Regulatory Requirements for Hazardous Materials

Other McGraw-Hill Books on Environmental Engineering

Regulatory Requirements for Hazardous Materials

Somendu B. Majumdar, Ph.D.

McGraw-Hill, Inc.

New York   San Francisco   Washington, D.C.   Auckland   Bogotá
Caracas   Lisbon   London   Madrid   Mexico City
Milan   Montreal   New Delhi   San Juan
Singapore   Sydney   Tokyo   Toronto

Library of Congress Cataloging-in-Publication Data

Majumdar, Somendu B.
    Regulatory requirements for hazardous materials / Somendu B. Majumdar.
       p.   cm.
    ISBN 0-07-039761-9
    1. Hazardous substances—Law and legislation—United States.
2. Hazardous wastes—Law and legislation—United States. I. Title.
KF3958.M335 1993
344.73′04622—dc20
[347.3044622]
                                          92-32345
                                            CIP

1 2 3 4 5 6 7 8 9 0 DOC/DOC 9 8 7 6 5 4 3 2

ISBN 0-07-039761-9

The sponsoring editor for this book was Gail Nalven, the editing supervisor was Jim Halston, and the production supervisor was Suzanne W. Babeuf. It was set in Times Roman by McGraw-Hill's Professional Book Group composition unit.

Printed and bound by R. R. Donnelley & Sons Company.

This book is dedicated to the loving memory of my mother

Contents

Preface

Today's world is complex and is enmeshed in a web of governmental regulations. They require the concerted efforts and expertise of several distinct and diverse professions. No single profession has all the answers to the confusing and often ambiguous questions that these regulations pose, particularly where a hazardous material is concerned.

This book aims at bringing together the skills and expertise of several fields in order to simplify today's murky world of hazardous material regulations. In this context, the term "hazardous material" has been used in the broadest sense in order to include hazardous and toxic substances, chemicals and pesticides as well as various types of wastes. The tripartite approach used in this book allows a comprehensive review, dissemination, and analytical scrutiny of various environmental statutes and the regulations promulgated thereunder through the eyes of a scientist/engineer, lawyer and corporate/business manager. It is hoped that the contents of the book will raise appropriate inquiries in order to handle environmental concerns and effectively manage hazardous materials in a most responsible fashion.

It must, however, be emphasized that the purpose of the book is not to turn every reader into an instant scientist, lawyer, or a business manager. The basic objective here is to bridge the gap that currently exists among the professionals of the scientific, legal, and corporate communities so that a full-fledged understanding can be developed with regard to hazardous material regulations. Finally, this book is not meant to be a substitute for the expertise of any particular profession. Appropriate counsel should be sought based on the cirumstances of a specific situation.

I have been extremely lucky to receive valuable comments and advice from experts and practitioners in three specific areas—engineering, law, and business management. In addition, the availability of so many excellent technical books and legal treatises made my task relatively easy. I am indeed indebted to these authors, many of whose books have been suggested as recommended reading, as well as those within the agencies who contributed to the interpretation of various rules and regulations.

On a more personal note, I am eternally grateful to some of my friends, teachers and colleagues—Mr. A. C. Tan for encouraging me to write this book as well as providing me with the excellent facilities of his home, Dr. Arjun Tan for his most ungrudging review and very helpful comments, Mr. S. P. Goel for ready assistance in finding appropriate government publications, Tracey McLean, Esq. for obtaining important case decisions and law review articles, Mark Arbon, Esq. and Andrew Leja, Esq. for skillfully guiding me through the complex formatting and indexing required for the manuscript. I am genuinely inspired by the kind comments and advice I received from Prof. Joseph Sweeney of the Fordham University School of Law and Kevin Brown, Esq. I am sincerely touched by the unhesitating willingness of Dean William Hadley, Ph.D. of the New Mexico State University and Mr. Glenn Batchelder, Vice President, Engineering of Groundwater Technology, Inc. for contributing two excellent and much needed chapters of the book. I am also appreciative of the most unselfish help and dedication of Ms. Sherry West who took upon herself to type the entire first draft of the manuscript. Of course, the book would have never seen its light of day without the constant support and confidence of the publishing and editorial staff of McGraw-Hill, particularly that of Ms. Gail Nalven, Senior Editor, and Mr. Jim Halston, Editing Supervisor.

Lastly, this book would not have been completed without the mental support, encouragement, and constant sacrifice by the members of my family who never allowed me to quit. I am fortunate to have a father as an inspiring guide and a wife and a son who understood the gravity of my task and the Herculean challenge that I was faced with. Hopefully, when Robin grows up, he will have some sympathy for my failure in not finding any time to play or read to him during the preparation of this book. Above all, I express my appreciation to the readers of this book for their time and commitment and sincerely hope that they will be able to forgive me for my inabilities and shortcomings, if any. I sincerely welcome their comments that may benefit my future endeavors.

Somendu B. Majumdar

Glossary of Important Terms, Acronyms, and Abbreviations

ACGIH	American Conference of Governmental Industrial Hygienists
ACL	Alternative Concentration Limit
ACM	Asbestos-Containing Material
ACO	Administrative Court Order
ADR	Alternative Dispute Resolution
AEA	Atomic Energy Act
AEC	Atomic Energy Commission
AHERA	Asbestos Hazard Emergency Response Act
AIP	Agreement in Principle
ANSI	American National Standards Institute
AOC	Area of Contamination
APA	Administrative Procedure Act
AQCR	Air Quality Control Region
ARARs	Applicable or Relevant and Appropriate Requirements
ARC	Alternate Remedial Contractor
ARCS	Alternative Remedial Contracting System
ASME	American Society of Mechanical Engineers
AST	Aboveground Storage Tank
ASTM	American Society for Testing of Materials
ATSDR	Agency for Toxic Substances and Disease Registry
BACT	Best Available Control Technology
BAT	Best Available Technology (Economically Achieveable)
BCT	Best Conventional Technology
BDAT	Best Demonstrated Available Technology
BHP	Biodegradation, Hydrolysis and Photolysis
BIFs	Boilers and Industrial Furnaces
BLS	Bureau of Labor Statistics
BMP	Best Management Practice
BMR	Baseline Monitoring Report
BNA	Base Neutral Acid

BOD	Biochemical Oxygen Demand
BOM	Bureau of Mines
BPJ	Best Professional Judgment
BPT	Best Practicable Technology (Currently Available)
CA	Consent Agreement
CAA	Clean Air Act
CAAA	Clean Air Act Amendments (of 1990)
CAG	Carcinogen Assessment Group
CAMU	Corrective Action Management Unit
CAS	Chemical Abstracts Service
CASRN	Chemical Abstracts Service Registry Number
CBEC	Concentration-based Exemption Criteria
CDC	Center for Disease Control
CEPA	Conscientious Employee Protection Act
CEQ	Council on Environmental Quality
CERCLA	Comprehensive Environmental Response, Compensation, and Liability Act (popularly known as "Superfund")
CERCLIS	Comprehensive Environmental Response, Compensation, and Liability Information Systems
CESQG	Conditionally Exempt Small Quantity Generator
CFC	Chlorofluorocarbon
CFR	Code of Federal Regulations
CGI	Combustible Gas Indicator
CGL	Comprehensive General Liability
CLP	Contract Laboratory Program
CO	Consent Order
COD	Chemical Oxygen Demand
CPI	Chemical Process Industries
CPSA	Consumer Product Safety Act
CPSC	Consumer Product Safety Commission
CRC	Community Relations Coordinator
CRP	Community Relations Plan
CSF	Cerebrospinal Fluid
CWA	Clean Water Act
DCA	Dangerous Cargo Act
DEIS	Draft Environmental Impact Statement
DHEW	Department of Health, Education and Welfare
DHHS	Department of Health and Human Services
DMR	Discharge Monitoring Report
DNA	Deoxyribonucleic Acid
DO	Dissolved Oxygen

DOA	Department of Agriculture
DOC	Department of Commerce
DOD	Department of Defense
DOE	Department of Energy
DOI	Department of Interior
DOJ	Department of Justice
DOL	Department of Labor
DOT	Department of Transportation
DRE	Destruction and Removal Efficiency
DSE	Domestic Sewage Exclusion
DTRE	Department of Treasury
EA	Environmental Assessment
EDF	Environmental Defense Fund
EIA	Environmental Impact Assessment
EIS	Environmental Impact Statement
EP	Extraction Procedure
EPA	Environmental Protection Agency
EPCRA	Emergency Planning and Community Right-to-Know Act
EQR	Environmental Quality Report
ERA	Expedited Response Action
ERC	Emergency Response Coordinator
ERDA	Energy Research and Development Administration
ERRIS	Emergency and Remedial Response Inventory System
ERT	Environmental Response Team
FAA	Federal Aviation Administration
FCLAA	Federal Cigarette Labeling and Advertising Act
FCMHSA	Federal Coal Mine Health and Safety Act
FCR	Field Change Request
FDA	Food and Drug Administration
FELA	Federal Employers' Liability Act
FEMA	Federal Emergency Management Agency
FEPCA	Federal Environmental Pesticide Control Act
FERC	Federal Energy Regulatory Commission
FFA	Federal Facilities Agreement
FFCA	Federal Facilities Compliance Agreement
FFDCA	Federal Food, Drug, and Cosmetic Act
FHSA	Federal Hazardous Substances Act
FHSLA	Federal Hazardous Substance Labeling Act
FHWAC	Federal Highway Commission
FIFRA	Federal Insecticide, Fungicide, and Rodenticide Act

FMIA   Federal Meat Inspection Act
FMSHAA   Federal Mine Safety and Health Amendments Act
FMSHRC   Federal Mine Safety and Health Review Commission
FOIA   Freedom of Information Act
FONSI   Finding of No Significant Impact
FOP   Field Operations Plan
FPC   Federal Power Commission
FR   Federal Register
FRA   Federal Railroad Administration
FRSA   Federal Railroad Safety Act
FS   Feasibility Study
FSAP   Field Sampling and Analysis Plan
FTC   Federal Trade Commission
FWPCA   Federal Water Pollution Control Act
FWQC   Federal Water Quality Criteria
GACT   Generally Achieveable Control Technology
GC   Gas Chromatograph
GC/MS   Gas Chromatograph/Mass Spectrometer
GLC   Ground Level Concentration
GWPS   Ground Water Protection Standard
HAP   Hazardous Air Pollutant
HASP   Health and Safety Plan
HAZMAT   Hazardous Material
HCFC   Hydrochlorofluorocarbon
HCS   Hazard Communication Standard
HDD   Halogenated Dibenzodioxin
HDF   Halogenated Dibenzofuran
HEAs   Health Effect Advisories
HH&E   Human Health & Environment
HLRW   High-Level Radioactive Waste
HMTA   Hazardous Materials Transportation Act
HMTUSA   Hazardous Materials Transportation Uniform Safety Act
HRS   Hazard Ranking System
HSWA   Hazardous and Solid Waste Amendments
HWM   Hazardous Waste Management
HWMU   Hazardous Waste Managment Unit
IAG   Inter-Agency Agreement
IARC   International Agency for Research on Cancer
ID No.   Identification Number (as provided by EPA)
IOC   Inorganic Chemical

IRIS	Integrated Risk Information System
IS	Interim Status
LAER	Lowest Achieveable Emission Rate
LCRS	Leachate Collection and Removal System
LC50	Lethal Concentration 50%
LD50	Lethal Dose 50%
LDRs	Land Disposal Restrictions
LEPC	Local Emergency Planning Committee
LPCs	Limiting Permissible Concentrations
LLRW	Low-Level Radioactive Waste
LLRWPA	Low-Level Radioactive Waste Policy Act
LLRWPAA	Low-Level Radioactive Waste Amendments Act
LUST	Leaking Underground Storage Tank
MACT	Maximum Achieveable Control Technology
MCD	Model Consent Decree
MCL	Maximum Contaminant Level
MCLG	Maximum Contaminant Level Goal
MEI	Maximum Exposed Individual
MESA	Mine Enforcement Safety Administration
mg/kg	milligram per kilogram
mg/L	milligram per liter
MOU	Memorandum of Understanding
MPRSA	Marine Protection, Research, and Sanctuaries Act
MSDS	Material Safety Data Sheet
MTBE	Methyl Tertbutylether
MTU	Mobile Treatment Unit
MWTA	Medical Waste Tracking Act
NAAQS	National Ambient Air Quality Standards
NAM	National Association of Manufacturers
NBAR	Nonbinding Allocations of Responsibility
NCP	National Oil and Hazardous Substances Pollution Contingency Plan (popularly known as "National Contingency Plan")
NCR	Nonconformance Report
NEA	National Energy Act
NEC	National Electrical Code
NEMA	National Electrical Manufacturer's Association
NEPA	National Environmental Policy Act
NESHAPs	National Emission Standards for Hazardous Air Pollutants
NFPA	National Fire Prevention Association
NHPA	National Historic Preservation Act

NIMBY	"Not in My Backyard" (Syndrome)
NIOSH	National Institute for Occupational Safety and Health
NOAA	National Oceanic and Atmospheric Administration
NOD	Notice of Deficiency
NOEL	No Observable Effect Level
NOI	Notice of Intent
NOV	Notice of Violation
NPDES	National Pollutant Discharge Elimination System
NPDWRs	National Primary Drinking Water Regulations
NPDWS	National Primary Drinking Water Standard
NPL	National Priorities List
NPRM	Notice of Proposed Rulemaking
NRC	a. Natinal Response Center, or b. Nuclear Regulatory Commission
NRDC	National Resources Defense Council
NRT	National Response Team
NSEQ	New Source Environmental Questionnaire
NSPS	New Source Performance Standard
NSR	New Source Review
NTIS	National Technical Information Service
NWPA	Nuclear Waste Policy Act
OECM	Office of Enforcement and Compliance Monitoring
O&M	Operation & Maintenance
OPA	Oil Pollution Act
OSC	On-Scene Coordinator
OSHA	Occupational Safety and Health Administration
OSH Act	Occupational Safety and Health Act
OSHRC	Occupational Safety and Health Review Commission
OSW	Office of Solid Waste
OSWER	Office of Solid Waste and Emergency Response
OTS	Office of Toxic Substances
OU	Operable Unit
OVA	Organic Vapor Analyzer
OWPE	Office of Waste Programs Enforcement
PA	Preliminary Assessment
PAC	Public Advisory Committee
PAH	Polynuclear Aromatic Hydrocarbon
PCB	Polychlorinated Biphenyl
PCE	Perchloroethylene (also known as tetrachloroethylene)
PCS	Permit Compliance System
PEIS	Programmatic Environmental Impact Statement

PEL	Permissible Exposure Limit
PIAT	Public Information Assist Team
PID	Photo Ionization Detector
PMN	Premanufacture Notification
POM	Polycyclic Organic Matter
POTWs	Publicly Owned Treatment Works
ppb	parts per billion
PPE	Personal Protective Equipment
ppm	parts per million
PRP	Potentially Responsible Party
PSD	Prevention of Significant Deterioration
PSES	Pretreatment Standards for Existing Sources
PSNS	Pretreatment Standards for New Sources
PVC	Polyvinyl Chloride
PWSA	Ports and Waterways Safety Act
QA	Quality Assurance
QC	Quality Control
RA	Remedial Action
RACT	Reasonably Available Control Technology
RAT	Radiological Assistance Team
RCC	Resource Conservation Committee
RCRA	Resource Conservation and Recovery Act
RD	Remedial Design
R&D	Research and Development
RDF	Refuse Derived Fuel
REM	Remedial Engineering Management (Control Program)
RFA	RCRA Facility Assessment
RFD	Reference Dose
RI	Remedial Investigation
RI/FS	Remedial Investigation/Feasibility Study
RMW	Radioactive Mixed Waste
ROD	Record of Decision
RPM	Remedial Project Manager
RQ	Reportable Quantity
RQG	Reduced Quantity Generator
RRC	Regional Response Center
RRT	Regional Response Team
RSPA	Research and Special Programs Administration
SAC	Support Agency Coordinator
SARA	Superfund Amendments and Reauthorization Act

SDWA	Safe Drinking Water Act
SERC	State Emergency Response Commission
SI	Site Investigation
SIC	Standard Industrial Classification
SIP	State Implementation Plan
SITE	Superfund Innovative Technologies Evaluation
SMCLs	Secondary Maximum Contaminant Levels
SMOA	Superfund Memorandum of Agreement
SMP	Site Management Plan
SNUR	Significant New Use Rule
SOC	Synthetic Organic Chemical
SPCC	Spill Prevention Control and Countermeasure
SPDES	State Pollutant Discharge Elimination System
SPHEH	Superfund Public Health Evaluation Manual
SRM	Standard Reference Manual
SSA	Sole Source Aquifer
SSC	Scientific Support Coordinator
STEL	Short Term Exposure Limit
SWDA	Solid Waste Disposal Act
SWMU	Solid Waste Management Unit
TAG	Technical Assistance Grant
TC	Toxicity Characteristic
TCA	1,1,1-Trichloroethane
TCB	Tetrachlorobenzene
TCDD	2,3,7,8-Tetrachlorodibenzo-p-dioxin
TCE	Trichloroethene (also known as Trichloroethylene)
TCL	Target Compound List
TCLP	Toxic Characteristic Leaching Procedure
TCP	2,4,5-Tricholorphenol
TDS	Total Dissolved Solid
TLV	Threshold Limit Value
TPH	Total Petroleum Hydrocarbon
TPQ	Threshold Planning Quantity
TSCA	Toxic Substances Control Act
TSD	Treatment, Storage or Disposal
TSDF	Treatment, Storage or Disposal Facility
TSS	Total Suspended Solid
TWA	Time Weighted Average
UAO	Unilateral Administrative Order

UCR	Unit Carcinogenic Risk
µg/kg	microgram per kilogram
µg/L	microgram per liter
UIC	Underground Injection Control (Program)
UL	Umbrella Liability
UMTRCA	Uranium Mill Tailings Radiation Control Act
USC	United States Code
USCG	United States Coast Guard
USDA	United States Department of Agriculture
USGS	United States Geological Survey
UST	Underground Storage Tank
UV	Ultraviolet (Ray)
VO	Volatile Organic
VOCs	Volatile Organic Compounds
WHP	Wellhead Protection (Program)
WMA	Wholesome Meat Act
WP	Work Plan
WPM	Work Plan Memorandum
WQ	Water Quality
WQA	Water Quality Act
WQC	Water Quality Criteria
WRC	Water Resources Council

Regulatory Requirements
for Hazardous Materials

1

Overview of Federal Statutes on Hazardous Materials

1.1 Introduction

With the advent of the industrial revolution, there has been an insidious proliferation of waste materials of virtually every kind. As the nation's appetite for industrial chemicals grows, so do the complexity and nature of waste and waste products. Proper handling and disposal of these wastes have become absolutely essential. It also makes good business sense to plan for the proper handling and disposal of the wastes that eventually result from these chemicals, through use, storage, further processing, or discarding. Unfortunately, a significant portion of such wastes is composed of hazardous and toxic substances which may generally be grouped as hazardous materials. Hence, today's hazardous materials require a comprehensive approach for their safe and effective handling and disposal.

With the passage of time, any plan for the safe handling and disposal of hazardous materials which may be in the form of toxic chemicals, hazardous substances, or hazardous wastes takes on a more complex form. Whenever industrial progress reaches another milestone, public concern about the proper handling and disposal of various hazardous materials generated by industries moves up another notch. For public health and safety and environmental protection, there is a continuing need to assess the nature of the hazardous materials that are directly or indirectly created by the nation's industries. Also as new knowledge is constantly gained through medical and scientific research, there can never be a final word as to their impact on public health and the environment. Hence, the status of all potentially harmful substances and waste products which are comprised of hazardous materials must be periodically assessed based on the best available information.

Such risk assessment should include not only those hazardous and toxic materials that are presently making an impact on public health and environment, but also those materials that may be commercially available in the future and have the potential to adversely affect public health or the environment. This knowledge is crucial to ameliorate the consequences of the improper release and disposal of hazardous materials. These hazardous materials may contain hazardous substances, toxic chemicals, hazardous wastes, and other harmful ingredients including radioactive wastes, polychlorinated biphenyls, vinyl chloride, trichloroethylene, methyl isocyanate, dioxin, mercury, lead, asbestos, and innumerable other industrial chemicals, pesticides, and natural and synthetic products.

Historically, in the United States, the 1960s brought greatly increased public sensitivity to environmental pollution. The 1970s marked the decade of congressional response to growing public concern and outcry against environmental pollution. The 1980s gave birth to several tough federal environmental laws.

Pollution prevention and control of hazardous materials are no longer voluntary. Congress has responded to public concern by passing several major laws dealing with toxic chemicals, hazardous substances, and hazardous waste. The current regulatory objective of waste minimization can be traced to several federal and state laws and local codes and ordinances.

At the federal level, Congress attempted to deal with the burgeoning problem of hazardous materials and perceived health and safety hazards by enacting pollution control laws that have their genesis in the common law principles. Historically, the nuisance laws were one of the early congressional attempts to curb environmental pollution by regulating land use and preventing widespread technological abuse. These laws are credited with the revolutionary concept that while one may use or enjoy one's land or property freely, such freedom may be restricted if that use or enjoyment affects others adversely. This simple legal concept has, by necessity, become extremely complex with the unprecedented growth of industrial products and chemical substances, the use or disposal of which can cause health or environmental damages if they remain unregulated or uncontrolled. It is no surprise that there are a host of federal statutes that deal with various hazardous and toxic materials which are created by humans or industrially processed from natural sources.

To appreciate the depth and breadth of the complex regulatory morass at the federal level, one has to be well versed in the federal statutes as well as various regulatory policies and programs that are based on the statutory authorities. In addition, it is a virtual requirement these days to check appropriate state and local laws and the regulations promulgated thereunder to comprehend and establish the full extent of current legal requirements. Gone are the days when understanding of a single statute, rule, or regulation could be construed as "diligent review." Today's environmental laws and regulations form an endless web as they not only overlap, but also, by necessity, relate to each other. Hence, following one law is never enough as it provides only one piece of a complex puzzle.

In this context, note that there are specific definitions for terms such as *hazardous substances, toxic substances, hazardous waste,* etc. These are all terms of art and must be fully understood in the context of their statutory or regulatory meanings and not merely limited to their plain English or dictionary meanings. It is absolutely imperative from a legal sense that each statute or regulation promulgated be read in conjunction with the terms defined in that specific statute or regulation. In this overview, however, any reference to hazardous materials conveys their plain meaning, and as such they include toxic chemicals, hazardous substances, dangerous chemical products and by-products, and hazardous wastes, including radioactive wastes, unless stated otherwise.

Also this overview is limited to certain federal statutes and regulations as they apply to hazardous materials, whether in the form of hazardous or toxic substances, harmful chemicals, pesticides, or hazardous wastes. This by no means suggests that these statutes are the only ones which may apply to a hazardous material in a particular situation. As indicated earlier, one must consult and review all applicable state and local laws, codes, rules, and regulations to ascertain the full depth and breadth of statutory and regulatory requirements.

Generally, at the federal level, there are numerous federal statutes and regulations which address various aspects of hazardous materials, including their use, reuse, processing, handling, storage, transportation, and ultimate disposal. While it is not very meaningful to provide a laundry list of all such statutes and regulations in this limited scope of the subject, a basic understanding of some of the most common and well-encompassing federal statutes and the regulations promulgated thereunder is almost a necessity. Some of these important federal statutes are the

- National Environmental Policy Act (NEPA)
- Clean Air Act (CAA)
- Clean Water Act (CWA)
- Toxic Substances Control Act (TSCA)
- Resource Conservation and Recovery Act (RCRA)
- Comprehensive Environmental Response, Compensation, and Liability Act (CERCLA)
- Superfund Amendments and Reauthorization Act (SARA)
- Occupational Safety and Health (OSH) Act
- Federal Insecticide, Fungicide, and Rodenticide Act (FIFRA)
- Hazardous Materials Transportation Act (HMTA)
- Federal Food, Drug, and Cosmetic Act (FFDCA)
- Safe Drinking Water Act (SDWA)
- Oil Pollution Act (OPA)
- Atomic Energy Act (AEA)

Since many of these federal statutes have been amended several times and some have progeny or second- and third-generation versions, they must be reviewed in their entirety, i.e., along with their amendments, progeny, and sister statutes. Anything less than that may create a treacherous gap in today's complex, murky world of environmental laws, policies, and programs.

To comprehend minimal legal concepts and requirements, a basic understanding of certain important federal statutes is essential. An overview of several important federal environmental statutes is provided below to give the reader a sense of statutory and regulatory duties and obligations. Several of these federal statutes and their regulatory requirements are also discussed in detail in later chapters as they have major implications with respect to various types and sources of hazardous materials. It is extremely important to understand the exact scope of the statutory provisions since regulations are interpretive rules based on statutes. As a result, regulations can be challenged in a court of law if they fail to interpret the statutes correctly.

1.2   National Environmental Policy Act

The National Environmental Policy Act (NEPA) of 1969 was actually signed into law in 1970. The purpose of NEPA is to provide a high priority to environmental questions whenever the federal government proposes any action that may affect the environment. Thus, it provides a national environmental policy while obligating federal agencies to consider environmental concerns before embarking on a federal project or program. Compliance with the provisions of NEPA is required if any proposed action or activity involves federal grants, assistance or review, and permits, licenses, or approvals.

NEPA does not create any substantive rights per se. It is, in the true sense, a procedural law that provides guidelines to federal agencies for consideration of various environmental matters. The federal agencies are required to consider the environmental consequences of their proposed actions at the planning stage. The reach of NEPA is limited and does not extend to the states or state agencies, although state versions of NEPA are very common. In essence, NEPA offers a statutory basis to everyone to force a review of federal decisions, especially when the federal agency involved does not have direct or distinct environmental responsibilities. Through NEPA, Congress established a national policy that requires all federal agencies, including defense or military agencies, to give proper and adequate consideration to environmental impacts of their decisions involving a proposed action or project.

The centerpiece of NEPA is the statutory requirement for an *Environmental Impact Statement* (EIS). An EIS is required to include a "detailed statement" depicting how a major "federal action" may significantly affect the quality of the environment. The draft EIS (DEIS) is to be submitted to local, state, and federal agencies that are interested (i.e., involved agencies) in the project in addition to making it available to the affected public in the community. The final EIS (FEIS) is required to address and include comments and responses made on the DEIS.

There are specific matters which must be addressed in an EIS by a federal agency:

- The environmental impact of the proposed action by the federal government
- Alternatives to the proposed action and their relative environmental impact
- All adverse environmental impacts due to the proposed action that cannot be avoided
- The relationship between short-term uses of the environment and the maintenance and growth of its long-term productivity
- All irreversible and irretrievable expenditure of natural resources in the event of implementation of the proposed action

All these considerations reflect a well-defined environmental policy under NEPA—to conduct all federal government activities in such a way as to promote the general welfare while encouraging productive and enjoyable harmony between people and the environment. In short, one goal of NEPA is to strike a "balance between population and resource use."

NEPA includes three subchapters or titles. Title I deals with the national environmental policy and specific obligations of the federal agencies. It incorporates six specific goals for the federal agencies:

- To fulfill the responsibilities of each generation as trustee of the environment for succeeding generations
- To assure all Americans of a safe, healthful, productive, and aesthetically and culturally pleasing surrounding
- To attain the widest range of beneficial uses of the environment without degradation, risk to health or safety, or other undesirable and unintended consequences
- To preserve important historic, cultural, and natural aspects of our national heritage and maintain, wherever possible, an environment which supports diversity and variety of individual choice
- To achieve a balance between population and resource use which will permit high standards of living and a wide sharing of life's amenities
- To enhance the quality of renewable resources and to approach the maximum attainable recycling of depletable resources

Title II of NEPA created the Council on Environmental Quality (CEQ) and gave it the authority to establish procedures and regulations that must be complied with by the federal agencies. CEQ, created in 1977, provides various rules that a federal agency must follow in order to determine its responsibilities and statutory requirements under NEPA. Section 204 of NEPA sets forth the specific statutory duty and functions of CEQ. Briefly, CEQ is required to "assist and advise" the President of the United States in the preparation of the Environmental Quality Report (EQR). Such a report not only covers the general issues of environmental conditions and quality but also deals with the adequacy and need for additional legislation to meet NEPA's stated goals. In essence, CEQ is responsible for assessing, developing, and recommending national environmental policies and legislative programs that may be required to protect and enhance the environmental quality of this nation.

Title III of NEPA established the Science Advisory Board (SAB). SAB is required to provide whatever scientific advice is requested by the EPA administrator, by the Committee on Environment and Public Works of the U.S. Senate, or by the Committee on Science and Technology, Energy and Commerce, or the Public Works and Transportation of the House of Representatives.

NEPA includes certain procedural provisions which are designed to achieve its stated environmental policy and goal. As indicated earlier, it does not per se create any new substantive rights. Nevertheless, since the enactment of NEPA, there has been considerable litigation involving either improper commission of an act or an omission to act adequately by the federal agencies, particularly with respect to the EIS requirements. The basic issue in the earlier cases was whether the statutory provisions in NEPA require a federal agency to consider certain specific environmental issues in all its decisions or only in certain situations. In a 1978 case, *Vermont Yankee Nuclear Power Corporation v. NRDC, Inc.,* the U.S. Supreme Court held that while NEPA sets forth "significant" and "substantial" goals, its mandate to the federal agencies is "essentially procedural." Two years later, in *Strycker's Bay Neighborhood Council v. Karlen,* the Supreme Court held that NEPA does not create any judicially enforceable substantive rights. Thus, once a federal agency has considered the environmental consequences of its proposed action, it has met the procedural requirements of NEPA and any decision taken by the agency thereafter falls well within its discretionary authority.

The decisions of the Supreme Court should not, however, be translated as an open invitation to a federal agency to reach its decisions without regard to the policies and goals of NEPA just because the agency has met the specific procedural requirements. All federal agencies are obligated to exercise their discretionary authority regarding any proposed federal action in a way which is consistent with the national environmental policy stated in Title I of NEPA. This requirement is so germane under NEPA that it even survived the Executive Order signed by President Reagan in 1981 under which the federal agencies are required to accept only those actions which result in least net cost and a net benefit to society.

As indicated earlier, CEQ acts as an adviser to the President of the United States on environmental issues. It is also charged with the responsibility of providing guidance to federal agencies on matters that arise from the statutory provisions of NEPA. Thus, CEQ is responsible for promulgating regulations under NEPA that are to be complied with by all federal agencies.

The first time CEQ promulgated such regulations was in 1979. They were last amended in 1986. These regulations are based on the central EIS provision of NEPA for "federal action." In this context, federal action includes all activities directly undertaken by any federal agency, including construction of facilities, operation of projects, and funding of programs which may be administered by others, such as state or local agencies and private businesses. As a general rule, the CEQ regulations require the preparation of an EIS for all "major federal actions." Specific actions that constitute major federal actions are set forth in 40 CFR (Code of Federal Regulations) §1508.18. They include projects financed, assisted, conducted, regulated, or approved by federal agencies including new or revised agency rules, regulations, plans, policies or procedures, and legislative proposals. However, this does not necessarily mean that environmental considerations may be bypassed by federal agencies if the proposed action does not constitute a major federal action or "significantly" affect the environmental quality. However, in such federal actions, the preparation of a formal EIS may not be necessary, and less vigorous consideration may be acceptable.

It is generally accepted that federal agencies have the right to make the preliminary decision as to whether an EIS should be prepared for a proposed action. If a federal agency finds, based on its preliminary environmental assessment of a proposed action, that there is "no significant impact" from its proposed action, it can bypass the EIS process altogether. However, a *finding of no significant impact* (FONSI) is not a mere cursory statement or decision. The federal regulations provide a specific definition of FONSI as set forth in 40 CFR §1508.13. Based on this, FONSI is a document which must be prepared by a federal agency and must include, even if briefly, the reasons why the proposed action will not have a "significant effect on the human environment" and for which an EIS will not be prepared. At a minimum, a FONSI must include either a summary or a full-fledged environmental assessment of the proposed federal action. If the assessment is included, a simple incorporation of it by reference may be adequate.

Section 102(C)(i) of NEPA requires that an EIS be prepared for any "major federal action" that "significantly" affects the environment. It is necessary for a federal agency, prior to its preparation of an EIS, to consult with and receive comments about the proposed action from other federal agencies which may have either jurisdiction or special expertise on environmental matters that may have to be addressed. Also, a draft EIS has to be prepared at an early stage of the proposed action, and it must be circulated for a period of not less than 45 days for comments from all involved or interested agencies, including state and local agencies which may be affected. The final EIS is prepared only after all comments and questions on the draft EIS are considered or resolved. The federal agency responsible for the proposed action has to make its decision on the issues at hand by careful review of the final EIS and the various alternatives presented by developing a *record of decision* (ROD).

The regulations, set forth in 40 CFR §1506.10, specify certain time sequences. As a general rule, no federal action will be undertaken at least for 90 days after the draft EIS or 30 days after the final EIS has been made available to CEQ and the public for review. The final agency proposal, to be submitted to the President, CEQ, and the public, must include the final EIS, comments, responses, and views expressed by other agencies.

In situations which require "major federal actions" by more than one federal agency, the regulations require that a "lead agency" determination be made by all such "involved" or interested agencies. The lead agency, once designated, is responsible for the preparation of the EIS for the proposed action. In case of disputes within the involved agencies as to which one is best suited to act as the lead agency in a proposed action, CEQ is responsible for designating the lead agency from the involved agencies in that federal action.

In summary, NEPA and its central provision, EIS, may have a profound impact on the decision-making process of federal agencies. In spite of the initiatives undertaken by the Reagan administration to focus federal agency attention on business conditions besides environmental quality for all their proposed actions, NEPA still remains a strong force in any decision making involving a federal action.

1.3   Clean Air Act

The Clean Air Act Amendments (CAAA) of 1990 have made significant changes in the basic Clean Air Act enacted in 1970. The statutory provisions of the 1990 amendments are extremely profound and in a way revolutionary, based on the depth and breadth of responsibilities imposed on the Environmental Protection Agency (EPA). While EPA is still working to promulgate regulations reflecting the 1990 amendments, the detailed statutory provisions have left very little room for EPA's discretionary rule making.

The basic Clean Air Act of 1970 and the 1977 amendments that followed consist of three titles. Title I deals with stationary air emission sources, Title II deals with mobile air emission sources, and Title III includes definitions of appropriate terms, provisions for citizen suits, and applicable standards for judicial review. The Clean Air Act Amendments of 1990, in sharp contrast to the previous federal clean air statutes, contain extensive provisions for control of hazardous air pollutants and acid rain. Some of these new or expanded titles in the 1990 amendments provide mandatory control requirements for several controversial topics, such as hazardous air pollutants and acid rain. At the same time, they provide new and added requirements for such original ideas as state implementation plans for attainment of the national ambient air quality standards and permitting requirements for the attainment and nonattainment areas. Title III now calls for a vastly expanded program to regulate *hazardous air pollutants* (HAPs) or the so-called air toxics. The previous national emission standards have been replaced by the new technology-based standards espoused in this newly amended federal statute.

Under the Clean Air Act Amendments of 1990, congressional mandate requires EPA to establish, during the first phase, technology-based *maximum achievable control technology* (MACT) emission standards. The MACT standards will apply to those major categories or subcategories of sources which emit at least 10 tons per year of any one or 25 tons per year of any combination of the listed 189 hazardous air pollutants. In addition, Title III provides for second-round, health-based standards to address the issue of residual risks due to air toxic emissions from the sources equipped with MACT. This congressional requirement will force EPA to determine whether the MACT standards can protect health with an "ample margin of safety."

In view of the 1990 amendments, it is perhaps correct to suggest that Congress is becoming more and more concerned about human exposure to various toxic substances that may be emitted into the atmosphere. The underlying concern is the risk of cancer along with other irreversible and incapacitating illnesses due to toxic air emissions. By enacting this new clean-air statute, Congress has sent a clear message that it is determined to provide technologically available means best suited to protect public health in view of the current and anticipated industrial growth in this nation.

Section 112 of the original Clean Air Act that dealt with hazardous air pollutants has been greatly expanded by the 1990 amendments. The list of HAPs has been increased many fold as the current list now includes 189 hazardous air pollutants from the previous group of seven hazardous air pollutants prior to the 1990 amendments. In addition, the standards for emission control have been tightened and raised to a very high level, referred to as the "best of the best," in order to reduce the risk of exposure to various hazardous air pollutants.

The Clean Air Act Amendments of 1990 contain specific deadlines and detailed timetables for major events. They require EPA to promulgate the MACT standards for forty source categories by November 15, 1992, along with a determination of its priorities for regulating other sources of HAPs. All EPA standards are required to be promulgated within 10 years of the passing of the 1990 amendments, i.e., by November 15, 2000. Implementation of these statutory requirements will be costly, particularly for smaller industries. This is all the more critical since a source subject to the MACT standard must achieve compliance within 3 years of promulgation of the MACT standard unless it receives an extension of up to 1 year by EPA.

The statute also offers a novel credit plan whereby full implementation may be delayed by up to 6 years. This will be allowed if there is more than 90 percent reduction (for particulates, 95 percent reduction) of hazardous air pollutants voluntarily achieved, using 1987 as the baseline, prior to EPA's rule promulgation under the Clean Air Act Amendments of 1990.

The statute established a new regulatory program for accidental releases. This reflects a congressional thinking similar to Title III of SARA, also known as the Emergency Planning and Community Right-to-Know Act (EPCRA) of 1986. Pursuant to this statutory scheme, EPA is required to publish a list of at least 100 extremely hazardous air pollutants by November 15, 1992. Interestingly, the statute itself identifies sixteen such pollutants, thus providing EPA with a head start.

1.4   Clean Water Act

The first comprehensive water-quality legislation put forth by Congress is the Federal Water Pollution Control Act (FWPCA) of 1972. Earlier laws, e.g., Water Pollution Control Act of 1948 and Water Quality Act of 1965, were generally limited to control of pollution of interstate waters and the adoption of water-quality standards by the states for interstate water within their borders. FWPCA is credited with the first comprehensive program for controlling and abating water pollution.

In 1977, Congress amended FWPCA, but in doing so, it renamed the earlier legislation. The new, amended statute became the Clean Water Act (CWA) of 1977. CWA was amended in 1978 to deal more effectively with spills of petroleum or oil and hazardous substances. Certain important amendments to CWA followed in 1987 under the new name of the Water Quality Act (WQA). WQA is largely aimed at improving water quality in those areas which lack or have insufficient compliance record with minimum national discharge standards. In all other respects, the current federal clean-water statute remains deeply entrenched in the statutory provisions of the 1977 enactment, which have been slightly modified or expanded by later amendments.

The federal clean-water statute makes a distinction between conventional and toxic pollutants. As a result, two standards of treatment are required prior to their discharge into the navigable waters of the nation. For conventional pollutants that generally include degradable nontoxic organics and inorganics, the applicable treatment standard is *best conventional technology* (BCT). For toxic pollutants, on the other hand, the required treatment standard is *best available technology* (BAT), which is a higher standard than BCT.

The statutory provisions of CWA have five major sections that deal with specific issues:

- Nationwide water-quality standards
- Effluent standards from certain specific industries
- Permit programs for discharges into receiving water bodies based on the National Pollutant Discharge Elimination System (NPDES)
- Discharge of toxic chemicals, including oil spills
- Construction grant program for *publicly owned treatment works* (POTW)

It must be emphasized that one of the primary congressional concerns that led to the passage of CWA is the actual and potential impact of water pollution through intentional or accidental discharge and release of hazardous and toxic pollutants into the waters of the nation. It is important to note here that Congress also passed a comparable law, the Marine Protection Research and Sanctuaries Act of 1972, which addresses the disposal of wastes, including hazardous and toxic pollutants, in the oceans that are considered as the dumping ground for these pollutants.

Originally, section 307(a) of CWA included exclusive statutory provisions for effluent limitations of listed toxic pollutants that may be discharged into receiving water bodies. As a result of settlement of a suit brought by the Natural Resources Defense Council (NRDC) against EPA in the form of a consent decree, EPA agreed to develop a toxics control strategy. Reacting to this, CWA included the technology-based aspects of the judicial mandate contained in the final NRDC consent decree. Several important elements are part of this judicial mandate:

- Adoption of a list of priority pollutants with primary emphasis on toxic substances
- Adoption of BAT standards for the discharge of each listed toxic substance
- EPA authorization to add any substance to or delete it from the list of toxic substances
- Provision for upgrading and enforcing pretreatment regulations for eventual discharges from the POTWs
- EPA authorization to adopt regulations that establish *best management practices* (BMPs) to control the discharge of toxic pollutants which emanate from surface runoff or other uncontrolled discharges by industrial facilities
- Promulgation of compliance requirements for BAT effluent limitations for toxic pollutants within 3 years of the establishment of such limitations

At present, the BAT effluent limitations focus primarily on those priority pollutants which were identified in the NRDC consent decree as well as certain other toxic pollutants that have been identified in section 307(a) of CWA. Additionally, EPA regulations as set forth in 40 CFR 112 are an attempt to prevent discharges of oil to surface waters as well as to control such discharges whenever they occur. One of the tools that EPA uses for this purpose is the regulatory requirement to prepare, maintain, and implement a *spill prevention control and countermeasure* (SPCC) plan.

In addition, section 311 of CWA includes elaborate provisions for regulating intentional or accidental discharges of oil and hazardous substances. Included there are response actions required for oil spills and the release or discharge of toxic and hazardous substances. Pursuant to its authority under sections 311(b)(2)(A) and 501(a) of CWA, EPA has designated certain elements and compounds as hazardous substances and developed an appropriate list, set forth in 40 CFR 116.4. The person in charge of a vessel or an onshore or offshore facility from which any designated hazardous substance is discharged, in quantities equal to or exceeding its reportable quality, must notify the appropriate federal agency as soon as such knowledge is obtained. Such notice should be provided in accordance with the procedures set forth in 33 CFR §153.203.

Note that after the enactment of the Comprehensive Environmental Response, Compensation, and Liability Act (CERCLA) in 1980, Congress significantly expanded EPA's authority to clean up and remediate contamination resulting from spills and other unpermitted releases of hazardous substances. Hence, there is an obvious overlap between CWA and CERCLA with respect to certain toxic pollutants. But they are not mutually exclusive since both statutes provide EPA with specific statutory authority and underlying duty. In fact, in the historical sense, section 311 of CWA is a precursor to the CERCLA provisions that set forth a *National Contingency Plan* (NCP) for cost recovery for necessary cleanup and remedial work and the imposition of strict liability on polluters responsible for spills or discharges of hazardous substances.

Section 311 of CWA specifically prohibits the discharge of "harmful quantities of oil and hazardous substances" unless such discharges are allowed under a valid NPDES permit. In the case of an oil spill, EPA regulations promulgated under CWA and set forth in 40 CFR §110.3 define what constitutes a "harmful quantity." It is set as one that violates applicable water-quality standards or causes "a film or sheen upon or discoloration of the surface" of the water or adjoining shorelines. For a hazardous substance, on the other hand, the designated *reportable quantity* (RQ) of the substance as specified by EPA is the threshold value.

Pursuant to EPA's regulations, as set forth in 40 CFR part 117, reportable quantities of hazardous substances are sorted in five categories (X, A, B, C, and D), each having a designated RQ which signifies the quantity that may be harmful if discharged into the navigable waters of the United States. For example, for a hazardous substance in category X, the harmful quantity is 1 pound. For hazardous substances belonging to categories A, B, C, and D, the respective harmful or reportable quantities are 10, 100, 1000, and 5000 pounds. EPA's Table 117.3, which provides the reportable quantities of the designated hazardous substance, is reproduced in Table 1.1.

It is important to note here that any violation of section 7003 of the Resource Conservation and Recovery Act (RCRA) due to the creation of a substantial and imminent endangerment by hazardous wastes generally causes a violation of CWA. This is due to the textual language in section 307 of CWA which also sets the limitations of toxic effluents. Such a violation of RCRA may additionally trigger a violation of section 402 of CWA which requires an effluent discharge only under a valid NPDES permit. Alternatively, the cause of action may be based on a permit violation under section 404 of CWA if it involves "dredge and spill" materials.

The effluent limitations of toxic pollutants under CWA that are most commonly encountered include several pesticides, such as aldrin, dieldrin, DDT, and endrin. Also included are mercury and several toxic chemicals and substances, such as polychlorinated biphenyls (PCBs) and toxaphene, as well as various chlorinated organics and organic solvents, including trichloroethylene (TCE) and xylenes.

1.5 Toxic Substances Control Act

In 1971, CEQ recommended that the EPA administrator be empowered "to restrict the use or distribution of any substance which he [or she] finds hazardous to human health or to the environment." Eventually, after several attempts by Congress, the Toxic Substances Control Act (TSCA) of 1976 was

TABLE 1.1   Reportable Quantities of Hazardous Substances Designated Pursuant to Section 311 of the Clean Water Act

Material	Category	RQ in pounds (kilograms)
Acetaldehyde	C	1,000 (454)
Acetic acid	D	5,000 (2,270)
Acetic anhydride	D	5,000 (2,270)
Acetone cyanohydrin	A	10 (4.54)
Acetyl bromide	D	5,000 (2,270)
Acetyl chloride	D	5,000 (2,270)
Acrolein	X	1 (0.454)
Acrylonitrile	B	100 (45.4)
Adipic acid	D	5,000 (2,270)
Aldrin	X	1 (0.454)
Allyl alcohol	B	100 (45.4)
Allyl chloride	C	1,000 (454)
Aluminum sulfate	D	5,000 (2,270)
Ammonia	B	100 (45.4)
Ammonium acetate	D	5,000 (2,270)
Ammonium benzoate	D	5,000 (2,270)
Ammonium bicarbonate	D	5,000 (2,270)
Ammonium bichromate	A	10 (4.54)
Ammonium bifluoride	B	100 (45.4)
Ammonium bisulfite	D	5,000 (2,270)
Ammonium carbamate	D	5,000 (2,270)
Ammonium carbonate	D	5,000 (2,270)
Ammonium chloride	D	5,000 (2,270)
Ammonium chromate	A	10 (4.54)
Ammonium citrate dibasic	D	5,000 (2,270)
Ammonium fluoborate	D	5,000 (2,270)
Ammonium fluoride	B	100 (45.4)
Ammonium hydroxide	C	1,000 (454)
Ammonium oxalate	D	5,000 (2,270)
Ammonium silicofluoride	C	1,000 (454)
Ammonium sulfamate	D	5,000 (2,270)
Ammonium sulfide	B	100 (45.4)
Ammonium sulfite	D	5,000 (2,270)
Ammonium tartrate	D	5,000 (2,270)
Ammonium thiocyanate	D	5,000 (2,270)
Amyl acetate	D	5,000 (2,270)
Aniline	D	5,000 (2,270)
Antimony pentachloride	C	1,000 (454)
Antimony potassium tartrate	B	100 (45.4)
Antimony tribromide	C	1,000 (454)
Antimony trichloride	C	1,000 (454)
Antimony trifluoride	C	1,000 (454)
Antimony trioxide	C	1,000 (454)
Arsenic disulfide	X	1 (0.454)
Arsenic pentoxide	X	1 (0.454)
Arsenic trichloride	X	1 (0.454)
Arsenic trioxide	X	1 (0.454)
Arsenic trisulfide	X	1 (0.454)
Barium cyanide	A	10 (4.54)
Benzene	A	10 (4.54)
Benzoic acid	D	5,000 (2,270)
Benzonitrile	D	5,000 (2,270)
Benzoyl chloride	C	1,000 (454)
Benzyl chloride	B	100 (45.4)
Beryllium chloride	X	1 (0.454)
Beryllium fluoride	X	1 (0.454)
Beryllium nitrate	X	1 (0.454)
Butyl acetate	D	5,000 (2,270)
Butylamine	C	1,000 (454)
n-Butyl phthalate	A	10 (4.54)
Butyric acid	D	5,000 (2,270)
Cadmium acetate	A	10 (4.54)
Cadmium bromide	A	10 (4.54)
Cadmium chloride	A	10 (4.54)
Calcium arsenate	X	1 (0.454)

TABLE 1.1 Reportable Quantities of Hazardous Substances Designated Pursuant to Section 311 of the Clean Water Act (*Continued*)

Material	Category	RQ in pounds (kilograms)
Calcium arsenite	X	1 (0.454)
Calcium carbide	A	10 (4.54)
Calcium chromate	A	10 (4.54)
Calcium cyanide	A	10 (4.54)
Calcium dodecylbenzenesulfonate	C	1,000 (454)
Calcium hypochlorite	A	10 (4.54)
Captan	A	10 (4.54)
Carbaryl	B	100 (45.4)
Carbofuran	A	10 (4.54)
Carbon disulfide	B	100 (45.4)
Carbon tetrachloride	A	10 (4.54)
Chlordane	X	1 (0.454)
Chlorine	A	10 (4.54)
Chlorobenzene	B	100 (45.4)
Chloroform	A	10 (4.54)
Chlorosulfonic acid	C	1,000 (454)
Chlorpyrifos	X	1 (0.454)
Chromic acetate	C	1,000 (454)
Chromic acid	A	10 (4.54)
Chromic sulfate	C	1,000 (454)
Chromous chloride	C	1,000 (454)
Cobaltous bromide	C	1,000 (454)
Cobaltous formate	C	1,000 (454)
Cobaltous sulfamate	C	1,000 (454)
Coumaphos	A	10 (4.54)
Cresol	C	1,000 (454)
Crotonaldehyde	B	100 (45.4)
Cupric acetate	B	100 (45.4)
Cupric acetoarsenite	X	1 (0.454)
Cupric chloride	A	10 (4.54)
Cupric nitrate	B	100 (45.4)
Cupric oxalate	B	100 (45.4)
Cupric sulfate	A	10 (4.54)
Cupric sulfate, ammoniated	B	100 (45.4)
Cupric tartrate	B	100 (45.4)
Cyanogen chloride	A	10 (4.54)
Cyclohexane	C	1,000 (454)
2,4-D Acid	B	100 (45.4)
2,4-D Esters	B	100 (45.4)
DDT	X	1 (0.454)
Diazinon	X	1 (0.454)
Dicamba	C	1,000 (454)
Dichlobenil	B	100 (45.4)
Dichlone	X	1 (0.454)
Dichlorobenzene	B	100 (45.4)
Dichloropropane	C	1,000 (454)
Dichloropropene	B	100 (45.4)
Dichloropropene-Dichloropropane (mixture)	B	100 (45.4)
2,2-Dichloropropionic acid	D	5,000 (2,270)
Dichlorvos	A	10 (4.54)
Dicofol	A	10 (4.54)
Dieldrin	X	1 (0.454)
Diethylamine	B	100 (45.4)
Dimethylamine	C	1,000 (454)
Dinitrobenzene (mixed)	B	100 (45.4)
Dinitrophenol	A	10 (45.4)
Dinitrotoluene	A	10 (4.54)
Diquat	C	1,000 (454)
Disulfoton	X	1 (0.454)
Diuron	B	100 (45.4)
Dodecylbenzenesulfonic acid	C	1,000 (454)
Endosulfan	X	1 (0.454)
Endrin	X	1 (0.454)
Epichlorohydrin	B	100 (45.4)
Ethion	A	10 (4.54)
Ethylbenzene	C	1,000 (454)
Ethylenediamine	D	5,000 (2,270)
Ethylenediamine-tetraacetic acid (EDTA)	D	5,000 (2,270)
Ethylene dibromide	X	1 (0.454)

TABLE 1.1   Reportable Quantities of Hazardous Substances Designated Pursuant to Section 311 of the Clean Water Act (*Continued*)

Material	Category	RQ in pounds (kilograms)
Ethylene dichloride	B	100 (45.4)
Ferric ammonium citrate	C	1,000 (454)
Ferric ammonium oxalate	C	1,000 (454)
Ferric chloride	C	1,000 (454)
Ferric fluoride	B	100 (45.4)
Ferric nitrate	C	1,000 (454)
Ferric sulfate	C	1,000 (454)
Ferrous ammonium sulfate	C	1,000 (454)
Ferrous chloride	B	100 (45.4)
Ferrous sulfate	C	1,000 (454)
Formaldehyde	B	100 (45.4)
Formic acid	D	5,000 (2,270)
Fumaric acid	D	5,000 (2,270)
Furfural	D	5,000 (2,270)
Guthion	X	1 (0.454)
Heptachlor	X	1 (0.454)
Hexachlorocyclopentadiene	A	10 (4.54)
Hydrochloric acid	D	5,000 (2,270)
Hydrofluoric acid	B	100 (45.4)
Hydrogen cyanide	A	10 (4.54)
Hydrogen sulfide	B	100 (45.4)
Isoprene	B	100 (45.4)
Isopropanolamine dodecylbenzenesulfonate	C	1,000 (454)
Kepone	X	1 (0.454)
Lead acetate	D	5,000 (2,270)
Lead arsenate	X	1 (0.454)
Lead chloride	B	100 (45.4)
Lead fluoborate	B	100 (45.4)
Lead fluoride	B	100 (45.4)
Lead iodide	B	100 (45.4)
Lead nitrate	B	100 (45.4)
Lead stearate	D	5,000 (2,270)
Lead sulfate	B	100 (45.4)
Lead sulfide	D	5,000 (2,270)
Lead thiocyanate	B	100 (45.4)
Lindane	X	1 (0.454)
Lithium chromate	A	10 (4.54)
Malathion	B	100 (45.4)
Maleic acid	D	5,000 (2,270)
Maleic anhydride	D	5,000 (2,270)
Mercaptodimethur	A	10 (4.54)
Mercuric cyanide	X	1 (0.454)
Mercuric nitrate	A	10 (4.54)
Mercuric sulfate	A	10 (4.54)
Mercuric thiocyanate	A	10 (4.54)
Mercurous nitrate	A	10 (4.54)
Methoxychlor	X	1 (0.454)
Methyl mercaptan	B	100 (45.4)
Methyl methacrylate	C	1,000 (454)
Methyl parathion	B	100 (45.4)
Mevinphos	A	10 (4.54)
Mexacarbate	C	1,000 (454)
Monoethylamine	B	100 (45.4)
Monomethylamine	B	100 (45.4)
Naled	A	10 (4.54)
Naphthalene	B	100 (45.4)
Naphthenic acid	B	100 (45.4)
Nickel ammonium sulfate	B	100 (45.4)
Nickel chloride	B	100 (45.4)
Nickel hydroxide	A	10 (4.54)
Nickel nitrate	B	100 (45.4)
Nickel sulfate	B	100 (45.4)
Nitric acid	C	1,000 (454)
Nitrobenzene	C	1,000 (454)
Nitrogen dioxide	A	10 (4.54)
Nitrophenol (mixed)	B	100 (45.4)
Nitrotoluene	C	1,000 (454)
Paraformaldehyde	C	1,000 (454)
Parathion	A	10 (4.54)

TABLE 1.1   Reportable Quantities of Hazardous Substances Designated Pursuant to Section 311 of the Clean Water Act (*Continued*)

Material	Category	RQ in pounds (kilograms)
Pentachlorophenol	A	10 (4.54)
Phenol	C	1,000 (454)
Phosgene	A	10 (4.54)
Phosphoric acid	D	5,000 (2,270)
Phosphorus	X	1 (0.454)
Phosphorus oxychloride	C	1,000 (454)
Phosphorus pentasulfide	B	100 (45.4)
Phosphorus trichloride	C	1,000 (454)
Polychlorinated biphenyls	X	1 (0.454)
Potassium arsenate	X	1 (0.454)
Potassium arsenite	X	1 (0.454)
Potassium bichromate	A	10 (4.54)
Potassium chromate	A	10 (4.54)
Potassium cyanide	A	10 (4.54)
Potassium hydroxide	C	1,000 (454)
Potassium permanganate	B	100 (45.4)
Propargite	A	10 (4.54)
Propionic acid	D	5,000 (2,270)
Propionic anhydride	D	5,000 (2,270)
Propylene oxide	B	100 (45.4)
Pyrethrins	X	1 (0.454)
Quinoline	D	5,000 (2,270)
Resorcinol	D	5,000 (2,270)
Selenium oxide	A	10 (4.54)
Silver nitrate	X	1 (0.454)
Sodium	A	10 (4.54)
Sodium arsenate	X	1 (0.454)
Sodium arsenite	X	1 (0.454)
Sodium bichromate	A	10 (4.54)
Sodium bifluoride	B	100 (45.4)
Sodium bisulfite	D	5,000 (2,270)
Sodium chromate	A	10 (4.54)
Sodium cyanide	A	10 (4.54)
Sodium dodecylbenzenesulfonate	C	1,000 (454)
Sodium fluoride	C	1,000 (454)
Sodium hydrosulfide	D	5,000 (2,270)
Sodium hydroxide	C	1,000 (454)
Sodium hypochlorite	B	100 (45.4)
Sodium methylate	C	1,000 (454)
Sodium nitrite	B	100 (45.4)
Sodium phosphate, dibasic	D	5,000 (2,270)
Sodium phosphate, tribasic	D	5,000 (2,270)
Sodium selenite	B	100 (45.4)
Strontium chromate	A	10 (4.54)
Strychnine	A	10 (4.54)
Styrene	C	1,000 (454)
Sulfuric acid	C	1,000 (454)
Sulfur monochloride	C	1,000 (454)
2,4,5-T acid	C	1,000 (454)
2,4,5-T amines	D	5,000 (2,270)
2,4,5-T esters	C	1,000 (454)
2,4,5-T salts	C	1,000 (454)
TDE	X	1 (0.454)
2,4,5-TP acid	B	100 (45.4)
2,4,5-TP acid esters	B	100 (45.4)
Tetraethyl lead	A	10 (4.54)
Tetraethyl pyrophosphate	A	10 (4.54)
Thallium sulfate	B	100 (45.4)
Toluene	C	1,000 (454)
Toxaphene	X	1 (0.454)
Trichlorfon	B	100 (45.4)
Trichloroethylene	B	100 (45.4)
Trichlorophenol	A	10 (4.54)
Triethanolamine dodecylbenzenesulfonate	C	1,000 (454)
Triethylamine	D	5,000 (2,270)
Trimethylamine	B	100 (45.4)
Uranyl acetate	B	100 (45.4)
Uranyl nitrate	B	100 (45.4)
Vanadium pentoxide	C	1,000 (454)

TABLE 1.1   Reportable Quantities of Hazardous Substances Designated Pursuant to Section 311 of the Clean Water Act (*Continued*)

Material	Category	RQ in pounds (kilograms)
Vanadyl sulfate	C	1,000 (454)
Vinyl acetate	D	5,000 (2,270)
Vinylidene chloride	B	100 (45.4)
Xylene (mixed)	C	1,000 (454)
Xylenol	C	1,000 (454)
Zinc acetate	C	1,000 (454)
Zinc ammonium chloride	C	1,000 (454)
Zinc borate	C	1,000 (454)
Zinc bromide	C	1,000 (454)
Zinc carbonate	C	1,000 (454)
Zinc chloride	C	1,000 (454)
Zinc cyanide	A	10 (4.54)
Zinc fluoride	C	1,000 (454)
Zinc formate	C	1,000 (454)
Zinc hydrosulfite	C	1,000 (454)
Zinc nitrate	C	1,000 (454)
Zinc phenolsulfonate	D	5,000 (2,270)
Zinc phosphide	B	100 (45.4)
Zinc silicofluoride	D	5,000 (2,270)
Zinc sulfate	C	1,000 (454)
Zirconium nitrate	D	5,000 (2,270)
Zirconium potassium fluoride	C	1,000 (454)
Zirconium sulfate	D	5,000 (2,270)
Zirconium tetrachloride	D	5,000 (2,270)

enacted. It was designed to understand the use or development of chemicals and to provide controls, if necessary, for those chemicals which may threaten human health or the environment.

For many, familiarity with TSCA generally stems from its specific reference to polychlorinated biphenyls, which raise a vivid, deadly characterization of the harm caused by them. But TSCA is not a statute that deals with a single chemical or chemical mixture or product. In fact, under TSCA, EPA is authorized to institute testing programs for various chemical substances that may enter the environment. Under TSCA's broad authorization, EPA may obtain data on the production and use of various chemical substances and mixtures so that it can protect public health and the environment from the effects of harmful chemicals. In actuality, TSCA supplements the appropriate sections dealing with toxic substances in other federal statutes, e.g., section 307 of the Clean Water Act and section 6 of the Occupational Safety and Health Act.

The objective of TSCA is to provide the necessary control before a chemical is allowed to be mass-produced and enter the environment. TSCA has a two-pronged regulatory framework. In the first stage, TSCA helps EPA in acquiring information sufficient to identify and evaluate potential hazards from chemical substances. In the second stage, it allows EPA to regulate the production, use, distribution, and disposal of chemical substances as and when necessary.

At the heart of TSCA is a *premanufacture notification* (PMN) requirement under which a manufacturer must notify EPA at least 90 days prior to the production of a new chemical. In this context, a "new chemical" is a chemical that is not listed in EPA's TSCA-based Inventory of Chemical Substances or is an unlisted reaction product of two or more chemicals. For chemicals already on this list, a notification is required if there is a new use that could significantly increase human or environmental exposure. No notification is required for chemicals that are manufactured in small quantities solely for scientific research and experimentation.

TSCA requires manufacturers and processors of chemicals to maintain and submit all appropriate records including data related to any adverse health or environmental effects. TSCA authorizes EPA to order changes in processing or production steps, product recalls, and, in certain situations, total product ban. To facilitate this program, TSCA authorizes inspections of facilities by authorized agency personnel. Furthermore, TSCA permits citizen suits by the public against the violators.

Under TSCA's import rule, which became effective on January 1, 1984, each chemical substance subject to TSCA regulations that is imported into the United States must also be listed on the TSCA Inventory of Chemical Substances (TSCA inventory). For this purpose, the exact chemical composition or identity is important. An importer of a chemical substance must provide a certification of compliance to the U.S. Customs Service indicating that the shipment is in compliance with all applicable TSCA rules or is exempt from TSCA under certain exemptions as specified in TSCA.

Sections 8(a) and 8(d) of TSCA require the EPA administrator to promulgate rules whereby any person who manufactures, processes, or distributes in commerce any chemical substance or mixture must maintain records and submit copies of health and safety studies that have been made regarding such chemicals. Small manufacturers are generally exempted from the provisions of section 8 of TSCA, but they may be subject to record-keeping and notification requirements pursuant to other applicable laws and regulations.

In September 1982, EPA issued a final rule requiring the submittal of specific information on forty chemicals and chemical categories, thirty-nine of which (including asbestos) were recommended for testing by the eight-agency Interagency Testing Committee. On September 27, 1990, EPA imposed a notification requirement on the manufacturers and importers of chemical substances that produce or import 10,000 pounds or more per year per facility of chemicals that are listed on the TSCA inventory. This notification requirement came soon after an August 1990 supplemental listing of 5000 new chemicals in the 1990 TSCA inventory by EPA. Taken together, these attempts by EPA demonstrate a planned tightening of EPA's control over a significant number of chemicals that are either produced or imported for use or sale in this country.

Section 6(e) of TSCA aims at creating stringent control over manufacturing, processing, distribution in commerce, and the use of PCBs, which also go by the names Aroclor, Inerteen, Pyranol, and generically as askarels. This congressional mandate was based on a widely held belief that PCBs pose a significant risk to public health and the environment. There are currently three regulatory classifications of PCBs. Anything less than 50 parts per million (ppm) of PCB is treated as a non-PCB class and is specifically exempt from the TSCA-authorized regulations.

Under TSCA, EPA has been given broad discretionary power to take any regulatory measures deemed necessary to restrict chemicals suspected of causing harm to humans and the environment. To date, PCB regulation has been the prime focus of TSCA. One should, however, assume that TSCA will cover other chemicals in the long run. The 1990 tightening of the regulatory scheme of EPA is only a warning.

1.6   Resource Conservation and Recovery Act

Since its enactment in 1976, the Resource Conservation and Recovery Act (RCRA) has been amended several times, and each amendment has placed additional burdens on manufacturing and other industrial operations to promote safer solid and hazardous waste management programs. Notwithstanding these amendments, subtitle C of RCRA, which deals with hazardous waste, remains the centerpiece of RCRA. Besides the regulatory requirements for hazardous waste management, it spells out the mandatory obligations of generators, transporters, and disposers of hazardous waste as well as those of owners and/or operators of hazardous waste *treatment, storage,* or *disposal* (TSD) facilities.

Before any action under RCRA is planned, it is essential to understand what constitutes a solid waste and what constitutes a hazardous waste. Section 1004(27) of RCRA defines *solid waste* as follows:

> The term "solid waste" means any garbage, refuse, sludge, from a waste treatment plant, water supply treatment plant, or air pollution control facility and other discarded material, including solid, liquid, semisolid, or contained gaseous material resulting from industrial, commercial, mining and agricultural operations and from community activities, but does not include solid or dissolved materials in domestic sewage, or solid or dissolved materials in irrigation return flows or industrial discharges which are point sources subject to permits under section 402 of the Federal Water Pollution Control Act, as amended (86 Stat. 880), or source, special nuclear, or byproduct material as defined by the Atomic Energy Act of 1954, as amended (68 Stat. 923).

This statutory definition of solid waste is indeed quite complicated. EPA's promulgation of rules defining what can constitute a solid waste was subsequently challenged. One case in 1987 compelled EPA to modify its definition of solid waste based on abandonment or random discarding. Pursuant to the court decision in *American Mining Congress v. EPA,* EPA proposed to exclude certain recycled materials that are part of a continuous manufacturing process from the definition of solid waste on January 8, 1988.

Pursuant to EPA regulations, a solid waste becomes a hazardous waste if it exhibits any one of four specific characteristics: ignitability, reactivity, corrosivity, and toxicity. Under EPA's regulations, certain types of solid wastes (e.g., household waste) are not considered to be hazardous wastes irrespective of their characteristics. Additionally, EPA has provided certain regulatory exemptions based on very specific criteria. For example, hazardous waste generated in a product or raw material storage tank, transport vehicle, or manufacturing processes and samples collected for monitoring and testing purposes are exempt from the regulations.

Besides these four characteristics of hazardous wastes, EPA has established three hazardous waste lists: hazardous wastes from nonspecific sources (e.g., spent nonhalogenated solvents), hazardous wastes from specific sources (e.g., bottom sediment sludge from the treatment of wastewaters from wood preserving), and discarded commercial chemical products and off-specification species, containers, and spill residues.

Pursuant to EPA's hazardous waste regulations promulgated under RCRA, the act of discarding materials is not limited to planned disposal only. Any accidental spill involving any one of the listed chemicals, chemical products, or residues will trigger EPA's hazardous waste management program. As a result, even though a facility does not discard or plan on discarding any such chemical in the foreseeable future, it must undertake an appropriate compliance program under RCRA in order to respond to any accidental occurrence.

Under RCRA, the hazardous waste management program is based on a "cradle-to-grave" concept so that all hazardous wastes can be traced and fully accounted for. Section 3010(a) of RCRA requires all generators and transporters of hazardous wastes as well as owners and operators of all TSD facilities to file a notification with EPA within 90 days after the promulgation of the regulations. The notification should state the location of the facility and include a general description of the activities as well as the identified and listed hazardous wastes being handled.

While EPA does provide certain regulatory exclusions and specific variances (upon appropriate demonstration) from its solid waste classification, Congress mandated EPA to establish a special regulatory scheme for industrial operations and to provide appropriate RCRA permits for industries dealing with hazardous waste. Since August 16, 1985, EPA has required even the small quantity generators (SQGs) to use a uniform manifest (commonly known as RCRA manifest, or simply manifest) for shipping any hazardous waste. This essentially closed the benefits of permit exemption accorded to the small quantity generators. As the possibility of new RCRA amendments by Congress looms large, it is safe to suggest that, in the near future, small quantity generators may no longer enjoy the full benefits of current permit exemption status.

Under RCRA, all regulated hazardous waste facilities must exist and/or operate under valid, activity-specific permits. Two categories of TSD facilities currently exist: interim-status facilities and permitted facilities. Interim-status facilities are those facilities which are currently operating without final RCRA permits. This is in accordance with a legislative decision that the issuance of permits to such facilities would be time-consuming and should be temporarily deferred. Nevertheless, all interim facilities must meet a three-part statutory test. They must

- Be in existence as of November 19, 1980
- Notify EPA of their hazardous waste management activities in accordance with section 3010(a) of RCRA
- File preliminary permit applications

All TSD facilities are required to satisfy certain minimum record-keeping requirements. In addition to using a manifest for the transportation of hazardous waste, the facilities must keep records of vital information about the waste, such as the results of waste analysis, trial tests, and inspections; a description and the quantity of each hazardous waste received; and methods and dates of treatment, storage, and disposal. Furthermore, all basic reports, including a biennial report of waste management activities, must be filed with the regional administrator of EPA.

Under RCRA, the states may be authorized to develop and carry out their own hazardous waste programs in lieu of the EPA-administered federal program. But such delegation of hazardous waste programs must ensure that the state programs are no less stringent than the federal programs as formulated by EPA.

1.7 Comprehensive Environmental Response, Compensation, and Liability Act

The Comprehensive Environmental Response, Compensation, and Liability Act (CERCLA), popularly known as the *Superfund,* was passed by Congress and signed into law on December 11, 1980. While RCRA deals basically with the management of hazardous wastes that are generated, treated, stored, or disposed of, CERCLA provides a response to the environmental release of various hazardous substances. This includes the release of various pollutants or contaminants into the air, water, or land. A CERCLA response or liability will be triggered by an actual release or the threat of a "hazardous substance or pollutant or contaminant" being released into the environment.

It is no surprise that these terms have been defined very broadly under CERCLA. Pursuant to section 101(14) of CERCLA, a *hazardous substance* is any substance requiring special consideration due to its toxic nature by EPA under the Clean Air Act, the Clean Water Act, or the Toxic Substances Control Act (TSCA) beside CERCLA itself. Also, by definition, hazardous substance includes the *hazardous waste* as defined under RCRA. Additionally, under section 101(33) of CERCLA, a *pollutant or contaminant* can be any other substance not necessarily designated or listed but that "will or may reasonably" be anticipated to cause any adverse effect in organisms and/or their offspring. It is therefore obvious that CERCLA has a much broader jurisdiction than RCRA, which applies only to solid and hazardous wastes.

As its popular name implies, CERCLA includes a federal funding program for the necessary cleanup work once releases of hazardous substances into the environment have taken place. At present, there are two funding programs covered under CERCLA and its progeny, the Superfund Amendments and Reauthorization Act (SARA): the Hazardous Substance Response Fund and the Postclosure Liability Trust Fund.

The central purpose of CERCLA is to provide a response mechanism for cleanup of any hazardous substance released, such as an accidental spill, or of a threatened release of a hazardous substance.

The responsibilities of implementing various action plans or responses prompted by such releases are shared by several federal agencies. While EPA has the overall responsibility for plan implementation, several other federal agencies and departments are earmarked for specific tasks. For example, all evacuation and relocation is handled by the Federal Emergency Management Agency. The U.S. Coast Guard, on the other hand, provides appropriate response actions in coastal zones, the Great Lakes, ports, and harbors. Also, under this federal scheme, the Department of Defense deals with releases from its own vessels and facilities. In addition, specific administrative functions are handled by the Departments of Transportation, Health and Human Services, Interior, Treasury, Justice, and so on.

This has resulted in not only specific interrelationships within several federal agencies and departments but also certain overlaps between CERCLA and other federal statutes, such as RCRA, TSCA, CAA, and CWA. As a result, the definition of *hazardous substance* in CERCLA seems to be well conceived, although extremely broad. Section 101(14) of CERCLA therefore defines a hazardous substance which by reference relates to several other federal environmental statutes as well as section 102 of CERCLA itself.

Section 102 of CERCLA is a catchall provision because it requires EPA to promulgate regulations to establish "that quantity of any hazardous substance the release of which shall be reported pursuant to Section 103" of CERCLA. Thus, under CERCLA, the list of *potentially responsible parties* (PRPs) can include all direct and indirect culpable parties who have either released a hazardous substance or violated any statutory provision. In addition to its own response action for an actual or potential release of a hazardous substance, pollutant, or contaminant, EPA can find certain responsible private parties liable for cleanup actions and/or costs as well as for reporting requirements. In this context, note that in *O'Neill v. Q.L.C.R.I. Inc.,* the court extended a lender's liability, using CERCLA's broad liability provisions for PRPs, to a PRP for the water pollution caused. Under this theory, lending institutions which help finance property acquisitions may be found liable for various types of environmental pollution, such as air pollution, wetlands degradation, and ocean dumping in addition to the well-publicized cases involving releases of hazardous substances into the environment.

The centerpiece for response, cleanup, and remedial action is the National Contingency Plan (NCP), which was originally developed under section 311(c)(2) of CWA but whose scope has been significantly broadened under CERCLA. Presently, it includes establishing "procedures and standards for responding to releases of hazardous substances, pollutants, and contaminants." In essence, all environmental releases to land, air, and water of hazardous substances, pollutants, and contaminants fall under the aegis of the NCP. In *United States v. Carolawn Co., Inc.,* the court agreed with EPA's position that any amount of a listed hazardous substance will trigger CERCLA's jurisdiction. This makes consistency with the NCP extremely important in any cost recovery action.

Besides the NCP, section 105(a)(8)(B) of CERCLA requires EPA to establish a *national priorities list* (NPL) of inactive hazardous waste disposal sites that release or threaten to release any hazardous substance. National attention is given to such sites for the purpose of removal of hazardous substances, pollutants, or contaminants based on their degree of seriousness. This is in addition to appropriate remedial actions at those locations for protecting human health and the environment. To determine the extent of a cleanup, EPA may proceed with a *remedial investigation/feasibility study* (RI/FS) before the actual remedy is selected by a record of decision (ROD).

CERCLA, as indicated earlier, achieves its goals through well-orchestrated funding mechanisms. The Hazardous Substance Response Trust Fund (Response Fund) receives 87.5 percent of its revenue from petroleum and chemical feedstock taxes; the balance comes from an appropriation process. For the Postclosure Liability Trust Fund (Postclosure Trust Fund), a tax is imposed on qualified facilities where hazardous substances have been disposed. Persons responsible for hazardous substance releases are also held liable for all costs incurred during the cleanup and remedial actions. If a particular defendant is unavailable, the Response Fund thus set up will absorb the costs of remediation of all released hazardous substances.

While section 111 of CERCLA spells out the details of how and when money in the Hazardous Substance Superfund (or simply the Superfund) will be provided by the federal government, it also provides specific limitations and restrictions. For example, CERCLA has no provisions that compensate for personal injuries or death.

As in the case of RCRA, the states are free to impose additional liability on wrongdoers for the release of hazardous substances. There is no federal preemption clause to inhibit state liabilities beyond federal liabilities. Furthermore, under the 1980 provisions of CERCLA, there is no specific statute of limitations for cost recovery actions by private parties that are not seeking reimbursement from the federal funding program (i.e., from the Superfund). Pursuant to the 1986 amendments by way of SARA, however, no claim for recovery of damages against the responsible parties may be brought after 3 years of the date of the discovery of the loss or the date of EPA's final regulations, whichever is later. Section 112(d) of CERCLA prohibits presentation of any claim for recovery of response costs from the Superfund after 6 years from the date of completion of all response actions. Pursuant to section 113(g)(2) of CERCLA, any private party seeking cost recovery for a removal action from a PRP must initiate its action within 3 years of completion of its removal action. If the action is for remedial action cost recovery, it must be instituted within 6 years after initiation of on-site physical construction of the remedial action.

1.8   Superfund Amendments and Reauthorization Act

The Superfund Amendments and Reauthorization Act (SARA), enacted on October 17, 1986, is a progeny of CERCLA. It addresses closed hazardous waste disposal sites that may release hazardous substances into any environmental medium. The most revolutionary part of SARA is the Emergency Planning and Community Right-to-Know Act (EPCRA), which is covered under Title III of SARA.

Title III of SARA has brought sweeping changes to the old ways of dealing with hazardous or toxic chemicals and substances following the tragedy in Bhopal, India, with its release of deadly methyl isocyanate. The statute mandated both planning for chemical emergencies and prevention programs for chemical spills or accidents. Under the community right-to-know provision of Title III of SARA, several deadlines have been set for specific implementation.

EPCRA includes three subtitles and four major parts: emergency planning, emergency release notification, hazardous chemical reporting, and toxic chemical release reporting. Subtitle A is a framework for emergency planning and release notification. Subtitle B deals with various reporting requirements for "hazardous chemicals" and "toxic chemicals." Subtitle C provides various dimensions of civil, criminal, and administrative penalties for violations of specific statutory requirements.

Title III of SARA also requires EPA to review emergency systems for monitoring, detecting, and preventing releases of extremely hazardous substances at facilities that produce, use, or store such substances.

Other provisions of SARA basically reinforce and/or broaden the basic statutory program dealing with the releases of hazardous substances under CERCLA. Section 313 of EPCRA contained in Title III of SARA deals with the toxic chemical release report. It requires owners and operators of certain facilities that manufacture, process, or otherwise use one of the 329 listed chemicals and chemical categories to report all environmental releases of these chemicals annually. This information about total annual releases of chemicals from the industrial facilities can be made available to the public. EPA's regulations contain specific provisions for dealing with trade secrets and confidential information that EPA may receive from the owners and operators of the facilities that manufacture, process, import, or use any of the listed chemicals.

SARA created a new section 310 in CERCLA to deal with citizen suits for bringing civil actions against any person, including the federal government and any other government agency, for a violation of a standard, regulation, condition, requirement, or order. As a result, CERCLA now allows citizen groups to bring suits against EPA and other federal agencies for failure to comply with their mandatory duties or obligations under CERCLA. In addition, citizen suits may be directed to the PRPs responsible for releases of hazardous substances and for the necessary cleanup and remediation.

1.9 Occupational Safety and Health Act

There is little dispute that EPA is the prime federal agency responsible for administering the statutory requirements of various environmental statutes. As an exception to this general understanding, the Occupational Safety and Health Administration (OSHA) is responsible for administering the Occupational Safety and Health (OSH) Act of 1970. OSHA is entrusted with the major responsibility for workplace safety and worker's health. It is also responsible for certain consultative services for management, employer-employee training, and inspection of workplaces to ensure compliance and enforcement of applicable standards under the OSH Act. OSHA is the federal agency responsible under this statute to promulgate various legally enforceable standards.

Historically, unlike EPA, which was created as an independent regulatory agency, OSHA is a division of the Department of Labor. As a result, it is much less independent. While OSHA primarily functions as an enforcement agency, OSHA and EPA, in certain situations, have overlapping functions because they deal with the common subjects of human health and safety.

OSH Act was carefully drafted with a well-meaning goal. The goal is that "no employee will suffer material impairment of health or functional capacity" due to a lifetime occupational exposure to chemicals and hazardous substances. The statute imposes a duty on the employers to provide employees with a safe workplace environment, free of known hazards that may cause death or serious bodily injury. The statute covers all employers and their employees in all the states and federal territories with certain exceptions. Generally, the statute does not cover self-employed persons, farms solely employing family members, and those workplaces covered under other federal statutes.

These federal safety and health standards under OSH Act are actually provided by the National Institute for Occupational Safety and Health (NIOSH). OSHA basically carries out the enforcement functions through federal and state inspectors. Additionally, as a labor-oriented agency, OSHA provides public education and consultation on matters related to occupational health and safety.

Under OSH Act, several health standards have been promulgated. As of now, the final health standards promulgated by OSHA apply to asbestos, several carcinogens (as listed), vinyl chloride, inorganic arsenic, lead, coke oven emissions, benzene, cotton dust, 1,2-dibromo-3-chloropropane, acrylonitrile, ethylene oxide, and formaldehyde.

This is by no means a final list. Under OSH Act, the list of chemicals and health standards will be reviewed and expanded as new findings are made about health and safety aspects of various chemicals.

Like the majority of federal environmental programs, the states have been encouraged to come up with their own occupational safety and health programs. This has not taken place to any significant degree. When it does, state programs will be the primary regulatory controls, and they will not be preempted by the federal program as long as they are judged to be "at least as effective as" the federal program.

Current OSHA standards impose certain record-keeping and reporting requirements on employers. There are also requirements about posting of certain specific materials at designated or prominent locations at the workplace for the purpose of keeping employees informed. OSH Act provides employees with several rights, including the right to information on hazards at the workplace, a request for inspection by NIOSH or OSHA to determine hazards, and protection of job safety for exercising the statutory rights. The employer is subject to workplace safety standards and is responsible for monitoring the workplace environment for purposes of compliance.

In most states, the workers' compensation law provides an exclusive remedy to an injured employee and acts as a bar to any other recovery by the injured employee unless the employer is found liable for intentional torts or fails to secure the appropriate compensation for the injured employee. However, OSHA, as the federal enforcement agency, is not restricted from bringing additional charges against such employer because of violation of workplace safety standards. In its present form, the hazard communication regulation (also known as the *worker right-to-know* rule) requires that all hazardous chemicals be labeled, that all *material safety data sheets* (MSDSs) for the chemicals with special reference to hazards be prepared, and that all affected workers and customers be informed of potential risks to health and safety stemming from exposure and handling of workplace chemicals. There is some overlap here with EPCRA because subtitle B of EPCRA requires reporting of "hazardous chemicals" for which material safety data sheets must be prepared or made available pursuant to OSHA regulations.

In October 1987, OSHA initiated the proposed federal cancer policy, which suggests further tightening of toxic chemical regulations. Three other agencies—EPA, the Food and Drug Administration (FDA), and the Consumer Product Safety Commission (CPSC)—are expected to follow suit.

1.10   Federal Insecticide, Fungicide, and Rodenticide Act

The Federal Insecticide, Fungicide, and Rodenticide Act (FIFRA) was enacted by Congress on October 21, 1972. Since that time, FIFRA has been amended several times, most notably by the Federal Environmental Pesticide Control Act (FEPCA) of 1982 and several FIFRA amendments in 1975, 1978, 1980, 1984, 1988, and 1990, of which the latter two are considered most significant.

While the increasing use of pesticides in this country has been credited with the bountiful crop harvests and significant reductions in crop infestation by pests, Congress has been aware for more than half a century through public outcry about the health and environmental hazards that are typically associated with them. It is generally undisputed that FIFRA was enacted to deal with toxicity that may be linked to these pesticides and related synthetic products.

These pesticides and like products are mostly manufactured. Because of their intended use, in most instances, they are toxic by design, to be effective. As a result, while the benefits of various pesticides, herbicides, and rodenticides are not overlooked, these chemicals require tight control. The statutory provisions of FIFRA and the regulations promulgated thereunder are aimed at doing just that—keeping a tight control on the production and application of various pesticides, fungicides, and rodenticides.

In this context, note that many of these regulated pesticides and herbicides, because of their primary use on agricultural land, enter the receiving waters either through point sources of waste stream or by surface runoff. As a result, such toxic effluents or discharges may also be regulated by the Clean Water Act or by its most recent amendment, the Water Quality Act, which specifically addresses the issue of stormwater discharges.

The heart of FIFRA is a pesticide registration program. In this respect, FIFRA resembles TSCA since TSCA also deals with manufactured chemical substances and mixtures that may be toxic. Pursuant to the current EPA regulations promulgated under FIFRA, all new pesticide products, with minor exceptions, must first be registered with EPA prior to their use, manufacture, or distribution in commerce. EPA also retains the authority to issue a suspension order or cancellation of any registered pesticide if there is a "substantial question of safety" or the pesticide itself presents an "imminent hazard" to people or the environment.

Under the 1988 FIFRA amendments, the states have been given greater regulatory and primary enforcement power over pesticides. As a result, while EPA still retains the overall responsibility and the ultimate veto power, states are assuming a greater role in pesticide application and use, including enforcement.

1.11   Hazardous Materials Transportation Act

The Hazardous Materials Transportation Act (HMTA) was enacted in 1974 and reauthorized by Congress in 1990. The new Hazardous Materials Transportation Uniform Safety Act (HMTUSA) of 1990 incorporates several new legislative pieces while retaining the major provisions of the original HMTA. The statute authorizes the Secretary of Transportation to establish and enforce hazardous material regulations for all modes of transportation by highway, water, and rail.

Under this statutory authority, the Secretary of Transportation has broad authority to determine what is a hazardous material, using the dual tools of quantity and type. By this two-part approach, any material that may pose an unreasonable risk to human health or the environment may be declared a hazardous material. Such a designated hazardous material obviously includes both the quantity and the form that make the material hazardous.

The basic purpose of HMTA is to ensure safe transportation of hazardous materials through the nation's highways, railways, and waterways. The basic theme of HMTA is to prevent any person from offering or accepting a hazardous material for transportation anywhere within this nation if that material is not properly classified, described, packaged, marked, labeled, and properly authorized for shipment pursuant to the regulatory requirements. Under the Department of Transportation (DOT) regulations, a *hazardous material* is defined as any substance or material, including a hazardous substance and hazardous waste, which the Secretary of Transportation deems capable of posing an unreasonable risk to health, safety, and property when transported in commerce. DOT thus has broad authority to regulate the transportation of hazardous materials which, by definition, include hazardous substances as well as hazardous wastes.

HMTUSA imposes specific restrictions on the packaging, handling, and shipping of hazardous materials. For shipping and receiving of hazardous chemicals, hazardous wastes, and radioactive materials, the appropriate documentation, markings, labels, and safety precautions are required. Bills of lading and manifests are required and must accompany all shipments. In addition, the drivers responsible for transportation of hazardous materials must meet certain minimum qualifications and use only those vehicles which are inspected and approved.

If a discharge or release of a hazardous substance or hazardous waste or any other hazardous material occurs during transportation, an official of the local, state, or federal government acting within the scope of official responsibility may authorize immediate removal of such material without the preparation of a manifest.

HMTUSA imposes stricter penalties than the original HMTA did. Both civil penalties and criminal sanctions may be imposed, depending on the nature of the violation.

1.12   Federal Food, Drug, and Cosmetic Act

The Federal Food, Drug, and Cosmetic Act (FFDCA) of 1938 was intended to protect U.S. consumers from harmful and potentially harmful food and food additives as well as drug and cosmetic products. This legislation was a marked departure from the original Food and Drug Act of 1906 which was predicated on voluntary inspection and very little regulation. Nevertheless, the original Federal Food, Drug, and Cosmetic Act of 1938 was ambiguous. Moreover, the statutory authority granted to the federal Food and Drug Administration (FDA), the agency responsible for regulation under this statute, was often embroiled in and challenged by various court actions. Congress responded to this by passing an amendment in 1953. The 1953 amendment not only made well-conceived provisions for the safety, purity, and wholesomeness of food and food products but also provided FDA with an armor of clear and unambiguous power to regulate them.

FDA is the prime regulatory agency authorized under FFDCA. In fact, all FDA-promulgated regulations to date have only been made pursuant to FFDCA. The current FDA regulation over food and food products reflects three areas of concern and the need for control. The first area of concern is with respect to the safety of food. The emphasis here is the need for an assurance that the food sold and consumed in this nation does not present any health hazard. The second area of concern deals indirectly with the safety of food so that any contamination or adulteration does not include any hazardous or harmful ingredients. Here the emphasis is on the conditions under which the food is processed, the types of material that are either deliberately or inadvertently introduced during the processing or production stages, and the general quality or purity levels of various food ingredients used. The third area of concern is the possibility of economic adulteration of food which degrades the general quality or nourishment value of the food but does not necessarily make it unwholesome.

By necessity, FDA regulations are exceedingly complex, particularly when the underlying judgments about the wholesomeness of food are not always scientific and are often based on personal knowledge. The reason for this is the statutory definition of adulteration. Section 402(a)(3) of FFDCA defines a food to be *adulterated* "if it consists in whole or in part of any filthy, putrid or decomposed substance, or it if is otherwise unfit for food." In *Young v. Community Nutrition Institute,* the Supreme Court upheld FDA's discretionary authority under the statute to promulgate or not to promulgate tolerance levels of harmful substances by any rational standards.

As of now, it is up to FDA to set a tolerance level for harmful, hazardous, and toxic substances present in foods or food products. In this context, note that EPA is the regulatory agency for pesticides which often can be found in foods. While EPA derives its general authority for regulating pesticides under FIFRA, EPA is also specifically authorized to set tolerance limits for pesticide residues in food under an agreement between EPA and FDA. Hence, the indirect role of EPA in the area of foods or food products should not be overlooked, particularly when it is the prime regulatory agency dealing with toxic and hazardous substances, pesticides, and other chemicals.

1.13 Safe Drinking Water Act

The Safe Drinking Water Act (SDWA), enacted in 1974, was amended several times by Congress in the 1970s and 1980s, the latest coming on June 19, 1986, to require EPA to set national drinking water standards. Thus the motion was set to implement various statutory provisions of SDWA at an early date while assuring the public about high-quality water supplies through public water systems.

SDWA is truly the first federal intervention to set the limits of contaminants in drinking water. The 1986 amendments came 2 years after the passage of the Hazardous and Solid Waste Amendments (HSWA) or the so-called RCRA amendments of 1984. As a result, certain statutory provisions were added to these 1986 amendments to reflect the changes made in the *underground injection control* (UIC) systems. In addition, the Superfund Amendments and Reauthorization Act (SARA) of 1986 set the groundwater standards the same as the drinking water standards for the purpose of necessary cleanup and remediation of an inactive hazardous waste disposal site.

Pursuant to the provisions of SDWA, EPA is required to undertake a two-prong program: to identify contaminants in drinking water that may have a deleterious effect on public health and to set a *maximum contaminant level* (MCL) for each such contaminant. In this context, the MCL means the maximum permissible level of a contaminant that can be present in water which is supplied through a public water system that has at least fifteen service connections or which regularly serves at least twenty-five individuals.

The 1986 amendments of SDWA included additional elements by requiring

EPA to establish *maximum contaminant level goals* (MCLGs) and national primary drinking water standards. The MCLG must be set at a level at which no known or anticipated adverse effects on human health occur, thus providing an "adequate margin of safety." Establishment of a specific MCLG depends on the evidence of carcinogenicity in drinking water or EPA's reference dose which is individually calculated for each specific contaminant. The MCL, an enforceable standard, however, must be set to operate as the *national primary drinking water standard* (NPDWS). Theoretically, the MCLs will be very close to the MCLGs whenever feasible. Furthermore, the 1986 amendments required EPA to identify those drinking water contaminants which may be comprised of volatile organic compounds, organics, inorganics, radionuclides, pesticides, and conventional pollutants including heavy metals and microorganisms. The EPA administrator is authorized to publish an MCLG and promulgate a national primary drinking water regulation for each identified contaminant.

In the context of hazardous and toxic substances, the priority list of drinking water contaminants is very important since it includes the contaminants known for their adverse effect on public health. Furthermore, most if not all are known or suspected to have hazardous or toxic characteristics that can compromise human health.

The statute requires EPA to update its priority list. In its first list, EPA is required to publish MCLs and MCLGs for twenty-five of the contaminants listed on the priority list of drinking water contaminants. At 3-year intervals, starting with the first publication of the MCLs and MCLGs for twenty-five priority contaminants, EPA is required to publish the MCLs and MCLGs for another twenty-five priority-list contaminants.

Table 1.2 provides the list of 83 contaminants which must be regulated under SDWA. In July 1987, EPA promulgated the MCLs and MCLGs for only nine contaminants. However, in January 1988, EPA published a priority list of drinking water contaminants to satisfy one of the mandates under the 1986 amendments of SDWA. On May 22, 1989, EPA proposed the MCLs and MCLGs for thirty-six contaminants. On January 30, 1991, EPA issued its final rule promulgating the MCLGs and the national primary drinking water regulations (NPDWRs) for twenty-six synthetic organic chemicals (SOCs) and seven inorganic chemicals (IOCs). The NPDWRs provide the MCLs or the treatment levels for SOCs and IOCs.

As indicated earlier, SDWA requires EPA to regulate UIC programs. This is done to protect and safeguard a vast number of aquifers which provide a source of drinking water to numerous people in this country, particularly in rural areas. Under SDWA, EPA is authorized to list all those states where there is a need for regulated UIC programs to prevent endangerment of drinking water quality. Pursuant to its statutory authority, EPA may implement and enforce a state UIC permit program. This is most likely to include cases where certain hazardous waste disposal, mining operations, and oil and gas production activities may make the drinking water supply system susceptible to underground injection, if unregulated. Under EPA's permitting program,

TABLE 1.2 Statutory List of 83 Contaminants for Regulation under the Safe Drinking Water Act

Inorganics	
1. Arsenic	13. Molybdenum*
2. Barium	14. Asbestos
3. Cadmium	15. Sulfate
4. Chromium	16. Copper
5. Lead	17. Vanadium*
6. Mercury	18. Sodium*
7. Nitrate	19. Nickel
8. Selenium	20. Zinc*
9. Silver*	21. Thallium
10. Fluoride	22. Beryllium
11. Aluminum*	23. Cyanide
12. Antimony	

Organics (pesticides/PCBs)	
1. Endrin	19. 1,1,2-Trichloroethane
2. Lindane	20. Vydate
3. Methoxychlor	21. Simazine
4. Toxaphene	22. PAHs
5. 2-4-D†	23. PCBs
6. 2,4,5-TP(Silvex)	24. Altrazine
7. Aldicarb	25. Phthalates
8. Chlordane	26. Acrylamide
9. Dalapon	27. 1,2-Dibromo-3-chloropropane (DBCP)
10. Diquat	28. 1,2-Dichloropropane
11. Endothall	29. Pentachlorophenol
12. Glyphosate	30. Pichloram
13. Carbofuran	31. Dinoseb
14. Alachlor	32. Ethylene dibromide (EDB)
15. Epichlorohydrin	33. Dibromomethane*
16. Toluene	34. Xylene
17. Adipates	35. Hexachlorocyclopentadiene
18. 2,3,7,8-TCDD (dioxin)	

Volatile organic compounds	
1. Trichloroethylene	8. Benzene
2. Tetrachloroethylene	9. Chlorobenzene
3. Carbon tetrachloride	10. Dichlorobenzene
4. 1,1,1-Trichloroethane	11. Trichlorobenzene
5. 1,2-Dichloroethane	12. 1,1-Dichloroethylene
6. Vinyl chloride	13. *Trans*-1,2-Dichloroethylene
7. Methylene chloride	14. *Cis*-1,2-Dichloroethylene

Microorganisms and turbidity	
1. Total coliforms	4. Viruses
2. Turbidity	5. Standard plate count
3. *Giardia lamblia*	6. *Legionella*

Radionuclides	
1. Radium 226 and 228	4. Gross alpha particle activity
2. Beta particle and photon radioactivity	5. Radon
3. Uranium	

*These seven contaminants have subsequently been substituted by aldicarb sulfone, aldicarb sulfoxide, ethylbenzene, heptachlor, heptachlor epoxide, nitrite, and styrene.
†Also known as 2,4-Dichlorophenoxy acetic acid.

all underground injection activities not only must comply with the drinking water standards, but also must meet specific permit conditions that are in unison with the provisions of the Clean Water Act. However, depending on the classification of underground wells, the regulatory requirements may vary.

To date, there are five broad classifications for underground wells:

1. Class I injection wells are used for industrial and municipal waste disposal, including storage and disposal of nuclear and hazardous waste.

2. Class II injection wells are used for brine injection or reinjection for oil and gas recovery and for the storage of liquid hydrocarbons at standard temperature and pressure.

3. Class III injection wells are used for extractions of materials for energy or minerals.

4. Class IV injection wells are used for the injection of hazardous or radioactive wastes into or above underground sources of drinking water.

5. Class V injection wells may be used for alternate energy development and in conjunction with other systems, such as septic systems and cesspools.

Under the "hammer" provisions of the Resource Conservation and Recovery Act, class IV injection wells are no longer permitted. In addition, the 1984 RCRA amendments contain several restrictions and provisions for the permitting of underground injection wells that may be used for storage and disposal of hazardous wastes.

SDWA provides pertinent provisions for state implementation plans. EPA may delegate administration of its regulatory program to a state as long as the state's program and the drinking water standards are "no less stringent" than the federal program and the appropriate national drinking water standards.

1.14 Oil Pollution Act

The Oil Pollution Act of 1924 was the first federal statute prohibiting pollution of waters strictly by oil. This statute, however, has been generally ineffective for lack of appropriate regulations and has been largely revamped by later federal statutes dealing with oil pollution of navigable waters of the United States.

As indicated earlier, the Federal Water Pollution Control Act (FWPCA) of 1972 provided a comprehensive plan for the cleanup of waters polluted by oil spills and intentional or accidental release of oil into the water. The subsequent laws, including the Clean Water Act of 1977 and with its later amendments, provide for regulation of pollution of waters by oil spills and other forms of unpermitted discharges. These legislations also incorporate certain provisions of the Rivers and Harbors Act of 1899, which was intended to prevent any obstruction to the use of navigable waters for interstate commerce. This statute, although quite old, has also been interpreted broadly by the U.S. Supreme Court in a 1966 case. In *United States v. Standard Oil Co.*, the Supreme Court held that the statutory provisions of the Rivers and Harbors Act bar obstruction of navigable waters by discharges of refuse or industrial waste, including oil. However, the invocation of this statute is mostly ceremonial now, in view of the specific provisions of CWA and its later amendments which prohibit pollution of waters by the discharge or spill of oil.

EPA regulations promulgated under the Federal Water Pollution Control Act of 1972 and the Deepwater Port Act of 1974, as set forth in 40 CFR part 110, also address the issue of discharge of oil into or upon navigable waters and adjoining shorelines. The applicable regulations for the prevention of oil pollution of navigable waters and adjoining shorelines by non-transportation-related onshore and offshore facilities are set forth in 40 CFR part 112.

The Oil Pollution Act (OPA) of 1990 is the latest of the federal statutes dealing with oil pollution of U.S. waters. OPA is the most comprehensive federal law dealing with both prevention of oil spills and the ensuing liability. However, unlike other laws dealing with oil pollution of navigable waters, OPA specifically deals with petroleum vessels and onshore and offshore facilities and their owners and operators.

OPA includes a panoply of operational requirements as well as a new liability and compensation scheme. It also imposes strict liability for oil spills on the owners and operators of petroleum vessels and facilities. In addition, OPA increases the liability limits for oil spills set under the Clean Water Act by more than tenfold. Pursuant to the statutory provisions, all claims for removal costs and damages under OPA, except specific exceptions, must first be presented to the responsible party or guarantor. If such parties are unavailable, the fund specifically set up pursuant to OPA will be responsible for compensation. For an oil spill at an onshore facility, a responsible party may be responsible for up to $350 million. This may be reduced up to $8 million at the discretion of EPA.

In the true sense, OPA reinforces as well as increases the current liability for any unpermitted discharge of oil, whether intentional or accidental, based on the Clean Water Act. But it does not replace CWA or any other prior federal law. As a result, it must be reviewed along with other applicable federal laws dealing with oil spills, e.g., the Ports and Waterways Safety Act, the Outer Continental Shelf Lands Act, and the Clean Water Act.

1.15 Atomic Energy Act and Other Federal Nuclear Waste Statutes

The Atomic Energy Act (AEA) of 1954 was truly the first comprehensive federal legislation to address the use of atomic energy for peaceful purposes. However, AEA does not provide any statutory guidance with regard to the handling and disposal of radioactive wastes.

Due to the prolific growth of both low-level and high-level radioactive wastes as a result of uranium mining, nuclear power plant operations, and production of nuclear weapons and bombs, the general provisions of AEA have very little impact today. Congress has attempted to address this problem and close the gap in existing federal legislations by enacting several new federal statutes for both high- and low-level radioactive wastes in recent years. However, the issue of handling and disposal of radioactive waste is not yet fully resolved.

Generally high-level radioactive wastes are generated during reprocessing of spent nuclear fuel and use of fuel rods. Low-level radioactive wastes, however, are generated from various nuclear power plants and medical, commercial, and nonmilitary uses of radioactive isotopes as well as the contamination of items by radioactive materials. Currently, the Department of Energy (DOE) is the regulatory agency responsible for locating and developing high-level radioactive or nuclear waste repositories. The Nuclear Regulatory Commission (NRC), on the other hand, is responsible for regulating nuclear power generation throughout the nation.

The Low-Level Radioactive Waste Policy Act (LLRWPA) of 1980 specifies that each state is responsible for providing appropriate disposal of the low-level (radioactive) waste (LLW) generated within its borders. In addition, LLRWPA encourages a regional concept, popularly known as *compacts,* for establishing regional low-level radioactive waste disposal sites.

The Nuclear Waste Policy Act (NWPA) of 1982 addresses the growing problem of high-level radioactive wastes. It provides a comprehensive high-level radioactive waste management plan and includes a funding program similar to the Superfund for the purpose of nuclear waste repository selection and construction. To date, only two nuclear or radioactive waste repositories, including one at Yucca Mountains, have been selected. But unfortunately no construction activities are planned at either location due to public opposition and growing controversies about the selection process.

This leaves the subject of entire radioactive wastes and their handling or disposal in suspension. Under the statutory authority, DOE is still the regulatory agency for "source," "special nuclear," and "byproduct material." However, in the absence of appropriate disposal sites, its role in radioactive waste disposal has been relegated to that of a toothless tiger.

In this context, it should be stressed that RCRA provisions include all types of radioactive waste except the above-referenced three types of high-level radioactive wastes. RCRA is also the federal statute with complete authority over "mixed waste" which is generated due to the contamination of hazardous waste by radioactive waste. Pursuant to RCRA, EPA has the ultimate say in the disposal of all hazardous wastes including mixed waste.

Finally, it is extremely important to note that both federal and state laws may apply to radioactive wastes and their proper disposal. The case laws generally suggest that there is no federal preemption in the case of nuclear or radioactive wastes.

1.16   Common Laws and Miscellaneous Statutes

In addition to the provisions included in the above federal statutes, private citizens may seek remedy against damages due to pollution from hazardous and toxic substances by invoking causes of action under common laws. These common law remedies can be found where the causes of action constitute a "tort" or "tortious conduct." Briefly, they include the following causes of action under the principles of tort:

1. *Private nuisance:* This arises when one interferes with the use and enjoyment of another's land or property.

2. *Public nuisance:* This arises when one's activities interfere with the rights of a community to enjoy reasonable property rights and where certain individual members of the community suffer damages not experienced by other members of the community.

3. *Trespass:* This arises when one actually enters upon another person's land without that person's invitation or permission and causes an interference with the landowner's use and enjoyment of the land.

4. *Strict trespass:* This arises due to an unprivileged and unauthorized entry by one into another's land. However, this remedy for pollution damages is available in only a few states.

5. *Negligence:* This arises when one's negligent action causes another to suffer personal injury or property damages.

In the context of these common law remedies, note that some of the causes of action require proof of fault and actual damages along with proximate or legal cause. But some are based on strict liability provisions whereas others may involve equitable relief in the form of a temporary or permanent injunction or temporary restraining order.

It is, however, obvious that any harm to human health or the environment by hazardous materials can be forcefully challenged under a number of federal statutes with specific liability provisions and also by common law principles. It is, therefore, a cardinal duty to fulfill one's statutory and regulatory obligations while handling hazardous materials, in the form of hazardous substances, toxic chemicals, hazardous wastes, or other dangerous and harmful pollutants or contaminants.

Finally, it is appropriate to emphasize that there are other lesser known federal statutes, not generally labeled as environmental statutes, which contain very explicit language or definition of what may be construed as a hazardous substance or material. For example, the Federal Hazardous Substances Act (FHSA) of 1960 defines a *hazardous substance* as any substance or mixture which may be toxic, corrosive, an irritant, a strong sensitizer, flammable or combustible, or generating pressure through decomposition, heat, or other means and which may cause substantial personal injury or illness. While the importance of these statutes has lessened due to the enactment of several major environmental statutes since the 1970s, they have not been repealed or otherwise obliterated. They can be invoked as good laws under the appropriate circumstances. As a result, nothing should be disregarded casually while one is dealing with any type of hazardous material, substance, or waste.

Recommended Reading

1. S. Majumdar, "Regulatory Requirements and Hazardous Materials," pp. 17–24, *Chemical Engineering Progress,* American Institute of Chemical Engineers, New York, May 1990.

2. J. C. Sweeney, "Protection of the Environment in the United States," *Fordham Environmental Law Report,* vol. 1, no. 1, Spring 1989.

3. J. G. Arbuckle, M. E. Bosco, D. R. Case, E. P. Laws, J. C. Martin, M. L. Miller, R. V. Randle, R. G. Stoll, T. F. P. Sullivan, T. A. Vanderver, Jr., P. A. J. Wilson, *Environmental Law Handbook,* 10th ed., Government Institutes, Inc., Rockville, Md., 1989.

4. P. Grad, *Treatise on Environmental Law,* vol. IA, Matthew Bender, New York, 1987.

5. W. P. Keaton, (ed.), *Prosser and Keaton on the Laws of Torts,* 5th ed., West Publishing Co., St. Paul, Minn., 1984.

6. Health Assessment Guidance Manual (draft), Agency for Toxic Substances and Disease Registry, U.S. Department of Health and Human Services, October 31, 1990.

Important Cases

1. *Young v. Community Nutrition Institute,* 106 S. Ct. 2360 (1986).

2. *Chemical Manufacturers Association v. Natural Resources Defense Council, Inc.,* 470 U.S. 116 (1985).

3. *Strycker's Bay Neighborhood Council v. Karlen,* 444 U.S. 223 (1980).

4. *Vermont Yankee Nuclear Power Corp. v. NRDC, Inc.,* 435 U.S. 519 (1978).

5. *United States v. Standard Oil Co.,* 384 U.S. 224 (1966).

6. *American Mining Congress v. EPA,* 824 F. 2d 1177 (D.C. Cir. 1987).

7. *O'Neill v. Q.L.C.R.I., Inc.,* No. 88-0704P (D.R.I. Oct. 29, 1990).

8. *U.S. v. Carolawn Co., Inc.,* 698 F. Supp. 616 (D.S.C. 1987).

Carcinogens and Their Regulatory Control

2.1 Introduction

Records show that since World War II, there has been a drastic surge in the production and development of chemicals, particularly organic chemicals. Most of these chemicals have generally played a significant role in our industrial progress and overall prosperity. However, they have also accounted for certain bitter consequences by increasing the risks to human health and the environment. Some of these chemicals have been found to be too hazardous or toxic. They have been confirmed or suspected of causing cancer in human beings and animals exposed to such chemicals.

Throughout the nation, there is an abundance of hazardous and toxic pollutants in all three environmental media (air, water, and land). This is an inevitable result of our industrial revolution. The cancer-causing chemicals, generally known as carcinogens, have been of great concern since they may cause cancer from exposure or physical contact. Most of the time, such deadly exposure or contact is the consequence of human confinement to an enclosed building or surrounding where carcinogenic chemicals are manufactured, processed, used, or stored.

The Occupational Safety and Health (OSH) Act of 1970 was passed to ensure "safe and healthful working conditions" in as great a degree as possible to the workers who may encounter safety and health risks within the working environment. Section 6(b)(5) of OSH Act states the statutory goal: to ensure that no worker will suffer "material impairment of health or functional capacity" even if there is an occupational exposure during the worker's entire working life.

OSH Act established the National Institute for Occupational Safety and Health (NIOSH) in the Department of Health, Education, and Welfare (DHEW) and the Occupational Safety and Health Administration (OSHA) in the Department of Labor (DOL). Pursuant to OSH Act, NIOSH has been given advisory responsibility while OSHA has been empowered with the overall administrative responsibility, including the setting of safety and health standards, enforcement of the statutory and regulatory provisions, and public education.

2.2 Cancer and Chemical Carcinogens

Cancer is a term used to describe not one specific disease but several diseases. In reality, cancer is an amalgamation of diseases which result from many causes, including environmental factors and hereditary or genetic makeup.

In general, cancer is typically characterized by an uncontrolled growth of cancerous or abnormal cells. In all cases of cancer, the structure of a deoxyribonucleic acid (DNA) undergoes a change, thereby causing the cell containing that DNA to function abnormally. This change in DNA may occur spontaneously or may be caused by an external factor. In any event, the change in DNA results in uncontrolled cell division or growth which frequently produces a mass of tissue characterized as tumor.

However, not all tumors are cancerous. When a tumor is self-contained, it is called a benign tumor. When a tumor is not self-contained, it is known as a malignant tumor. Cells from a malignant tumor separate and travel through the bloodstream to the lymph channels and to other parts of the body, thus setting up satellite tumors, known as metastases. These metastases invade other normal tissues, thereby resulting in their abnormal growth. The process of malignancy continues unless it can be stopped or reversed by external factors, including anticancer drugs, radiation or chemotherapy, and surgical removal of malignant tumors.

Different kinds of cancers grow at different rates. But all produce malignant tumors that displace normal cells, prevent them from getting nourishment, and eventually kill the normal or healthy cells. It is important to note that certain types of benign tumors may be early precursors for the development of malignant tumors. As a result, benign tumors should not be completely ignored.

There is no one reason or cause for contracting or developing cancer. While it is not considered contagious, it may be the result of hereditary or genetic makeup. A significant percentage of cancers result from cell mutations or spontaneous changes that take place in the cells. Nevertheless, a high percentage of cancers are believed to be caused by environmental factors, such as sunlight, radiation, chemicals, and viruses.

The environmental agents which produce cancers are collectively called *carcinogens*. Carcinogens include both manufactured (i.e., synthetic) and natural chemicals which may be present in air, water, and various foods and food products. In addition, exposure to certain environmental factors, such as x-rays and sunlight, can cause cancer.

A carcinogen is generally defined as a substance or condition which increases the incidence of irreversible benign or malignant tumors or produces unusual tumors in animals and human beings. A chemical carcinogen, as the term implies, is caused by a chemical, natural or synthetic, inorganic or organic, although the latter type is generally more common in inducing cancers. It is a well-documented fact that chemical carcinogens can enter the body through the skin, through the lungs, or through the digestive system.

Broadly categorized, there are two types of chemical carcinogens: one acts directly, and the other acts indirectly. A directly acting chemical carcinogen generally causes cancer at the site of exposure. For example, certain *polynuclear aromatic hydrocarbons* (PAHs), when inhaled, may cause cancer of the lungs, whereas emissions from coke ovens may cause skin cancer through contact with the skin.

The indirectly acting carcinogens are changed chemically to other substances, known as metabolites. This is brought about by the process of digestion or metabolism. The metabolites may cause changes in the lining of the urinary bladder, thus giving rise to bladder tumors and eventually bladder cancers.

Cancers can also develop through a combination of several carcinogens or an amalgamation of carcinogens and noncarcinogens. Furthermore, besides causing cancers, carcinogens may cause other health problems which may have short- or long-term effects.

2.3　Identification of Cancer-Causing Agents

Not every chemical or every environmental factor causes cancer. In fact, it is claimed that certain environmental factors that a human being encounters actually help prevent cancers. This is achieved through blocking or reversal of the effects of cancer-causing agents or carcinogens. In addition, it has been established that the majority of chemicals, even those which are otherwise toxic or harmful to human health or the environment, are noncarcinogenic.

In short, relatively few chemicals, natural or synthetic, cause cancers. However, it is not always easy to determine which chemical is a carcinogen and which is not. This is due to both the problem of tracking down the exposed population and the variability of the latency periods. For example, cancers linked to asbestos or tobacco smoke grow for many years before they can be detected. In addition, the problem of identification and detection of carcinogens is greatly aggravated by the indirect contribution of various environmental factors. This difficulty in isolating one chemical compound or substance often leads to a temporary listing of suspected carcinogens.

To date, some thirty agents or industrial processes have been identified as causing cancers in humans due to different forms of exposure. This identification also provides a very helpful insight into how humans may be exposed to these cancer-causing agents or processes. Table 2.1 is a list of cancer-causing agents, included in the National Toxicology Program. It also specifies likely means of human exposure to such agents based on available scientific data and epidemiological studies.

It is important to know the dose-and-effect relationship of carcinogens or suspected carcinogens. Generally speaking, the formation of a tumor is the re-

TABLE 2.1   Cancer-Causing Agents and Their Likely Exposure

Cancer-causing agent	Likely use and/or exposure route
1. 4-Aminobiphenyl	Used as an antioxidant in rubber and in the manufacture of dyes.
2. Arsenic and certain arsenic compounds	Used in pesticides, glass manufacturing, ceramics, and metal ore smelting. Also found in various foods and food products, including drinking water.
3. Asbestos	Present in asbestos-containing materials (ACMs) which are used in fire-resistant products, thermal insulation, and brake linings.
4. Auramine	Used in dye manufacturing.
5. Benzene	Used in the manufacturing of chemicals, plastics, paints, and adhesives. Also present in gasoline and fuel oils.
6. Benzidine	Used in dye manufacturing.
7. N,N-*Bis*-(2-chloroethyl)-2-naphthylamine	Used in drugs.
8. *Bis*-(chloromethyl) ether and chloromethyl ether (technical grade)	Used in the manufacture of chemicals and plastics.
9. Chlorambucil	Used in drugs.
10. Chromium and chromium compounds	Used in the manufacture of metal alloys, protective metal coatings, and paints. Also naturally present in drinking water and food.
11. Coke oven emissions	Released during coke manufacturing.
12. Cyclophosphamide	Used in drugs.
13. Diethylstilbestrol	Used in drugs.
14. Hematite ores	Present in underground iron ore mines and deposits.
15. Isopropyl alcohol	Manufacture of isopropyl alcohol using a strong acid process.
16. Melphalan	Used in drugs.
17. Mustard gas	Used as a chemical warfare agent.
18. 2-Naphthylamine	Used in dye manufacturing.
19. Nickel refining	Found in the refining process for nickel and nickel compounds.
20. Radiation and radioactive materials	Present in sunlight and in natural uranium deposits. Exposure is also likely during certain medical examination involving x-rays and various industrial operations involving radioactive materials including nuclear weapons.
21. Soots, tars, and mineral oils	Generated in coal tar and creosote manufacturing and in the manufacture of crude mineral oils, cutting oils, and shale oils.
22. Thorium oxide	Used in x-ray inquiry, manufacture of ceramics, incandescent lamps, magnesium alloys, nuclear reactors, and vacuum tubes.
23. Tobacco and tobacco smoke	Used and/or generated due to cigarettes, cigars, chewing tobacco, and snuff.
24. Vinyl chloride	Used in plastic manufacturing.

sult of a critical dosage or level of exposure to a carcinogen. However, no safe level of an actual or potential carcinogen has been established thus far.

Human cancers have been found to occur even at a very low level of exposure. A case in point is a cancer caused by secondary exposure, such as contraction of cancer by a member of an asbestos worker's family or by a nonsmoking member of a smoker's family. Based on existing information, however, the probability of contracting cancer is related to the degree or level of exposure related to a carcinogen, either directly or indirectly.

In this respect, it is also important to note that the degree of natural resistance to carcinogens may vary from person to person. This is where the genetic or hereditary makeup becomes a major factor. Another important consideration here is the individual lifestyle. Nevertheless, there is no disagreement within the scientific or medical community that zero or minimal level of exposure to carcinogens is the best available safeguard.

2.4 Occupational Carcinogens and Exposure Limits

There is no easy way to determine which chemicals are, or may be, carcinogens. Many chemicals that turned out to be occupational carcinogens because of their widespread use or existence in workplaces have been discovered through painstaking epidemiological studies. Generally, if there is a prevalence of cancer within a select group of workers who are exposed to a particular chemical or a group of chemicals with similar characteristics in the workplace as compared to others, it triggers an epidemiological concern. However, before any full-scale investigation or epidemiological study is begun, such select group of workers may have to be additionally screened on the basis of gender, race, age, personal habits, and the time, extent, and type of exposure to the chemical under study. In addition, other factors and variables have to be taken into consideration before a chemical can be suspected of being or labeled as a carcinogen.

The 1976 NIOSH Registry of Toxic Effects of Chemical Substances lists over 22,000 substances found to produce some form of toxic effect on test animals. However, not all these listed substances are officially labeled as human carcinogens. Nevertheless, almost all animal carcinogens, with the notable exception of arsenic, have been documented to be human carcinogens based on preliminary epidemiological studies. As a result, the animal carcinogens may be labeled as potential or suspected human carcinogens until adequate epidemiological studies can be carried out. These epidemiological studies may be either in vitro, which are short-term, or in vivo, which are long-term.

However, there is still disagreement among the medical and scientific groups as to which chemical or substance should be labeled as a carcinogen. Much of the confusion and debate arises from lack of a universal definition of cancer.

For regulatory purposes, OSH Act defines a chemical carcinogen as a chemical whose exposure has been found, through an appropriately designed study, to cause cancer in

- Humans
- Two different animal (mammal) species
- One animal species when results are implicated in separate studies

- One animal species if the results are supported by multitest evidence of mutagenicity

A *threshold limit value* (TLV) has been established by the American Conference of Governmental Industrial Hygienists for representing conditions under which almost all the workers may be exposed to a workplace chemical day after day without developing any adverse effect. The TLV actually represents the level of airborne concentration of a particular chemical or chemical substance. As a result, TLVs are different for different chemicals. Moreover, TLVs are based on the best available information about a particular chemical and include data from animal experimental studies, human epidemiological studies and other industrial or documented information.

Like all other scientific sources of information, TLVs typically undergo annual reviews. For regulatory purposes, OSHA promulgated a federal standard, set forth in 29 CFR 1910.1000, to prescribe the limits of maximum allowable exposures to air contaminants by workers that may result from various chemicals and chemical substances. These *permissible exposure limits* (PELs) can be the time-weighted average (TWA) of both the short-term exposure limit (STEL) and the ceiling limit as set by OSHA's final rule. However, OSHA has never suggested the safe level of exposure to a particular chemical, listed or unlisted.

In fact, to date, it has been the official position of NIOSH that it is not possible to demonstrate safe levels of exposure to chemical carcinogens. Thus the bottom line here is to reduce exposure to any known or suspected chemical carcinogen to the lowest level achievable. However, even then, no employer or industry is exempt from litigation or liability if cancer-causing chemicals are discovered and if there is supportive evidence of a tortious or wrongful conduct, including causal connection and the resulting damage from exposure to such chemicals.

2.5   Protection from Carcinogenic Exposure

Generally, four basic methods are used, either singularly or in combination, to reduce employee exposure to a known or suspected chemical carcinogen:

1. Use of an alternative material or chemical that is neither a known nor a suspected carcinogen. This may result in a change in the selection of raw materials, production processes, manufacturing or performance standards, by-products, economics, and planned attributes of the finished product.

2. Containment or isolation of the process. This is useful where a carcinogen is formed as a result of certain reactions of the noncarcinogenic chemicals during the processing cycles.

3. Complete or limited isolation of employees from exposure to chemical carcinogens. This method calls for the use of a closed control room that is isolated from areas where any chemical processing takes place.

4. Use of personal protective equipment. This may include various types of

protective gear, such as face masks, respirators, goggles, hand gloves, ear-plugs, protective clothing, and dermatological cream.

The maximum daily exposure to a suspected or known carcinogen may be further reduced by limiting the hours of exposure over a day or a week or a month or a year. In addition, it is extremely important to make routine inspection and undertake regular sampling to ensure effective control. Additionally, routine physical examination of the susceptible workers may be extremely useful in establishing the need for additional control as well as early detection and treatment of cancers.

2.6 Health Hazards and Regulatory Standards

Since its inception, OSHA has been largely preoccupied with the problem of health hazards caused by chemical carcinogens and the need for promulgation of standards that will render the workplace safer from exposure to occupational carcinogens. Such a task, no matter how laudable, is inherently complex because of the workplace variables and the large diversity of workers. It is further complicated by the need to institute medical surveillance, accurate and detailed record keeping, in-plant monitoring, and necessary cooperation from both employers and employees.

The federal publication *NIOSH Registry of Toxic Effects for Chemical Substances* contains the names of regulated chemicals. These regulated carcinogens are further referenced in 29 CFR sections 1910.1001 to 1910.1101. Appendix 2A contains OSHA's list of air contaminants and their exposure limits (both transitional and final rule limits).

Because of the complexity involved, there is no single federal standard developed to regulate carcinogenic chemicals. Instead, NIOSH is required to issue Current Intelligence Bulletins (CIBs) whenever there is confirmation of a human carcinogen. Section 6 of OSH Act requires OSHA to establish two standards—national consensus standards and federal standards—unless it is determined by the Secretary of Labor that promulgation of such standard will have no positive impact. Table 2.2 lists those chemicals and other substances for which federal standards have been developed by OSHA. Note, however, that not all the listed substances are known carcinogens. Rather, they have been included in using certain health criteria, including well-documented personal tragedy and court actions which raise serious concerns. Pursuant to section 6(c) of OSH Act, the Secretary of Labor is also required to set the requirement for an emergency temporary standard if it is determined that

- Employees are exposed to grave danger from exposure to toxic or physically harmful substances or agents.
- Such temporary standard is necessary to protect employees from such danger.

The OSHA standards set the minimum regulatory requirements to prevent the exposure of workers to carcinogens. These standards apply to all areas

TABLE 2.2   List of Chemicals and Substances with Final OSHA Health Standards

1. Asbestos, tremolite, anthophyllite, and actinolite
2. Currently regulated carcinogens
   a. 4-Nitrobiphenyl (CAS 92933)
   b. Alpha-naphthylamine (CAS 134327)
   c. Methyl chloromethyl ether (CAS 107302)
   d. 3,3'-Dichlorobenzidine (CAS 91941) and its salt
   e. bis-Chloromethyl ether (CAS 542881)
   f. Beta-naphthylamine (CAS 91598)
   g. Benzidine (CAS 92875)
   h. 4-Aminodiphenyl (CAS 92671)
   i. Ethyleneimine (CAS 151564)
   j. Beta-propiolactone (CAS 57578)
   k. 2-Acetylaminofluorene (CAS 53963)
   l. 4-Dimethylaminoazobenzene (CAS 60117)
   m. N-Nitrosodimethylamine (CAS 62759)
3. Vinyl chloride
4. Inorganic arsenic
5. Lead
6. Benzene
7. Coke oven emissions
8. Cotton dust
9. 1,2-Dibromo-3-chloropropane
10. Acrylonitrile
11. Ethylene oxide
12. Formaldehyde

where chemicals or substances are manufactured, processed, packaged, handled, stored, or released. These are specific regulations that apply to open vessels and closed or isolated systems of operation. Moreover, respiratory protection is required for employees engaged in the transfer of a carcinogen between containers, vessels, or like carriers.

In addition, an employer is required to establish a "regulated area" where any chemical carcinogen is manufactured, processed, used, repackaged, handled, stored, or released. A *regulated area* is defined as an area where entry and exit are restricted and controlled. Only authorized employees are to be allowed in the regulated areas. Moreover, entrances to regulated areas must be posted with signs stating:

CANCER-SUSPECT AGENT

AUTHORIZED PERSONNEL ONLY

Before an employee is authorized to enter a regulated area, such employee must undergo proper training. An acceptable level of training must include several specific topics at a minimum. Additionally, all authorized personnel working in a regulated area must have a prearranged physical examination and must be enrolled in a medical surveillance program.

Those workers who may come in direct contact with chemical carcinogens as a result of regular repair, maintenance, and decontamination activities, such as cleanups of spills, are required to wear clean, impervious, protective garments or coveralls. These protective clothings may include continuous air-supplied hoods, gloves, and boots. All protective garments and gears must be decontaminated prior to their removal. In addition, these workers are required to shower after removing the garments.

Entrances to areas where such maintenance and decontamination work is to be performed must be posted with signs stating

CANCER-SUSPECT AGENT EXPOSED IN THIS AREA.
IMPERVIOUS SUIT INCLUDING GLOVES, BOOTS, AND
AIR-SUPPLIED HOOD REQUIRED AT ALL TIMES.
AUTHORIZED PERSONNEL ONLY.

The contents of containers of carcinogenic chemicals or substances and contaminated equipment, if accessible to or handled by nonauthorized or untrained personnel, must be identified by their full chemical name and the Chemical Abstracts Service Registry Number (CASRN). Directly under or next to the contents' identification, a warning sign must be posted, stating

CANCER-SUSPECT AGENT

In addition, if the carcinogen or carcinogens in the container include any other element which may prove to be a health hazard, such as corrosivity, toxicity, or reactivity, it must be clearly indicated on the label.

OSHA has established a hazard communication requirement as set forth in 29 CFR section 1910.1200. The basic purpose of this regulatory requirement is to ensure evaluation of the hazards of all chemicals that are produced in or imported into the United States and to transmit such hazard information to both employers and employees. Pursuant to the regulatory requirement, the transmittal of such information must be through comprehensive hazard communication programs.

In summary, the regulatory requirements based on these general standards are designed to control possible exposure to the carcinogens and to afford maximum protection to the workers from possible exposure to known or suspected carcinogens. In this context, note that OSHA has promulgated specific standards which are much more detailed than these general standards for specific chemicals or substances. Such standards are set forth in 29 CFR part 1910.

2.7   Regulatory Dilemma over Carcinogens

Due to the unpredictable vagaries of suspected carcinogens, the regulators face a dilemma of how to regulate them since there are two major concerns over any regulation: the costs of doing business and the safety and welfare of human lives. Furthermore, there is a lack of scientific data to provide the existence of a reasonably safe threshold for exposures to known and suspected carcinogens. Under public pressure, Congress adopted Representative Delaney's amendment which prohibited the use of any food additive that has been found to induce cancer in humans or laboratory animals. Thus, under the so-called Delaney clause, the federal regulations are susceptible to adopting a zero-risk approach to all carcinogens.

However, such a decision tilts the scale to one side of the equation dealing with human health and safety. Any such blind adoption of a threshold quantity will have a catastrophic effect on industries and eventually on the economic well-being of the nation as well as its citizens. Thus, federal agencies are faced with a real dilemma, namely, how to strike a proper balance, particularly while dealing with toxic chemicals, hazardous substances, and the like which may contain a number of known and suspected carcinogens.

This intractable problem is compounded by the existence of less-than-perfect current scientific tools to detect cancer, particularly due to its long latency period. This has led to a greater reliance on laboratory animals, which in turn raises serious questions about the test protocols followed in such tests as well as the extrapolation of the test results to humans. As a result of this quagmire, the courts are generally endorsing the nonthreshold assumption for carcinogens. In *Natural Resources Defense Council v. EPA,* the court upheld OSHA's embracing of the no-threshold or zero-risk assumption for ethylene oxide.

However, any stringent regulation that calls for a zero-risk criterion has an opposing and somewhat paradoxical effect on effective enforcement. Certain courts have voiced their concerns about the unworkability and impracticability of the zero-risk approach. In *Public Citizen v. Young,* the court suggested in *dicta* that failure to employ a *de minimis* doctrine is directly contrary to the primary legislative goal. Nevertheless, this court was compelled to find that the zero-risk control is the applicable standard based on the enabling statute and its legislative history.

Congress has also debated this issue, particularly about the applicability of a zero-risk approach in connection with toxic chemicals, hazardous substances and hazardous waste. This has made risk assessment a primary tool for regulatory agencies before suggesting a threshold level for a known or suspected carcinogen. Unfortunately, most of the recent federal environmental statutes still embrace the loose concept of "an ample margin of safety."

There is inherently a vague standard in the absence of a proper definition of what constitutes an ample margin of safety. Nevertheless, this is credited as a workable solution on the implication that an ample margin of safety is a lower standard than the zero-risk standard. In *Industrial Union Department, AFL-CIO, v. American Petroleum Institute,* the Supreme Court rejected the plaintiff's argument that the clean-air statute was meant to completely eliminate any risk of serious harm with absolute certainty. In essence, the Supreme Court rejected the notion of zero risk. This seminal decision established a subfoundation for the federal regulation of carcinogens and other hazardous and toxic substances.

At present, most federal agencies are using the concept of significant risk in devising the acceptable threshold levels for hazardous and toxic substances. In recent federal statutes, the standard based on an "ample margin of safety" has been linked to a level of harm that affects 1 in 1 million. However, this standard is yet to be tested by the nation's courts. The only standard that is dead for all practical purposes is the earlier, often quoted standard based on zero risk.

2.8   Federal Cancer Policy

OSHA's federal cancer policy has identified three goals to date:

- To avoid repetitious scientific debate and ad hoc decisions on health standards
- To streamline OSHA's standard-setting process by prefabricating the essential elements of alternative versions
- To harmonize the policies of various health regulatory agencies

There are three federal health agencies currently involved in developing a federal cancer policy in addition to OSHA: the Environmental Protection Agency (EPA), the Food and Drug Administration (FDA), and the Consumer Product Safety Commission (CPSC). As of now, however, there has been no formal announcement of a comprehensive federal cancer policy in spite of OSHA's efforts since 1977. This has been attributed to the lack of a Presidential directive. This effort is further stymied due to a decision by the U.S. Supreme Court in *Industrial Union Division v. American Petroleum Institute v. Donovan.*

To combat against occupational and health hazards due to exposure to carcinogens and other toxic chemical substances, Congress passed several important laws including the OSH Act; the Federal Food, Drug, and Cosmetic Act; the Clean Air Act; the Clean Water Act; and the Toxic Substances Control Act. The regulatory implications of these federal statutes are discussed in later chapters.

2.9   Tort Actions and Regulatory Approach against Carcinogens

In recent years, advances in industrial technology have been blamed for various newer types of lung diseases, including cancers. The medical community has found a link between increased cases of thoracic neoplasm or new tumor-like growths and the increased environmental exposures to asbestos and several newer chemical compounds, particularly of the hydrocarbon group. Petrochemicals and chemicals using petroleum feedstock appear to have significant etiological relationships to various types of health disorders, including respiratory tract neoplasms and laryngeal cancers.

In several recent studies, polyvinyl chloride and vinyl chloride have been attributed to be the cause behind high human cancer mortality, particularly in those occupations where polymerization of chemical molecules is prevalent. Similarly, organic solvents have been blamed for a high level of lung cancer. In the past, chloroprene, benzoyl chloride, and chloromethyl methyl ether achieved notoriety for inducing lung cancers. Recently, with the massive discoveries of abandoned hazardous waste disposal sites and the prevalence of trichloroethylene, tricholoroethane, tetrachloroethylene, xylene, methylene chloride, and chloroethane in those wastes, attention has been focused on this newer group of organic solvents or hydrocarbons. Several medical researchers have associated them with pulmonary edema and other oncogenic or tumor-producing diseases in humans. Last, polychlorinated phenols, dioxins, polychlorinated and polybrominated biphenyls, and other halogenated hydrocarbons are suspected to be carcinogens because of the reported diseases that have been attributed to such chemicals.

As a result, a legal theory of causal connection linking these carcinogenic chemicals or suspected carcinogens has been taking shape. All human diseases, including cancers, that result from exposure to synthetic or manufactured chemicals are being labeled as personal injuries inflicted due to someone's tortious conduct. Hence, remedies are being sought by victims, using the traditional tort principles.

However, to seek an effective tort remedy, one has to overcome several difficulties which are seldom experienced in a typical personal-injury action. The first problem results from the long latency period of cancer symptoms and the fixed statutes of limitations that generally run for only a few years. The second problem arises from the difficulty of establishing a proximate cause due to limited scientific data and the lack of etiologic details and understanding.

This apparent legal difficulty and the functional objective of deterring human exposure to carcinogens have not been lost to the legislators. Congress has passed numerous statutes to grant authorities to various regulatory agencies to wage war on human cancer caused by exposure to environmental factors which are created or caused by humans.

At this time, there are six major federal agencies which are authorized under fifteen federal statutes to regulate the production, use, storage, treatment, and disposal of known carcinogens and other probable carcinogenic or toxic chemicals. These agencies, enumerated below, are responsible in varying degrees for regulating human exposure to various carcinogens, known and suspected.

1. The Environmental Protection Agency regulates the manufacture, use, storage, treatment, or disposal of carcinogens and toxic chemicals under six statutes and their specific provisions. These statutes include:

   - Toxic Substances Control Act
   - Federal Insecticide, Fungicide, and Rodenticide Act
   - Resource Conservation and Recovery Act
   - Clean Air Act
   - Clean Water Act
   - Safe Drinking Water Act

2. The Occupational Safety and Health Administration is empowered to regulate workplaces mostly through the Occupational Safety and Health (OSH) Act.

3. The Consumer Product Safety Commission exercises its regulatory authority over carcinogens under the Consumer Product Safety Act (CPSA) as well as the Federal Hazardous Substances Act (FHSA).

4. The Food and Drug Administration exercises its regulatory control generally under the statutory authority granted by the Federal Food, Drug, and Cosmetic Act.

5. The Department of Transportation regulates the handling and transportation of carcinogens under several federal statutes, e.g., the Hazardous Materials Transportation Act (HMTA), the Federal Railroad Safety Act (FRSA), the Dangerous Cargo Act (DCA), and the Ports and Waterways Safety Act (PWSA).

6. The Department of Agriculture (DOA) has only a limited regulatory authority to control carcinogens in meat and agricultural products under the Federal Meat Inspection Act (FMIA) and the Wholesome Meat Act (WMA).

Apart from the above, several other federal agencies play a limited role under certain statutory authority that is rooted in several unrelated federal statutes. These agencies include the Federal Trade Commission (FTC), the Nuclear Regulatory Commission (NRC), the Department of Energy, the Bureau of Mines (BOM), and the Mine Enforcement Safety Administration (MESA). They regulate carcinogens under several federal statutes, including the Federal Cigarette Labeling and Advertising Act (FCLAA), the Atomic Energy Act (AEA), the Federal Coal Mine Health and Safety Act (FCMHSA), and the Federal Mine Safety and Health Amendments Act (FMSHAA).

Note that in many instances the duties and responsibilities of these federal agencies overlap due to the specific authority granted by these statutes. Hence, one carcinogen may be regulated by several federal agencies. Also note that these agencies generally attempt to prescribe the limits of human exposure to various carcinogens and toxic chemicals. While such regulations reflect the general consensus that these carcinogens are dangerous and should only be used or handled with care, they do not ban such products outright. Moreover, an adherence to federal or any other legal standards and requirements establishes only good workplace safety programs, but it is by no means a warranty against legal actions. Any private remedy due to personal injury from regulated exposure to carcinogens may ultimately be sought by victims under traditional tort principles. Preventive remedies are rather difficult to achieve due to the inherent difficulties, discussed earlier, in establishing a dose-and-effect relationship.

Cancers that result from occupational and nonvoluntary exposures to various chemicals and chemical substances constitute tort actions based on the theory of personal injury. However, to recover damages, one has to meet the traditional burden of proof, showing both causal connection and the extent of damages resulting from such exposures. Prescription of safe limits of human exposure to known or suspected carcinogens makes good medical sense. However, it does not create a safe harbor in the absence of statutory or regulatory reforms. To date, there is no federal law or regulation that exempts one from the liability of a tort action that can be linked to a synthetic chemical irrespective of the level of exposure. Hence, in today's environment, any risk management for hazardous and toxic chemicals is an extremely difficult task.

Recommended Reading

1. Michael Dore, *Law of Toxic Torts,* vol. 2, Clark Boardman Callaghan, New York, 1991.

2. U.S. Department of Health and Human Services, *Everything Does Not Cause Cancer,* Public Health Service, National Institute of Health, NIH Publication 90-2039, March 1990.

3. U.S. Department of Health, Education, and Welfare, *Carcinogens—Regulation and Control,* Public Health Service, Center for Disease Control, National Institute for Occupational Safety and Health, Cincinnati, Ohio, 1977.

Important Cases

1. *Industrial Union Division v. American Petroleum Institute v. Donovan,* 100 S. Ct. 2478 (1981).

2. *American Textile Manufacturers Institute, Inc., v. Donovan,* 101 S. Ct. 2478 (1981).

3. *Whirlpool Corporation v. Marshall,* 444 U.S. 842 (1980).

4. *Industrial Union Department, AFL-CIO, v. American Petroleum Institute,* 448 U.S. 607 (1980).

5. *Marshall v. Barlow's Inc.,* 436 U.S. 307 (1978).

6. *Brock v. L.E. Myers Company,* 818 F. 2d 1270 (6th Cir. 1987).

7. *Natural Resources Defense Council v. EPA,* 824 F. 2d 1146 (D.C. Cir. 1987).

8. *Public Citizen v. Young,* 831 F. 2d 1108 (D.C. Cir. 1987).

Limits for Air Contaminants

APPENDIX 2A Limits for Air Contaminants

Substance	CAS No. (f)	Transitional limits PEL* ppm (a)	PEL* mg/m³ (b)	Skin desig-nation	Final rule limits** TWA ppm (a)	TWA mg/m³ (b)	STEL (c) ppm (a)	STEL (c) mg/m³ (b)	Ceiling ppm (a)	Ceiling mg/m³ (b)	Skin desig-nation
Acetaldehyde	75-07-0	200	360		100	180	150	270			
Acetic acid	64-19-7	10	25		10	25					
Acetic anhydride	108-24-7	5	20						5	20	
Acetone	67-64-1	1000	2400		750	1800	1000	2400			
Acetone[h]	67-64-1	1000	2400		750	1800	1000[h]	2400[h]			
2-Acetylaminofluorine; see 1910.1014	53-96-3										
Acetylene dichloride; see 1,2-Dichloroethylene											
Acetylene tetrabromide	79-27-6	1	14		1	14					
Acetylsalicylic acid (Aspirin)	50-78-2					5					
Acrolein	107-02-8	0.1	0.25		0.1	0.25	0.3	0.8			
Acrylamide	79-06-1		0.3	X		0.03					X
Acrylic acid	79-10-7				10	30					X
Acrylonitrile; see 1910.1045	107-13-1										
Aldrin	309-00-2		0.25	X		0.25					X
Allyl alcohol	107-18-6	2	5	X	2	5	4	10			X
Allyl chloride	107-05-1	1	3		1	3	2	6			
Allyl glycidyl ether (AGE)	106-92-3	(C)10	(C)45		5	22	10	44			
Allyl propyl disulfide	2179-59-1	2	12		2	12	3	18			
alpha-Alumina	1344-28-1										
Total dust						10					
Respirable fraction						5					
Aluminum (as Al) Metal	7429-90-5										
Total dust			15			15					
Respirable fraction			5			5					
Pyro powders						5					
Welding fumes***						5					
Soluble salts						2					
Alkyls						2					
4-Aminodiphenyl; see 1910.1011	92-67-1										
2-Aminoethanol; see Ethanolamine											
2-Aminopyridine	504-29-0	0.5	2		0.5	2					
Amitrole	61-82-5					0.2					
Ammonia	7664-41-7	50	35				35	27			
Ammonium chloride fume	12125-02-9					10		20			

Substance	CAS No.	Transitional Limits ppm	Transitional Limits mg/m³	Transitional Skin	Final Rule Limits ppm	Final Rule Limits mg/m³	Final Skin
Ammonium sulfamate	7773-06-0						
Total dust			15			10	
Respirable fraction						5	
n-Amyl acetate	628-63-7	100	525		100	525	
sec-Amyl acetate	628-38-0	125	650		125	650	
Aniline and homologs	62-53-3	5	19	X	2	8	X
Anisidine (o-,p-isomers)	29191-52-4		0.5	X		0.5	X
Antimony and compounds (as Sb)	7440-36-0		0.5			0.5	
ANTU (alpha Naphthylthiourea)	86-88-4		0.3			0.3	
Arsenic, organic compounds (as As)	7440-38-2		0.5			0.5	
Arsenic, inorganic compounds (as As); see 1910.1018	7440-38-2						
Arsine	7784-42-1	0.05	0.2		0.05	0.2	
Asbestos; see 1910.1001 and 1910.1101	Varies						
Atrazine	1912-24-9		5			5	
Azinphos-methyl	86-50-0		0.2	X		0.2	X
Barium, soluble compounds (as Ba)	7440-39-3		0.5			0.5	
Barium sulfate	7727-43-7						
Total dust			15			10	
Respirable fraction			5			5	
Benomyl	17804-35-2						
Total dust			15			10	
Respirable fraction			5			5	
Benzene; see 1910.1028 See Table Z-2 for the limits applicable in the operations or sectors excluded in 1910.1028[a]	71-43-2						
Benzidine; see 1910.1010							
p-Benzoquinone; see Quinone							
Benzo(a)pyrene; see Coal tar pitch volatiles	92-87-5						
Benzoyl peroxide	94-36-0		5			5	
Benzyl chloride	100-44-7	1	5		1	5	
Beryllium and beryllium compounds (as Be)	7440-41-7	Tbl. Z-2	Tbl. Z-2		Tbl. Z-2	Tbl. Z-2	
Biphenyl; see Diphenyl							
Bismuth telluride, Undoped	1304-82-1						
Total dust			15			15	
Respirable fraction			5			5	
Bismuth telluride, Se-doped						5	
Borates, tetra, sodium salts							
Anhydrous	1330-43-4					10	
Decahydrate	1303-96-4					10	
Pentahydrate	12179-04-3					10	
Boron oxide	1303-86-2						
Total dust			15			10	
Boron tribromide	10294-33-4				1	10	
Boron trifluoride	7637-07-2	(C)1	(C)3		(C)1	(C)3	
Bromacil	314-40-9				1	10	

53

APPENDIX 2A Limits for Air Contaminants (Continued)

Substance	CAS No. (f)	Transitional limits PEL* ppm (a)	PEL* mg/m³ (b)	Skin desig-nation	Final rule limits** TWA ppm (a)	TWA mg/m³ (b)	STEL (c) ppm (a)	STEL (c) mg/m³ (b)	Ceiling ppm (a)	Ceiling mg/m³ (b)	Skin desig-nation
Bromine	7726-95-6	0.1	0.7		0.1	0.7	0.3	2			
Bromine pentafluoride	7789-30-2				0.1	0.7					
Bromoform	75-25-2	0.5	5	X	0.5	5					X
Butadiene (1,3-Butadiene)	106-99-0	1000	2200		1000	2200					
Butane	106-97-8				800	1900					
Butanethiol; see Butyl mercaptan											
2-Butanone (Methyl ethyl ketone)	78-93-3	200	590		200	590	300	885			
2-Butoxyethanol	111-76-2	50	240	X	25	120					X
n-Butyl-acetate	123-86-4	150	710		150	710	200	950			
sec-Butyl acetate	105-46-4	200	950		200	950					
tert-Butyl acetate	540-88-5	200	950		200	950					
Butyl acrylate	141-32-2				10	55					
n-Butyl alcohol	71-36-3	100	300						50	150	X
sec-Butyl alcohol	78-92-2	150	450		100	305	150	450			
tert-Butyl alcohol	75-65-0	100	300		100	300					
Butylamine	109-73-9	(C)5	(C)15	X					5	15	X
tert-Butyl chromate (as CrO₃)	1189-85-1		(C)0.1	X						0.1	X
n-Butyl glycidyl ether (BGE)	2426-08-6	50	270		25	135					
n-Butyl lactate	138-22-7				5	25					
Butyl mercaptan	109-79-5	10	35		0.5	1.5					
o-sec-Butylphenol	89-72-5				5	30					X
p-tert-Butyltoluene	98-51-1	10	60		10	60	20	120			
Cadmium fume (as Cd)	7440-43-9		Tbl Z-2			0.1				0.3	
Cadmium dust (as Cd)	7440-43-9		Tbl Z-2			0.2				0.6	
Calcium carbonate	1317-65-3										
Total dust			15			15					
Respirable fraction			5			5					
Calcium cyanamide	156-62-7					0.5					
Calcium hydroxide†	1305-62-0					5†					
Calcium oxide†	1305-78-8		5			5†					
Calcium silicate	1344-95-2										
Total dust			15			15					
Respirable fraction			5			5					

Substance	CAS No.	Transitional ppm	Transitional mg/m³	Transitional Skin	Final TWA ppm	Final TWA mg/m³	Final STEL ppm	Final STEL mg/m³	Final Skin
Calcium sulfate	7778-18-9								
Total dust			15			15			
Respirable fraction			5			5			
Camphor, synthetic	76-22-2		2			2			
Caprolactam	105-60-2								
Dust						1		3	
Vapor					5	20	10	40	
Captafol (Difolatan®)	2425-06-1					0.1			
Captan	133-06-2		5			5			
Carbaryl (Sevin®)	63-25-2		5			5			
Carbofuran (Furadan®)	1563-66-2					0.1			
Carbon black	1333-86-4		3.5			3.5			
Carbon dioxide	124-38-9	5000*	9000		10,000	18,000	30,000	54,000	
Carbon disulfide	75-15-0	Tbl. Z-2			4	12	12	36	X
Carbon monoxide	630-08-0	50	55		35	40	200	229	
Carbon tetrabromide	558-13-4				0.1	1.4	0.3	4	
Carbon tetrachloride	56-23-5	Tbl. Z-2			2	12.6			
Carbonyl fluoride	353-50-4				2	5	5	15	
Catechol (Pyrocatechol)	120-80-9	5			5	20			X
Cellulose	9004-34-6								
Total dust			15			15			
Respirable fraction			5			5			
Cesium hydroxide	21351-79-1					2			
Chlordane	57-74-9		0.5	X		0.5			X
Chlorinated camphene	8001-35-2		0.5	X		0.5		1	X
Chlorinated diphenyl oxide	55720-99-5		0.5			0.5			
Chlorine	7782-50-5	(C)1			0.5	1.5	1	3	
Chlorine dioxide	10049-04-4	0.1			0.1	0.3	0.3	0.9	
Chlorine trifluoride	7790-91-2	(C)0.1			(C)0.1	(C)0.4			
Chloroacetaldehyde	107-20-0	(C)1			(C)1	(C)3			
a-Chloroacetophenone (Phenacyl chloride)	532-27-4	0.05			0.05	0.3			
Chloroacetyl chloride	79-04-9				0.05	0.2			
Chlorobenzene	108-90-7	75	350		75	350			
o-Chlorobenzylidene malononitrile	2698-41-1	0.05	0.4		(C)0.05	(C)0.4			X
Chlorobromomethane	74-97-5	200	1050		200	1050			
2-Chloro-1,3-butadiene; see b-Chloroprene									
Chlorodifluoromethane	75-45-6	1000			1000	3500			
Chlorodiphenyl (42% Chlorine) (PCB)	53469-21-9		1	X		1			X
Chlorodiphenyl (54% Chlorine) (PCB)	11097-69-1		0.5	X		0.5			X
1-Chloro,2,3-epoxypropane; see Epichlorohydrin									
2-Chloroethanol; see Ethylene chlorohydrin									
Chloroethylene; see Vinyl chloride									
Chloroform (Trichloromethane)	67-66-3	(C)50	(C)240		2	9.78			
bis(Chloromethyl) ether; see 1910.1008	542-88-1								

APPENDIX 2A Limits for Air Contaminants (*Continued*)

Substance	CAS No. (f)	Transitional limits PEL* ppm (a)	PEL* mg/m³ (b)	PEL* Skin designation	Final rule limits** TWA ppm (a)	TWA mg/m³ (b)	STEL (c) ppm (a)	STEL (c) mg/m³ (b)	Ceiling ppm (a)	Ceiling mg/m³ (b)	Skin designation
Chloromethyl methyl ether; see 1910.1006	107-30-2										
1-Chloro-1-nitropropane	600-25-9	20	100		2	10					
Chloropentafluoroethane	76-15-3				1000	6320					
Chloropicrin	76-06-2	0.1	0.7		0.1	0.7					
beta-Chloroprene	126-99-8	25	90	X	10	35					X
o-Chlorostyrene	2039-87-4				50	285	75	428			
o-Chlorotoluene	95-49-8				50	250					
2-Chloro-6-trichloro-methyl pyridine	1929-82-4										
Total dust			15			15					
Respirable fraction			5			5					
Chlorpyrifos	2921-88-2					0.2					
Chromic acid and chromates (as CrO₃)	7440-47-3		Tbl. Z-2							0.1	X
Chromium (II) compounds (as Cr)	7440-47-3		0.5			0.5					
Chromium (III) compounds (as Cr)	7440-47-3		0.5			0.5					
Chromium metal (as Cr)	7440-47-3		1			1					
Chrysene; see Coal tar pitch volatiles											
Clopidol	2971-90-6										
Total dust			15			15					
Respirable fraction			5			5					
Coal dust (less than 5% SiO₂). Respirable fraction			Tbl. Z-3			2					
Coal dust (greater than or equal to 5% SiO₂), Respirable quartz fraction.			Tbl. Z-3			0.1					
Coal tar pitch volatiles (benzene soluble fraction), anthracene, BaP, phenanthrene, acridine, chrysene, pyrene.	65966-93-2		0.2			0.2					
Cobalt metal, dust, and fume (as Co)	7440-48-4		0.1			0.05					
Cobalt carbonyl (as Co)	10210-68-1					0.1					
Cobalt hydrocarbonyl (as Co)	16842-03-8					0.1					
Coke oven emissions; see 1910.1029	—										
Copper	7440-50-8										
Fume (as Cu)			0.1			0.1					
Dusts and mists (as Cu)			1			1					
Cotton dust (raw)			1			1					

This 8-hour TWA applies to respirable dust as measured by a vertical elutriator cotton dust sampler or equivalent instrument. The time-weighted average applies to the cotton waste processing operations of waste recycling (sorting, blending, cleaning and willowing) and garnetting. See also 1910.1043 for cotton dust limits applicable to other sectors.

Substance	CAS No.	Trans. ppm	Trans. mg/m³	Trans. Skin	Final TWA ppm	Final TWA mg/m³	Final STEL/Ceiling ppm	Final STEL/Ceiling mg/m³	Final Skin
Crag herbicide (Sesone)	136-78-7								
Total dust			15			10			
Respirable fraction			5			5			
Cresol, all isomers	1319-77-3	5	22	X	5	22			X
Crotonaldehyde	123-73-9; 4170-30-3	2	6		2	6			
Crufomate	299-86-5		5			5			
Cumene	98-82-8	50	245	X	50	245			X
Cyanamide	420-04-2					2			
Cyanides (as CN)	Varies with compound		5			5			
Cyanogen	460-19-5				10	20			
Cyanogen chloride	506-77-4						(C)0.3	(C)0.6	
Cyclohexane	110-82-7	300	1050		300	1050			
Cyclohexanol	108-93-0	50	200		50	200			X
Cyclohexanone	108-94-1	50	200		25	100			X
Cyclohexene	110-83-8	300	1015		300	1015			
Cyclohexylamine	108-91-8				10	40			
Cyclonite	121-82-4					1.5			X
Cyclopentadiene	542-92-7	75	200		75	200			
Cyclopentane	287-92-3				600	1720			
Cyhexatin	13121-70-5					5			
2,4-D (Dichlorophenoxyacetic acid)	94-75-7		10			10			
Decaborane	17702-41-9	0.05	0.3		0.05	0.3	0.15	0.9	X
Demeton (Systox)	8065-48-3		0.1	X		0.1			X
Dichlorodiphenyltrichloroethane (DDT)	50-29-3		1	X		1			X
Dichlorvos (DDVP)	62-73-7		1	X		1			X
Diacetone alcohol (4-Hydroxy-4-methyl-2-pentanone)	123-42-2	50	240		50	240			
1,2-Diaminoethane; see Ethylenediamine									
Diazinon	333-41-5		0.1			0.1			X
Diazomethane	334-88-3	0.2	0.4		0.2	0.4			
Diborane	19287-45-7	0.1	0.1		0.1	0.1			
1,2-Dibromo-3-chloropropane; see 1910.1044	96-12-8								
2-N-Dibutylaminoethanol	102-81-8	2	14	X	2	14			X
Dibutyl phosphate	107-66-4	1	5		1	5	2	10	
Dibutyl phthalate	84-74-2		5			5			
Dichloroacetylene	7572-29-4						(C)0.1	(C)0.4	
o-Dichlorobenzene	95-50-1	(C)50	(C)300				(C)50	(C)300	
p-Dichlorobenzene	106-46-7	75	450		75	450	110	675	
3,3'-Dichlorobenzidine; see 1910.1007	91-94-1								

APPENDIX 2A Limits for Air Contaminants (*Continued*)

Substance	CAS No. (f)	Transitional limits PEL* ppm (a)	mg/m³ (b)	Skin desig-nation	Final rule limits** TWA ppm (a)	mg/m³ (b)	STEL (c) ppm (a)	mg/m³ (b)	Ceiling ppm (a)	mg/m³ (b)	Skin desig-nation
Dichlorodifluoromethane	75-71-8	1000	4950		1000	4950					
1,3-Dichloro-5,5-dimethyl hydantoin	118-52-5		0.2			0.2		0.4			
1,1-Dichloroethane	75-34-3	100	400		100	400					
1,2-Dichloroethylene	540-59-0	200	790		200	790					
Dichloroethyl ether	111-44-4	(C)15	(C)90	X	5	30	10	60			X
Dichloromethane; see Methylene chloride											
Dichloromonofluoromethane	75-43-4	1000	4200		10	40					
1,1-Dichloro-1-nitroethane	594-72-9	(C)10	(C)60		2	10					
1,2-Dichloropropane; see Propylenedichloride											
1,3-Dichloropropene	542-75-6				1	5					X
2,2-Dichloropropionic acid	75-99-0				1	6					
Dichlorotetrafluoroethane	76-14-2	1000	7000		1000	7000					
Dicrotophos	141-66-2					0.25					X
Dicyclopentadiene	77-73-6				5	30					
Dicyclopentadienyl iron	102-54-5										
Total dust			15			10					
Respirable fraction			5			5					
Dieldrin	60-57-1		0.25	X		.025					X
Diethanolamine	111-42-2				3	15					
Diethylamine	109-89-7	25	75		10	30	25	75			
2-Diethylaminoethanol	100-37-8	10	50	X	10	50					X
Diethylene triamine	111-40-0				1	4					X
Diethyl ether; see Ethyl ether											
Diethyl ketone	96-22-0				200	705					
Diethyl phthalate	84-66-2					5					
Difluorodibromomethane	75-61-6	100	860		100	860					
Diglycidyl ether (DGE)	2238-07-5	(C)0.5	(C)2.8		0.1	0.5					
Dihydroxybenzene; see Hydroquinone											
Diisobutyl ketone	108-83-8	50	290		25	150					
Diisopropylamine	108-18-9	5	20	X	5	20					X
4-Dimethylaminoazobenzene; see 1910.1015	60-11-7										
Dimethoxymethane; see Methylal											
Dimethyl acetamide	127-19-5	10	35	X	10	35					X
Dimethylamine	124-40-3	10	18		10	18					

58

Substance	CAS No.	TWA ppm	TWA mg/m³	Skin	TWA ppm	TWA mg/m³	STEL ppm	STEL mg/m³	Skin
Dimethylaminobenzene; see Xylidine									
Dimethylaniline (N-Dimethyl-aniline)	121-69-7	5	25	X	5	25			X
Dimethylbenzene; see Xylene									
Dimethyl-1,2-dibromo-2,2-dichloroethyl phosphate	300-76-5		3			3			X
Dimethylformamide	68-12-2	10	30	X	10	30			X
2,6-Dimethyl-4-hepta-none; see Diisobutyl ketone									
1,1-Dimethylhydrazine	57-14-7	0.5		X	0.5	1			X
Dimethylphthalate	131-11-3		5			5			
Dimethyl sulfate	77-78-1	1	5	X	0.1	0.5			X
Dinitolmide (3,5-Dinitro-o-toluamide)	148-01-6					5			
Dinitrobenzene (all isomers)	(alpha-) 528-29-0; (meta-) 99-65-0; (para-) 100-25-4		1	X		1			X
Dinitro-o-cresol	534-52-1		0.2	X		0.2			X
Dinitrotoluene	25321-14-6		1.5	X		1.5			X
Dioxane (Diethylene dioxide)	123-91-1	100	360	X	25	90			X
Dioxathion (Delnav)	78-34-2					0.2			X
Diphenyl (Biphenyl)	92-52-4	0.2	1		0.2	1			
Diphenylamine	122-39-4					10			
Diphenylmethane diisocyanate; see Methylene bisphenyl isocyanate									
Dipropylene glycol methyl ether	34590-94-8	100	600	X	100	600	150	900	X
Dipropyl ketone	123-19-3				50	235			
Diquat	85-00-7					0.5			
Di-sec octyl phthalate (Di-2-ethylhexyl-phthalate)	117-81-7		5			5		10	
Disulfiram	97-77-8					2			
Disulfoton	298-04-4					0.1			X
2,6-Di-tert-butyl-p-cresol	128-37-0					10			
Diuron	330-54-1					10			
Divinyl benzene	1321-74-0	10			10	50			
Emery	112-62-9								
Total dust			15			10			
Respirable fraction			5			5			
Endosulfan	115-29-7					0.1			X
Endrin	72-20-8		0.1	X		0.1			X
Epichlorohydrin	106-89-8	5	19	X	2	8			X
EPN	2104-64-5		0.5	X		0.5			X
1,2-Epoxypropane; see Propylene oxide									
2,3-Epoxy-1-propanol; see Glycidol									
Ethanethiol; see Ethyl mercaptan									
Ethanolamine	141-43-5	3	6		3	8	6	15	

APPENDIX 2A Limits for Air Contaminants (*Continued*)

Substance	CAS No. (f)	Transitional limits — PEL* ppm (a)	PEL* mg/m³ (b)	PEL* Skin desig-nation	Final rule limits** — TWA ppm (a)	TWA mg/m³ (b)	STEL (c) ppm (a)	STEL (c) mg/m³ (b)	Ceiling ppm (a)	Ceiling mg/m³ (b)	Skin desig-nation
Ethion	563-12-2					0.4					X
2-Ethoxyethanol	110-80-5	200	740	X	200	740					X
2-Ethoxyethyl acetate (Cellosolve acetate)	111-15-9	100	540	X	100	540					X
Ethyl acetate	141-78-6	400	1400		400	1400					
Ethyl acrylate	140-88-5	25	100	X	5	20	25	100			X
Ethyl alcohol (Ethanol)	64-17-5	1000	1900		1000	1900					
Ethylamine	75-04-7	10	18		10	18					
Ethyl amyl ketone (5-Methyl-3-heptanone)	541-85-5	25	130		25	130					
Ethyl benzene	100-41-4	100	435		100	435	125	545			
Ethyl bromide	74-96-4	200	890		200	890	250	1110			
Ethyl butyl ketone (3-Heptanone)	106-35-4	50	230		50	230					
Ethyl chloride	75-00-3	1000	2600		1000	2600					
Ethyl ether	60-29-7	400	1200		400	1200	500	1500			
Ethyl formate	109-94-4	100	300		100	300					
Ethyl mercaptan	75-08-1	(C)10	(C)25		0.5	1					
Ethyl silicate	78-10-4	100	850		10	85					
Ethylene chlorohydrin	107-07-3	5	16	X					1	3	X
Ethylenediamine	107-15-3	10	25		10	25					
Ethylene dibromide	106-93-4	Tbl. Z-2	Tbl. Z-2	Tbl. Z-2	Tbl. Z-2	Tbl. Z-2	Tbl. Z-2	Tbl. Z-2	Tbl. Z-2	Tbl. Z-2	Tbl. Z-2
Ethylene dichloride	107-06-2	Tbl. Z-2	Tbl. Z-2	Tbl. Z-2	1	4	2	8			
Ethylene glycol	107-21-1								50	125	
Ethylene glycol dinitrate ¹	628-96-6	(C)0.2	(C)1	X				*0.1			X
Ethylene glycol methyl acetate; see Methyl cellosolve acetate											
Ethyleneimine; see 1910.1012	151-56-4										
Ethylene oxide; see 1910.1047	75-21-8										
Ethylidene chloride; see 1,1-Dichloroethane											
Ethylidene norbornene	16219-75-3								5	25	
N-Ethylmorpholine	100-74-3	20	94	X	5	23					X
Fenamiphos	22224-92-6					0.1					X
Fensulfothion (Dasanit)	115-90-2					0.1					X
Fenthion	55-38-9					0.2					X
Ferbam	14484-64-1										

Substance	CAS No.	Transitional Limits ppm	Transitional Limits mg/m³	Final Rule Limits TWA ppm	Final Rule Limits TWA mg/m³	Final Rule Limits STEL ppm	Final Rule Limits STEL mg/m³	Skin
Total dust			15		10			
Ferrovanadium dust	12604-58-9		1		1		3	
Fluorides (as F)	Varies with compound		2.5		2.5			
Fluorine	7782-41-4	0.1		0.1	0.2			
Fluorotrichloromethane (Trichlorofluoromethane)	75-69-4	1000	5600	1000	5600			
Fonofos	944-22-9				0.1			X
Formaldehyde; see 1910.1048; See Table Z-2 for operations or sectors excluded from 1910.1048 or for which limit(s) is(are) stayed.	50-00-0							
Formamide	75-12-7			20	30	30	45	
Formic acid	64-18-6	5	9	5	9			
Furfural	98-01-1	5	20	2	8			X
Furfuryl alcohol	98-00-0	50	200	10	40	15	60	X
Gasoline	8006-61-9			300	900	500	1500	
Germanium tetrahydride	7782-65-2			0.2	0.6			
Glutaraldehyde	111-30-8				0.2		0.8	
Glycerin (mist)	56-81-5							
Total dust			15		10			
Respirable fraction			5		5			
Glycidol	556-52-5	50	150	25	75			
Glycol monoethyl ether; see 2-Ethoxyethanol								
Grain dust (oat, wheat, barley)					10			
Graphite, natural respirable dust	7782-42-5		Tbl Z-3		2.5			
Graphite, synthetic								
Total dust			15		10			
Respirable fraction			5		5			
Guthion[a]; see Azinphos methyl								
Gypsum	13397-24-5							
Total dust			15		15			
Respirable fraction			5		5			
Hafnium	7440-58-6		0.5		0.5			
Heptachlor	76-44-8		0.5		0.5			X
Heptane (n-Heptane)	142-82-5	500	2000	400	1600	500	2000	
Hexachlorobutadiene	87-68-3			0.02	0.24			
Hexachlorocyclo-pentadiene	77-47-4			0.01	0.1			
Hexachloroethane	67-72-1	1	10	1	10			X
Hexachloronaphthalene	1335-87-1		0.2		0.2			X
Hexafluoroacetone	684-16-2			0.1	0.7			X
n-Hexane	110-54-3	500	1800	50	180			
Hexane isomers	Varies with compound			500	1800	1000	3600	
2-Hexanone (Methyl n-butyl ketone)	591-78-6	100	410	5	20			X
Hexone (Methyl isobutyl ketone)	108-10-1	100	410	50	205	75	300	

APPENDIX 2A Limits for Air Contaminants (*Continued*)

Substance	CAS No. (f)	Transitional limits			Final rule limits**						
		PEL* ppm (a)	mg/m³ (b)	Skin desig-nation	TWA ppm (a)	mg/m³ (b)	STEL (c) ppm (a)	mg/m³ (b)	Ceiling ppm (a)	mg/m³ (b)	Skin desig-nation
sec-Hexyl acetate	108-84-9	50	300		50	300					
Hexylene glycol	107-41-5								25	125	
Hydrazine	302-01-2	1	1.3	X	0.1	0.1					X
Hydrogenated terphenyls	61788-32-7				0.5	5					
Hydrogen bromide	10035-10-6	3	10						3	10	
Hydrogen chloride	7647-01-0	(C)5	(C)7						5	7	
Hydrogen cyanide	74-90-8	10	11	X			4.7	5			X
Hydrogen fluoride (as F)	7664-39-3	Tbl. Z-2			3		6				
Hydrogen peroxide	7722-84-1	1	1.4		1	1.4					
Hydrogen selenide (as Se)	7783-07-5	0.05	0.2		0.05	0.2					
Hydrogen sulfide	7783-06-4	Tbl. Z-2			10	14	15	21			
Hydroquinone	123-31-9		2			2					
2-Hydroxypropyl acrylate	999-61-1				0.5	3					X
Indene	95-13-6				10	45					
Indium and compounds (as In)	7440-74-6					0.1					
Iodine	7553-56-2	(C)0.1	(C)1						0.1	1	
Iodoform	75-47-8				0.6	10					
Iron oxide fume	1309-37-1		10			10					
Iron pentacarbonyl (as Fe)	13463-40-6				0.1	0.8	0.2	1.6			
Iron salts (soluble) (as Fe)	Varies with compound					1					
Isoamyl acetate	123-92-2	100	525		100	525					
Isoamyl alcohol (primary and secondary)	123-51-3	100	360		100	360	125	450			
Isobutyl acetate	110-19-0	150	700		150	700					
Isobutyl alcohol	78-83-1	100	300		50	150					
Isooctyl alcohol	26952-21-6				50	270					X
Isophorone	78-59-1	25	140		4	23					
Isophorone diisocyanate	4098-71-9				0.005		0.02				X
2-Isopropoxyethanol	109-59-1				25	105					
Isopropyl acetate	108-21-4	250	950		250	950	310	1185			
Isopropyl alcohol	67-63-0	400	980		400	980	500	1225			
Isopropylamine	75-31-0	5	12		5	12	10	24			

Substance	CAS No.	Transitional ppm	Transitional mg/m³	Trans. Skin	Final ppm	Final mg/m³	Final STEL ppm	Final STEL mg/m³	Final Skin
N-isopropylaniline	768-52-5				2	10			X
Isopropyl ether	108-20-3	500	2100		500	2100			
Isopropyl glycidyl ether (IGE)	4016-14-2				50	240	75	360	
Kaolin									
Total dust			15			10			
Respirable fraction			5			5			
Ketene	463-51-4	0.5	0.9		0.5	0.9	1.5	3	
Lead inorganic (as Pb); see 1910.1025.	7439-92-1								
Limestone	1317-65-3								
Total dust			15			15			
Respirable fraction			5			5			
Lindane	58-89-9		0.5	X		0.5			X
Lithium hydride	7580-67-8		0.025			0.025			
L.P.G. (liquefied petroleum gas)	68476-85-7	1000	1800		1000	1800			
Magnesite	546-93-0								
Total dust			15			15			
Respirable fraction			5			5			
Magnesium oxide fume	1309-48-4								
Total particulate			15			10			
Malathion	121-75-5		15	X		10			X
Total dust									
Maleic anhydride	108-31-6	0.25	1		0.25	1			
Manganese compounds (as Mn)	7439-96-5		(C)5					(C)5	
Manganese fume (as Mn)	7439-96-5		(C)5			1		3	
Manganese cyclopenta-dienyl tricarbonyl (as Mn)	12079-65-1					0.1			X
Manganese tetroxide (as Mn)	1317-35-7					1			
Marble	1317-65-3								
Total dust			15			15			
Respirable fraction			5			5			
Mercury (aryl and inorganic) (as Hg)	7439-97-6		Tbl. Z-2			0.1			X
Mercury (organo) alkyl compounds (as Hg)	7439-97-6		Tbl. Z-2			0.01		0.03	X
Mercury (vapor) (as Hg)	7439-97-6		Tbl. Z-2			0.05			
Mesityl oxide	141-79-7	25	100		15	60	25	100	X
Methacrylic acid	79-41-4				20	70			X
Methanethiol; see Methyl mercaptan									
Methomyl (Lannate)	16752-77-5					2.5			
Methoxychlor	72-43-5								
Total dust			15			10			
2-Methoxyethanol; see Methyl cellosolve									
4-Methoxyphenol	150-76-5		5			5			
Methyl acetate	79-20-9	200	610		200	610	250	760	
Methyl acetylene (Propyne)	74-99-7	1000	1650		1000	1650			
Methyl acetylene-propadiene mixture (MAPP)		1000	1800		1000	1800	1250	2250	
Methyl acrylate	96-33-3	10	35	X	10	35			X

63

Substance	CAS No. (f)	Transitional limits			Final rule limits**						
		PEL*			TWA		STEL		Ceiling		Skin desig-nation
		ppm (a)	mg/m³ (b)	Skin desig-nation	ppm (a)	mg/m³ (b)	ppm (a)	mg/m³ (c) (b)	ppm (a)	mg/m³ (b)	
Methylacrylonitrile	126-98-7				1	3					X
Methylal (Dimethoxy-methane)	109-87-5	1000	3100		1000	3100					
Methyl alcohol	67-56-1	200	260		200	260	250	325			X
Methylamine	74-89-5	10	12		10	12					
Methyl amyl alcohol; see Methyl isobutyl carbinol											
Methyl n-amyl ketone	110-43-0	100	465		100	465					
Methyl bromide	74-83-9	(C)20	(C)80	X	5	20					X
Methyl butyl ketone; see 2-Hexanone											
Methyl cellosolve (2-Methoxyethanol)	109-86-4	25	80	X	25	80					X
Methyl cellosolve acetate (2-Methoxyethyl acetate)	110-49-6	25	120	X	25	120					X
Methyl chloride	74-87-3	Tbl. Z-2			50	105	100	210			
Methyl chloroform (1,1,1-Trichloroethane)	71-55-6	350	1900		350	1900	450	2450			
Methyl 2-cyanoacrylate	137-05-3				2	8	4	16			
Methyl cyclohexane	108-87-2	500	2000		400	1600					
Methylcyclohexanol	25639-42-3	100	470		50	235					
o-Methylcyclohexanone	583-60-8	100	460	X	50	230	75	345			X
Methylcyclopentadienyl manganese tricarbonyl (as Mn)	12108-13-3					0.2					X
Methyl demeton	8022-00-2					0.5					X
4,4'-Methylene bis (2-chloroaniline) (MBOCA)	101-14-4				0.02	0.22					X
Methylene bis(4-cyclohexylisocyanate)	5124-30-1								0.01	0.11	X
Methylene chloride	75-09-2	Tbl. Z-2			Tbl. Z-2		Tbl. Z-2		Tbl. Z-2		
Methyl ethyl ketone peroxide (MEKP)	1338-23-4								0.7	5	
Methyl formate	107-31-3	100	250		100	250	150	375			
Methyl hydrazine (Monomethyl hydrazine)	60-34-4	(C)0.2	(C)0.35	X					0.2	0.35	X
Methyl iodide	74-88-4	5	28	X	2	10					X
Methyl isoamyl ketone	110-12-3				50	240					
Methyl isobutyl carbinol	108-11-2	25	100	X	25	100	40	165			X
Methyl isobutyl ketone; see Hexone											
Methyl isocyanate	624-83-9	0.02	0.05	X	0.02	0.05					X
Methyl isopropyl ketone	563-80-4				200	705					
Methyl mercaptan	74-93-1	(C)10	(C)20		0.5	1					
Methyl methacrylate	80-62-6	100	410		100	410					
Methyl parathion	298-00-0					0.2					X

Methyl propyl ketone; see 2-Pentanone

Substance	CAS No.	Transitional Limits ppm	mg/m³	Skin	Final Rule TWA ppm	mg/m³	STEL ppm	mg/m³	Skin
Methyl silicate	681-84-5	1	6		1	6			
alpha-Methyl styrene	98-83-9	(C)100	(C)480		50	240	100	485	
Methylene bisphenyl isocyanate (MDI)	101-68-8	(C)0.02	(C)0.2		0.02	0.2			
Metribuzin	21087-64-9		5			5			
Mica; see Silicates									
Molybdenum (as Mo)	7439-98-7								
Soluble compounds			5			5			
Insoluble compounds									
Total dust			15			10			
Monocrotophos (Azodrin*)	6923-22-4		0.25			0.25			
Monomethyl aniline	100-61-8	2	9	X	0.5	2			X
Morpholine	110-91-8	20	70	X	20	70	30	105	X
Naphtha (Coal tar)	8030-30-6	100	400		100	400			
Naphthalene	91-20-3	10	50		10	50	15	75	
alpha-Naphthylamine; see 1910.1004	134-32-7								
beta-Naphthylamine; see 1910.1009	91-59-8								
Nickel carbonyl (as Ni)	13463-39-3	0.001	0.007		0.001	0.007			
Nickel, metal and insoluble compounds (as Ni)	7440-02-0		1			1			
Nickel, soluble compounds (as Ni)	7440-02-0		1			0.1			
Nicotine	54-11-5		0.5	X		0.5			X
Nitric acid	7697-37-2	2	5		2	5	4	10	
Nitric oxide	10102-43-9	25	30		25	30			
p-Nitroaniline	100-01-6	1	6	X		3			X
Nitrobenzene	98-95-3	1	5	X	1	5			X
p-Nitrochlorobenzene	100-00-5		1	X		1			X
4-Nitrodiphenyl; see 1910.1003	92-93-3								
Nitroethane	79-24-3	100	310		100	310			
Nitrogen dioxide	10102-44-0	(C)5	(C)9				1	1.8	
Nitrogen trifluoride	7783-54-2	10	29		10	29			
Nitroglycerin[1]	55-63-0	(C)0.2	(C)2	X				0.1	X
Nitromethane	75-52-5	100	250		100	250			
1-Nitropropane	108-03-2	25	90		25	90			
2-Nitropropane	79-46-9	25	90		10	35			
N-Nitrosodimethylamine; see 1910.1016	62-75-9								
Nitrotoluene	88-72-2	5	30	X	2	11			X
o-isomer	88-72-2								
m-isomer	99-08-1								
p-isomer	99-99-0								
Nitrotrichloromethane; see Chloropicrin									
Nonane	111-84-2	200	1050		200	1050			
Octachloronaphthalene	2234-13-1		0.1	X		0.1			X
Octane	111-65-9	500	2350		300	1450	375	1800	

Substance	CAS No. (f)	Transitional limits PEL* ppm (a)	mg/m³ (b)	Skin desig-nation	Final rule limits** TWA ppm (a)	mg/m³ (b)	STEL (c) ppm (a)	mg/m³ (b)	Ceiling ppm (a)	mg/m³ (b)	Skin desig-nation
Oil mist, mineral	8012-95-1		5			5					
Osmium tetroxide (as Os)	20816-12-0		0.002		0.0002	0.002	0.0006	0.006			
Oxalic acid	144-62-7		1			1		2			
Oxygen difluoride	7783-41-7	0.05	0.1						0.05	0.1	
Ozone	10028-15-6	0.1	0.2		0.1	0.2	0.3	0.6			
Paraffin wax fume	8002-74-2		2			2					
Paraquat, respirable dust	1910-42-5 4685-14-7 2074-50-2		0.5	X		0.1					X
Parathion	56-38-2		0.1	X		0.1					X
Particulates not otherwise regulated											
Total dust			15			15					
Respirable fraction			5			5					
Pentaborane	19624-22-7	0.005	0.01		0.005	0.01	0.015	0.03			
Pentachloronaphthalene	1321-64-8		0.5	X		0.5					X
Pentachlorophenol	87-86-5		0.5	X		0.5					X
Pentaerythritol	115-77-5										
Total dust			15			10					
Respirable fraction			5			5					
Pentane	109-66-0	1000	2950		600	1800	750	2250			
2-Pentanone (Methyl propyl ketone)	107-87-9	200	700		200	700	250	875			
Perchloroethylene (Tetrachloroethylene)	127-18-4	Tbl. Z-2			25	170					
Perchloromethyl mercaptan	594-42-3	0.1	0.8		0.1	0.8					
Perchloryl fluoride	7616-94-6	3	13.5		3	14	6	28			
Perlite											
Total dust			15			15					
Respirable fraction			5			5					
Petroleum distillates (Naphtha)		500	2000		400	1600					
Phenol	108-95-2	5	19	X	5	19					X
Phenothiazine	92-84-2		5	X		5					X
p-Phenylene diamine	106-50-3		0.1	X		0.1					X
Phenyl ether, vapor	101-84-8	1	7		1	7					
Phenyl ether-biphenyl mixture, **vapor**		1	7		1	7					

Substance	CAS No.	OSHA TWA ppm	OSHA TWA mg/m³	OSHA skin	NIOSH TWA ppm	NIOSH TWA mg/m³	NIOSH STEL/C ppm	NIOSH STEL/C mg/m³	skin
Phenylethylene; see Styrene									
Phenyl glycidyl ether (PGE)	122-60-1	10	60		1	6			
Phenylhydrazine	100-63-0	5	22		5	20			
Phenyl mercaptan	108-98-5	0.5	2	X	0.5	2		0.25	X
Phenylphosphine	638-21-1				0.05	0.25			
Phorate	298-02-2		0.1			0.05		0.2	X
Phosdrin (Mevinphos®)	7786-34-7		0.1	X	0.01	0.1	0.03	0.3	X
Phosgene (Carbonyl chloride)	75-44-5	0.1	0.4		0.1	0.4			
Phosphine	7803-51-2	0.3	0.4		0.3	0.4	1	1	
Phosphoric acid	7664-38-2		1		1	1		3	
Phosphorus (yellow)	7723-14-0		0.1			0.1			
Phosphorus oxychloride	10025-87-3				0.1	0.6			
Phosphorus pentachloride	10026-13-8		1			1			
Phosphorus pentasulfide	1314-80-3		1			1		3	
Phosphorus trichloride	7719-12-2	0.5	3		0.2	1.5	0.5	3	
Phthalic anhydride	85-44-9	2	12			6			
m-Phthalodinitrile	626-17-5					5			
Picloram	1918-02-1								
Total dust		15				10			
Respirable fraction		5				5			
Picric acid	88-89-1		0.1	X		0.1			X
Piperazine dihydro-chloride	142-64-3					5			
Pindone (2-Pivalyl-1,3-indandione)	83-26-1		0.1			0.1			
Plaster of Paris	26499-65-0								
Total dust		15				15			
Respirable fraction		5				5			
Platinum (as Pt)	7440-06-4								
Metal						1			
Soluble salts	65997-15-1	0.002				0.002			
Portland cement									
Total dust		Tbl. Z-3				10			
Respirable fraction		Tbl. Z-3				5			
Potassium hydroxide	1310-58-3								
Propane	74-98-6	1000	1800		1000	1800			
Propargyl alcohol	107-19-7				1	2		2	
beta-Propriolactone; see 1910.1013	57-57-8								X
Propionic acid	79-09-4				10	30			
Propoxur (Baygon)	114-26-1					0.5			X
n-Propyl acetate	109-60-4	200	840		200	840	250	1050	
n-Propyl alcohol	71-23-8	200	500		200	500	250	625	
n-Propyl nitrate	627-13-4	25	110		25	105	40	170	
Propylene dichloride	78-87-5	75	350		75	350	110	510	
Propylene glycol dinitrate	6423-43-4				0.05	0.3			

APPENDIX 2A Limits for Air Contaminants (*Continued*)

Substance	CAS No. (f)	Transitional limits PEL* ppm (a)	Transitional limits PEL* mg/m³ (b)	Transitional Skin desig- nation	Final rule limits** TWA ppm (a)	Final rule limits** TWA mg/m³ (b)	Final rule limits** STEL (c) ppm (a)	Final rule limits** STEL (c) mg/m³ (b)	Final rule limits** Ceiling ppm (a)	Final rule limits** Ceiling mg/m³ (b)	Final rule limits** Skin desig- nation
Propylene glycol monomethyl ether	107-98-2				100	360	150	540			
Propylene imine	75-55-8	2	5	X	2	5					X
Propylene oxide	75-56-9	100	240		20	50					
Propyne; see Methyl acetylene											
Pyrethrum	8003-34-7		5			5					
Pyridine	110-86-1	5	15		5	15					
Quinone	106-51-4	0.1	0.4		0.1	0.4					
Resorcinol	108-46-3				10	45	20	90			
Rhodium (as Rh), metal fume and insoluble compounds	7440-16-6		0.1			0.1					
Rhodium (as Rh), soluble compounds	7440-16-6		0.001			0.001					
Ronnel	299-84-3		15			10					
Rosin core solder pyrolysis products, as formaldehyde						0.1					
Rotenone	83-79-4		5			5					
Rouge											
Total dust						10					
Respirable fraction						5					
Selenium compounds (as Se)	7782-49-2		0.2			0.2					
Selenium hexafluoride (as Se)	7783-79-1	0.05	0.4		0.05	0.4					
Silica, amorphous, precipitated and gel	112926-00-8		Tbl Z-3			6					
Silica, amorphous, diatomaceous earth, containing less than 1% crystalline silica	61790-53-2		Tbl Z-3			6					
Silica, crystalline cristobalite (as quartz), respirable dust	14464-46-1		Tbl Z-3			0.05					
Silica, crystalline quartz (as quartz), respirable dust	14808-60-7		Tbl Z-3			0.1					
Silica, crystalline tripoli (as quartz), respirable dust	1317-95-9		Tbl Z-3			0.1					
Silica, crystalline tridymite (as quartz), respirable dust	15468-32-3		Tbl Z-3			0.05					
Silica, fused, respirable dust	60676-86-0		Tbl Z-3			0.1					
Silicates (less than 1% crystalline silica)											
Mica (respirable dust)	12001-26-2		Tbl Z-3			3					
Soapstone, total dust			Tbl Z-3			6					
Soapstone, respirable dust			Tbl Z-3			3					
Talc (containing asbestos): use asbestos limit. See 29 CFR 1910.1001			Tbl Z-3								
Talc (containing no asbestos), respirable dust	14807-96-6		Tbl Z-3			2					
Tremolite			Tbl Z-3								

See 29 CFR 1910.1101

Substance	CAS No.	Transitional ppm	Transitional mg/m³	Transitional Skin	Final TWA ppm	Final TWA mg/m³	STEL ppm	STEL mg/m³	Ceiling ppm	Ceiling mg/m³	Final Skin
Silicon	7440-21-3										
Total dust			15			10					
Respirable fraction			5			5					
Silicon carbide	409-21-2										
Total dust			15			10					
Respirable fraction			5			5					
Silicon tetrahydride	7803-62-5	5			5	7					
Silver, metal and soluble compounds (as Ag)	7440-22-4		0.01			0.01					
Soapstone; see Silicates											
Sodium azide	26628-22-8										
(as HN₃)									0.1		X
(as NaN₃)										0.3	X
Sodium bisulfite	7631-90-5					5					
Sodium fluoroacetate	62-74-8		0.05	X		0.05					X
Sodium hydroxide	1310-73-2		2							2	
Sodium metabisulfite	7681-57-4					5					
Starch	9005-25-8										
Total dust			15			15					
Respirable fraction			5			5					
Stibine	7803-52-3	0.1	0.5		0.1	0.5					
Stoddard solvent	8052-41-3	500	2900		100	525					
Strychnine	57-24-9		0.15			0.15					
Styrene	100-42-5	Tbl. Z-2			50	215	100	425			
Subtilisins (Proteolytic enzymes)	9014-01-1							0.00006 (60 min.)*			
Sucrose	57-50-1										
Total dust			15			15					
Respirable fraction			5			5					
Sulfur dioxide	7446-09-5	5	13		2	5	5	13			
Sulfur hexafluoride	2551-62-4	1000	6000		1000	6000					
Sulfuric acid	7664-93-9		1			1					
Sulfur monochloride	10025-67-9	1	6						1	6	
Sulfur pentafluoride	5714-22-7	0.025	0.25						0.01	0.1	
Sulfur tetrafluoride	7783-60-0								0.1	0.4	
Sulfuryl fluoride	2699-79-8	5	20		5	20	10	40			
Sulprofos	35400-43-2					1					
Systox®, see Demeton											
2,4,5-T	93-76-5		10			10					
Talc; see Silicates											
Tantalum, metal and oxide dust	7440-25-7		5			5					
TEDP (Sulfotep)	3689-24-5		0.2	X		0.2					X
Tellurium and compounds (as Te)	13494-80-9		0.1			0.1					
Tellurium hexafluoride (as Te)	7783-80-4	0.02	0.2		0.02	0.2					

Substance	CAS No. (f)	Transitional limits PEL* ppm (a)	mg/m³ (b)	Skin designation	Final rule limits** TWA ppm (a)	mg/m³ (b)	STEL (c) ppm (a)	mg/m³ (b)	Ceiling ppm (a)	mg/m³ (b)	Skin designation
Temephos	3383-96-8										
Total dust			15			10					
Respirable fraction			5			5					
TEPP	107-49-3		0.05	X		0.05					X
Terphenyls	26140-60-3	(C)1	(C)9						0.5	5	
1,1,1,2-Tetrachloro-2,2-difluoroethane	76-11-9	500	4170		500	4170					
1,1,2,2-Tetrachloro-1,2-difluoroethane	76-12-0	500	4170		500	4170					
1,1,2,2-Tetrachloroethane	79-34-5	5	35	X	1	7					X
Tetrachloroethylene; see Perchloroethylene											
Tetrachloromethane; see Carbon tetrachloride											
Tetrachloronaphthalene	1335-88-2		2	X		2					X
Tetraethyl lead (as Pb)	78-00-2		0.075	X		0.075					X
Tetrahydrofuran	109-99-9	200	590		200	590	250	735			
Tetramethyl lead, (as Pb)	75-74-1		0.075	X		0.075					X
Tetramethyl succinonitrile	3333-52-6	0.5	3	X	0.5	3					X
Tetranitromethane	509-14-8	1	8		1	8					
Tetrasodium pyrophosphate	7722-88-5					5					
Tetryl (2,4,6-Trinitrophenyl-methyl-nitramine)	479-45-8		1.5	X		1.5					X
Thallium, soluble compounds (as Tl)	7440-28-0		0.1	X		0.1					X
4,4'-Thiobis(6-tert, Butyl-m-cresol)	96-69-5										
Total dust			15			10					
Respirable fraction			5			5					
Thioglycolic acid	68-11-1				1	4					X
Thionyl chloride	7719-09-7								1	5	
Thiram	137-26-8		5			5					
Tin, inorganic compounds (except oxides) (as Sn)	7440-31-5		2			2					
Tin, organic compounds (as Sn)	7440-31-5		0.1			0.1					X
Tin oxide (as Sn)	21651-19-4					2					
Titanium dioxide	13463-67-7										
Total dust		Tbl. Z-2	15			10					
Toluene	108-88-3				100	375	150	560			
Toluene-2,4-disocyanate (TDI)	584-84-9	(C)0.02	(C)0.14		0.005	0.04	0.02	0.15			
m-Toluidine	108-44-1				2	9					X
o-Toluidine	95-53-4	5	22	X	5	22					X

Substance	CAS No.	Transitional ppm	Transitional mg/m³	Transitional Skin	Final ppm	Final mg/m³	Final STEL ppm	Final STEL mg/m³	Final Skin
p-Toluidine	106-49-0				2	9			X
Toxaphene; see Chlorinated camphene									
Tremolite; see Silicates									
Tributyl phosphate	126-73-8		5		0.2	2.5			
Trichloroacetic acid	76-03-9				1	7			
1,2,4-Trichlorobenzene	120-82-1						5	40	
1,1,1-Trichloroethane; see Methyl chloroform									
1,1,2-Trichloroethane	79-00-5	10	Tbl. Z-2	X	10	45			X
Trichloroethylene	79-01-6		Tbl. Z-2		50	270	200	1080	
Trichloromethane; see Chloroform									
Trichloronaphthalene	1321-65-9		5	X		5			X
1,2,3-Trichloropropane	96-18-4	50	300		10	60			
1,1,2-Trichloro-1,2,2-trifluoroethane	76-13-1	1000	7600		1000	7600	1250	9500	
Triethylamine	121-44-8	25	100		10	40	15	60	
Trifluorobromomethane	75-63-8	1000	6100		1000	6100			
Trimellitic anhydride	552-30-7				0.005	0.04			
Trimethylamine	75-50-3				10	24	15	36	
Trimethyl benzene	25551-13-7				25	125			
Trimethyl phosphite	121-45-9				2	10			
2,4,6-Trinitrophenyl; see Picric acid									
2,4,6-Trinitrophenylmethyl nitramine; see Tetryl									
2,4,6-Trinitrotoluene (TNT)	118-96-7		1.5	X		0.5			X
Triorthocresyl phosphate	78-30-8		0.1			0.1			X
Triphenyl amine	603-34-9					5			
Triphenyl phosphate	115-86-6		3			3			
Tungsten (as W)	7440-33-7								
Insoluble compounds						5		10	
Soluble compounds						1		3	
Turpentine	8006-64-2	100	560		100	560			
Uranium (as U)	7440-61-1								
Soluble compounds			0.05			0.05		0.2	
Insoluble compounds			0.25			0.2		0.6	
n-Valeraldehyde	110-62-3				50	175			
Vanadium	1314-62-1								
Respirable dust (as V_2O_5)			(C)0.5			0.05			
Fume (as V_2O_5)			(C)0.1			0.05			
Vegetable oil mist									
Total dust			15			15			
Respirable fraction			5			5			
Vinyl acetate	108-05-4				10	30	20	60	
Vinyl benzene; see Styrene									
Vinyl bromide	593-60-2	5			5	20			
Vinyl chloride; see 1910.1017									
Vinylcyanide; see Acrylonitrile									
Vinyl cyclohexene dioxide	106-87-6				10	60			X

APPENDIX 2A   Limits for Air Contaminants (*Continued*)

Substance	CAS No. (f)	Transitional limits PEL* ppm (a)	Transitional limits PEL* mg/m³ (b)	Transitional limits Skin designation	Final rule limits** TWA ppm (a)	Final rule limits** TWA mg/m³ (b)	Final rule limits** STEL (c) ppm (a)	Final rule limits** STEL (c) mg/m³ (b)	Final rule limits** Ceiling ppm (a)	Final rule limits** Ceiling mg/m³ (b)	Final rule limits** Skin designation
Vinylidene chloride (1,1-Dichloroethylene)	75-35-4				1	4					
Vinyl toluene	25013-15-4	100	480		100	480					
VM & P Naphtha	8032-32-4				300	1350	400	1800			
Warfarin	81-81-2		0.1			0.1					
Welding fumes (total particulate)***						5					
Wood dust, all soft and hard woods, except Western red cedar.						5		10			
Wood dust, Western red cedar	1330-20-7					2.5					
Xylenes (o-, m-, p- isomers)	1477-55-0	100	435		100	435	150	655			
m-Xylene alpha, alpha'- diamine										0.1	X
Xylidine	1300-73-8	5	25	X	2	10					X
Yttrium	7440-65-5		1			1					
Zinc chloride fume	7646-85-7		1			1		2			
Zinc chromate (as CrO₃)	Varies with compound	Tbl. Z-2	Tbl. Z-2	Tbl. Z-2						0.1	
Zinc oxide fume	1314-13-2		5			5		10			
Zinc oxide	1314-13-2										
Total dust			15			10					
Respirable fraction			5			5					
Zinc stearate	557-05-1										
Total dust			15			10					
Respirable fraction			5			.5					
Zirconium compounds (as Zr)	7440-67-7		5			5		10			

*The transitional PELs are 8-hour TWAs unless otherwise noted; a (C) designation denotes a ceiling limit.

**Unless otherwise noted, employers in General Industry (i.e., those covered by 29 *CFR* 1910) may use any combination of controls to achieve these limits until Dec. 31, 1992 as set forth in 29 *CFR* 1910.1000(f).

***As determined from breathing-zone air samples.

(a) Parts of vapor or gas per million parts of contaminated air by volume at 25°C and 760 torr.

(b) Approximate milligrams of substance per cubic meter of air.

(c) Duration is for 15 minutes, unless otherwise noted.

(d) The final benzene standard in 1910.1028 applies to all occupational exposures to benzene except some subsegments of industry where exposures are consistently under the action level (i.e., distribution and sale of fuels, sealed containers and pipelines, coke production, oil and gas drilling and production, natural gas processing, and the percentage exclusion for liquid mixtures); for the excepted subsegments, the benzene limits in Table Z–2 apply.

(e) Exposures under 10,000 ppm to be cited de minimus.

(f) The CAS number is for information only. Enforcement is based on the substance name. For an entry covering more than one metal compound measured as the metal, the CAS number for the metal is given—not the CAS numbers for the individual compounds.

(g) Compliance with the subtilisins PEL is assessed by sampling with a high volume sampler (600–800 liters per minute) for at least 60 minutes.

(h) The acetone STEL does not apply to the cellulose acetate fiber industry. It is in effect for all other sectors.

(i) The Final Rule Limit of 5 mg/m³ is not in effect as a result of reconsideration. Calcium hydroxide is covered by the exposure limits for particulates not otherwise regulated of 5 mg/m³ respirable dust and 15 mg/m³ total dust.

(j) The Final Rule Limit TWA of 5 mg/m³ is not in effect as a result of reconsideration. The calcium oxide Transitional Limit of 5 mg/m³ remains in effect and employee exposures shall be kept below that level pursuant to the methods of compliance specified in 29 CFR 1910.1000(e).

(k) The Final Rule Limit STEL of 0.1 mg/m³ is not in effect as a result of reconsideration for the industrial sector of civilian manufacture and distribution of explosives and propellants for civilian use. The Final rule limits skin designation and the Transitional limits ceiling limit of 1 mg/m³ remain in effect for this sector until completion of the reconsideration.

(l) The Final Rule Limit STEL of 0.1 mg/m³ is not in effect as a result of reconsideration for the industrial sector of civilian manufacture and distribution of explosives and propellants for civilian use. The Final rule limits skin designation and the Transitional limits ceiling limit of 2 mg/m³ remains in effect for this sector until completion of the reconsideration.

3

Basic Principles of Hazardous Waste Toxicology*

3.1 Introduction

Objectives

The basic objectives of this chapter are

- To introduce the reader to the characteristics of hazardous waste which may make it toxic (animal and environmental characteristics)
- To introduce the reader to the principles of toxicology which would increase understanding of why hazardous waste may produce a toxic response
- To use specific examples of hazardous chemicals in illustrating basic toxicology principles
- To introduce the reader to the characteristics of an organism that may make it susceptible to hazardous chemicals
- To introduce the reader to organ systems that may be attacked by hazardous wastes and to indicate mechanisms

History and scope

Paracelsus (1493–1541) indicated that all substances are poisons and there is none that is not a poison. According to him, it is the dose that differentiates a poison from a remedy.

This means that no substance is inherently either safe or toxic. It is only the way that it is used or its improper disposal that leads it to being toxic. In fact, in the case of waste, it is only defined as hazardous after improper disposal. Under the Resource Recovery and Conservation Act (RCRA), waste is consid-

---

*This chapter has been authored by Dr. William M. Hadley, Dean, College of Pharmacy, The University of New Mexico, Albuquerque, New Mexico.

ered to be a potentially valuable material and only dangerous when disposal is improper.

In this chapter, some of the major characteristics which lead a waste to being toxic after improper disposal are considered. Also why these characteristics are important in terms of basic principles of toxicology is discussed. Applying the information will allow an understanding of why a waste may be toxic if its disposal is improper. The information provided is introductory, and the reader should refer to the reference section for access to more extensive information.

What is toxicology?

Toxicology is a science with its roots in antiquity. Much of the early toxicity information was developed as humans tried new foods, examined substances for use in treatment of disease (drugs), or developed new weapons (poisons). The toxicity of substances was systematically examined in animals, with Mattieu Joseph Bonaventura Orfila (1787–1853) often cited as a pioneer. He also developed a number of analytical techniques that allowed the detection of poisons in tissue.

Toxicology as a science is rapidly developing and expanding in scope. Historically, it tended to be a descriptive science. This is based on the theory that a material is present in a specific concentration or dose and produces an observable effect. In the early days of toxicology, the steps between exposure and effect were usually unknown, and unraveling them was not possible then.

Now toxicology is becoming a mechanistically oriented science. As more sophisticated analytical instrumentation, biochemical assays, and molecular biology techniques have been developed, toxicologists have applied them to understand the steps between exposure and effect.

Toxicologists draw on many disciplines in assessing and predicting toxicity. It is anticipated that as our understanding of biological and environmental systems becomes more complete and as more mechanisms of toxicity are known and understood, better modeling and prediction of toxicity will be possible.

Epidemiology and risk assessment are rapidly expanding interests of toxicologists. This expansion in interest reflects the increasing importance of modeling and the resulting prediction of toxicity to humans and to the environment.

What is hazardous waste?

Hazardous waste has many definitions. One that is commonly used is the following, from RCRA:

> The term "hazardous waste" means a solid waste or combination of solid wastes, which because of its quantity, concentration, or physical, chemical, or infectious characteristics may
>
> (a) cause or significantly contribute to an increase in mortality or an increase in serious irreversible or incapacitating reversible illness, or

(b) pose a substantial present or potential hazard to human health or the environment when improperly treated, stored, transported, or disposed of, or otherwise managed.

Even though this definition focuses on solid and hazardous wastes, the same toxicology principles are applicable to any waste. Analysis of the above definition leads to many of the factors that must be considered in determining when a waste becomes hazardous. The list of factors is not all-inclusive.

The statutory definition of hazardous waste, however, reflects the intent of Congress and shapes the policy of agencies such as EPA. The characteristics of a waste that make it hazardous in EPA's regulations include one or more of the following:

1. Quantity
2. Concentration
3. Physical characteristics
4. Chemical characteristics
5. Infectious characteristics
6. Causes serious irreversible disease
7. Causes incapacitating reversible disease
8. Poses a substantial threat to human health
9. Poses a substantial threat to the environment
10. Is improperly used

Under this definition, effects on both the environment and human health are considered important, but interestingly only agents that cause serious irreversible or incapacitating reversible disease are considered hazardous. Agents that produce diseases that do not fall in these categories are not considered hazardous in this case. The focus is on certain specific and important characteristics of agents, e.g., quantity, concentration, physical characteristics, chemical characteristics, and infectious characteristics. Why these and other characteristics may be important in determining the toxicity of hazardous wastes in humans or the environment is addressed later in this chapter.

3.2 Toxic Effects in Animals

A very concentrated hazardous waste may produce a high local concentration of certain chemicals or chemical substances in the waste. The environmental impact due to such concentration will decline as the substance moves away from the disposal site and may drop to the point where no damage occurs. However, if the quantity of waste is large, the concentrations of chemicals or their ingredients throughout the environment may be high enough to cause widespread toxic effects. The polychlorinated biphenyls (PCBs) and DDT are good examples of chemicals that have been spread in large quantities throughout the environment.

Toxic materials may produce local damage at the site of exposure, with the intensity of effect related to the concentration of the toxic material at the site and the damage threshold of the tissue. However, most toxic materials in addition are able to cause effects at sites distant from the exposure site. For a toxic substance to cause damage at a distant site, it must reach the site in concentrations sufficient to exceed the damage threshold there.

Toxicokinetics

Toxicokinetics is the science which examines the factors that influence the absorption of a substance from the environment into an organism and the factors that influence the distribution, storage, biotransformation, and excretion of a substance after it has been absorbed. Ultimately, these factors in summation determine the concentration at the site in the organism where a toxic effect is produced.

Absorption.   Absorption is the process by which a substance passes from outside the organism to inside. Absorption may occur through the skin, through the gastrointestinal epithelium, through the respiratory tract epithelium, or through the eyes. The membranes in these organs or organ systems present barriers to entry. The effectiveness of the barrier varies from tissue to tissue. The primary barrier is the plasma membrane of the cell.

Epithelial membranes such as the skin are composed of closely associated cells. In general, tight junctions exist between the cells so that for a substance to pass through the epithelial membrane, it must pass through the plasma membranes of the individual cells. The basic structure of the cell membrane is that of a lipid bilayer which is modified by other molecules such as proteins and carbohydrates. To pass the barrier posed by the plasma membrane, the toxic substance must usually dissolve in the membrane and diffuse down a concentration gradient to the extracellular fluid and eventually reach the blood or the lymph. Since the membrane is primarily lipid in nature, the toxic substance must have lipid solubility if it is to dissolve in the membrane. Many toxic substances are lipid-soluble and may easily diffuse through membranes; organic chemicals in particular are often lipid-soluble.

The proteins and carbohydrates may provide entry to certain substances by forming pores through the membrane or by serving as carriers for active transport. The pores are very small (approximately 4 Angstroms) and constitute a small portion of the membrane surface. Only very small molecules or ions are able to pass through them. Most toxic substances are not able to pass through pores. For active transport by a carrier to occur, the toxic substance must fit the structural requirements of the carrier. Since this characteristic is uncommon in toxic substances, active transport is not an important route for most molecules.

The skin is the largest organ and presents a large surface for absorption. With such a large absorbing surface available, even if the rate of penetration per unit area is small, the total quantity absorbed may be great. On the surface of the skin, the membranes of dead cells are layered in the stratum coreum to provide a barrier to absorption. However, many substances readily penetrate the skin. The substances dissolve in the membranes and diffuse down a concentration gradient to the extracellular fluid surrounding the skin cells and eventually reach the lymph or the blood. The skin varies in composition with the skin on some areas, presenting lesser or greater barriers to penetration. The thickness of the skin in the area and the blood supply to the area are factors that influence the absorption rate. Areas of the skin such as the palms of the hand and the soles of the feet may present greater barriers to entry because of layers of callus. The blood supply to the skin varies from area to area with the absorption rate increasing in areas of greater blood supply.

Modification of the skin structure or dissolution of the agent in an oil or organic solvent will increase the rate of penetration. When the skin is abraded or hydrated, the rate of penetration will increase. This may be a problem in workers who continually expose their hands or other parts of their skin to water.

The structures in the nasal cavity and the lungs have large surface areas and ample blood supplies. In addition, the barriers to entry are fewer than those in the skin. In the lung, there may only be one cell thickness between the inhaled air and the extracellular fluid or blood. A mucus coat is also present. However, it does not present a barrier to penetration for most organic substances. Since it traps particles, it may serve to hold the substance at the absorbing surface for a longer period with a possible increase in absorption.

Inhaled substances often produce local damage to the nasal or lung tissues. In addition, they readily reach the general circulation. Highly water-soluble substances such as ammonia or formaldehyde usually do not penetrate into the small lung spaces, but instead are trapped by the mucus coat and may cause significant local damage. Less water-soluble materials such as vapors of organic solvents will penetrate the smallest lung spaces and rapidly pass to the general circulation. The rate of penetration is such that equilibrium between the air and the blood is reached very rapidly.

Smokers who work with toxic substances must be concerned with not only direct inhalation of vapors or gases but also the inhalation of substances that they have transferred to the cigarette through hand contact. Both the original substance and its pyrolytic products will enter the lungs and may be absorbed.

If a chemical or a particle is trapped by the mucus coat in the ciliated parts of the respiratory tract, much of it is moved by the cilia and eventually passes down into the gastrointestinal tract. Further local damage or absorption may occur as it passes through.

The gastrointestinal tract from the mouth to the anus offers a ready entry route for toxic substances. A chemical that is taken orally either intentionally or by inadvertent means, such as a food contaminant, may cause local damage to the gastrointestinal tract or reach the general circulation through absorption. There are tight functions between the cells lining the gastrointestinal tract; however in most areas the cell layer is only one cell thick. The epithelium is covered by a mucus coat, but it does not serve as an effective barrier to entry for many substances. The blood supply and the lymphatic circulation are ample, and many substances rapidly penetrate the epithelium and enter the blood or lymph.

The eyes may serve as an absorption site for toxic substances. The eyes have an ample blood supply which is readily accessible to substances entering the eye. Since the surface area of the eye is not large in comparison to that of the skin, lungs, or gastrointestinal tract, the quantity absorbed through the eyes is often relatively small. A more serious problem, though, is the possibility of direct local damage to the epithelium of the eye.

Distribution.   Once the toxic chemical reaches the bloodstream or the lymphatic circulation, it is rapidly distributed to all tissues. Some tissues, such as the brain, have barriers to entry, but in general the distribution to most body sites occurs within a matter of minutes. The blood carries the chemical to the tissue, and in most tissues relatively large pores exist between the capillary endothelium cells. Most chemicals are able to pass between the cells and readily enter the tissue.

In the blood or lymph, the toxic chemical either may be in solution or may be bound to a blood component such as albumin or a cell. Those substances that are poorly water-soluble are often transported through the blood bound to albumin or some other blood component. A few chemicals are actively taken up by blood cells. For example, lead is actively taken up by red blood cells. Over time the distribution between the various blood components changes as equilibrium is reached.

In the brain and in some other organs such as the testes, barriers to entry exist. In those tissues, tight junctions exist between the capillary endothelium cells, and to gain entry, a substance must pass through the plasma membrane of the individual cells. In the brain there is a layering of plasma membranes of accessory cells around the neurons and blood vessels presenting a further barrier to access. See Fig. 3.1. In addition, the cerebrospinal fluid (CSF) is continually formed at the choroid plexus and flows through the ventricles and back to the bloodstream at the arachnoid villi. The CSF is continuous with the extracellular fluid surrounding the neurons, and as it flows out of the brain, it carries substances from the brain back to the blood. Organic ions may be transported out of the brain at the choroid plexus or removed with the CSF flow. If the rate of removal by all mechanisms is fast in comparison to the rate of entry, the brain concentration of a chemical will be low. In general, only those substances that are lipid-soluble and able to penetrate plasma membranes rapidly reach significant concentrations in the brain. A lipid-soluble substance may reach a concentration in the brain which is much higher than that in the blood. This accumulation occurs because the brain contains a high percentage of lipid and the solubility of the lipophilic substance is much higher in the brain than in blood.

Storage.   Certain chemicals are retained in the tissues for prolonged periods. Storage in a tissue occurs because the chemical has high affinity or solubility

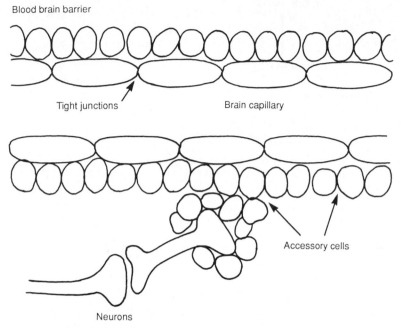

Blood brain barrier

Tight junctions   Brain capillary

Accessory cells

Neurons

Figure 3.1   Blood brain barrier.

in the tissue. Storage of lipid-soluble chemicals often occurs in those tissues that contain relatively high concentrations of lipid. As already mentioned, highly lipid-soluble substances rapidly enter the brain where they are concentrated. The blood supply to the adipose tissue is less than that to the brain and to many other tissues. As a result, the rate of delivery of the substance to the adipose tissue is lower than that to the brain. Initially, a substance enters the brain tissue, but over time it redistributes to the adipose tissue where it is stored. The chemical in the adipose storage sites is usually inert. If the chemical remaining in the blood is removed by excretion or biotransformation, it is replaced by diffusion of chemical from the storage site. If exposure is not continuous, all the chemical in the storage site may be removed over time. However, if exposure continues and biotransformation or excretion is slower than entry, the concentration in the storage site will increase until the rate of entry is balanced by elimination.

Chemicals may specifically bind to tissue components and be stored. For example, lead is stored in bone by replacing calcium in the bone matrix. See Fig. 3.2. The lead stored in bone was once considered inert; however, this has been questioned recently. Lead in the bone matrix is released when calcium is released. Another example of a chemical class that binds to a specific tissue component is the binding of the polycyclic aromatic hydrocarbons (PAHs) to DNA. The binding is covalent, and the PAH remains in the tissue until DNA repair occurs or the cell dies. This is another case where the substance is not inert since the DNA is altered and the cell function may be changed.

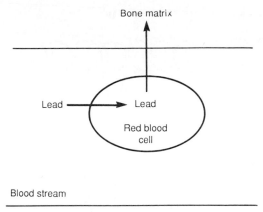

Figure 3.2   Binding of lead.

Biotransformation.   To excrete a chemical, the body may have to biotransform it to a metabolite that is less easily reabsorbed in the kidney. For reabsorption, chemicals must be lipid-soluble or structurally similar to one of the very limited number of biochemicals that are actively recovered from the urine. Metabolites are usually less lipid-soluble than the parent compound and, as a result, are reabsorbed more slowly. This leads to an increase in the excretion rate.

The enzyme system that is responsible for the biotransformation of many chemicals is the microsomal mixed-function oxidase system that is primarily localized in the smooth endoplasmic reticulum of a variety of cells. A major site of localization of this system is the parenchymal cells of the liver. The enzyme family in the system that catalyzes most of the biotransformations is called cytochromes P450. This is not a single enzyme, but instead a group of isoforms with similar characteristics. As a group, these isoforms are not very selective as to the substrates they biotransform. Frequently, these enzymes biotransform chemicals to more polar and usually more water-soluble forms through an oxidation mechanism. Some typical reactions are shown in Table 3.1.

A metabolite or a metabolic intermediate may be more toxic than the parent molecule. A hydroxyl group is introduced into the chemical molecule, or a hydroxyl intermediate is present that spontaneously decomposes to a more stable form. The metabolite or, in many cases, the metabolic intermediate is a very reactive molecule. Reaction of the intermediate with tissue components may occur spontaneously and produce toxicity. An example is the damage produced in the liver by the metabolic intermediates of carbon tetrachloride.

Biological systems have developed mechanisms to protect against the toxicity of reactive metabolites. For example, glutathione commonly combines with reactive metabolites. However, these protective mechanisms may be overloaded, and tissue damage occurs. In other cases, no protective mechanism exists. The microsomal mixed-function oxidase system therefore is a mixed blessing. It is necessary for excretion of many chemicals, but in the process of converting chemicals to an excretable form, it makes some of them many times more toxic.

TABLE 3.1 Selected Reactions Catalyzed by Cytochromes P450

N- and O- Dealkylations

$R\text{-}NHCH_2\text{-}CH_3 \text{---}^{[o]}\text{----} > RNH_2 + CH_3CHO$

$R\text{-}OCH_3 \text{-----}^{[o]}\text{----} > ROH + CH_2O$

Side Chain (Aliphatic) and Aromatic Hydroxylations

$RCH_2CH_3 \text{----}^{[o]}\text{----} > RCH(OH)\ OH_3$

N-Oxidation and N-Hydroxylation

$(R)_3N \text{----}^{[o]}\text{----} > R_3N\text{->}O$

$RNHR' \text{----}^{[o]}\text{----} RN(OH)R'$

Desulfuration

$R_3P\text{=}S \text{----}^{[o]}\text{----} > R_3 = O + SO_4^{-2} + PO_4^{-3}$

Deamination of Amines

$RCH_2NH_2 \text{----}^{[o]}\text{----} > RCHO + NH_3$

Other metabolic conversions occur in the body, but only one other biotransformation, the conjugation reaction, is of general importance for a variety of chemicals. In the group of reactions termed *conjugations,* the chemical is bonded to another molecule such as glucuronic acid, an amino acid, or a sulfate. In general, the conjugation reaction increases water solubility which results in decreased reabsorption in the kidney and faster excretion. These reactions are catalyzed by a variety of enzymes localized in many tissue sites. These enzymes typically exhibit broad substrate structural specificity, and since there are a number of them, a wide variety of structures may be conjugated. The toxicity of the conjugated metabolite is usually less than that of the parent chemical.

The tissue content of biotransforming enzymes varies. The presence of a biotransforming enzyme in a tissue sometimes leads to tissue-specific toxicity. Particularly in the case of the cytochromes P450 family of enzymes, the isoform content varies from tissue to tissue. Certain isoforms are unique and found only in a specific tissue. If a toxic metabolite is produced by that isoform, toxicity may occur in only that tissue or in a specific cell type of that tissue.

Induction and inhibition of the biotransforming enzymes may occur. Toxicity may increase or decrease as a result of induction or inhibition. The cytochrome P450 family of enzymes is inducible on exposure to a chemical that acts as an inducing agent. The induction usually requires several hours to days. Many chemicals may act as inducing agents, and a chemical may induce its own metabolism. The halogenated hydrocarbons such as the PCBs are good examples of the types of compounds that produce induction. When induction occurs, the isoform pattern in a tissue changes. The content of some isoforms may decline while that of others may actually increase, the net result being an overall increase in activity. Other biotransforming enzymes such as glucuronyl transferases are also inducible. Inhibition of metabolism may occur and potentially may have an effect on toxicity. For example, piperonyl butoxide is included in insecticides because it is a cytochrome P450 inhibitor. In that case, the biotransformation of other components of the insecticide is slowed, and the toxicity is increased.

Excretion.   In most animals, the major site of excretion of chemicals is the kidney. Excretion in the bile is also important for some chemicals. The lungs serve as the excretion route for many gases. In fish, exchange of chemicals from the gills into the surrounding water is often important. Most chemicals pass from the blood into the urine by the process of filtration in Bowman's capsule within the kidney proximal tubule. See Fig. 3.3. A few chemicals are actively secreted into the proximal tubule, but this is not common because of the structural specificity required by the carrier.

Once the chemical is filtered in Bowman's capsule, it may pass with the urine out of the body, or if it is sufficiently lipid-soluble, it may be reabsorbed back into the blood passing through the tubule cells. Chemicals that are sufficiently lipid-soluble to be absorbed through the gastrointestinal epithelium or the skin are also probably sufficiently lipid-soluble to be reabsorbed in the kidney. If a compound is completely reabsorbed in the kidney, it will never be

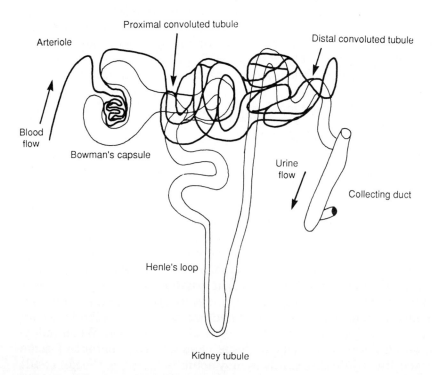

Figure 3.3 Filtration process in Bowman's capsule.

excreted. The chemical remains in the organism unless changed by metabolism to a more polar form that is not readily reabsorbed.

Biliary excretion is less common and is an active process. Molecular size and structure are important determinants of biliary excretion. Some heavy metals such as cadmium are also excreted through the bile.

Breast milk is not usually a quantitatively important route of excretion. However, in the nursing mother, the quantity of a chemical transferred to the child may be important. Lipid-soluble chemicals easily enter the milk and concentrate there. The polychlorinated biphenyls are examples of chemicals that move rapidly from the mother to the nursing infant. In one report, one mother's entire body load of polychlorinated biphenyls moved from the mother to the infant. The quantity of material transferred was sufficient to produce a toxic effect in the infant.

Certain substances such as arsenic are concentrated in the skin. When the epidermis is sloughed, the substance is excreted. The same is true of hair. In the case of hair, the period of exposure to the substance may be determined by assuming that the hair grows at a constant rate and by doing analysis of segments of the hair for the content of the substance.

In summary, if a chemical is lipid-soluble, it will be rapidly absorbed and widely distributed in tissue. Initially, it will reach highest concentrations in tissues that have high blood flow and high lipid content such as the brain. Later it will be stored in the adipose tissue and only slowly released for metabolism and excretion. Unless the chemical is converted by metabolism to a more water-soluble form, the excretion rate will be minimal and accumulation may occur.

Toxicodynamics

The toxic effect produced by a chemical may occur as the result of a nonspecific interaction with a component of the organism or through interaction with a highly specific receptor. Most chemicals, if not all, have more than one toxic action with the effect produced being dependent on the concentration of the chemical and the site of action. A chemical may produce toxic effects by both specific and nonspecific interactions. All the interactions follow the law of mass action. A second-order equation describes the interaction:

$$T + C \rightleftharpoons TC \rightarrow \rightarrow \text{toxic effect}$$

Here T is the tissue where the chemical produces the toxic effect, and C is the chemical that produces the toxic effect. The chemical interacts with the tissue, and a toxic effect occurs. A number of intermediate effects may occur in the tissue before the observable toxic effect is produced. For example, a chemical might be biotransformed to a toxic metabolite that then would dissolve in a brain neuron plasma membrane, causing a fluidity change. This fluidity change might then lead to a change in membrane permeability which then might lead to a change in function for the neuron and eventually through a continuing series of changes to the observable toxic effect of central nervous system depression. The number and complexity of the intermediate steps vary from chemical to chemical, and in most cases only a portion of the intermediate steps is known or understood.

Many chemicals appear to produce toxic effects through a nonspecific interaction with tissue. In some cases the interaction may actually be nonspecific whereas in others the state of knowledge may not have advanced to the point where a specific site of action has been identified. An example of a toxic effect, which is thought to result from a nonspecific interaction of the chemical with the target tissue, is the depressant effect on the brain of many lipid-soluble substances. The substances dissolve in brain cell plasma membranes. When this happens, the fluidity of the membrane changes, and in concert the membrane function changes. The depressant effect results from a nonspecific interaction.

Increasingly, specific tissue components that chemicals interact with in producing toxic effects are being identified. These specific chemical components are called *receptors*. Receptors may be any tissue component. Commonly, they are proteins, enzymes, or nucleic acids. An example of a highly specific interaction with a known receptor is the interaction of polycyclic aromatic hydrocarbon compounds with a protein called the Ah receptor. The quantity of the Ah receptor present in a given organism is genetically determined. The higher the concentration of the receptor present, the greater the toxicity of polycyclic aromatic hydrocarbons in that organism.

In most cases, whether an effect is produced by specific or nonspecific interaction is not known with a high degree of certainty. Often knowledge of the specific site of action is limited to a whole organ or a section of an organ. Even if it is known that a compound interacts with a specific receptor, the events that occur between the compound's interacting with the receptor and the occurrence of the toxic effect are unknown. For example, as mentioned above, the polycyclic aromatic hydrocarbons are thought to produce their toxic effects by interacting with the Ah receptor. However, some compounds produce a number of effects such as induction of P450 cytochromes, liver necrosis, and tumors. The steps between the responsible interaction and the ultimate effect are not known with any certainty. Many toxicologists are currently working to elucidate mechanisms of toxic effects, with much of the effort directed at the cellular and molecular level.

Information about the intervening steps between receptor interaction and toxic effect is slowly appearing as the research progresses. The rapid expansion of knowledge about cell function and structure at the biochemical and molecular level has provided the toxicologist with the techniques and information needed to examine the toxic effects of chemicals at the macromolecular level. Use of computer models of receptors and the study of the interaction of chemicals with receptors are in their infancy. As more is learned about the mechanisms of toxicity, the production of safer chemicals will be possible. Chemicals will be modified to reduce the interaction with the specific receptor or the nonspecific site, and the result will be less toxicity.

Dose-response relationship

The toxic effect that occurs and the intensity of the effect are determined by the concentration of the chemical in the tissue where the toxic effect is manifest and more specifically at the site of action or receptor. In the intact organism or cell, the concentration of chemical at the site of action is usually unknown and may only be inferred by measuring a concentration at a distant site such as the blood and then extrapolating to the concentration at the site of action.

The relationship between the toxic effect produced and the concentration required to produce the effect varies from chemical to chemical. The relationship between the concentration and the effect may be expressed mathematically with the expression derived from the mass action equation. The relationship is often expressed graphically, as shown in Fig. 3.4.

When concentration is expressed in arithmetic units, the concentration response curve normally has the shape of a rectangular hyperbola.

If the concentration is expressed in logarithmic units, the curve normally has an S or sigmoid shape. Deviation from the normal curve shape is an indication that the response is being influenced by other factors. For example, if the blood concentration is measured but entry barriers exist, the blood concentration may not reflect the concentration at the site of action; or if the series of responses between the interaction of the chemical with tissue and the observable response is complex, then the curve shape will be an integration of all the factors and may not be normal. The curve shape may be used to examine the question of whether factors other than concentration are influencing the response.

Before the advent of computers, the relationships were often expressed graphically to allow better visualization. The mathematical relationship established in the mass action equation was expressed in a variety of ways to allow conversion of the graphical representation to a linear format. In toxicology, data were often expressed graphically in probit units as a way of converting the graphical representation to a linear format. A probit is equivalent to one standard deviation. See Fig. 3.5.

In toxicology, a concentration-response relationship is not always evident. An example is the case of the allergic response. The amount of material needed to

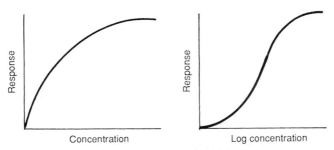

Figure 3.4 Concentration-response relationship.

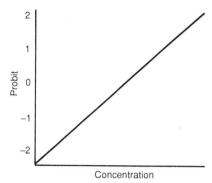

Figure 3.5   Probit-concentration graph.

elicit a response may be very small. Once the threshold for the response is exceeded, a maximum response is produced with little or no further exposure.

In biological systems, a chemical must usually exceed a threshold concentration before an observable response is seen. This suggests that a sufficient activation of receptors must occur for a response to take place and that activation below that level produces no observable response. In the case of nonspecific interaction with plasma membranes, the threshold will be exceeded when the fluidity of the membrane is changed to a point so that its function is affected and an observable change occurs. In other cases, such as damage to DNA, the threshold might be exceeded only when the ability to repair the damage before expression is exceeded. Below the threshold, no dose-response relationship exists.

Repeated exposure to a chemical may result in an accumulation of damage or an accumulation of the chemical to the point where an observable effect is seen. Accumulation of damage occurs when the rate of repair is exceeded by the rate of damage. When accumulation of damage occurs, the concentration of chemical required to produce an observable toxic effect is often less than that required after a single-dose exposure. When accumulation of the chemical occurs, the concentration is reached at the site of action, and the toxicity usually resembles that of an effect due to a single exposure.

Factors affecting response

The effect of genotype on the response to chemicals has been extensively studied in laboratory animals. Significant variation is seen even in inbred strains of animals that appear phenotypically identical and that have been raised in closely controlled laboratory environments.

The genotype of humans varies tremendously. This variation is expressed in an equally wide variety of phenotypes. The human response to toxic chemicals often varies according to a normal distribution, with the magnitude of the variation being large. For example, a more than twentyfold variation is seen in the rate of biotransformation of some chemicals. Any observed variation in response is often attributed to a variation in a toxicokinetic parameter such as absorption, biotransformation, or excretion. However, an equally great variation is probably present in pharmacodynamic parameters. Less information is available for the variation in toxicodynamic parameters.

Only a portion of the variation seen from one human to the next may be attributed to genetic differences. Nearly all toxicodynamic and toxicokinetic parameters can be influenced by the environment. Many environmental factors such as chemical exposure, diet, smoking, stress, and alcohol use contribute to the variation. For example, if a person works or lives in an environment where there is an exposure to polychlorinated biphenyls which induce cytochromes P450, the person will biotransform both polychlorinated biphenyls and many other chemicals more rapidly. Toxic responses to the chemicals may be increased or decreased depending on whether the metabolites are toxic or nontoxic. The degree of variation is different for each person, as would be expected in a normally distributed population. Some individuals are unusually resistant or sensitive to environmental effects.

The age of an animal or a human may influence the toxic effect of a chemical. A chemical that shows a large variation in response with age is methyl mercury. Methyl mercury, even in very low concentrations, produces irreversible central nervous system damage in the fetus. The degree of damage to the central nervous system decreases with age. Adult humans show no effect at concentrations that will irreversibly damage the fetus.

Often the variation in toxic response with age may be traced back to a difference in a toxicokinetic parameter. For example, biotransformation and excretion may not be completely developed in the neonate and may decline rapidly as the person reaches late middle age and old age. In general, the variation in toxic responses increases as the population ages. Some individuals continue to show the response of a much younger adult while others show much less resistance to toxic effects. Genetics, environmental exposure, and disease all contribute to the age-related variation. Little information is available on toxicodynamic changes with age.

Laboratory rats show a sex difference in the toxic response to many chemicals. Much of the variation may be attributed to a major difference in biotransformation. The levels of individual P450 cytochromes are different in male and female rats. Growth hormone and sex hormones play a central role in the difference. Sex-related differences have been seen in the toxic response to chemicals in a variety of species. Gender differences have also been reported in humans, but they have not been studied extensively.

Whether the presence of combinations of chemicals in the environment or the organism enhances or decreases toxic responses has only recently received much attention. Depending on the specific combination, the toxic effects will be enhanced or attenuated. Extensive research is needed to explore this highly relevant area.

Previous or concurrent disease will change the toxic response to a chemical. The disease state often removes the physiological reserve normally present in the tissue. For example, the person who has fibrotic lung disease has little physiological reserve available. Inhaled chemical may produce toxic effects which are unexpectedly severe for the exposure concentration. Another good example is the person who has cirrhosis of the liver. If cirrhosis is severe, the biotransformation enzyme levels in the liver will be lower than normal and toxic effects of compounds that are biotransformed in the liver will be changed. The same relationship is likely to extend to other organisms and, ultimately, the environment as a whole. However, few studies have been done.

Many biological systems vary in function in a cyclic pattern over time. The cyclic patterns may recur daily or on longer time scales. As the systems vary in function, the toxic effect of chemicals also changes. A good example of daily variation is the response to agents that depress the central nervous system. Compounds produce a greater depression when the animal is in the sleep cycle part of the day and a lesser depression when the animal is in the active cycle part of the day. Biotransformation enzyme activities also vary over a 24-hour period. The cycle appears to be controlled by light and is related to cyclic changes that occur in the endocrine system. In particular, the adrenal gland produces a varying amount of corticosteroids on a 24-hour cycle. When a toxic effect is examined in any organism, the presence of cycles should be considered, or else the effect may be obscured or enhanced by the normal variation.

Stress has been observed to influence the toxic effect of chemicals. Many biological systems appear to change as the level of stress changes. In particular, the immune system and the endocrine system show dramatic changes during stress. In general, the ability of a stressed organism to mount a defense against chemical toxicants is reduced. When the immune system's function is changed, the organism does not respond normally to infection and may have less ability to scavenge abnormal cells. Tumors may result. Many biological systems are under endocrine control. As the endocrine system function changes, other activities such as biotransformation and the immune response change. When these changes occur, the organism does not respond normally to toxic insult.

3.3   Toxic Effects in the Environment

Commercial mixtures

Chemicals usually enter the environment not as pure substances, but as commercial products that are mixtures of chemicals or as components of waste. In either case, the toxic effect produced is a combination of the effects of the various chemical components. Each of the components will produce different toxicities at different concentrations. As the substances remain in the environment, they will move and change.

An example of a commercial product that may be produced in many different mixtures is the PCBs. There are 209 isomers of the PCBs, and each commercial mixture contains a different mixture of these isomers. In addition, some of the commercial mixtures contain polychlorinated dibenzofurans as contaminants. The commercial mixtures also vary in isomer and contaminant content from batch to batch.

Environmental movement

Chemicals rapidly move throughout the environment. Chemicals easily move from one environmental location to another with equilibrium occurring when the rate of addition equals the rate of breakdown. Movement in both water and air is rapid. By injecting chemicals that are not naturally occurring into the air, the worldwide distribution of substances in a matter of days can be demonstrated. DDT and the PCBs have been found in the body fat of organisms in Antarctica, which is hundreds of miles from where these substances might have been used.

Breakdown products

The chemical composition of waste changes as the chemicals move through the environment. Photolysis, oxidation, hydrolysis, and biotransformation by organisms are some of the modifying processes. Some chemicals will be resistant to modification, and others will be modified almost immediately. The mixture present in the environment will consist of those chemicals that resist modification and the products of the more easily modified chemicals. If the original mixture is used in studying toxic effects, the toxicity seen may be different from that produced by the chemicals after modification.

Persistence

Depending on the conditions, substances may persist in the environment for very long periods. Most substances will either degrade or move in the environment if water, oxygen, or sunlight is present. However, if a substance is tightly bound to a soil particle, it may not be available for movement into the water and may be chemically unreactive while bound. The substance will persist for an abnormally long period, and accumulation may occur if release continues. Since modern landfills are designed to prevent the entry of water and to exclude oxygen and sunlight, even materials such as paper and food which would be expected to degrade rapidly reportedly persist for long periods. If water enters the landfill at a later time, the accumulated substances will leach from the landfill and potentially contaminate the groundwater.

In areas of low rainfall and high soil pH, the accumulation of heavy metals in the topsoil layer may occur. The concentration may increase to the point where environmental and human health effects may occur. Lead in street surface dirt in arid areas and cities has been found to reach concentrations as high as several milligrams per gram of soil. When this dirt is blown by the wind, it may be inhaled by humans and the lead may be absorbed. Little information is available about organics in soil in arid areas.

Bioconcentration and bioaccumulation

Lipid-soluble chemicals may concentrate in the lipid of an organism. On continuing exposure to the chemical, bioaccumulation may occur in the organism's lipid. If the organism is consumed by another organism, its accumulation of chemical will be passed on to the lipid of the consuming organism. Over time, the consuming organism will concentrate the chemical content from the lipid of the many organisms it consumes, and bioconcentration will occur. The concentration of a chemical in the body fat of an animal at the top of a food chain may be several orders of magnitude higher than that found in the environment.

Theories of bioaccumulation and bioconcentration are used to explain how a chemical such as DDT can cause a decline in pelican populations when the environmental concentration of hazardous substances or toxic chemicals is far below that required to produce acute toxicity. The pelicans consume fish that have consumed many smaller organisms. The smaller organisms accumulated DDT from the environment, and the fish concentrated DDT in body fat as the many smaller organisms were consumed. The pelicans then consumed many fish, further concentrating the DDT until it reached toxic concentrations.

3.4   Toxicity Testing

Lack of testing of mixtures

Most toxicity testing has been done on individual chemicals instead of on commercial mixtures. The use of data from using individual chemicals for the purpose of making toxicity predictions for commercial chemicals is problematic. The toxicity of the commercial mixture is a combination of the effects of the compounds present. As indicated above, commercial mixtures may vary in their components from batch to batch; the predicted toxicity will also vary. In addition, the components will vary with the residence time of the chemical in the environment. Further, the rate of change will be influenced by the microenvironments with the chemical moving from one to another over time. For example, if a chemical is bound to a clay particle and buried in an anaerobic environment, the rate of change may be much different from that for the same chemical which is in solution and exposed to oxygen and sunlight.

Species used in testing

Toxicity testing is done in a wide variety of species, but in many cases the species used may not be well suited to predicting effects in the environment. Rats and mice are readily available and inexpensive. As a result, most toxicity testing is done in inbred mice and rats. The physiology and biochemistry of these species are quite different from those of humans and of many other animals. Some important differences are the length of lifespan, biotransformation enzymes, and bacterial flora. These differences and many others make the prediction of toxicity in human and other organisms in the environment difficult and uncertain.

Extrapolation

Frequently, there is little toxicity information available on a chemical. High-dose and short-term exposure data obtained in rats or mice are frequently all that is available. In the case of the new chemical, entity information may be completely lacking. The prediction of toxicity in these cases requires extrapolation. Extrapolation may require the use of structure-action data obtained by using other similar chemicals. It may often require the prediction of long-term and low-dose toxicity in humans or other organisms from high-dose and relatively short-term exposures in mice or rats. Because of the many unknown and uncontrollable factors and the imprecise nature of the models used in making such predictions, the extrapolation is uncertain.

In assessment of risk, concentration-response curves are often used in extrapolating from high-concentration animal exposures to low-concentration human exposures. This is particularly true of carcinogenicity studies and is necessitated by the lack of information on human exposures and toxicities. A number of models are used in making the extrapolation. The commonly used model assumes that the concentration response is linear from high to low concentrations. Use of this model as well as other models generates controversy. Depending on the critic, the models are described as either over- or underestimating the risk at low concentrations. The possible existence of a threshold for damage is often ignored. Even very low doses are assumed to produce some effect. Because of the uncertainty of the risk estimate for any model, a safety margin is often used in establishing the concentration which is considered to be generally safe. The size of the safety margin varies with the toxic effect. A safety margin of a factor of 1000 is commonly used in carcinogen risk assessment, and lesser safety margins are used in risk assessment of other toxic effects. Obviously, better models are needed.

Increasingly, computer models are being used in extrapolation. This is particularly true for the prediction of toxicity from the structure for the new chemical entity. A variety of computerized methods are available for extrapolating from animal data to humans. The methods vary in sophistication and use different models. Little has been done in the use of computer models to predict environmental effects.

Threshold

The models used in extrapolation often ignore the concept of threshold. The *threshold* is the lowest concentration of chemical that produces an observable change. The concept is based on the presence of a physiological reserve in most biological systems and the repair of damage that may occur before any observable toxicity is produced. Thresholds probably exist for most effects produced by a chemical.

By assuming there is no threshold and that even very low concentrations of a chemical may produce some damage, the estimated "no-effect" concentration is made even lower. All the assumptions are biased in the direction of safety. Controversy therefore exists over the appropriate model to use and whether the safety bias is too great.

The lethal concentration 50 percent

The concentration of a chemical which kills 50 percent of a population of organisms is known as the *lethal concentration 50 percent* or the LC50. In animals, it is sometimes referred to as the lethal dose 50 percent or the LD50. In the past, the LD50 was routinely determined for chemicals. The information was then used in comparing the relative toxicity of chemicals. As the sensitivity to the use of animals in toxicity studies has increased, questions have been raised about the value of the data and the determination of the LD50 has decreased. There is little reason to continue to determine the LD50. The determination of mechanisms and the no-effect level are far more useful.

Mechanism of toxicity

During the last several years, there has been a shift from descriptive toxicity studies to mechanistic studies. The shift has occurred because of the importance of understanding the mechanism through which a compound causes its toxic effect. If the mechanism is known, the data are much more useful in extrapolations involving new compounds, in the development of safer chemicals, and in the development of better treatments.

Alternatives to toxicity testing in animals

Animal rights groups have been increasingly active in criticizing the use of animals in toxicity testing. Many groups object to the use of animals in toxicity tests that cause pain without proper relief. In some cases, laboratories have been damaged and animals removed. Several states have now enacted legislation that makes interfering with the use of animals in research a crime.

Alternatives to the testing of chemicals in animals are being extensively investigated. In vitro testing of chemicals in cell culture has increased dramatically. As more information is gained on tissue biochemistry and structure and on how chemicals interact with them in producing toxicity, more prediction of toxicity by using computer models may be possible. However, the continuing use of animals in toxicity testing is essential since the toxicity testing of chemicals in vitro and the use of computer models are still in their infancy.

3.5   Organ-Specific Toxicity

Many chemicals produce organ-specific toxicity. This is often the case because the organ is exposed to high concentrations of a chemical or because of the biotransformation of a chemical to a toxic metabolite. In the following sections, some examples of organ-specific toxicity are presented.

Skin

The skin is a very large organ in terms of both size and surface area. It is also exposed directly to many environmental chemicals. It is protected from damage in a number of ways. For example, the layers of dead keratinized cells that make up the stratum corneum of the epidermis are relatively impermeable to most chemicals. The skin contains a variety of biotransforming enzymes including the cytochrome P450 system.

Chemicals diffusing through the skin may be delivered to distant sites through the blood or lymph. Since the blood and lymph flow vary from one skin area to another, the rate at which the chemicals reach the general circulation varies depending on the skin area exposed.

The skin is able to respond to injury. When the skin is continually abraded, the surface layer thickens to form calluses; and when injury from sunlight occurs, specific skin cells, called melanocytes, respond by increasing the production of melanin, which provides protection. The large blood supply to the skin quickly delivers the immune system components when injury occurs. In addition, the large quantities of biotransforming enzymes may quickly detoxify some chemicals, and in some cases the enzymes are inducible.

Chemicals may cause damage to the skin through either direct or indirect action. Corrosive chemicals such as acids and bases may cause damage through direct chemical action to destroy the skin structure. Others, such as some of the polycyclic aromatic hydrocarbons (PAHs), cause damage through metabolites. Cytochrome P450 isoforms present in the skin cells produce metabolites of the PAHs which then interact with DNA in the skin cells. Tumors may ultimately occur at the site of exposure. Sensitizers such as nickel may cause local reactions by activation of the immune system.

Liver

Chemicals which are swallowed may be absorbed from the gastrointestinal tract and pass with the blood through the portal vein to the liver. Through this process, the liver is often exposed to high concentrations of chemicals. The liver contains a wide variety of biotransforming enzymes which may transform the chemical to a form that may be more readily excreted through the kidney or the bile.

The liver is sometimes damaged by the chemicals during their transit through the organ. The damage may occur directly through chemical action or indirectly through toxic metabolites. Cell-specific damage is often associated with the presence of a discrete biotransforming enzyme. Because of the presence of large quantities of biotransforming enzymes in the parenchymal cells, they are commonly the site of toxicity. Often the toxic damage is greatest in those parenchymal cells surrounding the central vein. Carbon tetrachloride is a chemical which demonstrates this effect. It is biotransformed by cytochrome P450 isoforms to a toxic metabolite with the toxicity focused on the parenchymal cells closest to the central vein. Other tissues may also be damaged by the toxic metabolites produced in the liver and delivered through the blood to other tissues.

Respiratory tract

The respiratory tract consists of a number of components including the nasal cavity, the trachea, and the lungs. The inhaled chemical may produce local toxicity, may readily pass into the blood, or may be trapped and then swallowed, with toxicity or further absorption occurring in the gastrointestinal tract.

Each component of the respiratory tract may be the site of toxicity for an inhaled substance. The water solubility of the inhaled chemical and the particle size are important parameters in determining the site of toxicity. If a chemical is very water-soluble or if it has a particle size greater than 5 micrometers, then it will primarily produce effects in the nasal cavity, the trachea, the lining of the tissues, and the larger lung spaces. The lower the water solubility of the substance and the smaller the particle size, the farther the chemical will penetrate into the lungs. Chemical particles that are smaller than 0.1 micrometer will penetrate into the alveoli.

The nasal and lung surfaces are covered with a mucus coat. Particles and water-soluble substances may be trapped in the mucus coat. If they are trapped in a ciliated section, they will be moved by ciliary action to the esophagus where they will be swallowed. In the smaller lung spaces, where the epithelium is not ciliated, the chemical may be absorbed into the blood or, in the case of insoluble particles, engulfed by the macrophages and moved to the regional lymph nodes. In most sections of the respiratory tract, the chemical has to pass through only a single cell layer to reach the blood.

The respiratory tract contains a large number of different cell types. Each varies in its response to a toxic chemical. Chemicals may produce damage by direct corrosive action, by indirect stimulation of inflammatory responses by the immune system, or by biotransformation to toxic metabolites. If a substance causes direct tissue damage, the lungs often fill with fluid. They may become ineffective in oxygenating the blood. The same effect may occur when a substance causes an inflammatory response. If the direct damage or the inflammatory response continues over an extended period, the lung cells may be replaced by fibrous tissue and the lungs may again be ineffective in oxygenation of the blood. Another potential result of continuing direct damage or inflammation is the production of tumors.

In the lung, the biotransformation enzymes are localized in specific cell types. The enzymes present vary from one cell type to another. This high degree of localization of the biotransformation enzymes often results in the toxicity's being localized in specific cell types.

The nasal cavity is the site of toxicity of many inhaled chemicals. As in the lungs, chemicals have to pass only a single cell layer to reach the blood with systemic toxicity as a possible result. The damage to the nasal tissue may be produced by direct corrosive damage to the tissues, an inflammatory response mediated by the immune system, or a toxic effect produced by a metabolite. The nasal epithelium has a mucus coat, but it is not an effective barrier in preventing substances from reaching the nasal epithelium.

The sustentacular cell, which is found surrounding the olfactory receptor cells, appears to contain large amounts of biotransforming enzymes, particularly the P450 cytochromes. Other cells in the nasal cavity epithelium such as Bowman's glands and Stensen's glands also contain the P450 cytochromes and other xenobiotic biotransforming enzymes. Some of the cytochromes P450 isoforms are specific to the nasal cavity. Certain chemicals appear to cause site-specific toxicity in the nasal cavity because of biotransformation to toxic metabolites. A number of nasal carcinogens appear to require activation by P450 cytochromes mediated demethylation, with the toxic metabolite being formaldehyde. The tumors often arise in the ethmoturbinate area of the nasal cavity which is the site of the sustentacular and olfactory receptor cells.

Gastrointestinal tract

The gastrointestinal system extends from the mouth to the anus and consists of a number of components that vary in their characteristics. A chemical may enter the gastrointestinal tract either through the mouth or in secretions from the respiratory tract. A number of substances such as saliva, hydrochloric acid, and bile are secreted into the gastrointestinal tract. Each substance may interact with the chemical to influence its toxicity. Biotransforming enzymes are found throughout the epithelium lining of the gastrointestinal tract with the activity localized in specific cells. In addition, the bacterial flora of the gastrointestinal tract biotransform many ingested chemicals.

In the mouth, the chemical is exposed to enzymes present in saliva. Chemicals may either bind to the enzymes and be inactivated or be biotransformed by the enzymatic activity. Biotransforming enzymes are found in epithelium of the tongue and the mouth and in the bacterial flora. Biotransformation in the mouth has been implicated in the carcinogenicity of chemicals producing tumors of the mouth and tongue.

The esophageal epithelium contains biotransforming enzymes which have been implicated in the carcinogenicity of ingested chemicals. The esophagus is a highly vascularized organ. It is an indirect site of toxicity for ethanol. Ethanol has been implicated as a causative agent in cirrhosis of the liver. In cirrhosis, the veins of the esophagus may become engorged with blood and ultimately ulcerate, causing life-threatening bleeding into the esophagus. Chemicals that directly corrode the epithelium of the esophagus may also cause bleeding or perforation followed by infection. Direct injury to the esophagus may be followed by fibrous tissue formation and stricture.

The stomach epithelium is coated with mucus. Chemicals readily penetrate the mucus barrier and reach the epithelial cells. Biotransforming enzymes are found in stomach epithelial cells with a toxic effect sometimes produced by a metabolite. Substances that cause a depletion of the mucus coat may produce toxicity by exposing the epithelium to erosion by the hydrochloric acid and digestive enzymes normally found in the stomach.

The intestinal epithelium varies from segment to segment. Variation is also seen in the bacterial flora. The intestinal epithelium is coated with mucus, but the coat is not an effective barrier to chemicals. Biotransformation of chemicals occurs both in the intestinal epithelium and in the bacterial flora. The metabolites may produce toxic effects locally or pass through the blood to produce toxicity systemically.

Chemicals sometimes follow complex pathways in producing toxicity either in the intestine or systemically. For example, a chemical might be absorbed from the intestine, biotransformed in the liver, excreted back into the intestine in the bile, further biotransformed by the bacterial flora, reabsorbed, and excreted through the kidney. Toxicity might occur at any time during the circuitous path from absorption to excretion.

Kidney

The kidney is often the site of chemical-induced toxicity. As part of its normal excretion function, the kidney concentrates chemicals in the tubules. This causes the tubular cells to be exposed to relatively high concentrations of chemicals. The kidney tubular cells also contain a variety of biotransforming enzymes. Metabolism of chemicals to toxic metabolites may occur in the tubular cells. The metabolites may cause direct damage to the tubular cells or may diffuse into the blood and cause toxic effects at other sites. Since the kidney is very important in maintaining homeostasis in the body, damage that results in a reduction in kidney function may have adverse effects on many other body functions.

Immune system

The immune system is increasingly being identified as a target of chemical toxicity. Because the immune system is involved in a wide variety of functions such as control of infections, abnormal-cell removal, and inflammatory responses, if its function is changed, severe toxicity may result. The understanding of the wide variety of functions and diseases that are immune-system-mediated is just developing. For example, if function is reduced, routine infections may become life-threatening and tumors may occur. If function is enhanced, autoimmune diseases such as arthritis may be induced or accentuated.

Brain

In general, the biotransformation activity in the brain cells is relatively low. A chemical that penetrates the blood-brain barrier may be retained in the brain for prolonged periods. Also, since the brain has a high lipid content and high blood flow, lipid-soluble chemicals may rapidly enter the brain and be concentrated there. Toxicity sometimes occurs rapidly and may be irreversible since neurons are not able to divide and repair may be slow. Later the chemical may be redistributed to the adipose tissue for inert storage, but this usually occurs after the chemical has reached the brain and produced toxicity.

3.6   The Future

Toxicology is rapidly moving from its historical descriptive approach to a future that will be dominated by mechanistic approaches. New techniques in molecular biology, biochemistry, and computer modeling are being applied to understanding the mechanisms of toxicities. By charting the mechanism underlying a toxic effect, the toxicologist will be able to generalize the information to the prediction of toxicity for new chemicals. With the advent of molecular modeling using supercomputers and as increased information is available on molecular targets of chemical toxicants, the prediction of specific toxicity without toxicity testing in animals or the environment could become routine. It will be possible to predict the environmental stability and toxicity of a chemical and to modify its structure to reduce its stability. Thereby, the persistence of the chemical as well as its structure may be modified to reduce toxicity. These techniques will allow development of alternative chemicals that are environmentally safe.

Similar techniques may be applied to develop genetically engineered organisms to help in elimination of preexisting environmental contaminants. For example, an existing or specially designed biotransformation enzyme could be inserted into an organism. The organism could then be released into the contaminated area and the contaminant eliminated by biotransformation to nontoxic chemicals. The organism will be engineered to have a short survival time in the environment to alleviate public concerns. This approach has already been used in oil spill cleanup.

Using these and other tools, the toxicologist will be able to reduce the potential dangers associated with the necessary use of chemicals. Methods of dealing with inadvertent chemical spills and hazardous wastes will become more specific and effective.

Recommended Reading

1. Alfred Goodman Gilman, Theodore W. Rall, Alan S. Nies, and Palmer Taylor (eds.), *The Pharmacological Basis of Therapeutics,* 8th ed., Pergamon Press, New York, 1990.

2. Curtis D. Klassen, Mary O. Amdur, and J. Doull (eds.), *Toxicology: The Basic Science of Poisons,* 3d ed., Macmillan, New York, 1986.

3. William B. Pratt and Palmer Taylor (eds.), *Principles of Drug Action: The Basis of Pharmacology,* 3d ed., Churchill Livingstone, New York, 1990.

4. U.S. Congress, Office of Technology Assessment, *Identifying and Controlling Immunotoxic Substances—Background Paper,* OTA-BP-BA-75, Government Printing Office, Washington, April 1991.

Air Toxics under the Clean Air Act

4.1 Introduction

In a wave of environmental reform, Congress passed four amendments to the Clean Air Act of 1963: the Air Quality Act of 1967, the Clean Air Act Amendments of 1970, the Clean Air Act Amendments of 1977, and the Clean Air Act Amendments (CAAA) of 1990. Taken together, they comprise the federal framework to regulate air emissions and airborne pollutants that may be emitted from stationary and mobile air emission sources. With the passing of the second, third, and fourth of these federal air statutes, Congress played its role cautiously and, like a good neighbor, listened to the public comments and expectations much more attentively.

During the 1950s and 1960s, public attention and congressional inquiries and interests were largely limited to visible sources of pollution. As a result, during these periods, the controllability of existing air pollution included only those mobile and stationary sources which were found to be aesthetically unappealing. The first effort to list invisible air pollutants or contaminants was initiated by the Environmental Protection Agency (EPA) pursuant to the Clean Air Act (CAA) of 1970, which has been labeled as the first major federal environmental act. Armed with its broad congressional mandate, EPA established the primary and secondary *national ambient air quality standards* (NAAQSs) for the protection of public health and welfare.

Thus, under statutory authority for the first time EPA declared what pollutants must be controlled under CAA of 1970. It initially selected seven *hazardous air pollutants* (HAPs) for the purpose of regulating them with an ample margin of safety to protect public health. However, very little attention was paid to the other hazardous or toxic air pollutants by EPA at the preliminary stage of regulation.

Nevertheless, there was a tacit recognition around that time that certain suspected carcinogens are emitted into the air and that they need to be controlled. This concern was not totally lost to the members of Congress. Eventually, Congress included specific provisions in CAA for the control of hazardous or toxic air pollutants. As a result of this legislative decision, section 112 of the 1970 clean air statute not only defined a hazardous air pollutant but also required the EPA administrator to publish a list of hazardous air pollutants within 90 days after the date of enactment of the statute.

There was an extensive amendment to the Clean Air Act in 1977. The Clean Air Act Amendments of 1977 revised section 112 and authorized EPA to establish emission standards for the hazardous air pollutants. EPA was also mandated to establish reliable work practice, design of control equipment, operational standards, or some combination of these to protect public health.

However, EPA's first major step in regulating hazardous air pollutants did not go much beyond its setting of the national emission standards for hazardous air pollutants (NESHAPs). Although several major environmental laws were passed during the 1980s to control or regulate hazardous substances, hazardous waste, and toxic chemicals, there was virtually no movement on the issue of hazardous air pollutants or air toxics. Hence, there was no regulatory mechanism for their control. EPA was unexplainably slow despite an almost unanimous recognition of the inadequacy of existing EPA program to control air toxics or hazardous air pollutants.

EPA's dismal record in controlling toxic or hazardous air pollutants was known to lawmakers. They were further disturbed by calls from environmental groups and industry associations for statutory modifications of section 112. After several failed attempts, the Clean Air Act Amendments (CAAA) of 1990 were finally passed in both the Senate and the House of Representatives and signed into law on November 15, 1990.

The 1990 CAAA has been truly revolutionary in terms of its far-flung approach as well as its specific and detailed textual description of the statutory provisions, compared to the previous air statutes. CAAA contains several separate and distinct titles addressing a multitude of regulatory programs and policies. It has left very little room for regulatory discretion although it addresses several matters for the first time. The difference in the level of public concern and attitude during the 1960s and the 1990s is well reflected in this new, comprehensive air statute which has been given the distinction of the succinct street name *the devil is in the detail*. However, the ambitious goals outlined in CAAA can be achieved only through effective and expeditious implementation by EPA and state and local governments. Right now, the focus is on EPA as it toils to promulgate its regulations under the CAAA of 1990.

One of the major thrusts of the CAAA of 1990 is the control of hazardous air pollutants. The old section 112 has undergone major modifications. The thoroughly revised section 112 currently is one of the most significant statutory sections. The extent of the modification can be fathomed from the simple fact that the list of hazardous air pollutants was expanded from a mere 7 under the last statutory and regulatory framework to 189 by virtue of this latest clean-air statute. Moreover, pursuant to the statutory requirements, EPA must promulgate new control standards for most of the sources that may emit these HAPs.

Indeed, the passing of CAAA forcefully brought home the issue of HAPs or air toxics. As indicated earlier, Congress left very little room for creative maneuvers or regulatory discretion by EPA. EPA has yet to promulgate all regulations under CAAA. However, this can lead to only one thing: The regulatory framework will have a very definite shape because of the well-delineated footprint of the statute. It is therefore clear that the regulations for hazardous air pollutants will be both extensive and somewhat draconian in comparison to past regulatory approaches. In short, EPA's listing of major source categories for air toxic emissions is expected to be broad and sweeping. In addition, under the statutory guideline, EPA will regulate such sources by imposing technology-based rather than risk-based control standards.

4.2 Definitions

The passage of the Clean Air Act Amendments of 1990 culminated from a process that spanned several years in both the House and the Senate. Several new terms and phrases were coined and interpreted during these lengthy deliberations by lawmakers. Many eventually found their way into CAAA. As a result, they require special understanding since they have specific statutory meanings and thus important legal ramifications. Section 111 of the statute defines these important terms.

CAAA defines *major source* as any stationary source that emits or has the potential to emit, after considering air pollution controls, at least 10 tons per year of any listed hazardous air pollutant or at least 25 tons per year of any combination of the listed hazardous air pollutants. Section 112(a) provides that a group of stationary sources may be considered as one major source if these stationary sources are within a contiguous area and under common control while meeting these threshold emission levels. The statute authorizes the EPA administrator to establish a lesser threshold quantity for effects of bioaccumulations, or different criteria for certain pollutants, e.g., radionuclides. Note that no judicial review of such administrative or regulatory decisions is permissible until a final standard is promulgated by EPA.

Area source is defined as any stationary source of toxic or hazardous air pollutant that is not a major source. In essence, area sources include small emission point sources as well as fugitive sources. The definition of *area source,* however, excludes motor or other vehicles that are to be regulated under Title II of CAAA for mobile sources.

New source means a source whose construction or reconstruction begins after EPA first proposes its regulations under section 112 of the statute to establish an emission standard for such source.

Stationary source, as defined in section 111(a), means any building, structure, facility, installation, or any unit thereof that emits or may emit any of the listed toxic or hazardous air pollutants.

Hazardous air pollutant means a pollutant that is listed in section 112(b). The statute includes 189 hazardous air pollutants of which 172 are individual substances and the other 17 are actually families of pollutants, rather than an individual chemical or chemical substance. However, this is an initial list of hazardous air pollutants. As indicated earlier, the EPA administrator is authorized by the statute to add air pollutants to or delete air pollutants from this list under certain specific criteria.

A *modification* is defined as any physical change in the method of operation of a major source if such change results in an increase in actual emissions of any hazardous air pollutant by an amount that exceeds the *de minimis* quantity.

Owner or operator means any person who owns, leases, operates, controls, or supervises a stationary source.

Adverse environmental effect means any threat of significant and widespread adverse effects on wildlife, aquatic life, or other natural resources. This includes any disruption of local ecosystem, significant or adverse impact on endangered or threatened species, or significant damage or degradation of environmental quality over large natural areas.

Carcinogenic effect conveys the meaning as stated by the EPA administrator in "Guidelines for Carcinogenic Risk Assessment" as of the date of enactment of the statute. The existing guidelines on carcinogenic risk assessment may be revised by the EPA administrator, but only after an appropriate notice and an opportunity for comment by the public.

4.3 Hazardous Air Pollutants

As indicated earlier, CAAA lists 189 hazardous air pollutants or air toxics in section 112(b). Table 4.1 lists the hazardous air pollutants specifically mentioned in this latest clean air legislation. However, the statute emphasizes regulation of categories and subcategories of sources that emit such HAPs rather than regulation of individual pollutants. As a result, the statute requires EPA to establish a list of major source categories that emit the listed HAPs.

The cornerstone of Title III of the act is the imposition of the *maximum achievable control technology* (MACT) standards for all such listed source categories using a well-planned schedule. The MACT standard is to be based on the best demonstrated control technology that offers the maximum degree of emission control for the listed hazardous air pollutants whenever achievable. While it is a technology-based standard, EPA is required to consider costs and other tangible factors or criteria for each category of sources that will be subject to this new standard. At the same time, EPA is empowered to prohibit emissions altogether, if warranted, because of extremely adverse health or environmental effects.

As indicated earlier, the list provided in section 112(b) is strictly a preliminary list of hazardous air pollutants. The statute authorizes the EPA administrator to add to, delete from, or modify this list at any time by appropriate rule making. The basis for such revision, however, is a determination that a pollutant or substance under review presents or may present a threat of adverse human health or environmental effects. However, there cannot be any addition of criteria pollutants or their precursors to this list of hazardous air pollutants. A precursor may be added only if it independently meets the statutory criteria for addition to the list. On the other hand, any substance regu-

TABLE 4.1   List of Hazardous Air Pollutants

Acetaldehyde	1,3-Dichloropropene
Acetamide	Dichlorvos
Acetonitrile	Diethanolamine
Acetophenone	*N,N*-Diethyl aniline (*N,N*-Dimethylaniline)
2-Acetylaminofluorene	Diethyl sulfate
Acrolein	3,3-Dimethoxybenzidine
Acrylamide	Dimethyl aminoazobenzene
Acrylic acid	3,3'-Dimethyl benzidine
Acrylonitrile	Dimethyl carbamoyl chloride
Allyl chloride	Dimethyl formamide
4-Aminobiphenyl	1,1-Dimethyl hydrazine
Aniline	Dimethyl phthalate
o-Anisidine	Dimethyl sulfate
Asbestos	4,6-Dinitro-*o*-cresol, and salts
Benzene (including benzene from gasoline)	2,4-Dinitrophenol
Benzidine	2,4-Dinitrotoluene
Benzotrichloride	1,4-Dioxane (1,4-Diethyleneoxide)
Benzyl chloride	1,2-Diphenylhydrazine
Biphenyl	Epichlorohydrin (1-Chloro-2,3-epoxypropane)
Bis(2-ethylhexyl)phthalate (DEHP)	1,2-Epoxybutane
Bis(chloromethyl)ether	Ethyl acrylate
Bromoform	Ethyl benzene
1,3-Butadiene	Ethyl carbamate (Urethane)
Calcium cyanamide	Ethyl chloride (Chloroethane)
Caprolactam	Ethylene dibromide (Dibromoethane)
Captan	Ethylene dichloride (1,2-Dichloroethane)
Carbaryl	Ethylene glycol
Carbon disulfide	Ethylene imine (Aziridine)
Carbon tetrachloride	Ethylene oxide
Carbonyl sulfide	Ethylene thiourea
Catechol	Ethylidene dichloride (1,1-Dichloroethane)
Chloramben	Formaldehyde
Chlordane	Heptachlor
Chlorine	Hexachlorobenzene
Chloroacetic acid	Hexachlorobutadiene
2-Chloroacetophenone	Hexachlorocyclopentadiene
Chlorobenzene	Hexachloroethane
Chlorobenzilate	Hexamethylene-1,6-diisocyanate
Chloroform	Hexamethylphosphoramide
Chloromethyl methyl ether	Hexane
Chloroprene	Hydrazine
Cresols/Cresylic acid (isomers and mixture)	Hydrochloric acid
o-Cresol	Hydrogen fluoride (Hydrofluoric acid)
m-Cresol	Hydrogen sulfide
p-Cresol	Hydroquinone
Cumene	Isophorone
2,4-D, salts and esters	Lindane (all isomers)
DDE	Maleic anhydride
Diazomethane	Methanol
Dibenzofurans	Methoxychlor
1,2-Dibromo-3-chloropropane	Methyl bromide (Bromomethane)
Dibutylphthalate	Methyl chloride (Chloromethane)
1,4-Dichlorobenzene(p)	Methyl chloroform (1,1,1-Trichloroethane)
3,3-Dichlorobenzidene	Methyl ethyl ketone (2-Butanone)
Dichloroethyl ether (Bis(2-chloroethyl)ether)	Methyl hydrazine

TABLE 4.1   List of Hazardous Air Pollutants (*Continued*)

Methyl iodide (Iodomethane)	Tetrachloroethylene (Perchloroethylene)
Methyl isobutyl ketone (Hexone)	Titanium tetrachloride
Methyl isocyanate	Toluene
Methyl methacrylate	2,4-Toluene diamine
Methyl tert butyl ether	2,4-Toluene diisocyanate
4,4-Methylene bis(2-chloroaniline)	*o*-Toluidine
Methylene chloride (Dichloromethane)	Toxaphene (chlorinated camphene)
Methylene diphenyl diisocyanate (MDI)	1,2,4-Trichlorobenzene
4,4'-Methylenedianiline	1,1,2-Trichloroethane
Naphthalene	Trichloroethylene
Nitrobenzene	2,4,5-Trichlorophenol
4-Nitrobiphenyl	2,4,6-Trichlorophenol
4-Nitrophenyl	Triethylamine
2-Nitropropane	Trifluralin
N-Nitroso-*N*-methylurea	2,2,4-Trimethylpentane
N-Nitrosodimethylamine	Vinyl acetate
N-Nitrosomorpholine	Vinyl bromide
Parathion	Vinyl chloride
Pentachloronitrobenzene (Quintobenzene)	Vinylidene chloride (1,1-Dichloroethylene)
Pentachlorophenol	Xylenes (isomers and mixture)
Phenol	*o*-Xylenes
p-Phenylenediamine	*m*-Xylenes
Phosgene	*p*-Xylenes
Phosphine	Antimony Compounds
Phosphorus	Arsenic Compounds (inorganic including arsine)
Phthalic anhydride	Beryllium Compounds
Polychlorinated biphenyls (Aroclors)	Cadmium Compounds
1,3-Propane sultone	Chromium Compounds
beta-Propiolactone	Cobalt Compounds
Propionaldehyde	Coke Oven Emissions
Propoxur (Baygon)	Cyanide Compounds[1]
Propylene dichloride (1,2-Dichloropropane)	Glycol ethers[2]
Propylene oxide	Lead Compounds
1,2-Propylenimine (2-Methyl aziridine)	Manganese Compounds
Quinoline	Mercury Compounds
Quinone	Fine mineral fibers[3]
Styrene	Nickel Compounds
Styrene oxide	Polycylic Organic Matter[4]
2,3,7,8-Tetrachlorodibenzo-*p*-dioxin	Radionuclides (including radon)[5]
1,1,2,2-Tetrachloroethane	Selenium Compounds

NOTE: For all listings above which contain the word "compounds" and for glycol ethers, the following applies: Unless otherwise specified, these listings are defined as including any unique chemical substance that contains the named chemical (i.e., antimony, arsenic, etc.) as part of that chemical's infrastructure.

[1]'X'CN where X = H' or any other group where a formal dissociation may occur. For example KCN or Ca(CN)$_2$.

[2]Includes mono- and di-ethers of ethylene glycol, diethylene glycol, and triethylene glycol R–(OCH2CH2)$_n$–OR' where

n = 1, 2, or 3

R = alkyl or aryl groups

R' = R, H, or groups which, when removed, yield glycol ethers with the structure: R–(OCH2CH)$_n$–OH. Polymers are excluded from the glycol category.

[3]Includes mineral fiber emissions from facilities manufacturing or processing glass, rock, or slag fibers (or other mineral derived fibers) of average diameter 1 micrometer or less.

[4]Includes organic compounds with more than one benzene ring, and which have a boiling point greater than or equal to 100°C.

[5]A type of atom which spontaneously undergoes radioactive decay.

SOURCE: Section 112, Clean Air Act Amendments of 1990.

lated for acid rain under Title IV of the act cannot be added to this list of HAPs solely on the basis of adverse environmental effects.

Section 112, included in Title III of the act, provides a specific "add or delete" provision. Pursuant to this, any person may send a petition to the EPA administrator to add or delete substances any time after 6 months of the enactment of the act. The statute requires the EPA administrator to either grant or deny such a request within 18 months of receiving that petition. If the petition is denied, the administrator has to publish a written explanation for such denial. Any denial on the grounds of inadequate time or resources available for review is, however, not acceptable.

The act authorizes EPA to prevent or reduce hazardous air pollutant emissions through a number of available means. EPA can require certain changes in the manufacturing processes, or restrict the use of certain raw materials, or undertake an appropriate recycling program, or change existing designs of equipment, operating methods, and the like. In short, to achieve a MACT standard, EPA can be very imposing.

Generally, any air pollutant can be added to the list of HAPs if such a pollutant causes or can be reasonably anticipated to cause adverse human health or environmental effects due to ambient concentrations, bioaccumulation, or deposition. A substance may be deleted from the list of HAPs if there are adequate data to demonstrate that its emission is not reasonably anticipated to cause adverse human health or environmental effects because of ambient concentrations, bioaccumulation, or deposition.

The act also includes "low-risk categories" of carcinogens. The EPA administrator is authorized to delete from the list of source categories any category of sources that emit carcinogens in quantities that cause a lifetime risk of carcinogenic effects in the range of 1 in 1 million or less. The EPA administrator may also delete a hazardous air pollutant from the list when such a pollutant may result in some adverse human health or environmental effects but not cancer. This is allowed, provided the emission level of such an air pollutant is kept low enough to protect public health with an "ample margin of safety."

In summary, section 112 of the act is a basic overhaul of the original section 112 of the Clean Air Act. The earlier criterion was to include a hazardous air pollutant in section 112 if it were reasonably anticipated to result in an increase in mortality or an increase in serious irreversible or incapacitating illnesses. Furthermore, pursuant to the new clean-air legislation, once EPA lists an air pollutant as a hazardous air pollutant under section 112, EPA must also establish standards for those sources that cause such an emission. Under the original statutory mandate, the emission standards simply reflected a threshold required to protect the public health with an "ample margin of safety."

Using this as a core criterion, EPA adopted specific national standards, i.e., NESHAPs, for only seven hazardous air pollutants until the recent past: inorganic arsenic, asbestos, beryllium, benzene, mercury, vinyl chloride, and radionuclides. EPA used a broad but vague yardstick for regulating these hazardous air pollutants to levels that it determined to be adequate for the protection of public health with an "ample margin of safety." This caused waves of litigation and resulted in considerable controversy as to what was precisely EPA's mandate to set NESHAPs and its obligation to regulate known probable or suspected carcinogens. In *Natural Resources Defense Council v. EPA,* the court held that the statutory language of "ample margin of safety" does not establish a "risk-free" level and that EPA could consider costs and technological feasibility after establishing a safe level as to what constitutes an ample margin of safety. However, since the term *ample margin of safety* was not precisely defined in the statute, its interpretation by EPA and the regulated community was largely a matter of judicial vagaries and subject to review on a case-by-case basis.

Title III of CAAA has attempted to eliminate these controversies by adopting a different approach altogether. Instead of developing a regulatory framework by using a pollutant-by-pollutant approach, it requires EPA to regulate categories of hazardous air pollutant emission sources by using a less controversial technology-based standard. For example, section 112(b) of CAAA does not require any *prevention of significant deterioration* (PSD) review for limited pollutants. Additionally, section 112(c) authorizes EPA to exempt *de minimis* sources in a category. As a result, while the list of hazardous air pollutants has greatly expanded, the new statute gives EPA a shorter leash for using its discretion as to what constitutes an ample margin of safety for hazardous air pollutants.

The "hammer" provision in the new clean-air legislation can be found in the appropriate permitting requirements for an air emission source. All major sources must obtain permits that reflect the MACT standards. Under the new statutory scheme, the MACT standards are to be based initially on the performance of 12 percent of the most tightly controlled existing facilities. After 8 years (or 9 years for the first source categories) of EPA's promulgation of the MACT standards, EPA is required to promulgate the residual risk standards based on risk to human health in order to provide an ample margin of safety.

Note that pursuant to Title III of CAAA, the first phase of emission standards and requirements is technology-based for the purpose of controlling emission of HAPs. Title III of CAAA also endorses a second phase of emission standards and requirements which is risk-based. This risk-based control will be required if there is any residual risk to public health or the environment from HAP emissions even after the incorporation of the MACT. In essence, if the MACT eliminates the residual risk, there will be no need to provide any risk-based control to reduce or eliminate the emission of HAPs.

4.4   Emission Standards and Compliance Requirements

The CAAA of 1990, as discussed earlier, provides a strict definition of a major source. However, this does not limit the EPA to keeping its regulatory focus only on major sources. The statute requires EPA to examine other sources that fall within the area-source designation for appropriate regulation. In fact, the statute sets a very definite regulatory framework within which EPA is required to operate. For a better understanding of this apparent bootstrap, it is important to understand several standards provided in the statute.

MACT standards

Pursuant to the act, EPA is required to publish a list of all major categories and subcategories of sources of hazardous air pollutants. In addition, EPA is mandated to develop a MACT standard for each major source category.

This MACT standard is a new standard altogether. Simplistically, it means maximum degree of reduction of a hazardous air pollutant that is technologically achievable. However, this is a quantum leap from EPA's earlier approach based on the *best available control technology* (BACT). The BACT standard is not entirely technology-based since it includes factors pertaining to energy use and economics in addition to environmental considerations. The introduction of the MACT standard gives EPA a free rein to develop emission standards on the basis of control technology. However, the legislative intent behind the new clean air legislation clearly indicates that EPA must also consider the cost of such technology before promulgating the MACT standard. In this respect, the MACT standard may be considered a higher standard than the BACT standard, but they are not very different in terms of their development criteria.

In this context, it is important that the statute authorizes EPA to institute alternative area source control programs, particularly for nonmajor sources, where the MACT standards will extract too high a price. Such sources may be regulated through a lesser standard, e.g., the *generally available control technology* (GACT) standard, or by acceptable management practices. Under the statutory schedule, regulations for area sources must be promulgated by November 15, 2000.

CAAA also requires EPA to promulgate the MACT standards based on a specific schedule. By November 15, 1991 (i.e., a year after the enactment of the 1990 amendments), EPA is required to publish a list of all major source categories and subcategories of hazardous air pollutants. This list, under the statutory mandate, must be revised at least once every 8 years. On June 21, 1991, EPA published a preliminary draft list of categories of major and area sources of hazardous air pollutants. Table 4.2 is a reproduction of this draft list.

In addition, the EPA administrator is required to list each category or subcategory of area sources that could present a threat to human health or to the environment. Also, within 5 years of the enactment of CAAA, the EPA administrator is required to list categories or subcategories of area sources to ensure that those area sources which represent 90 percent of the top 30 hazardous air pollutants in urban areas are appropriately regulated. Regulations that are to be geared for this result must be effective within 10 years of the enactment of the act, i.e., by November 15, 2000. However, for seven specific pollutants [e.g., alkylated lead compounds, polycyclic organic matter (POM), hexachlorobenzene, mercury, polychorinated biphenyls (PCBs), 2,3,7,8-tetrachlorodibenzofurans, and 2,3,7,8-tetrachlorodibenzo-p-dioxin], the EPA administrator is required to publish a list of categories and subcategories of sources within 5 years (i.e., by November 15, 1995) to ensure at least 90 percent reduction in their aggregate emissions.

TABLE 4.2   Preliminary Draft List of Categories of Major and Area Sources of Hazardous Air Pollutants

Industry Group—Fuel Combustion

Category Name

Industrial External Combustion
  Boilers
Institutional External Combustion
  Boilers
External Combustion Space
  Heaters
Industrial Electric Generation
  Turbines
Industrial Reciprocating IC
  Engines
Commercial/Institutional Turbines
Commercial Reciprocating IC
  Engines
Test Engine Aircraft
Test Engines—Turbine*
Test Engines—Reciprocating
Process Heaters
Secondary Metals Process Heaters
Petroleum Industry Process
  Heaters
Oil and Gas Steam Generation
Industrial In-Situ Fuel Use
Prescribed Burning
Residential Boilers
Residential Wood Combustion—
  Fireplaces
Residential Wood Combustion—
  Woodstoves

Industry Group—Metallurgical
Industry: Nonferrous Metals

Category Name

Aluminum Production*
Primary Lead Smelting
Primary Metals—Miscellaneous*
Secondary Aluminum*
Secondary Copper*

Industry Group—Metallurgical
Industry: Nonferrous Metals

Category Name

Battery Manufacturer: Non-Lead
  Types
Cadmium Refining
Lead Acid Battery Manufacturing
Non-Ferrous Alloys Production
Primary Copper Smelters
Secondary Metals—Miscellaneous
Zinc Smelting

Industry Group—Metallurgical
Industry: Ferrous Metals

Category Name

Ferroalloys Production
Iron & Steel Manufacturing
Gray Iron Foundries
Steel Foundry
Coke Byproduct Plants
Coke Ovens
Metal Shredding (Recycling)
Steel Pickling

Industry Group—Mineral
Products Processing and Use

Category Name

Taconite Iron Ore Processing*
Asphalt Concrete Manufacture
Brick Manufacturing
Cement Kilns*
Glass Manufacture*
Stone Quarries
Mining Operation—Sand/Gravel*
Metal Pipe Coating Asphalt/Coal
  Tar*
Asbestos Fabricating
Asbestos Manufacturing
Asbestos Milling
Asbestos Removal: Demolitions
Asbestos Removal: Renovations
Asbestos Waste Disposal:
  Demolitions
Asbestos Waste Disposal:
  Renovations
Construction: Spraying and
  Insulation
Asphalt Paving and Roofing
  Operations
Asphalt Processing
Automotive Transmission Plates
  Manufacturing
Brake Parts Manufacturing
Ceiling Tile Manufacturing
Friction Material Manufacturing
Mineral Dryers/Calciners
Mineral Wool Production
Ore Flotation
Refractories Production
Talc Manufacturing
Vermiculite Manufacturing
Wool Fiberglass Manufacturing

Industry Group—Petroleum
Refineries

Category Name

Petroleum Refining

Industry Group—Petroleum and
Gasoline Production and
Marketing

Category Name

Oil and Gas Production
Gasoline/Petroleum Storage
Petroleum Marketing (With Bulk
  Terminals and Plants)
Manganese Fuel Additives
Natural Gas Storage/Transmission
Oil Shale Retorting

Industry Group—Surface Coating
Processes

Category Name

Fabric Printing
Surface Coating Operations—Gen-
  eral Solvent Uses
Fabric Coating*
Paper Coating*
Large Appliance*
Magnet Wire*
Auto and Light Duty Truck
Metal Can*
Metal Coil*
Wood Furniture*
Metal Furniture*
Flat Wood Products*
Plastic Part*
Large Ship*
Large Aircraft*
Printing/Publishing
Architectural
Magnetic Tapes

Industry Group—Waste Treatment and Disposal

Category Name

Solid Waste Disposal–Open
  Burning
Sewage Sludge Incineration*
Municipal Landfills*
Groundwater Cleaning
Hazardous Waste Incineration

TABLE 4.2   Preliminary Draft List of Categories of Major and Area Sources of Hazardous Air Pollutants *(Continued)*

Industrial Group—Waste Treatment and Disposal

Tire Burning
Tire Pyrolysis
Cooling Water Chlorination–
  Steam Electric Generators
Wastewater Treatment Systems
Water Treatment Purification
Water Treatment–Boilers

Industry Group—Agricultural Chemicals Production and Use

Category Name

2,4-D Salts and Esters Production
4,6-Dinitro-*o*-Cresol Production
4-Chloro-2-methylphenoxyacetic
  Acid Production
Baygon℗ Production
Captafol℗ Production
Captan℗ Production
Carbamate Insecticides Production
Chlorthalonil Production
Dacthal℗ Production
Dichlorodiphenyltrichloroethane
  Production
Fumigation Use
Grain Fumigation Production
Metribuzin Production
Parathion Use
Pentachloronitrobenzene
  Production
Pentachlorophenol Production
R-11 (Butadiene Furfural-
  Cotrimer) Production
Sodium Pentachlorophenate
  Manufacture
Soil Fumigant Use
Space Fumigant Use
Substituted Phenyl Ureas
  Production
Thiocarbamates Production
Tordon Acid Production

Industry Group—Fibers Production Processes

Category Name

Acrylic Fibers/Modacrylic Fibers
Nylon Fibers
Rayon
Spandex
Triacetate Fibers

Industry Group—Food and Agriculture Industry

Category Name

Bakers Yeast Manufacturer
Coffee Roasting
Cotton Ginning
Prepared Food Manufacturing

Industry Group—Pharmaceutical Production Processes

Category Name

Pharmaceuticals Production

Industry Group—Polymers and Resins Production

Category Name

Acetal Resins Production
Acrylonitrile-Butadiene-Styrene/
  Styrene/Acrylonitrile
Alkyd Resins Production
Butyl Rubber Production
Carboxymethylcellulose
  Production
Cellophane Production
Cellulose Ethers
Epichlorohydrin Elastomers
Epoxy Resins
Foamed Plastics
Formaldehyde Resins
  Production
Hypalon℗ Production
Maleic Copolymers Production
Methyl Methacrylate-
  Acrylonitrile-Butadiene-Styrene
Methyl Methalcrylate-Butadiene
  Styrene Terpolymers
Methylcellulose Production
Neoprene Production
Nitrile Butadiene Rubber
  Production
Nylon Plastics Production
Phenolic Resins Production
Polybutadiene Rubber
  Production
Polycarbonates Production
Polyester Plastics
Polyester Resins Production
Polyether Polyols Production
Polyethylene Terephthalate Production

Polymerization of Vinylidene
  Chloride
Polymethyl Methacrylate
  Resins Production
Polystyrene Production
Polyurethane Foam
Polyurethane Production
Polyvinyl Acetate Emulsions
Polyvinyl Alcohol Production
Polyvinyl Butyral Production
Polyvinyl Chloride and Copoly-
  mers Production
Reinforced Plastics
Styrene Butadiene Rubber and
  Latex Production

Industry Group—Production and Use of Inorganic Chemicals

Category Name

Aluminum Chloride
Aluminum Fluoride
Ammonium Phosphates
Ammonium Sulfate
Calcium Oxide Production
Carbon Black
Charcoal
Chemical Intermediate
Chlorine
Chromium Chemicals
  Manufacturer
Cyanuric Chloride
Detergent
Fertilizer Formulation and Use
Fluorides
Hydrochloric acid
Hydrogen cyanide
Hydrogen fluoride
Isopropanolamines
Manganese chemicals
Phosphate fertilizers
Phosphoric acid
Phosphorus pentasulfide
  production
Phosphorus production
Phosphorus trichloride/oxychloride
  production
Quaternary ammonium
  compounds production
Rocket engine fuel
Sodium cyanide production
Uranium hexafluoride
  production

TABLE 4.2   Preliminary Draft List of Categories of Major and Area Sources of Hazardous Air Pollutants (*Continued*)

Industry Group—Production of
Synthetic Organic Chemicals

Category Name

Acenaphthene production
Acetaldehyde production
Acetaldol production
Acetamide production
Acetanilide production
Acetic acid production
Acetic anhydride production
Acetoacetanilide production
Acetone cyanohydrin production
Acetone production
Acetonitrile production
Acetophenone production
Acetyl chloride production
Acrolein production
Acrylamide production
Acrylic acid production
Acrylonitrile production
Adiponitrile production
Alizarin production
Alkyl anthraquinones production
Alkyl naphthalene sulfonates
  production
Allyl alcohol production
Allyl chloride production
Allyl cyanide production
Aminophenol (*p*-isomer)
  production
Aminophenol sulfonic acid
  (*o,p*-isomer) production
Ammonium thiocyanate
  production
Aniline hydrochloride production
Aniline production
Anisidine (o-isomer) production
Anthracene production
Anthraquinone production
Azobenzene production
Benzaldehyde production
Benzene production
Benzenedisulfonic acid
  production
Benezenesulfonic acid production
Benzil production
Benzilic acid production
Benzoic acid production
Benzoin production
Benzophenone production
Benzoyl chloride production
Benzyl acetate production
Benzyl alcohol production
Benzyl benzoate production

Benzyl chloride production
Benzyl dichloride production
Biphenyl production
Bis (chloromethyl) ether
  production
Bisphenol A production
Bromobenzene production
Bromoform production
Bromonaphthalene production
Butadinene (1,3-isomer) produc-
  tion
Butanediol (1,4-isomer)
  production
Butyl acrylate (N-isomer) produc-
  tion
Butylamine (N-isomer)
  production
Butylamine (S-isomer)
  production
Butylamine (T-isomer)
  production
Butylbenzyl phthalate
  production
Butylene glycol (1,3-isomer) pro-
  duction
Butyrolactone production
Caprolactam production
Carbaryl production
Carbazole production
Carbon Disulfide production
Carbon Tetrabromide
  production
Carbon Tetrachloride production
Carbon Tetrafluoride production
Chloral production
Chloracetic acid production
Chloroacetophenone (2-isomer)
  production
Chloroaniline (o-isomer) produc-
  tion
Chloroaniline (p-isomer) produc-
  tion
Chlorobenzaldehyde production
Chlorobenzene production
Chlorodifluoroethane production
Chlorodifluoromethane
  production
Chloroform production
Chloronaphthalene production
Chloronitrobenzene (1,3-isomer)
  production
Chloronitrobenzene (o-isomer)
  production
Chloronitrobenzene (p-isomer)
  production

Chlorophenols production
Chloroprene (2-chloro-1,3-
  butadiene) production
Chlorosulfonic acid production
Chlorotoluene (m-isomer)
  production
Chlorotoluene (o-isomer)
  production
Chlorotoluene (p-isomer)
  production
Chlorotrifluoromethane
  production
Chrysene production
Cresol (m-isomer) production
Cresols (o-isomer) production
Cresols (p-isomer) production
Cresols cresylic acid (mixed) pro-
  duction
Crotonaldehyde production
Cumene production
Cumene hydroperoxide
  production
Cyanamide production
Cyanoacetic acid production
Cyanoformamide production
Cyanogen chloride production
Cyanuric chloride production
Cyclohexane production
Cyclohexanol production
Cyclohexanone production
Cyclohexylamine production
Cyclooctadiene (1,5-isomer)
  production
Cyclooctadiene production
Decahydronaphthalate
  production
Di (2-methoxyethyl) phthalate
  production
Di-o-tolyguanidine production
Diacetoxy-2-butene (1,4-isomer)
  production
Diallyl phthalate production
Diaminophenol hydrochloride pro-
  duction
Dibromomethane production
Dibutoxyethyl phthalate
  production
Dichloro-1-butene (3,4-isomer) pro-
  duction
Dichloro-2-butene (1,4-isomer) pro-
  duction
Dichloroaniline (all isomers) pro-
  duction
Dichlorobenzene (1,4-isomer) (p-
  isomer) production

TABLE 4.2   Preliminary Draft List of Categories of Major and Area Sources of Hazardous Air Pollutants (*Continued*)

Industry Group—Production of Synthetic Organic Chemicals

Dichlorobenzene (*m*-isomer) production
Dichlorobenzene (*o*-isomer) production
Dichlorodifluoromethane production
Dichloroethane (1,2-isomer) production
Dichloroethyl ether production
Dichloroethylene (1,2-isomer) production
Dichlorophenol (2,4-isomer) production
Dichloropropene (1,3-isomer) production
Dichlorotetrafluoroethane production
Dicyanidiamide production
Diethanolamine production
Diethyl phthalate production
Diethylamine production
Diethylaniline (2,6-isomer) production
Diethylene glycol dibutyl ether production
Diethylene glycol diethyl ether production
Diethylene glycol dimethyl ether production
Diethylene glycol monobutyl ether production
Diethylene glycol ether monobutyl acetate production
Diethylene glycol monoethyl ether production
Diethylene glycol monoethyl ether acetate production
Diethylene glycol monohexyl ether production
Diethylene glycol monomethyl ether production
Diethylene glycol monomethyl ether acetate production
Diethylene glycol production
Diisodecyl phthalate production
Diisooctyl phthalate production
Dimethyl benzidine (3,3-isomer) production
Dimethyl either-*N,N* production
Dimethyl formamide (*N,N*-isomer) production
Dimethyl hydrazine (1,1-isomer) production

Dimethyl phthalate production
Dimethyl sulfate production
Dimethyl terephthalate production
Dimethylamine production
Dimethylaminoethanol (2-isomer) production
Dimethylaniline-*N,N* production
Dinitrobenzenes production
Dinitrophenol (2,4-isomer) production
Dinitrotoluene (2,4-isomer) production
Dioxane production
Dioxilane production
Diphenyl methane production
Diphenyl oxide production
Diphenyl thiourea production
Diphenylamine production
Dipropylene glycol production
Dodecyl benzene (branched) production
Dodecyl phenol (branched) production
Dodecylaniline production
Dodecylbenzene (N-isomer) production
Dodecylphenol production
Epichlorohydrin production
Ethanolamines (all isomers) production
Ethyl acetate production
Ethyl acrylate production
Ethyl benzene production
Ethyl chloride production
Ethyl chloroacetate production
Ethyl orthoformate production
Ethylamine production
Ethylaniline (N-isomer) production
Ethylaniline (O-isomer) production
Ethylcellulose production
Ethylcyanoacetate production
Ethylene dibromide production
Ethylene glycol diacetate production
Ethylene glycol dibutyl ether production
Ethylene glycol diethyl ether production
Ethylene glycol dimethyl ether production
Ethylene glycol monoacetate production

Ethylene glycol monobutyl ether production
Ethylene glycol monobutyl ether acetate production
Ethylene glycol monoethyl ether production
Ethylene glycol monoethyl ether acetate production
Ethylene glycol monohexyl ether production
Ethylene glycol monomethyl ether production
Ethylene glycol monomethyl ether acetate production
Ethylene glycol monooctyl ether production
Ethylene glycol monophenyl ether production
Ethylene glycol monopropyl ether production
Ethylene glycol production
Ethylene imine production
Ethylene oxide production
Ethylenediamine production
Ethylenediamine tetraacetic acid production
Ethylhexyl acrylate (2-isomer) production
Ethylnaphthalene (2-isomer) production
Fluoranthene production
Formaldehyde production
Formamide production
Formic acid production
Fumaric acid production
Glutaraldehyde production
Glyceraldehyde production
Glycerol dicholorohydrin production
Glycerol production
Glycine production
Glyoxal production
Guanidine nitrate production
Guanidine production
Hexachlorobenzene production
Hexachlorobutadiene production
Hexachlorocyclopentadiene production
Hexachloroethane production
Hexadiene (1,4-isomer) production
Hexamethylenetetramine production
Hexanetriol (1,2,6-isomer) production

TABLE 4.2   Preliminary Draft List of Categories of Major and Area Sources of Hazardous Air Pollutants (*Continued*)

Industry Group—Production of
Synthetic Organic Chemicals

Hydrogen cyanide production
Hydroquinone production
Hydrooxyadipaldehyde production
Iminodiethanol (2,2-isomer)
 production
Isobutyl acrylate production
Isobutylene production
Isophorone nitrile production
Isophorone production
Isophthalic acid production
Isopropylphenol production
Lactic acid production
Lead phthalate production
Linear alkylbenzene production
Maleic anhydride production
Maleic hydrazide production
Malic acid production
Metanilic acid production
Methacrylic acid production
Mathanol production
Methionine production
Methyl acetate production
Methyl acrylate production
Methyl bromide production
Methyl chloride production
Methyl ethyl ketone production
Methyl formate production
Methyl hydrazine production
Methyl isobutyl carbinol
 production
Methyl isobutyl ketone production
Methyl isocyanate production
Methyl mercaptan production
Methyl methacrylate production
Methylnaphthalenes production
Methyl phenyl carbinol production
Methyl tert butyl ether production
Methylamine production
Methylaniline (N-isomer)
 production
Methylcyclohexane production
Methylcyclohexanol production
Methylene chloride production
Methylene dianiline (4,4-isomer)
 production
Methylene diphenyl diisocyanate
 production
Methylionones (A-isomer)
 production
Methylpentynol production
Methylstyrene (A-isomer)
 production

N-vinyl-2-pyrrolidine
 production
Naphthalene production
Naphthalene sulfonic acid
 (A-isomer) production
Naphthalene sulfonic acid
 (B-isomer) production
Naphthol (A-isomer) production
Naphthol (B-isomer)
Naphtholsulfonic acid (1-isomer)
 production
Naphthylamine (1-isomer)
 production
Naphthylamine (2-isomer)
 production
Naphthylamine sulfonic acid (1,4-
 isomer) production
Naphthylamine sulfonic acid (2,1-
 isomer) production
Nitrilotriacetic acid production
Nitroaniline (M-isomer)
 production
Nitroaniline (O-isomer)
 production
Nitroanisole (O-isomer)
 production
Nitoranisole (P-isomer)
 production
Nitrobenzene production
Nitrophthalene (1-isomer)
 production
Nitrophenol (4-isomer)
 (P-isomer) production
Nitrophenol (O-isomer)
 production
Nitropropane (2-isomer)
 production
Nitrotoluene (2-isomer)
 (O-isomer) production
Nitrotoluene (3-isomer)
 (M-isomer) production
Nitrotoluene (4-isomer)
 (P-isomer) production
Nitrotoluene production
Nitroxylene production
Nonylbenzene (branched)
 production
Nonylphenol production
Octene-1 production
Octylphenol production
P-tert-butyl toluene production
Paraformaldehyde production
Paraldehyde production
Pentachlorophenol production
Pentaerythritol production

Perchloroethylene production
Perchloromethyl mercaptan
 production
Phenanthrene production
Phenetidine (P-isomer)
 production
Phenol production
Phenolphthalein production
Phenolsulfonic acids (all isomers)
 production
Phenylenediamine (P-isomer) pro-
 duction
Phyloroglucinol production
Phosgene production
Phthalic acid production
Phthalic anhydride production
Phthalimide production
Phthalonitrile production
Picoline (B-isomer) production
Polyethylene glycol production
Polypropylene glycol production
Propiolactone (B-isomer)
 production
Propionaldehyde production
Propionic acid production
Propyl chloride production
Propylene carbonate production
Propylene dichloride production
Propylene glycol monomethyl
 ether production
Propylene glycol production
Propylene oxide production
Pyrene production
Pyridine production
Quinone production
Resorcinol production
Salicylic acid production
Sodium chloroacetate
 production
Sodium cyanide production
Sodium methooxide production
Sodium phenate production
Stilbene production
Styrene production
Succinic acid production
Succinonitrile production
Sulfanilic acid production
Sulfolane production
Tartaric acid production
Terephthalic acid production
Tert-butylbenzene production
Tetrabromophthalic anhydride
 production
Tetrachlorobenzene (1,2,4,5-
 isomer) production

TABLE 4.2   Preliminary Draft List of Categories of Major and Area Sources of Hazardous Air Pollutants (*Continued*)

Industry Group—Production of Synthetic Organic Chemicals

Tetrachloroethane (1,1,2,2-isomer) production
Tetrachlorophthalic anhydride production
Tetraethyl lead production
Tetraethylene glycol production
Tetraethylenepentamine production
Tetrahydronaphthalene production
Tetrahydrophthalic anhydride production
Tetramethylenediamine production
Tetramethylethylenediamine production
Thiocarbanilide production
Thiourea production
Toluene 2,4-diamine production
Toluene 2,4-diisocyanate production
Toluene diisocyanate (mixture) production
Toluene production
Toluenesulfonic acids (all isomers) production
Toluenesulfonyl chloride production
Toluidine (O-isomer) production
Trichloroaniline (2,4,6-isomer) production
Trichlorobenzene (1,2,4-isomer) production
Tricholoroethane (1,1,2-isomer) production
Tricholoroethylene production
Tricholorofluoromethane production
Trichlorophenol (2,4,5-isomer) production
Tricholorotrifluoroethane (1,2,2-1,1,2-isomer) production
Triethanolamine production
Triethylamine production
Triethylene glycol dimethyl ether production
Triethylene glycol monomethyl ether production
Triethylene glycol production
Trimethylamine production
Trimethylcyclohexanol production
Trimethylcyclohexanone production
Trimethylcyclohexylamine production

Trimethylopropane production
Trimethylpentane (2,2,4-isomer) production
Tripropylene glycol production
Vinyl acetate production
Vinyl chloride production
Vinyl toluene production
Vinylcyclohexene (4-isomer) production
Vinylidene chloride production
Xanthates (potassium ethyl xanthate) production
Xylene (M-isomer) production
Xylene sulfonic acid production
Xylenes (mixed) production
Xylenes (O-isomer) production
Xylenes (P-isomer) production
Xylenol production

Industry Group—Miscellaneous

Category Name

Asphalt roofing manufacture
Pulp & paper production
Plywood/particle board manufacture*
Sawmill operations*
Tire production
Dry cleaning (petroleum solvent)
Dry cleaning (chlorinated solvents)—coin operation plant
Dry cleaning (chlorinated solvents)—coin operation self
Dry cleaning (chlorinated solvents)—commercial
Dry cleaning (chlorinated solvents)—industrial
Cold degreasing
Fabric dyeing*
Solvent extraction processes
Acrylic sheeting production
Aerosols production
Anesthetics
Benzyltrimethylammoniumchloride production
Boat building
Butadiene cylinders, lab testing
Butadiene dimers production
Chelating agents production
Chlorinated paraffins production
Chloroneb production
Chromium electroplating
Comfort cooling towers
Commercial sterilization facilities
Conveyorized degreasing

Deodorant production
Disinfectants production
Dodecanedioic acid production
Dyes and pigments production
Electric wiring
Electronics manufacture
Ethylidene norborne production
Explosives production
Flame retardant production
Hospital sterilizers
Hydrazine production
Industrial cooling towers
Industrial process aids—enhanced oil recovery
Ion exchange resins production
Jet fuel deicer use
Leather tanning
Lube oil additives
Lube oil dewaxing
Moth repellent
Oil/gas well acidizing
Open top vapor degreasing
Other electroplating
Paint removers use
Paints, coatings & adhesives: manufacture & use (other than surface coating)
Phosphate esters production
Photographic chemicals manufacture
Photographic film processing
Phthalate plasticizers production
Polymerization inhibitors use
Resins catalyst production
Rubber antioxidants production
Rubber cement manufacturing
Rubber chemicals production
Semiconductors manufacturing
Surface active agents production
Symmetrical tetrachloropyridine process
Synthetic tanning agents production
Vinylidene chloride copolymer fabrication
Wood preservation—direct use

Industry Group—Production and Use Activities (TRIS)

Category Name

1,1,2,2-Tetrachloroethane
1,1-Dimethyl hydrazine
1,2,4-Trichlorobenzene
1,2-Epoxybutane

TABLE 4.2   Preliminary Draft List of Categories of Major and Area Sources of Hazardous Air Pollutants (*Continued*)

Industry Group—Production and Use Activities (TRIS)		
1,2-Propylenimine	Bis (2-ethylhexyl) phthalate	Hexachlorobenzene
1,3-Dichloropropene	Bis (chloromethyl) ether	Hexachlorobutadiene
1,3-Propane sultone	Calcium cyanamide	Hexachlorocyclopentadiene
1,4-Dioxane	Captan	Hexachloroethane
2,4,6-Trichlorophenol	Carbaryl	Lindane (all isomers)
2,4-D Salts and esters	Carbonyl sulfide	Mercury compounds
2,4-Dinitrophenol	Catechol	Methoxychlor
2,4-Dinitrotoluene	Chloramben	Methyl hydrazine
2,4-Toluene diamine	Chlordane	Methyl iodide
2-Chloroacetophenone	Chlorobenzilate	Methyl isocyanate
3,3-Dichlorobenzidene	Chloromethyl methyl ether	N,N-Diethylaniline
4,4-Methylene bis	Cobalt compounds	O-Anisidine
4,6-Dinitro-o-cresol, and salts	Cresols/cresylic acid (isomers and mixture)	O-Toluidine
4-Aminobiphenyl	Dibenzofurans	P-Phenylenediamine
4-Nitropropane	Dichloroethyl ether	Pentachloronitrobenzene
Acetamide	Dichlorvos	Polychlorinated biphenyls (aroclors)
Acetonitrile	Diethanolamine	Propoxur
Allyl chloride	Diethyl sulfate	Quinone
Antimony compounds	Dimethyl sulfate	Styrene oxide
Benzotrichloride	Ethyl carbamate	Titanium tetrachloride
Beryllium compounds	Ethylene imine	Trifluralin
	Ethylene thiourea	Vinyl bromide
	Heptachlor	Vinylidene chloride

*These source categories and subcategories are included due to their speciation profiles of relatively poor quality ranking.

First-round standards

The statute permits phasing of emission reductions over a period of 10 years (i.e., by November 15, 2000). Section 112 of CAAA requires the EPA administrator to set a first round of technology-based emission standards for each category and subcategory of listed sources.

Pursuant to the statutory framework, EPA will have to regulate 40 source categories plus coke ovens by November 15, 1992, and 25 percent of the remaining source categories of HAPs within 4 years. Of the remaining source categories, an additional 25 percent will be brought under regulatory control within 7 years. The remaining 50 percent will have to be regulated by EPA by the end of the 10-year targeted deadline. If EPA fails to promulgate appropriate standards based on this schedule written into the statute, all sources that would have been subject to an EPA-developed standard will have to be permitted incorporating a level of control that is certified as the best control by an independent engineer. This "hammer" provision in the statute becomes effective within 18 months of EPA's scheduled promulgation date.

The first-round standards to be promulgated by EPA may be distinguished on the basis of classes, types, and size of sources in a category or subcategory. The first-round standards will have to incorporate a general standard developed on the basis of maximum emissions reduction achievable given the economics of implementation, energy requirements, and health and environmental impacts. Such a general standard will apply to both existing and new sources. The emission reductions thus required may be achieved by any means available. This may include substitution of raw materials used, changes in manufacturing processes, development of appropriate control equipment, and special operator training.

For existing sources of air emissions, the first-round standards must be as stringent as

- The average emissions of the best performing 12 percent of sources situated in a similar category or subcategory, except certain sources that have achieved the lowest achievable emission rate (LAER), for 30 or more sources   or
- The average emissions of the five best performing sources in a category or subcategory that includes less than 30 sources

The existing sources, however, may not have to meet the high standards for emission reduction set for the new sources. The act requires that the emission reduction standards for existing area sources be at least as stringent as the most stringent emission level already achieved by any source in the same category or subcategory. Thus, CAAA authorizes EPA to promulgate a fluid standard rather than a fixed standard.

For area sources, the EPA administrator is authorized to promulgate emission reduction requirements based on the GACT standard or appropriate management practices. However, requirements based on the GACT standard may have to be cost-effective.

Residual-risk standards

There is a tacit recognition by the lawmakers that the first-round standards based on control technology may not be enough to reduce emissions of HAPs to a level that offers an ample margin of safety. As a result, the statute incorporates a second round of emission standards based on residual risks. Section 112(f) of CAAA requires the EPA administrator to consult with the Surgeon General and report to Congress on residual risk involving emissions of hazardous air pollutants. Such report must be made within 6 years of the enactment of the act and must include

- The methods of calculating residual risk
- The risk to public health remaining or likely to remain after the application of the MACT standards
- The public health significance of such estimated remaining risks
- The available means or methods and costs of reduction of such risks
- Recommendations for legislation regarding such risks

In the absence of appropriate legislation, EPA is required to promulgate residual risk standards within 8 years of issuance of the MACT standards (or 9 years in the case of first-round standards) for a particular category or subcategory of sources. These residual risk standards for the hazardous air pollutants must provide

- An ample margin of safety to protect public health within the meaning of section 112 included in earlier statutes preceding the enactment of CAAA in 1990

- Protection against any adverse environmental effect

The statute allows EPA to determine if a residual risk standard is necessary. However, EPA has no such discretion if a pollutant is a carcinogen and the risk to the most exposed individual is 1 in 1 million or more. Currently, EPA is reviewing California's air toxics law ("Hot Spots" Information and Assessment Act of 1987) that lists approximately 330 toxic substances in order to develop appropriate risk assessments and a threshold level for "significant risks." Reportedly, there are two levels of standards, 1 in 10,000 and 1 in 1 million, which are being seriously reviewed by EPA for its eventual promulgation of threshold quantities of HAPs using the theory of an ample margin of safety.

These standards are to be promulgated for those hazardous air pollutants which are probable or suspected human carcinogens provided that the cancer risks to the most exposed individual are not increased to a level of more than 1 in 1 million. The residual risk standards will become effective upon their promulgation. The EPA administrator, however, may grant a waiver, allowing up to 2 additional years of compliance in the event a delay is required to install necessary controls and to take additional steps during the period of waiver to ensure protection of public health from imminent danger.

Under the new strategic scheme, the residual risk standards for each source category or subcategory, except area sources, must reflect a MACT standard as the baseline standard. The EPA administrator has been given 8 or 9 years to determine what additional standards beyond the first-round standards are necessary to reduce the risk of airborne carcinogens to the most exposed individual at a level less than 1 in 1 million.

Thus, it is likely that certain categories or subcategories of sources may not meet the regulatory standard by incorporating control technology based on the MACT standards. These categories may be subject to additional regulatory requirements if the remaining emissions, after installation of necessary controls, based on the MACT standards, contain HAPs at a level considered harmful to the public.

Miscellaneous provisions

The act includes a specific provision for coke oven emissions. The EPA administrator is required to set emission standards for coke ovens. Such standards are not allowed to provide more than

- 3 percent of leaking doors (5 percent of leaking doors for 6-meter batteries)
- 1 percent of leaking lids

- 4 percent of leaking offtakes
- 16 seconds of visible emissions per charge

The statute sets a compliance date of December 31, 1995, for emission standards from existing coke ovens.

The act also makes special reference to radionuclides. The EPA administrator, however, is not authorized to set emission standards for radionuclides that emit from a facility licensed by the Nuclear Regulatory Commission (NRC). This is, however, restricted to only those cases where the EPA administrator, after consultation with NRC, determines that there is an ample margin of safety within the NRC's regulatory program to protect public health. This special provision, however, does not bar any state or a political subdivision of a state from setting its own standards for emissions of radionuclides.

CAAA requires valid operating permits for all major sources. As indicated earlier, a hammer provision is included in CAAA which requires the owner or operator of any major source to submit a permit application, using the MACT standards, within 18 months of operation of that major source. This requirement is triggered only if EPA fails to promulgate a standard for a category or subcategory of a major source within the stipulated time frame.

Compliance schedule

Based on the statutory framework, after the promulgation of appropriate standards for various source categories by EPA, all sources subject to EPA regulation must comply with regulatory standards within 3 years of their promulgation. This is a general requirement for all sources, new and old.

CAAA includes a special rule that waives compliance with EPA-promulgated standards for a new source (i.e., a source that begins construction or reconstruction after the standard has been proposed but before its promulgation). This waiver may be allowed for up to 3 years from the date of promulgation if the promulgated standard is more stringent than the proposed standard. In such a case, however, the new source must comply with the proposed standard during this 3-year waiver period.

In addition, there are other compliance schedules. For example, for existing sources, the compliance date for any category or subcategory of sources may be as late as 3 years after the effective date of the EPA-promulgated standard. Extensions of up to 1 year may be allowed by EPA or an authorized state responsible for administering the permit program for the purpose of full compliance requiring installation of certain emission control equipment. For mining operations, an additional extension of up to 3 years may be granted if a 4-year compliance deadline is determined to be insufficient to reduce the level of emission of hazardous air pollutants.

In addition, there is a "Presidential exemption" built into the compliance schedule. It allows the President of the United States to exempt any stationary source from compliance requirements for up to 2 years if the President determines that the technology required to implement the standard is not available and that an extension is appropriate for national security reasons. This exemption, under certain situations, may be further extended by 4 years.

Table 4.3 includes the current EPA schedule for various activities associated with hazardous air pollutants as planned by EPA under the act. Table 4.4 provides EPA's compliance schedule under section 112(i).

4.5   Accidental Releases of Extremely Hazardous Air Pollutants

CAAA requires EPA to identify at least 100 extremely hazardous air pollutants by November 15, 1992. These substances are specially recognized because they may cause serious bodily injury or other serious adverse effects on human health, including death, and damages to the environment. Furthermore, EPA is required to promulgate appropriate regulations for their control.

An independent Chemical Safety and Hazard Identification Board (Safety Board) will be established within EPA to investigate all accidental releases of hazardous air pollutants. In this context, an *accidental release* means an unanticipated emission of a regulated substance or other extremely hazardous substance from a stationary source. The Safety Board will be required to make appropriate recommendation to the EPA administrator with respect to these accidental releases and appropriate safety measures against such releases.

The initial list of extremely hazardous substances includes chlorine, anhydrous ammonia, methyl chloride, ethylene oxide, vinyl chloride, methyl isocyanate, hydrogen cyanide, ammonia, hydrogen sulfide, toluene diisocyanate, phosgene, bromine, anhydrous hydrogen chloride, hydrogen fluoride, anhydrous sulfur dioxide, and sulfur trioxide. Pursuant to the statutory requirements, the owner or operator of a facility that handles such extremely hazardous substances must operate the facility safely. Moreover, the owner of the facility is required to perform an engineering analysis that identifies public health hazards and incorporates a risk management plan for preventing accidents.

The EPA administrator is required to review and revise the list of extremely hazardous substances at least once every 5 years. When a substance is added to this list, a threshold limit must be set for that substance. Within 3 years after the enactment of CAAA, EPA is required to promulgate regulations for prevention, detection, and minimization of accidental releases of such extremely hazardous substances.

4.6   State Programs

Using a flexible scheme provided in the act, a state with an approved permit program under Title V may issue a permit that authorizes a source to comply with alternative emission limitations in lieu of EPA-promulgated standards. However, this may be allowed only under certain specific situations, which include

- Achieving at least 90 percent reduction in emissions of hazardous air pollutants or 95 percent reduction for particulates by an existing (i.e., pre-MACT) source prior to the EPA proposal for the MACT standards   or

TABLE 4.3   EPA's Time Schedule for Title III Hazardous Air Pollutants

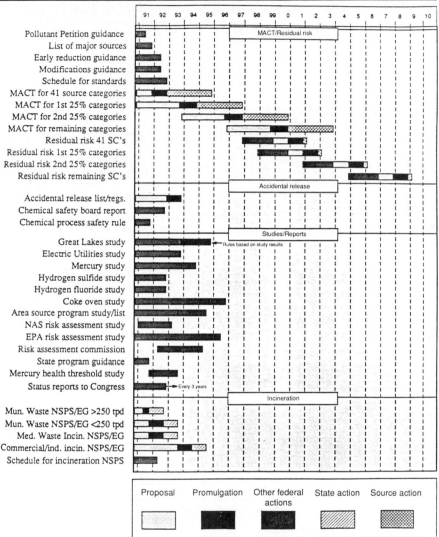

- Achieving early reductions by January 1, 1994 (i.e., post-MACT) and making an enforceable commitment to achieve the required MACT standards

The state may submit its programs for state implementation and enforcement of emission standards for hazardous air pollutants to EPA. These submittals may include appropriate provisions for the prevention of accidental releases of hazardous air pollutants.

On May 10, 1991, EPA proposed a new 40 CFR part 70 which, if accepted finally, will require the states to develop and submit to EPA programs for issuing operating permits to major stationary sources including the major sources for hazardous air pollutants under Title V of CAAA. By statutory pro-

TABLE 4.4   Schedule of Compliance under Section 112(i)

Sources	Technology-based standards	Health-based standards
New and/or reconstructed sources	Effective immediately	Effective immediately
Exceptions or extensions for new or reconstructed sources	3-year "special rule" extension for certain new sources	3-year "special rule" extension for certain new sources 10-year extension for certain new sources
Existing sources	As expeditiously as practicable, but no later than 3 years*†	90 days*
Extensions for existing sources	6-year extension for voluntary reductions	Waiver of up to 2 years
	5-year extension from date of installation of BACT or LAER	5-year extension from date of installation of BACT or LAER

*Compliance dates are from the date the standard is effective. Both health- and technology-based standards are effective upon promulgation.
†EPA or a state with a permit program may grant to an existing source an additional year to comply with the technology-based standards if necessary for installation of controls. An additional 3 years may be granted for drying and covering mining waste.

visions, EPA is required to promulgate such regulations within 12 months of the enactment of the statute. On July 21, 1991, EPA issued its final rule.

Pursuant to the statutory provisions in section 112(e) of CAAA, EPA may delegate partial or complete authority to qualified states. However, such a delegation cannot include an authority to set less stringent standards than those promulgated by EPA. The EPA administrator is also required to maintain a clearinghouse for hazardous air pollutants and to provide information on control technology, health and ecological risk assessment, risk modeling, and emissions monitoring and measurement.

4.7   Chemical Process Safety Management

The CAAA of 1990 requires the Occupational Safety and Health Administration (OSHA) to promulgate a safety standard for chemical processes within 12 months of enactment. This safety standard must take into consideration a list of extremely hazardous chemicals and the safety issues that reflect the health and safety impact due to exposure.

At a minimum, several specific requirements must be included in such an OSHA standard:

- Development and maintenance of written safety information
- Review of hazard assessments
- Assessment of workplace hazards
- Establishment of a system responding to hazard assessment findings
- Consultation with employees and their representatives regarding hazard assessments
- Implementation of written operating procedures

- Access to written safety and operating information for employees
- Training of employees, contractors, and contractors' agents who may be engaged
- Providing contractors and contractors' agents with appropriate information
- Establishment of a quality assurance program for process equipment and spare parts
- Establishment of appropriate maintenance systems for critical process equipment
- Arrangement for pre-start-up safety reviews
- Implementation of written procedures
- Investigation of incidents that result or have the potential to result in a major accident

4.8 Judicial Review

As indicated earlier, CAAA of 1990 provides very limited scope for judicial review. No preenforcement judicial review (i.e., review by a court of law) is allowed other than for administrative orders or penalties. Also, no judicial reviews will be available regarding addition of a pollutant to or deletion of a pollutant from the list of hazardous air pollutants or the final listing of source categories by the EPA administrator. Such actions, however, may be reviewed under section 307(b) of CAAA only after EPA issues emission standards for the affected pollutant or the source category itself.

4.9 Course of Action

In the absence of promulgation of regulations by EPA pursuant to the act, there is speculation as to how EPA will eventually tighten its control over hazardous air pollutants. However, the broad statutory authority given to EPA, while not unprecedented, makes it clear that EPA's role in reduction of emissions of hazardous air pollutants will be nothing short of remarkable and conspicuous.

As indicated earlier, EPA is required to define the MACT standard for each category or subcategory of a major source. Several factors will have to be included in this MACT definition, e.g., capital and operating costs, energy requirements, secondary environmental impacts, and industrywide economic impacts. While common sense suggests that gathering of this rudimentary but considerable information will be time-consuming, one cannot take much comfort from this assessment. This is because EPA has been at work for some time, anticipating many of the issues covered under both the House and the Senate bills. Hence, the regulated community may have to respond to the MACT standards in a hurry in order to be in full compliance.

In developing a proper course of action, it is important to recognize that the requirements of Title III for industrial facilities may have some overlap with other provisions of the statute. This is particularly important in cases that may involve ozone nonattainment areas along with possible emissions of *volatile organic compounds* (VOCs). In such cases, it is likely that the MACT standards may have to be replaced by a more stringent standard based on the *lowest achievable emission rate* (LAER) program.

Table 4.5 lists important statutory deadlines for various regulatory programs. These time schedules can be an excellent guide for planning purposes by the affected corporations and other business entities.

4.10   Penalties and Sanctions

The act increases the liability for violations considerably compared to the earlier clean air statutes. The criminal sanction under the act increases from a misdemeanor to a felony for a criminal offense as defined in the act.

This will bring unlawful air pollution in line with the criminal sanctions

TABLE 4.5   Important Deadlines under CAAA of 1990 for Regulatory Programs for Air Toxics

Activity description	Statutory deadline
Publication of a list of categories and subcategories of major and area sources of 189 listed hazardous air pollutants	November 15, 1991
Determination and/or prioritization of source categories for establishing the schedule for MACT standards for source categories to be regulated	November 15, 1992
Promulgation of MACT standards for the first 40 listed source categories	November 15, 1992
Promulgation of a list of, and establishing threshold quantities for, at least 100 extremely hazardous air pollutants posing greatest risk from accidental releases	November 15, 1992
Promulgation of regulations to implement risk management plans for prevention and detection of accidental releases	November 15, 1993
Promulgation of MACT standards for 25 percent of listed categories	November 15, 1994
First deadline for compliance with MACT standards (i.e., 3 years after promulgation of first MACT standard)	November 15, 1995*
Listing of area source categories emitting 30 or more hazardous air pollutants that present the greatest public health threat	1994–1995
Implementation of risk management plans by operators of stationary sources subject to accidental-release regulations	1986
Submission of EPA's report to Congress on residual risk determination	November 15, 1986
Promulgation of MACT standards for an additional 25 percent of the listed source categories	November 15, 1987
Promulgation of MACT standards for an additional 25 percent and all other listed source categories	November 15, 2000
Promulgation of residual risk standards for the first 40 listed source categories to provide an ample margin of safety	2001 and later†

*Most likely time line.

†The residual risk standards are to be promulgated 9 years after the promulgation of the MACT standards in the case of first regulated source categories and 8 years after for all other source categories.

imposed for unlawful hazardous waste disposal. This is the first time that one will be facing a possible felony prosecution for an air pollution violation.

Such an eventuality does not bode well for violators of environmental compliance requirements, particularly if they include corporations and their directors, officers, and key employees. It is most alarming to note that a corporation may be found guilty even if one employee fails to follow the direction of a corporate officer or director. Under the act, a corporation may face a criminal conviction even if its employees fail to make a required report (say, an accidental release of a reportable quantity of an air pollutant) to EPA.

There are also severe consequences of a criminal conviction. For a corporation, a felony conviction may mean a hefty fine plus disqualification from participation in federal government contracts. Under the new legislative initiative, the Department of Justice (DOJ) may want to prosecute all the so-called responsible corporate officers. By using this expansive theory, a corporate officer may be criminally liable if the officer simply supervises an employee who commits the so-called environmental crime.

4.11 Possible Prescription

It would be prudent for an individual, industry, or commercial operation to first take stock of its current air emissions from both emission points and fugitive sources. The next approach should include what is the best way to reduce and control such emissions through appropriate changes in raw materials selection and/or manufacturing processes. Then, to stay ahead of the game, appropriate control technology should be identified for the pollutants captured, and their removal or destruction efficiencies should be fully established. Last but not least, costs to achieve and maintain controlled emissions should be appropriately factored into the budget for all long-term operations and planning.

Although there are no set standards yet for developing work plans, there are several common denominators which should be included in a well-planned environmental strategy for shielding corporate officers as well as corporations from broad environmental liability from unwanted air emissions:

- Comprehensive environmental audit
- Detailed corporate policy statements
- Current operation manuals and prompt maintenance of control equipment
- Environmental training and education
- Complete record keeping and good housekeeping
- Prompt disclosure of any environmental releases
- Corporate awareness of the need for complete environmental compliance

Note that facilities can get credit for early reductions through a voluntary program. Hence, there is a way to reap some benefits from an early compliance program. EPA has offered a 6-year extension in meeting the MACT standards if the industry voluntarily commits to a 90 percent reduction of toxic air emissions prior to EPA's announcement of its proposed regulations. This is worth considering in spite of a bottleneck that has surfaced in this credit program. This problem arises because of a fundamental difficulty relating to the definition of *source.* EPA plans to solve this problem through a proposed rule soon. As a result, it is in the best interest of industries to plan for reduction of hazardous air pollutants as soon as practicable, but prior to the promulgation of EPA's standards so that the statutory credit can be locked in.

Recommended Reading

1. American Law Network Satellite Seminar, "Implementing the 1990 Clean Air Act," American Bar Association, Section of Natural Resources, Energy, and Environmental Law, February 21, 1991.

2. CQ, "Clean Air Act Amendments", November 24, 1990.

3. American Law Institute–American Bar Association Course of Study, "Implementing the 1990 Clean Air Act Amendments," Committee on Continuing Professional Education, Washington, November 8–9, 1990.

Important Cases

1. *Wisconsin Electric Power Co. v. Reilly (WEPCO),* 893 F. 2d 901 (7th Cir. 1990).

2. *Puerto Rican Cement Co. v. EPA,* 889 F. 2d 292 (1st Cir. 1989).

3. *Natural Resources Defense Council v. EPA,* 824 F. 2d 1146 (D.C. Cir. 1987).

Hazardous Wastes under RCRA

5.1 Introduction

One of the unfortunate and inescapable by-products of the industrial revolution is the prolific generation of solid, liquid, and gaseous wastes. Many such wastes, because of their chemical and physical characteristics, are hazardous and can cause serious problems to both people and the environment unless they are properly managed and disposed of.

Perhaps one of the most disconcerting geopolitical aspects of today's hazardous waste generation is that no country is totally immune from the ultimate fallout of such wastes. Even countries blessed with pristine land, clean water, and fresh air due to adherence to the old way of life can be susceptible. This happens when countries responsible for generation of hazardous wastes send their hazardous wastes to the nongenerating countries for ultimate storage and/or disposal. Unfortunately, countries in the latter category seldom have the technical expertise or resources to carry out a safe, scientifically acceptable hazardous waste management program. This has caused an uncontrolled and widespread global littering of hazardous waste with serious consequence to humans and nature.

The United States is perhaps one of the first industrialized nations that not only understood the dangerous ramifications of hazardous waste generation, but also felt an immediate need to map out a comprehensive hazardous waste management strategy. Strong public opinion throughout the nation ultimately forced the U.S. Congress to undertake a legislative response, if not a total cure, to bring the multifaceted ills of hazardous wastes under a certain degree of control.

However, this did not happen overnight. After several attempts by Congress to legislate various issues and aspects of hazardous waste, particularly those of industrial origin, Congress came up with its most potent weapon by passing the Resource Conservation and Recovery Act (RCRA) of 1976. In enacting RCRA, Congress attempted to address the issues surrounding both solid wastes and hazardous wastes. By doing this, Congress addressed and apparently plugged the loophole that existed under the guise of land disposal.

Since 1976, however, RCRA has been amended several times. Each of these amendments has placed additional burden on manufacturing and other industrial operations to provide safer solid and hazardous waste management programs.

While RCRA is by no means the only legislative act to deal with hazardous waste at the federal level, it still remains one of the most significant and comprehensive legislative controls, providing "cradle to grave" accounting for hazardous waste management nationally. As a result, it is extremely important to understand both the procedural and the substantive requirements of RCRA and their application to solid and hazardous wastes.

5.2 What Constitutes Solid and Hazardous Wastes

To comprehend what a hazardous waste is, one must first take into account the statutory definition of solid waste. Section 1004(27) of RCRA defines solid waste as follows:

> The term "solid waste" means any garbage, refuse, sludge from a waste treatment plant, water supply treatment plant or air pollution control facility and other discarded material, including solid, liquid, semisolid, or contained gaseous materials resulting from industrial, commercial, mining and agricultural operations and from community activities, but does not include solid or dissolved material in domestic sewage, or solid or dissolved materials in irrigation return flows or industrial discharges which are point sources subject to permits under section 402 [of the Federal Water Pollution Control Act, as amended (86 Stat. 880)], or source, special nuclear, or byproduct material as defined by the Atomic Energy Act of 1954, as amended (68 Stat. 923).

Thus, this statutory definition of solid waste goes far beyond the physical form since it includes liquid, semiliquid, and gaseous materials. The regulatory definition of solid waste which followed later was not simple either, for it embraced any discarded material that was not specifically excluded as solid waste. In addition, the EPA administrator is empowered to list other materials within the category of solid wastes according to certain specific criteria.

As a matter of fact, EPA was compelled by the court to modify its original definition of solid waste under RCRA. Pursuant to the decision in *American Mining Congress v. EPA,* which held that Congress did not intend EPA to regulate in-process secondary materials as solid waste under RCRA, EPA proposed to exclude certain recycled materials that are part of a continuous manufacturing process from the definition of solid waste on January 8, 1988. In its deliberations of *American Mining Congress,* the court interpreted the term *discarded materials* to include materials extracted from the manufacturing process even when they are later reused in that process. The court held that the sludges generated in wastewater treatment impoundments are subject to the hazardous waste regulations even though these sludges may be discarded in future. Due to the holding of the court, EPA is under pressure to define solid waste somewhat narrowly, particularly for certain secondary materials that are reintroduced or recycled into the manufacturing processes at the same location.

Section 1004(5) of RCRA defines the term *hazardous waste* as a

> Solid waste, or combination of solid wastes, which because of its quantity, concentration, or physical, chemical or infectious characteristics may:
>
> (A) cause, or significantly contribute to an increase in mortality or an increase in serious irreversible, or incapacitating reversible, illness; or
>
> (B) pose a substantial present or potential hazard to human health or the environment when improperly treated, stored, transported or disposed of, or otherwise managed.

Under this statutory construction, if a waste does not meet the definition of solid waste, it is not subject to RCRA regulations. Also, before a waste can be characterized as a hazardous waste, it must be a solid waste.

As discussed earlier, EPA's definition of solid waste under section 1004(27) of RCRA includes "any discarded material." Such discarded materials, under EPA's regulations, can be (1) abandoned materials, (2) recycled materials, and (3) inherently wastelike materials. Abandoned materials comprise those materials that have been disposed of, burned or incinerated, accumulated, stored, or treated before or in lieu of disposal, burning, or incineration.

Recycled materials fall into four categories:

- Materials that are applied to or placed on the land in a manner that constitutes disposal

- Materials that are burned as fuel or otherwise for the purpose of energy recovery

- Materials that are reclaimed (i.e., regenerated or processed to recover a usable product)

- Materials that are accumulated speculatively

EPA has also classified certain wastes (which bear hazardous waste numbers F020, F021, F022, F023, F026, and F028), set forth in 40 CFR section 261.31, to be inherently wastelike. It also set forth certain criteria under which other wastes may also be found inherently wastelike.

Subtitle C has always been the centerpiece of RCRA, in spite of numerous amendments, including the Hazardous and Solid Waste Amendments (HSWA) of 1984 (also known as the RCRA amendments of 1984). Subtitle C deals with the hazardous waste management in general. It also includes specific items, such as identification and listing of hazardous waste. In addition to wastes listed by EPA as hazardous and set forth in 40 CFR §§261.31, 261.32, and 261.33, residues from the *treatment, storage, or disposal* (TSD) facilities of listed hazardous wastes are also considered hazardous.

EPA regulations exclude certain wastes from the definition of solid waste, and thus from hazardous waste, by way of variance. A "discarded material" which is abandoned, recycled, or inherently wastelike may qualify for such a variance. However, if abandoned materials are disposed of, burned or incinerated, accumulated, stored, or treated (but not recycled), they will not qualify for a variance and thus will be treated as solid waste.

In summary, pursuant to the statutory definition of a hazardous waste, first, it must be a solid waste as defined and not excluded by a variance or otherwise by the regulatory agencies. Second, such a solid waste must pose a substantial present or potential threat to human health or to the environment. Third, a solid waste, if specifically exempted by EPA from being a hazardous waste, should not be treated as a hazardous waste irrespective of other considerations and characteristics.

5.3 Regulatory Exclusions

EPA's regulations exclude several wastes from the definitions of solid and hazardous wastes. Certain materials are excluded from the definition of solid waste under subtitle C of RCRA. They are set forth at 40 CFR §261.4 and include

- Domestic sewage that passes untreated through a sewer
- Secondary materials which are recycled in a particular manner, reflecting a normal production operation
- Secondary recycled materials that are used in normal production operations and fall in one of three conditional exemptions: the ingredient exemption, the commercial product substitution exemption, and the closed-loop production exemption
- Materials in an in situ mining operation that are not removed from the ground
- Pulping liquors (i.e., black liquor) which are reclaimed in a pulping liquor recovery furnace and reused in the pulping process
- Spent sulfuric acid which is used to make virgin sulfuric acid
- Spent wood-preserving solutions that have been reclaimed and are reused
- Certain solid or dissolved materials present in industrial wastewaters that are regulated under the National Pollutant Discharge Elimination System (NPDES) permit program
- Solid and dissolved materials present in irrigation return flows
- Radioactive wastes as defined by the Atomic Energy Act

Apart from these, EPA's regulations exclude certain solid wastes from the definition of hazardous wastes:

- Household wastes that are produced in households but not in office buildings, restaurants, shopping centers, retail stores, or house demolition debris
- Certain agricultural wastes generated from the raising of animals, including animal manures, and from the growing and harvesting of crops

- Mining overburden that (1) consists of materials overlaying a mineral deposit with economic value, (2) is removed for the purpose of gaining access to the mineral deposit, and (3) is used for the purpose of mine reclamation

- Wastes which fail the Toxicity Characteristic test because of chromium present exclusively or nearly exclusively as trivalent chromium, and not for any other constitutent

- Fossil-fuel wastes generated from the burning of such fuel

- Drilling fluids, produced waters, and other wastes associated with the exploration, development, or production of natural gas, crude oil, or geothermal energy

- Solid wastes resulting from the extraction, beneficiation, and processing of ores and minerals

- Certain discarded wood products containing arsenic

- Petroleum-contaminated wastes which fail the Toxicity Characteristic test and are subject to the corrective action regulations.

- Injected groundwater exhibiting the Toxicity Characteristic only

- Used chlorofluorocarbon refrigerants from totally enclosed equipment and systems

- Cement kiln dust waste except when it is burned or processed as hazardous waste

In addition, there are hazardous wastes which are exempted from certain EPA regulations. They include samples for testing at laboratories and facilities.

Note that on May 20, 1992, EPA published its proposed rule in response to a petition by the Chemical Manufacturers Association (CMA). If accepted, EPA will follow concentration-based exemption criteria (CBEC) for listed hazardous wastes. Additionally, it may establish "characteristic" levels for listed hazardous wastes.

5.4   Identification of Hazardous Wastes

Section 3001 of RCRA deals with the identification of hazardous wastes. EPA's regulations promulgated under this section of RCRA and set forth in 40 CFR part 261 identify hazardous wastes based on certain specific characteristics as well as regulatory listing of hazardous wastes.

Characteristics of hazardous waste

EPA regulations define a solid waste to be a hazardous waste if it exhibits any one of four specific characteristics: ignitability, corrosivity, reactivity, or toxicity.

Ignitability. The characteristic of ignitability is based on whether the solid waste is capable of causing fire during routine handling or exacerbating the dangers once a fire is started. A solid waste is deemed to exhibit the characteristic of ignitability if it meets any one of the following four criteria set forth in 40 CFR §261.21, which are based on the physicochemical nature of the solid waste:

- It is a liquid, but not an aqueous solution containing less than 24 percent alcohol (by volume), and has a flash point of not less than 140°F to be determined by certain specific test methods.

- It is a nonliquid which can cause fire under normal conditions through friction, moisture absorption, or spontaneous chemical changes, and it burns vigorously upon ignition, thus creating a hazard.

- It is a compressed gas which is ignitable and so defined by the regulations promulgated by the Department of Transportation (DOT) as set forth in 49 CFR §173.300.

- It is an oxidizer as defined by the DOT regulations included in 49 CFR §173.151.

While there are specific tests of ignitability for liquids, EPA regulations do not suggest any specific test for determining the ignitability of solids. Instead, one has to consider whether such solids are capable of causing any spontaneous fire that can create a hazard. For ignitable compressed gases and oxidizers, EPA simply refers to the DOT regulations and the specific test methods described there.

Any solid waste that exhibits any of these ignitability characteristics is designated by EPA number D001.

Corrosivity. The test for corrosivity is based on two prime considerations: (1) the ability of a solid waste to corrode metal, particularly steel, and (2) whether the pH of the solid waste is highly acidic or highly alkaline so that it can cause harm to human tissue and aquatic life, react with other wastes, and accelerate the migration of toxic wastes.

Pursuant to EPA's regulation as set forth in 40 CFR §261.22, a solid waste is deemed to be corrosive if it is

- A liquid and corrodes steel (SAE 1020) at a rate greater than 6.35 millimeters (0.250 inch) per year at a temperature of 55°C (130°F)   or

- Aqueous and has a pH less than or equal to 2.0 or greater than or equal to 12.5

It thus appears that this characteristic of corrosivity applies to only liquids and those solids and gases that can be liquefied under a certain temperature, pressure, or both.

Any solid waste that exhibits the corrosivity characteristic is designated by EPA number D002.

Reactivity. A solid waste is considered reactive if it exhibits any of the following tendencies, as set forth in 40 CFR §261.23:

- It is normally unstable and reacts violently or explodes if mixed with water.

- It reacts violently when heated.

- It generates toxic fumes when mixed with water in quantities which present a danger to human health or the environment.

- It is either a cyanide- or sulfur-bearing waste which generates toxic fumes if exposed to certain pH conditions.

- It can detonate or explode when exposed to heat or pressure.

- It is a forbidden explosive under 49 CFR §173.51 or falls in the category of class A or class B explosive as set forth in 49 CFR §173.53 or 49 CFR §173.88.

Unlike for other characteristics, EPA does not provide specific test conditions or protocols for determining reactivity. Instead, EPA relies heavily on narrative tests that should be taken into account to determine reactivity.

Reactive hazardous wastes are designated by EPA number D003.

Toxicity.   Initially, EPA developed a test protocol for the toxicity characteristic on the basis of EP toxicity. However, EP toxicity, in real scientific parlance, is not a characteristic, but a test procedure based on extraction of certain toxic constituents. Thus, an EP toxicity test of a solid waste under EPA's requirement basically measures the concentrations of certain toxic constituents (heavy metals and pesticides) that are released during the extraction procedure.

Effective September 25, 1990, EPA has replaced this EP toxicity test by a *toxicity characteristic* (TC) test. At present, EPA requires a *toxicity characteristic leaching procedure* (TCLP) test to determine the toxicity of a solid waste. As a result, if any of the constituents in a solid waste exceeds the specific regulatory level for constituents analyzed in a TCLP test, it is labeled as a hazardous waste. Table 5.1 lists these toxic constituents and their regulatory levels which, if equaled or exceeded, will confirm the EPA-designated toxicity characteristic. As of now, however, the TCLP test is not required for EPA's underground storage tank program.

All hazardous wastes that exhibit the toxicity characteristic are designated by EPA number D004.

Listed hazardous wastes

Besides establishing four specific characteristics of hazardous wastes, EPA has developed specific lists of hazardous wastes which can be grouped in three broad categories: characteristic wastes, acute hazardous wastes, and toxic wastes.

The characteristic wastes are listed by EPA based on the four waste characteristics discussed earlier. Their listing is an attempt by EPA to simplify inclusion of certain wastes in the hazardous waste category without undertaking all the individual tests prescribed for ignitability, corrosivity, reactivity, and toxicity.

TABLE 5.1 **Maximum Concentration of Contaminants for the Toxicity Characteristic**

EPA hazardous waste level no.*	Contaminant	CAS no.†	Regulatory level, mg/l
D004	Arsenic	7440-38-2	5.0
D005	Barium	7440-38-3	100.0
D018	Benzene	71-43-2	0.5
D006	Cadmium	7440-43-9	1.0
D019	Carbon tetrachloride	56-23-5	0.5
D020	Chlordane	57-74-8	0.03
D021	Chlorobenzene	108-90-7	100.00
D022	Chloroform	67-86-3	6.0
D007	Chromium	7440-47-3	5.0
D023	o-Cresol	95-18-7	200.00‡
D024	m-Cresol	108-39-4	200.00‡
D025	p-Cresol	106-44-5	200.00‡
D026	Cresol	—	200.00‡
D016	2,4-D	94-75-7	10.0
D027	1,4-Dichlorobenzene	106-46-7	7.5
D028	1,2-Dichloroethane	107-00-2	0.5
D029	1,1-Dichloroethylene	75-35-4	0.7
D030	2,4-Dinitrotoluene	121-14-2	0.13§
D012	Endrin	72-20-8	0.02
D031	Heptachlor (and its epoxide)	76-44-8	0.008
D032	Hexachlorobenzene	116-74-1	0.13†
D033	Hexachlorobutadiene	87-68-3	0.5
D034	Hexachloroethane	67-72-1	3.0
D008	Lead	7438-82-1	5.0
D013	Lindane	58-88-9	0.4
D009	Mercury	7439-97-8	0.2
D014	Methoxychlor	72-43-5	10.0
D035	Methyl ethyl ketone	78-83-3	200.0
D038	Nitrobenzene	98-95-3	2.0
D037	Pentachlorophenol	87-85-5	100.0
D038	Pyridine	110-85-1	5.0†
D010	Selenium	7792-49-2	1.0
D011	Silver	7440-22-4	5.0
D039	Tetrachloroethylene	127-18-4	0.7
D015	Toxaphene	8001-35-2	0.5
D040	Trichloroethylene	79-01-6	0.5
D041	2,4,5-Trichlorophenol	95-35-4	400.0
D042	2,4,6-Trichlorophenol	88-06-2	2.0
D017	2,4,5-TP (Silvex)	93-72-1	1.0
D043	Vinyl chloride	75-01-4	0.2

*Hazardous waste number.
†Chemical Abstracts Service number.
‡If o-, m-, and p-cresol concentrations cannot be differentiated, the total cresol (D026) concentration is used. The regulatory level of total cresol is 200 mg/L.
§Quantitation limit is greater than the calculated regulatory level. The quantitation limit therefore becomes the regulatory level.

EPA's listing of acute hazardous wastes is based on the theory that not all wastes that may have potential harmful acute or chronic effect on human health can be detected by using the test protocols developed for four hazardous waste characteristics. However, due to their characteristics or nature, these wastes conform to the statutory definition of hazardous wastes since they can cause substantial actual or potential threat to human health or to the environment. In short, they may result in an increase in mortality or serious irreversible or incapacitating illness.

Under the category of listed toxic wastes, EPA has included all those wastes which have exhibited toxic, carcinogenic, mutagenic, or teratogenic effects on living beings, both humans and animals. As a result, these toxic wastes are not restricted to any listing based solely on the TCLP test for toxicity. Rather, this listing of toxic wastes is based on a presumption of potential toxic, carcinogenic, mutagenic, or teratogenic effects.

Using these three broad concepts, EPA has officially documented and listed certain wastes as hazardous. The current listing of hazardous wastes by EPA includes three specific groups:

- Hazardous wastes from nonspecific sources as set forth in 40 CFR section 261.31

- Hazardous wastes from specific sources as set forth in 40 CFR section 261.32

- Discarded commercial chemical products as set forth in 40 CFR section 261.33

The last one is different from the first two since it includes two groups: acutely hazardous wastes and toxic wastes.

Appendix 5A provides the EPA list for these three specific groups of hazardous waste. These wastes are listed by EPA as hazardous because they typically and frequently exhibit one or more of the hazardous waste characteristics identified in subpart C of 40 CFR part 261 or meet the criteria for listing contained in 40 CFR section 261.11(a)(2) or 261.11(a)(3). Needless to say, since EPA's first promulgation of its interim and final regulations implementing section 3001 of RCRA on January 16, 1981, EPA's list of hazardous wastes from nonspecific and specific sources has been amended several times. Hence, the current list of hazardous wastes is not necessarily a fixed or final list.

Based on EPA's current format, all hazardous wastes from nonspecific sources are designated by the prescript F in addition to a specific hazardous waste number. Hazardous wastes from specific sources are designated by the prescript K while the discarded commercial chemical products are designated by the prescript P if they are deemed toxic and U if they are considered acutely hazardous.

Note that EPA deferred the listing of certain wastes at the time it released its official list of hazardous wastes because of the pressure of a court order. This happened in spite of the fact that these wastes fall within the statutory definition of hazardous wastes. Hence, the current EPA list for hazardous waste is by no means all-inclusive, even by today's standard.

It is also appropriate to note that polychlorinated biphenyls (PCBs), because of their toxicity, should have been regulated by EPA as hazardous wastes. Instead, EPA currently regulates PCBs under the Toxic Substances Control Act (TSCA) because TSCA-based regulations were promulgated prior to RCRA-based regulations. However, this may change as a result of the legislative comments expressed prior to the passage of the 1984 RCRA Amendments. Interestingly, several states, e.g., New York, which have been delegated the authority for administration of RCRA-based regulations by EPA have incorporated and are regulating PCBs as hazardous wastes.

The mixture rule

Pursuant to EPA's regulations, any mixture of a listed hazardous waste and a nonhazardous solid waste must be considered and handled as a hazardous waste, irrespective of whether the mixture exhibits any hazardous characteristics. This is the *mixture rule* which seriously discourages as well as prohibits any manipulation of the EPA lists, regardless of the relative quantities of hazardous wastes and nonhazardous solid wastes, unless the mixture qualifies for an exemption. This also effectively prohibits dilution of hazardous wastes by various means. This is schematically presented in a diagram by EPA which is reproduced in Fig. 5.1.

Notwithstanding, exemptions from hazardous waste regulations will apply if

- The hazardous waste included in the mixture is listed solely because of its exhibition of one or more hazardous characteristics while the mixture itself does not exhibit any such characteristics.

- The mixture is a discarded commercial chemical product and truly reflects a *de minimis* loss during normal manufacturing operations.

- The mixture is comprised of a nonhazardous, nonlisted wastewater and certain hazardous wastes in low concentrations, the discharge of which is not prohibited by the provisions of the Clean Water Act.

These considerations make EPA's position against waste dilution very clear. The hazardous waste management rules as promulgated by EPA under the statutory authority of RCRA cannot be circumvented by any nonroutine procedure or external means that results in sufficient dilution to prevent triggering of any of the four hazardous waste characteristics. However, EPA's regulations promulgated under RCRA do not preempt any procedural relief to seek variance or delisting of any solid waste from being treated as a hazardous waste.

On December 6, 1991, in a decision involving *Shell Oil v. EPA*, the Court of Appeals for the District Court of Columbia invalidated the mixture rule on the ground that EPA did not provide an opportunity to comment on this rule prior to its adoption. However, the court left the door open for EPA to consider reenacting the mixture rule in whole or in part on an interim basis under the "good cause" provision of section 553 of the Administrative Procedure Act.

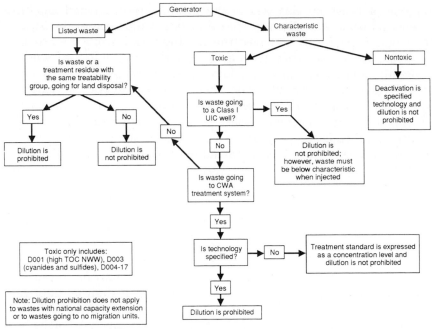

Figure 5.1   Dilution and its implications. (*Federal Register,* "Rules and Regulations," Vol. 56, No. 21, January 31, 1991.)

The derived-from rule

Under this rule, any waste that is generated due to the treatment, storage, or disposal activities of a hazardous waste must be considered as a hazardous waste unless specifically exempted. To apply this rule properly, three scenarios should be considered. First, if the waste is derived while one is dealing with a hazardous waste that has been so determined due to a listed characteristic, it will be treated as a hazardous waste only if it exhibits that particular characteristic. Second, if the waste is derived from a listed hazardous waste, it is a hazardous waste unless the parent waste is delisted. Third, if a reclaimed solid waste that has a beneficial use, and is thus exempted, is burned as fuel or disposed of, it is no longer exempted as a solid waste and will be subject to the full statutory definition of hazardous waste.

As in the case of the mixture rule, the derived-from rule has been vacated by the court's decision in *Shell Oil v. EPA* because of procedural irregularities. It is not known whether EPA will re-enact this rule on an interim basis to avoid undue disruption of its hazardous waste management program.

The Bevill Amendment

In October 1990, Congress incorporated a statutory amendment to exempt certain "special wastes" from EPA's regulations promulgated under subtitle C of RCRA. The purpose of this exclusion of special wastes from the regulatory requirements is to protect high-volume, low-hazard wastes from the burdensome impact of RCRA-based regulations.

Most notable of these special wastes are mining, milling, and smelting wastes. Their exclusion is credited to their sponsor, Congressman Bevill. The Bevill Amendment originally removed the mining waste from the scope of hazardous waste listing temporarily and subjected it to additional studies. Later, the Bevill Amendment included other types of special wastes.

To date, the exclusion of mining waste under the Bevill Amendment and that of other special wastes are well and alive. However, EPA has attempted to narrow the scope of exclusion of these special wastes through a 1985 reinterpretation and relisting proposal to Congress. If accepted, this will reduce the original scope of the Bevill Amendment. Congress has not yet responded to EPA's proposal.

Used oil

On September 23, 1991, EPA released its proposal containing three options for the treatment, storage, or disposal of used or waste oils. Under option 1, all used oils have to be treated as hazardous, thus subjecting them to full TSD requirements pursuant to subtitle C of RCRA. Under option 2, only those used oils that are "typically and frequently" found hazardous at their point of generation and 50 percent or more of which exhibit toxic characteristic will qualify as hazardous wastes. Under option 3, EPA will not list any used oil as hazardous but will instead follow a comprehensive set of management standards based on basic good housekeeping requirements. This may still subject any used oil to the EPA regulations promulgated under subtitle C of RCRA if the used oil is going to be disposed rather than recycled.

Also, under the 1984 RCRA Amendments, EPA proposed certain special requirements covering notification and record keeping for the transportation of used oil that will be recycled. On September 23, 1991, EPA issued a supplemental notice proposing certain used-oil management standards for recycled oil under section 3014 of RCRA. This was in direct response to the Used Oil Recycling Act (UORA) of 1980 which requires EPA to make a determination of the hazardousness of used oil. Also, in a 1988 case entitled *Hazardous Waste Treatment Council v. EPA*, the District of Columbia Court of Appeals held that EPA must determine whether used oils must be listed based on RCRA's technical criteria. EPA's proposal included amendment of 40 CFR §261.32 by adding four waste streams from the reprocessing and re-refining of used oil to the list of hazardous wastes from specific sources.

On May 20, 1992, EPA issued its final rule on the listing of used oil based upon the technical criteria of §§1004 and 3001 of RCRA and 40 CFR §§261.11(a)(1) and (a)(3). EPA's current position is not to list used oil destined for disposal as hazardous waste. However, EPA is yet to make a final decision on listing of used oil destined for recycling.

5.5 Variances

EPA's regulations specify three situations under which it may grant variances from the requirements of hazardous waste. Specifically, these variances are from the regulatory definition of solid waste. Under the statutory definition of hazardous waste as discussed earlier, if a waste no longer qualifies as a solid waste, it cannot be a hazardous waste.

Under the current EPA scheme, variances may be granted for the following wastes or waste products:

- Certain accumulated materials of which at least 75 percent is likely to be recycled in the following year

- Secondary materials which are reclaimed and reused in the original primary production process from which they were generated

- Materials resulting from an initial but significant and substantial reclamation operation which makes the reclaimed materials more like a product than a waste

For each of the specific conditions under which a variance is sought, EPA has a checklist of factors that must be considered for either granting or denying a variance. For example, through a variance procedure, EPA may allow certain combustion units to qualify as boilers even though they do not meet the regulatory definition of a boiler.

Pursuant to EPA's regulations, regional administrators of EPA as well as authorized state agencies may grant these variances. To obtain a variance under any one of the above situations, the application must include the standards under which the variance is to be considered as well as the reasons why such a variance is proper.

5.6 Delisting of Hazardous Waste

Individual waste streams may vary depending on raw materials, industrial processes, and other operating variables. Thus, absent necessary analysis, if a waste is initially considered hazardous under EPA's regulations, it may not always turn out to be hazardous. It is possible that a specific waste from an individual facility meeting EPA's listing description may not actually be hazardous.

For this reason, EPA's regulations, as set forth in 40 CFR §260.22, also provide for an exclusion or a delisting procedure allowing persons to demonstrate why a specific waste should not be regulated as a hazardous waste. Note that a delisting petition will not be considered by EPA for a hazardous waste that is listed because of its hazardous characteristics as defined in 40 CFR §§261.21, 261.22, 261.23, and 261.24 or because it is an acutely hazardous waste. But under this delisting scheme, a listed hazardous waste may be delisted based on its inconsequential acute or chronic effects and toxicity. To have these wastes excluded, petitioners must show that the wastes, as generated at their individual facilities, do not meet any of the criteria for which such wastes were originally listed. Additionally, in view of HSWA, the petitioners are required to demonstrate that the waste under question does not exhibit any of the four specific hazardous waste characteristics. Moreover, they must present sufficient information for EPA to determine whether the waste contains any other toxic constituents at hazardous levels. The exclusion, if granted, will be issued for a particular generating, storage, treatment, or disposal facility.

Most of the delisted hazardous wastes to date were originally listed as specific and nonspecific source hazardous wastes. In 1990, EPA granted petitions to delist several common types of wastes: wastewater treatment sludges from electroplating operations, wastewater treatment sludges from the manufacture of lead-based initiating compounds, wastewater treatment sludges from the chemical conversion coating of aluminum, spent nonhalogenated solvents and still bottoms from recovery of such solvents, wastes from the production of tri- or tetrachlorophenol, and spent pickle liquor generated by steel finishing operations.

During delisting determinations, EPA uses several "fate and transport" models to predict the concentrations of hazardous constituents that may be released from disposed wastes. On July 18, 1991, EPA issued a proposal for use of EPA's composite model for landfills (EPACML) to evaluate a delisting petition. If a petitioned waste is delisted by EPA, the waste is no longer subject to federal hazardous waste regulations, but still must be managed and disposed of in accordance with state and local laws and regulations. In this context, note that the substantive standards for delisting a treatment residue or a mixture are the same as those for listed hazardous wastes.

5.7   Generators of Hazardous Waste

EPA's regulations as set forth in 40 CFR part 261 define *generator* of hazardous waste as

> [A]ny person, by site, whose act or process produces hazardous waste identified or listed in Part 261 of this chapter or whose act first causes hazardous waste to become subject to regulation.

A *person* is defined by EPA regulations in broad terms and includes an individual, trust, firm, joint stock company, federal agency, corporation, partnership, association, state, municipality, commission, political subdivision of a state, or any interstate body.

In view of the evolution of the definition of hazardous waste, any generator of wastes must first determine whether such wastes are solid wastes and whether any of the generated solid wastes falls within the definition of hazardous waste. For the latter part, one must first determine whether it is listed as a hazardous waste and, if not, whether it triggers any of the four hazardous waste characteristics of the solid waste.

The generator requirements promulgated by EPA pursuant to section 3002 of RCRA are set forth in 40 CFR part 262. Figure 5.2 is a reproduction of EPA's flowchart that depicts the generator requirements. Pursuant to the EPA regulations, hazardous waste generators must identify the hazardous waste in terms of both quantity and constituents. In addition, a generator is responsible for adhering to certain requirements and standards set forth in

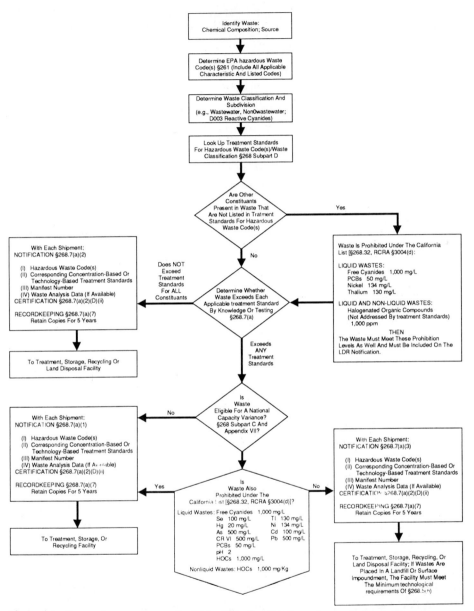

Figure 5.2   Generator requirements. *Note:* This flowchart can be used as a guide only.

these regulations, to facilitate proper tracking of all hazardous wastes from their source to final disposal. At a minimum, a generator, after it has identified the hazardous wastes, must

- Notify EPA about the generation of hazardous wastes.
- Obtain an EPA identification number as a hazardous waste generator.
- Label and package all hazardous waste containers used for either storage or transportation.
- Provide information pertaining to the chemical composition (e.g., chemical profile) of the waste to the transporters pursuant to the DOT regulations as well as EPA regulations for hazardous waste treatment, storage, or disposal facilities.
- Use a manifest system to track the waste movement after its generation and identification, starting with its transportation.
- Maintain complete records of the analytical tests performed as well as the results for at least 3 years after the wastes have been shipped off-site for treatment, storage, or disposal.
- Submit biennial reports to EPA describing the quantity, nature, and disposition of hazardous waste generated within the reporting period; the steps taken to reduce the volume and toxicity of the wastes; and the success achieved in minimizing them.
- Certify the manifests and sign a certificate stating that the "hazardous waste minimization program" under the 1984 RCRA Amendments has been adopted.

It is important to prepare the appropriate number of copies of the waste manifest so that everyone involved in RCRA's cradle-to-grave scenario is fully informed. The generator is responsible for maintaining the final copy when it is returned from the TSD facility.

In this context, note that a generator is responsible for preparing a proper manifest as well as the control and transport document that must accompany the shipment of hazardous waste from the generating facility. Also, in the event a generator does not receive the manifest or receives an incomplete or improper manifest, the generator is required to file an "exception report" with EPA or with the state agency, if the latter has been delegated such authority by EPA. A copy of the Uniform Hazardous Waste Manifest (commonly known as uniform manifest or simply manifest) as developed by EPA is shown in App. 5B.

Occasionally, there can be multiple generators of hazardous waste. Such a situation typically arises when a tank and related equipment are involved during manufacturing or on-site storage and the owners of the tank and equipment do not own the contents and are not engaged in the removal of waste residues from the tank. For such a situation, EPA may conclude that all those who are responsible for the tank and related equipment, including the contents and residue, qualify as generators and may bring appropriate enforcement action. Therefore, as a safeguard, all possible "multiple" generators should fulfill the requirements of a generator even if the possibility of unnecessary duplication exists.

There are additional requirements for a generator who exports hazardous wastes. Section 3017 of the RCRA Amendments of 1984 includes certain definite and stringent requirements irrespective of whether EPA or a delegated state agency is responsible for the RCRA program. In essence, these requirements include notification by the primary exporter to EPA, acceptance of the shipment by the appropriate governments of the transit and receiving countries, attachment of written consent by the foreign country to the manifest that accompanies the shipment, shipment conformance to the terms of consent, and written explanation if there are any significant discrepancies from the manifest.

EPA regulations, not surprisingly, include certain exemptions from the broad RCRA requirements. For example, a generator is allowed to accumulate its own hazardous wastes on-site without an RCRA permit for a TSD facility under certain specific circumstances:

- A generator can store up to 55 gallons of hazardous wastes or 1 quart of acutely hazardous waste as listed in 40 CFR §261.33(e) in containers at or near their points of generation that qualify as "satellite accumulation areas" indefinitely, without an RCRA permit or interim status, provided such containers are in good condition, compatible with the waste stored and appropriately marked or identified.

- A generator can store hazardous waste in any quantity on-site prior to its shipment off-site for a period not exceeding 90 days.

Under both circumstances, however, the generator must meet the general requirements of a generator, discussed earlier. The 90-day storage without obtaining an interim status or an RCRA permit for a TSD facility may be extended by EPA for up to 30 days in situations which involve unforeseen, temporary, and uncontrollable events or circumstances. Also, a generator who generates more than 100 kilograms but less than 1000 kilograms of hazardous waste in a calendar month may accumulate hazardous waste up to 180 days without an RCRA permit or interim status provided the quantity of accumulated hazardous waste on-site never exceeds 6000 kilograms.

5.8 Small Quantity Generator Exemptions

Pursuant to the 1984 Amendments of RCRA, EPA promulgated special regulations for small quantity generators. Under the regulatory definition, a *small quantity generator* is defined as a person who produces no more than 100 kilograms (220 pounds) of hazardous waste in a calendar month. Such a generator, defined as a *conditionally exempt small quantity generator* (CESQG), is exempt from most of the RCRA requirements that are otherwise applicable provided certain specific conditions are met:

■ The generator produces no more than 1 kilogram (2.2 pounds) of acutely hazardous waste during the month for which it qualifies as a small quantity generator.

■ The generator complies with all applicable requirements, e.g., identification of the waste, shipment restrictions, and no on-site accumulation of more than 1000 kilograms (2200 pounds) of hazardous waste at any time.

■ The generator treats or disposes of its hazardous waste at an RCRA-permitted on-site or off-site TSD facility or a facility which beneficially uses, reuses, recycles, or reclaims the waste in conformance with applicable EPA regulations.

Pursuant to EPA's regulations promulgated under RCRA, the conditionally exempt small quantity generators are not subject to 40 CFR parts 124, 262 to 266, 268, and 270. The manifest used by a conditionally exempt small quantity generator includes a modified certificate of waste minimization.

Note that the generators of no more than 1000 kilograms of hazardous waste per month also qualify for certain exemptions from the RCRA requirements, including on-site storage for 180 or 270 days without a TSD facility permit in certain specific situations. However, these exemptions are not as broad as those allowed for conditionally exempt small quantity generators. To bridge the gap between these two groups, EPA has provided certain specific exemptions for generators of 100 to 1000 kilograms per month of hazardous waste. These exemptions, as expected, are more comprehensive than those for conditionally exempt small quantity generators.

Generators of more than 1000 kilograms of hazardous waste may also be exempted from certain RCRA permit requirements if they store hazardous wastes for a short term. All other generators, unless specifically exempted, are regulated as large or nonexempt generators and are subject to complete RCRA-based regulation.

However, if the hazardous waste must be shipped over 200 miles to an off-site TSD facility, the on-site storage without a TSD facility permit may be extended up to 270 days for a generator of more than 100 kilograms but less than 1000 kilograms of hazardous waste in a calendar month. This is a recognition of practical problems that arise for disposal of hazardous wastes.

Last, a special exemption is allowed to farmers for on-site storage and disposal of waste pesticides. However, this is contingent on the storage of such waste in proper containers. At a minimum, these containers or their liners must be triple-rinsed with an appropriate solvent. Also the disposal of the waste must be in a manner consistent with the instructions provided on the pesticide label.

5.9 Transporters of Hazardous Waste

A *transporter* is defined as a person engaged in the off-site transportation of hazardous waste. Any on-site transportation of hazardous waste by generators, owners, or operators of permitted TSD facilities is not subject to federal regulations under RCRA. Pursuant to the requirements of section 3003 of

RCRA, EPA promulgated regulations for the transportation of hazardous wastes. These are set forth in 40 CFR part 263. In these regulations, EPA has expressly adopted certain regulations promulgated by DOT for the transportation of hazardous materials. These regulations essentially apply to the transporters of all types and quantities of hazardous waste.

Off-site transportation can be interstate or intrastate or both. However, any transportation within the site of hazardous waste generation is outside the scope of off-site transportation and thus is not subject to RCRA-based transporter requirements.

Pursuant to EPA's regulations promulgated under RCRA, the term *on-site* in connection with hazardous waste generation means

- The same or geographically contiguous property which may be divided by a public or private right-of-way under certain specific conditions   or
- Noncontiguous properties owned by the same person but connected by a right-of-way controlled by the same person and to which there is no public access

Pursuant to EPA regulations, a transporter may ship only that hazardous waste which has been properly recorded on a manifest. Moreover, the transporter is required to deliver the waste only to the destination as shown on the manifest, unless the shipment is made by a small quantity generator who has valid exemptions. In addition, a transporter must comply with certain specific requirements, summarized here:

- A transporter must obtain an EPA identification number prior to transporting any hazardous waste.
- A transporter must sign and date the manifest, acknowledging acceptance of the specific waste, and return one copy to the generator before leaving the generator's facility.
- A transporter must keep the manifest at all times while transporting hazardous waste. When the waste is delivered to another transporter or to the designated TSD facility, the transporter must obtain the signature of the next transporter or the operator of the TSD facility, as the case may be.
- A transporter must retain one copy of the manifest for his or her own records and give the remaining copies to the person receiving the waste.
- A transporter, if unable to deliver the waste according to the manifest, must contact the generator for instructions and revise the manifest, as required.
- A transporter must keep the executed copy of the manifest for at least 3 years.
- A transporter must notify the appropriate regulatory authorities and undertake prompt cleanup if there is any discharge, spill, or release of hazardous waste during the transport.

The transporter of hazardous waste may hold the waste for up to 10 days at a transfer facility or transfer station without obtaining an RCRA permit for storage. However, pursuant to EPA's regulations, a hazardous waste transporter may be subject to the hazardous waste generator requirements as set forth in 40 CFR part 260 if

- The transporter mixes hazardous wastes of different DOT shipping descriptions by placing them in a single container.
- The transporter imports hazardous waste from abroad.

A transporter of bulk shipments by water is exempted from the manifest requirements covered in EPA's regulations promulgated under RCRA. However, to qualify for such an exemption, the following requirements must be met:

- The facility to which the hazardous waste is shipped is so designated on the manifest.
- The shipping documents contain all the information on the manifest, with the exception of the EPA identification number; the generator's waste minimization certification and signature; and EPA's acknowledgment of consent if the waste is scheduled for export. They also include all such documents that accompany the waste.
- The transporter's signature and the delivery date are obtained by the person delivering the hazardous waste to the first bulk transporter by water.
- Each bulk transporter by water retains a copy of the shipping documents or the manifest.

Similar exemptions from the manifest requirements are also made to transporters via rail shipments. Not surprisingly, to qualify for such exemptions, a rail transporter must satisfy several procedures:

- The first rail transporter who accepts hazardous waste from the generator or from a nonrail transporter must sign and date the manifest, acknowledging acceptance of the waste, return a signed copy of the manifest to the person delivering the waste, and forward at least three copies of the manifest to one of several designated persons.
- The rail transporter must ensure that the hazardous waste is accompanied at all times by appropriate shipping documents required on the manifest (with the exception of the EPA identification number), the generator's waste minimization certification and signature, and the EPA acknowledgment of consent if the waste is being exported.
- The rail transporter must obtain the signature of the owner or operator of the designated facility on the manifest or on the shipping documents if the manifest has not been received at the facility, along with the delivery date.
- The rail transporter must retain a copy of the signed manifest or the shipping documents and must keep them for at least 3 years.

- A nonrail transporter must sign and date the manifest and provide a copy to the rail transporter if she or he accepts the hazardous waste from a rail transporter.

It is important to note that pursuant to the 1984 RCRA Amendments, railroads are statutorily shielded from citizen suits and imminent hazard enforcement provisions if the railroad is a mere transporter of hazardous waste under a contractual agreement and meets the due-care requirement.

Generally, two types of hazardous waste are exempt from the transporter requirements as set forth in 40 CFR part 263:

- Hazardous wastes that qualify as a recycling material, e.g., certain types of used and reclaimed oil; used batteries, scrap metals, petroleum coke, and certain fuels produced in refining; coke and coal tar operations in steel manufacturing and industrial ethyl alcohol production

- Hazardous waste shipped from a conditionally exempt small quantity generator who generates no more than 1 kilogram of acute hazardous waste or 100 kilograms of other categories of hazardous waste and who does not store more than 1000 kilograms of hazardous waste on-site at any time

Section 3003(b) of RCRA requires those EPA regulations promulgated under RCRA to interface and be consistent with the applicable provisions of the Hazardous Materials Transportation Act (HMTA) and the DOT regulations promulgated thereunder. The most important of these DOT regulations for transportation of hazardous materials are set forth in 49 CFR parts 172 and 174 through 177.

5.10   Treatment, Storage, and Disposal Facilities

Section 3004 of RCRA requires EPA to promulgate regulations covering certain specific performance standards for hazardous waste treatment, storage, or disposal facilities with the objective of protecting human health and the environment. Additionally, section 3005 of RCRA requires all TSD facilities to require appropriate permits unless they are specifically exempt.

Definitions

RCRA and the EPA regulations promulgated thereunder provide very specific definitions which must be taken into account in any discussion of their applicability in the context of TSD facilities.

A *facility* includes all contiguous land, structures, other appurtenances, and improvements on the land used for treating, storing, or disposing of hazardous waste. Thus a facility may include one or more TSD units providing treatment, storage, or disposal of hazardous waste separately or all together.

A facility can qualify as a *treatment facility* if one or more methods of treatment are carried out by the operator at such a facility. In this context, the term *treatment* signifies

[A]ny method, technique, or process, including neutralization, designed to change the physical, chemical, or biological character or composition of any hazardous waste so as to neutralize such waste or so as to render such waste nonhazardous, safer for transport, amenable for recovery, amenable for storage, or reduced in volume...

Thus, treatment encompasses any activity or processing that brings about a change in the physical form or chemical composition of the hazardous waste and renders it nonhazardous or reduces its original volume.

A *storage facility* is defined as a facility which is used for the containment or holding of hazardous waste either on a temporary basis or for a period of years, at the end of which the hazardous waste is treated, disposed of, or removed elsewhere. Hence, any such holding of hazardous waste will trigger this definition whether such waste is held in drums, tanks, and containers or is kept in stockpiles. However, waste stored in totally enclosed treatment units such as pipes or being held within the manufacturing units such as reactor vessels during production operations is exempt from the storage facility requirements until it is removed from such enclosed treatment or manufacturing process units.

A hazardous waste *disposal facility* is a facility at which the hazardous waste is deliberately brought into contact with air, land, or water by undertaking one or more of the following activities that comprise disposal:

[T]he discharge, deposit, injection, dumping, spilling, leaking, or placing of any...hazardous waste into any land or water...so that such...hazardous waste or any constituent thereof may enter the environment or be emitted into the air or discharged into any waters, including ground waters.

By this definition, any accidental spill of hazardous waste onto land or water and any unintentional release of hazardous waste into the air could conceivably constitute a disposal activity, thereby making the affected facility a TSD facility. However, EPA regulations specify that only intentional activities are included within this definition; thus, any accidental spilling or release of hazardous waste does not make the facility a disposal facility in the context of RCRA. Nonetheless, any accidental or unintentional spill or release of hazardous waste may trigger certain reporting requirements, as discussed in Chap. 10.

Exemptions

EPA regulations offer certain facilities both full and partial exemptions from the requirements of TSD facilities. The exempt activities and qualified TSD facilities include

- Facilities disposing hazardous waste in the ocean pursuant to a permit issued under the Marine Protection, Research, and Sanctuary Act (MPRSA)

- Disposal of hazardous waste by underground injection pursuant to the underground injection control (UIC) program of the Safe Drinking Water Act (SDWA)

- A Publicly Owned Treatment Works (POTW) having a valid National Pollutant Discharge Elimination System (NPDES) permit issued under the Clean Water Act (CWA) which complies with the conditions of the permit and specific operating and record-keeping requirements

- Facilities regulated under authorized state programs pursuant to section 3006 of RCRA

- Facilities handling certain recyclable materials which are exempt from the definition of hazardous waste because of their recycling programs

- Facilities used for temporary on-site storage for (1) up to 1 quart of acutely hazardous waste and 55 gallons of other hazardous waste for an indefinite period and (2) up to 90 days of accumulation for any quantity of hazardous waste

- Facilities accepting hazardous waste only from conditionally exempt small quantity generators

- Facilities consisting of elementary neutralization units and wastewater treatment units

- Facilities engaged in treatment or storage activities solely due to an emergency response procedure

- Facilities placing all the hazardous waste in a container of absorbent material to reduce the amount of free liquids

Permitting program

Unless a TSD facility is exempt, the owner or operator of the facility must obtain an RCRA permit. Section 3005 of RCRA sets forth the permitting program for the TSD facilities.

Pursuant to the EPA regulations, the RCRA-related permit requirements can be satisfied in several ways depending on the permitting status sought. Currently, there are two categories of TSD facilities: interim status facilities and permitted facilities. By virtue of their interim status, certain facilities that existed as of November 19, 1980, can continue operations until their RCRA permit applications are reviewed and permits are issued. However, these interim-status facilities are required to notify EPA about their hazardous waste management activities as well as to submit a preliminary permit application. When an interim-status facility receives the final RCRA permit, it loses its interim status and becomes a permitted facility.

A permitted facility, however, is a facility that has obtained an RCRA permit from EPA or from the state agency issuing permits under an EPA-delegated authority. Because of the statutory requirements set forth in section 3005(e) of RCRA, a specific TSD facility operating after November 19, 1980, is required to secure an RCRA permit prior to the beginning of any construction work unless it satisfies other qualifications or fulfills certain specific statutory or regulatory requirements that may shift the effective date to a new date.

A TSD facility may also operate without an RCRA permit if it satisfies the requirements for a permit by rule. This is allowed if the TSD facility is regulated under other valid environmental permit programs and if the facility is in full compliance with such permit requirements including general and specific terms and conditions. In addition, a facility operating under the permit by rule must meet certain performance standards relating to the hazardous waste management operations.

Performance standards

EPA's regulations, which are promulgated under section 3004 of RCRA and provide the standards for owners and operators of both interim-status and permitted TSD facilities, are set forth in 40 CFR parts 264 and 265. There are separate standards for interim-status facilities and permitted facilities. Moreover, EPA regulations provide both standards of general and specific applicability.

Standards of general applicability.   Standards of general applicability, as the term implies and as set forth in 40 CFR part 264, subpart B, apply to all TSD facilities. Pursuant to these, an operator of a TSD facility must

- Obtain an EPA identification number.
- Conduct or arrange for a detailed physical and chemical analysis of a representative sample of the hazardous waste prior to its acceptance.
- Notify the regulatory agency (EPA or authorized state agency) about the hazardous waste activities at the facility.
- Install a security system to prevent trespassing and unauthorized entries.
- Train facility personnel appropriately for their respective tasks.
- Maintain operating records until closure and comply with specific record-keeping requirements.
- Inspect all equipment and the facility on a regular basis.
- Document precautions to be exercised and steps to be taken for safe management of ignitable, reactive, or incompatible wastes.
- Install safety equipment and communication system, including alarm and other appropriate devices, for emergencies and accidents due to spill, fire, or explosion.
- Implement measures for groundwater protection.
- Write a closure plan and schedule and, in appropriate cases, postclosure procedures; have them approved by the regulatory agency.
- Establish financial responsibility requirements by ensuring availability of adequate funds for closure and postclosure care, corrective action, and third-party liability.

The 1984 RCRA Amendments also require health risk assessment for certain land-based TSD facilities. However, EPA has yet to promulgate regulations specifying the scope of such a risk assessment.

Standards of specific applicability. Besides the general standards, EPA has promulgated specific design, construction, and operating standards for eleven types of TSD facility. They cover tanks; containers; surface impoundments; waste piles; land treatment units; landfills; incinerators; thermal treatment units; chemical, physical, and biological treatment units; underground injection wells; and miscellaneous units. Their respective requirements are set forth in 40 CFR parts 264 through 267.

On February 21, 1991, EPA promulgated its final rule for the burning of hazardous waste in *boilers and industrial furnaces* (BIFs). This rule, commonly known as the *BIF rule,* took effect on August 21, 1991. By promulgating this BIF rule under the RCRA authority, EPA has expanded its control of hazardous waste combustion in order to regulate air emissions from the burning of hazardous waste in boilers and industrial furnaces.

In general, the BIF rule establishes control emissions of toxic organic compounds, toxic metals, hydrogen chloride, chlorine gas, and particulate matter from boilers and industrial furnaces burning hazardous wastes. The BIF rule requires boilers and industrial furnaces used to burn hazardous wastes to comply with the same *destruction and removal efficiency* (DRE) that currently applies to hazardous waste incinerators. The DRE criterion incorporates two minimum standards: 99.9999 percent DRE for dioxin-listed waste and 99.99 percent DRE for all other hazardous wastes.

To date, the BIF rule has established emission limits for ten toxic metals. These limits are based on projected inhalation health risks to a hypothetical *maximum-exposed individual* (MEI). Based on current federal policy on carcinogens, the standards for four carcinogenic metals—arsenic, beryllium, cadmium, and chromium—limit the increased lifetime cancer risk of the MEI to a maximum of 1 in 100,000. The standards for the noncarcinogenic metals—antimony, barium, lead, mercury, silver, and thallium—are based on *reference doses* (RFDs) below which no adverse health effects have been observed.

EPA's regulations include a specific small quantity on-site burner exemption. The BIF rule conditionally exempts the following:

- Boilers and industrial furnaces that burn small quantities of hazardous waste as fuel and operate under prescribed conditions
- Smelting, melting, and refining furnaces that process hazardous waste solely for metal reclamation
- Coke ovens, if the only hazardous waste they process consists of K087 (i.e., decanter tank tar sludge from coking operations) as an ingredient to produce coke

Owners and operators of facilities that burn hazardous waste in an on-site boiler or industrial furnace are exempt from the BIF rule, as codified in 40 CFR part 266, subpart H, if the hazardous waste burned in such a device does not exceed certain specified limits set by EPA. Table 5.2 reproduces the EPA table that relates this small-quantity burner exemption to the terrain-adjusted effective stack height.

TABLE 5.2   **Exempt Quantities for Small-Quantity Burner Exemption**

Terrain-adjusted effective stack height of device (meters)	Allowable hazardous waste burning rate (gallons/month)	Terrain-adjusted effective stack height of device (meters)	Allowable hazardous waste burning rate (gallons/month)
0 to 3.9	0	40.0 to 44.9	210
4.0 to 5.9	13	45.0 to 49.9	260
6.0 to 7.9	18	50.0 to 54.9	330
8.0 to 9.9	27	55.0 to 59.9	400
10.0 to 11.9	40	60.0 to 64.9	490
12.0 to 13.9	48	65.0 to 69.9	610
14.0 to 15.9	59	70.0 to 74.9	680
16.0 to 17.9	69	75.0 to 79.9	760
18.0 to 19.9	76	80.0 to 84.9	850
20.0 to 21.9	84	85.0 to 89.9	960
22.0 to 23.9	93	90.0 to 94.9	1,100
24.0 to 25.9	100	95.0 to 99.9	1,200
26.0 to 27.9	110	100.0 to 104.9	1,300
28.0 to 29.9	130	105.0 to 109.9	1,500
30.0 to 34.9	140	110.0 to 114.9	1,700
35.0 to 39.9	170	115.0 or greater	1,900

Disposal ban and restrictions.   The RCRA Amendments of 1984 prohibit land disposal of certain hazardous wastes in order to encourage advanced treatment and recycling of waste. This resulted in the banning of land disposal of bulk or noncontainerized liquid hazardous wastes and all those hazardous wastes that contain any free liquid. In addition, EPA's regulations prohibit the creation of a landfill of biodegradable or liquid-releasing absorbents that are used to remove the free liquids in containerized hazardous waste.

The restrictions thus imposed by EPA for land disposal of liquid hazardous waste are popularly known as the *hammer provisions*. All dioxin-containing hazardous wastes are banned from land disposal effective November 1986. In addition, EPA banned the land disposal of certain types of hazardous wastes effective July 1987, unless they are pretreated.

The hammer provisions also resulted in an EPA ranking of the hazardous wastes based on their intrinsic hazard and quantities along with a determination of the extent of the restriction and/or ban for land disposal. The EPA ranking and its schedule for land disposal restrictions and specific prohibitions against various hazardous wastes are set forth in 40 CFR part 268.

By virtue of the hammer provisions, if EPA fails to adopt restrictions by the statutory deadlines, certain restrictions set forth in section 3004 of RCRA will automatically become effective. Furthermore, EPA is also required to establish treatment standards for each restricted waste based on the *best demonstrated available technology* (BDAT).

EPA's land disposal restrictions as set forth in 40 CFR part 268 apply to generators and transporters of hazardous waste as well as to owners and operators of TSD facilities. However, wastes which were disposed of before November 8, 1986, need not be removed from the land where they were disposed.

Thus far, EPA issued three rules proclaiming its land disposal restrictions. The first, known as the *first-third rule,* was published by EPA on August 17,

1988. This rule contains provisions for only one-third of the listed hazardous wastes under EPA's RCRA program.

The *second-third* and the *third-third rules* were subsequently published by EPA on June 8, 1989, and June 1, 1990, respectively. Taken together, these three rules, which address three distinct and separate groups of hazardous wastes based on toxicity and volume considerations, encompass those hazardous wastes covered in 40 CFR part 268. EPA also provides appropriate treatment standards for these listed hazardous wastes by establishing its three rules prohibiting and/or restricting land disposal. These treatment standards reflect the corrective action program for hazardous waste management under RCRA.

Two distinct "soft hammer" and "hard hammer" provisions have been developed to restrict disposal of hazardous wastes if EPA fails to promulgate appropriate standards by the statutory deadlines. Figure 5.3 reproduces EPA's soft-hammer waste disposal restrictions whereas Fig. 5.4 provides the soft-hammer notification, certification, and demonstration requirements pursuant to EPA's current program. Unlike the soft-hammer provisions, the hard-hammer provisions prohibit all land disposal of hazardous wastes. Figure 5.5 reproduces EPA's hard-hammer deadlines.

In this context, note that the basic framework of EPA's land disposal restriction includes implementation of a BDAT standard for the majority of the listed hazardous wastes. EPA has also attempted to link the BDAT standards with specific technologies.

State authority.   Section 3006 of RCRA gives EPA the authority to delegate administration of the federal hazardous waste program to the states. The state programs can be more stringent, but under no circumstances are they to be less stringent than the federal program administered by EPA. Also, they must be consistent with the federal and other, already authorized state programs.

To receive such a delegation, a state must include several mechanisms in its program: a permitting plan, a manifest system, a demonstration of administrative resources to undertake a comprehensive program, and an adequate enforcement program.

Because of the dual administration of the RCRA program, it is important for a TSD facility to ensure compliance with the general provisions of RCRA and the requirements of the authorized program, whether federal or state.

5.11   RCRA Corrective-Action Requirement

The RCRA Amendments of 1984 empowered EPA with the authority to require certain TSD facilities to take actions to correct various environmental problems. This "corrective-action" authority does not extend to the non-TSD facilities that may deal with certain aspects of hazardous waste management and are subject to only some of the provisions of RCRA.

The focus of the corrective-action program is on those TSD facilities from which releases of contaminants are likely. This parallels the cleanup goals set

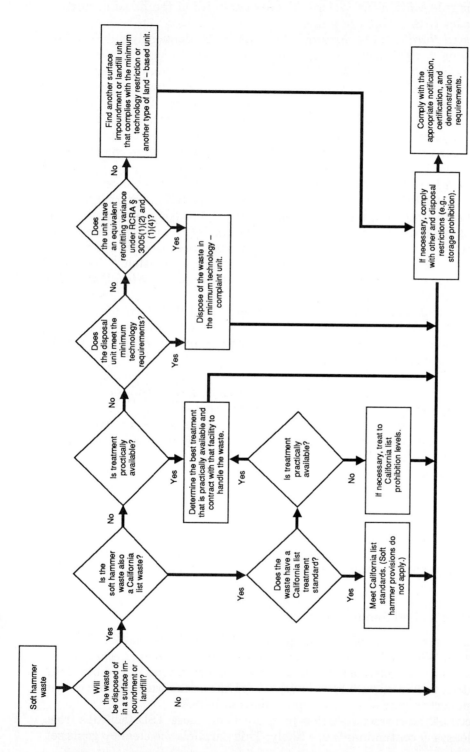

Figure 5.3 Identifying soft-hammer waste restrictions.

Requirement	Sent to	When	Required information
If Land Disposal Occurs in Surface Impoundment or Landfill Units			
Notification (off-site only)	Treatment or disposal facility receiving waste	With each waste shipment	Notification that the waste is a soft hammer waste. Specific information includes: - EPA hazardous waste number; - Any applicable prohibitions (e.g., soft hammer provision); - Manifest number associated with shipment of waste; and - Waste analysis data, where available.
Certification— If treatment is *not* practically available (off-site only)	EPA Regional Administrator and Disposal facility receiving waste	At time of first waste shipment and copy with each waste shipment	Certification should appear as follows: "EPA certifies under penalty of law that the requirements of 40 CFR 268.8(a)(1) have been met and that disposal in a landfill or surface impoundment is the only practical alternative to treatment currently available. EPA believes that the information submitted is true, accurate, and complete. EPA is aware that there are significant penalties for submitting false information, including the possibility of fine and imprisonment."
Certification— If treatment *is* practically available (off-site only)	EPA Regional Administrator and Treatment facility receiving waste	At time of first waste shipment and copy with each waste shipment	Certification should appear as follows: "EPA certifies under penalty of law that the requirement of 40 CFR 268.8(a)(1) have been met and that the agency has contracted to treat its waste (or will otherwise provide treatment) by the practically available technology which yields the greatest environmental benefit, as indicated in its demonstration. EPA believes that the information submitted is true, accurate, and complete. EPA is aware that there are significant penalties for submitting false information, including the possibility of fine and imprisonment."
Demonstration—If *no* treatment is available (off-site and on-site)	EPA Regional Administrator	At time of first waste shipment	List of facilities and facility officials contacted, addresses, telephone numbers, and contact dates. Also, a written discussion of when treatment or recovery is not practical for the waste.
Demonstration—If treatment *is* available (off-site and on-site)	EPA Regional Administrator	At time of first waste shipment	List of facilities and facility officials contacted, addresses, telephone numbers, and contact dates. Provide information on the chosen treatment technology selected because it provides the greatest environmental benefit.

Figure 5.4 Soft-hammer notification, certification, and demonstration requirements.

Waste	Hard-hammer statutory deadline
Solvent & dioxin wastes	November 8, 1986
California list wastes	July 8, 1987
CERCLA/RCRA corrective action soil and debris contaminated with solvent and dioxin and California list wastes	November 8, 1988
Scheduled wastes (1st Third, 2nd Third, and 3rd Third wastes)	May 8, 1990

Figure 5.5   Hard-hammer deadlines.

under the Comprehensive Environmental Response, Compensation, and Liability Act (CERCLA). However, an RCRA-based corrective-action program is different from the CERCLA program for the simple reason that the corrective action applies to only a TSD facility as defined in RCRA, whereas the CERCLA program applies to any facility where there is an actual release or a threat of a release of a hazardous substance. Hence, EPA's corrective-action authority under RCRA is narrower than EPA's remedial authority under CERCLA.

EPA has responded in kind to its authority under the corrective-action program of RCRA. It has thus far attempted to scope its corrective-action strategy and undertake an appropriate plan pursuant to its statutory authority. To date, EPA has made several announcements proposing how this corrective-action program will be formulated. EPA's proposed regulations, issued on July 27, 1990, contained some important concepts. While these are not yet final, it is important to understand them to appreciate the future direction of EPA's corrective-action programs.

EPA's corrective-action authority will cover all TSD facilities irrespective of whether they are operating under full RCRA permit or under the interim-status provisions of RCRA. Furthermore, under EPA's regulations, a facility where hazardous waste is generated without a TSD permit under an RCRA permitting program or under the interim-status provisions may be subject to the corrective-action program. This can happen when the 90-day on-site storage limit for a hazardous waste generator is exceeded. Hence, to avoid corrective-action provisions altogether, all generators of hazardous waste must limit their on-site storage of hazardous wastes strictly to 90 days.

EPA's proposed regulation has introduced a new term: *corrective-action management unit* (CAMU). According to EPA's plan, it may designate an area containing contaminated waters or soil as a CAMU. Furthermore, EPA wants to manage all activities within a CAMU outside of RCRA's permitting process for the purpose of simplification and expediency of correction efforts. This will apparently facilitate the closing operations or closure of a TSD facility when no further treatment, storage, or disposal is contemplated.

However, EPA's proposed rule places its corrective-action jurisdiction over the owner or operator of a TSD facility. Pursuant to its statutory authority, EPA may direct the owner or operator of a TSD facility to clean up actual releases of contaminants or to undertake appropriate steps to eliminate or reduce the impact of such potential releases. Moreover, using its corrective-action authority, EPA can place the entire financial burden of necessary investigation and cleanup due to a release of a contaminant from a TSD facility on the owner or operator of such a facility. It is important to understand here that unlike with CERCLA, there is no government funding mechanism set up to pay for any corrective-action response costs.

EPA's proposed regulations of 1990 contain some hitherto unknown concepts. For corrective-action programs, EPA favors a broader definition of hazardous waste than what is statutorily defined. EPA's current position is to go beyond those wastes which are either listed or identified as hazardous wastes under RCRA. Hence, if EPA prevails, corrective-action programs will apply to all those "hazardous wastes" included within the broad definition of RCRA's textual language, i.e., any solid waste which may cause serious or incapacitating illness or pose a substantial present or potential hazard to human health or to the environment. This can result in an application of the corrective-action program whenever any waste containing a hazardous constituent, chemical, or toxic substance that may adversely affect human health or the environment is released or is threatened to be released from a TSD facility. This can happen even if such a waste is neither listed nor defined as hazardous waste under EPA's RCRA regulation based on the four specific characteristics.

In the absence of any final EPA rule, it is difficult to predict how EPA's proposed corrective-action program will be finally shaped or implemented. It is also quite likely, in view of the expected financial and operational impact of any corrective-action plan, that the courts may ultimately decide the legal propriety as well as the applicability of EPA's final corrective-action activities.

5.12   Underground Storage Tank Systems

Subtitle I of RCRA requires EPA to develop regulations for a comprehensive program covering the *underground storage tank* (UST) systems. RCRA defines a UST as follows:

> Any one or combination of tanks (including underground pipes connected thereto) which is used to contain an accumulation of regulated substances, and the volume of which (including the volume of the underground pipes connected thereto) is 10 percent or more beneath the surface of the ground.

EPA's final rule on the technical requirements for UST became effective on December 2, 1988. This final rule conforms with section 9003 of RCRA under which EPA is required to establish requirements for release or leak detection, leak prevention, financial responsibility, and corrective action for all USTs containing regulated substances.

A UST is subject to EPA's regulations promulgated under RCRA if it contains a regulated substance. *Regulated substances* are those "substances defined as hazardous under the Comprehensive Environmental Response, Compensation, and Liability Act of 1980 (CERCLA), except hazardous wastes regulated under subtitle C of RCRA, and petroleum." Thus, regulated substances include all the hazardous substances listed in CERCLA and any form or class of petroleum, but not hazardous waste as defined in RCRA.

In this context, the term *petroleum* has a distinct definition. It includes crude oil or any fraction thereof that is liquid at standard temperature and pressure (60°F and 14.7 pounds per square inch absolute).

In the proper analysis, the UST program is a culmination of the provisions of two federal statutes, RCRA and CERCLA, rather than one particular statute. However, subtitle I of RCRA contains several basic and major provisions for the regulation of the UST systems. Section 9002 of RCRA requires the owner of a UST system to notify the appropriate state of the existence of such a system and, whenever appropriate, to go through a formal tank registration program.

Section 9003 of RCRA requires EPA to promulgate necessary regulations, applicable to all owners and operators of UST systems, for protecting human health and the environment against releases. A checklist of the minimum RCRA requirements includes

- Meeting of proper standards for design, construction, and installation of new tanks (i.e., tanks installed on or after May 8, 1986)
- Conformance with appropriate standards for upgrading existing UST systems
- Instituting of standards for the maintenance of a leak detection system, an inventory control system along with tank testing or a comparable system, and relevant record keeping for all UST systems
- Reporting of releases of regulated substances from any UST system along with corrective actions taken in response
- Maintaining of records of any monitoring, leak detection or inventory control system, or tank testing or comparable system
- Fulfilling of necessary closure procedures for an existing UST system to prevent future releases of regulated substances
- Demonstration of evidence of financial responsibility for undertaking corrective actions and compensating third parties for bodily injury and property damage

Under EPA's current program scope and interim prohibition, as indicated in 40 CFR part 280, subpart A, UST systems may be installed for storing regulated substances after May 8, 1985, only if such a system is

- Protected cathodically against corrosion, constructed of noncorrodible material, or designed to prevent any release of the stored substance
- Adequate to prevent releases due to corrosion or structural failure during the operational life of the tank
- Constructed of or lined with material compatible with the stored substance

A UST system without corrosion protection may also be installed at a site that a corrosion expert determines to be not corrosive enough to cause a release due to corrosion during its operating life.

As of December 22, 1988, these UST systems operating under the interim prohibition must meet EPA's requirements for new tank standards, except those tanks for which EPA has decided to defer its regulatory action.

Exclusions

Subtitle I excludes the following tanks and facilities from the definition of USTs:

- Farm or residential tanks of 1100-gallon capacity or less which store motor fuel for noncommercial purposes
- Tanks storing heating oil for consumption on the premises where stored
- Surface impoundments, pits, ponds, and lagoons
- Stormwater or wastewater collection systems
- Flow-through process tanks
- Liquid traps or associated collection lines directly related to oil or gas production and collection operations
- Pipeline facilities and collection lines regulated under the Natural Gas Pipeline Safety Act of 1968, the Hazardous Liquid Pipeline Act of 1979, or comparable state laws
- Storage tanks situated on or above the underground floor areas

In addition, EPA's regulations exclude the following UST systems from the requirements of subtitle I:

- UST systems with storage capacities of 110 gallons or less
- UST systems containing a *de minimis* concentration of regulated substances
- UST systems containing hazardous wastes that fall within subtitle C of RCRA or a mixture of such hazardous wastes and other regulated substances
- Emergency spill and overflow containment units emptied expeditiously after use
- Equipment or machinery containing regulated substances, e.g., hydraulic-lift tanks and electrical equipment tanks
- Wastewater treatment tanks which are an integral part of a wastewater treatment facility regulated under the Clean Water Act

Unfortunately, EPA's regulations under RCRA do not provide a definition of *de minimis* concentration. Instead, EPA has adopted a policy of case-by-case analysis.

Existing tanks

All existing tanks that qualify as UST systems must meet the requirements set forth by EPA in 40 CFR part 280. Appendix 5C provides the EPA form for notification for all underground tanks that have been used to store regulated substances since January 1, 1974. Within 10 years of December 22, 1988, all existing USTs either must meet the new tank performance standards or specific upgrade requirements or must be closed down permanently. Appendix 5D is a reproduction of EPA's final approach to upgrading of all existing tanks by the 1988 deadline.

To meet such standards or requirements, the owners and operators of existing tanks must meet at least the following regulatory requirements:

- Follow proper tank-filling practices to prevent releases due to spills and overfills.
- Use devices that prevent overfills and control and contain spills.
- Install a release detection system.
- Repair tanks in accordance with nationally recognized industry codes, such as the National Fire Prevention Association (NFPA) codes.
- Report suspected releases to the implementing agency within 24 hours, and perform release investigation and confirmation tests.
- Begin corrective action when a release is confirmed.

Note that even temporarily closed tanks are subject to these regulatory requirements. To initiate a permanent closure, first the tanks must be emptied, cleaned, and either removed from the ground or filled with an inert solid material.

Last, all requirements that apply to the tanks under this UST program also apply to the connected piping system which is an integral part of the tanks. Hence, if an existing tank is upgraded to meet the new tank performance standards, such an undertaking must also include the tank's piping, which is considered part of the tank.

New tanks

Section 9002 of RCRA requires owners of new tanks to give notice to the designated state or local agency within 30 days of bringing a new tank into use. As part of this notification process, the owners of new USTs must certify that the new tank has been installed in accordance with the regulations and that it complies with all regulatory requirements for a new tank.

EPA's regulations promulgated under section 9003(c) establish the technical performance standards for all underground storage tanks. As set forth in 40 CFR section 280.20(a), each new tank must generally be "properly designed and constructed, and any portion underground that routinely contains product must be protected from corrosion" in accordance with a code of practice developed by a nationally recognized association, such as NFPA, or by an independent testing laboratory. The owners and operators of new USTs, at a minimum, must certify that proper installation procedures were followed as well as identify and account for how the installation was accomplished.

EPA has specifically authorized the installation of four types of new tanks:

- Tanks constructed of fiberglass-reinforced plastic
- Tanks constructed of steel-fiberglass-reinforced plastic composite
- Tanks constructed of steel and cathodically protected
- Tanks constructed of metal when an expert has determined that the installation site is not too corrosive to result in a release

The major elements of EPA's regulations promulgated on September 23, 1988, which became effective on December 22, 1988, include the following requirements for new USTs:

- Tanks must be designed and constructed to retain their structural integrity for their operating life.
- Tanks and attached piping used to deliver the stored product must be protected from external corrosion.
- Any cathodic protection used must be monitored and maintained to keep the UST free of corrosion.
- Tanks must be equipped with devices to prevent overfills and to control or contain spills.
- Release detection must be instituted.

Any associated or interconnected piping of a new UST that contains regulated substances and is in contact with the ground must meet the same standards as the tank insofar as the design, construction, and corrosion protection are concerned. In addition, release detection is required for such piping. For new pressurized lines, continuous monitoring or mechanical detectors are also required.

All new-tank owners and operators must report suspected releases within 24 hours and undertake appropriate investigation and testing to confirm such releases. If there is a leak, owners and operators of USTs must undertake immediate corrective action.

In this context, note that the *aboveground storage tanks* (ASTs) are not currently subject to any special EPA regulations under RCRA. There is a simple reason for this seeming indifference: Most of the AST systems are part of the TSD operations and hence are regulated under subtitle C of RCRA. A number of states, however, have specific regulations for ASTs, and such regulatory requirements must be complied with to avoid violations. Also, the status quo at the federal level may change if EPA's proposed regulations of September 23, 1991, are adopted. If they are adopted, the owners and operators of ASTs will be subject to certain RCRA-based standards for storing hazardous waste in ASTs.

Effective January 4, 1989, petroleum UST owners and operators must demonstrate their capability to meet certain specific financial responsibilities including third-party liability. Details of these financial responsibility requirements are set forth in 40 CFR sections 280.90 to 280.111. Compliance with the financial responsibility requirements has been phased over a few years based on the number of USTs and the net worth of the owner or operator.

5.13 Medical Waste

The definition of medical waste specifically precludes it from the definition of hazardous waste. Nevertheless, it is considered a subpart of solid waste. Pursuant to EPA's definition, *medical waste* means any solid waste which is generated in the diagnosis, treatment, or immunization of human beings or animals. Furthermore, it includes those solid wastes which are generated during research, production, and testing of biological parts or components.

Subtitle J of RCRA, also known as the Medical Waste Tracking Act (MWTA) of 1988, requires EPA to set up a demonstration program for tracking the shipment and disposal of medical wastes in a few specific states. Unlike the solid waste program which has been delegated to the states by EPA pursuant to applicable RCRA provisions, there has been no such broad delegation for medical wastes. However, as set forth in 40 CFR section 259.1(c), generators, transporters, and owners or operators of intermediate handling or disposal facilities involved in transporting or managing regulated medical waste generated in a covered state (i.e., participating in the demonstration medical waste tracking program) must comply with the standards for its tracking and management even if such transport or management occurs in a noncovered state. As of now, Connecticut, New Jersey, New York, Rhode Island, and Puerto Rico are the only covered states.

A "regulated medical waste," as defined in 40 CFR §259.30, is any solid waste generated in the diagnosis, treatment, or immunization of human beings or animals. Hazardous waste identified or listed in 40 CFR part 261 is not regulated medical waste.

Pursuant to RCRA, EPA is required to promulgate regulations that cover certain types of medical waste:

- Waste blood and blood products including blood plasma, blood platelets, red or white corpuscles, and interferon
- Cultures and stocks of infectious agents and associated biologicals
- Discarded or wasted body fluids and secretions
- Pathological wastes, including tissues, organs, and body parts
- Contaminated animal carcasses and parts
- Wastes from surgery or autopsy that were in contact with infectious agents
- Laboratory wastes that were in contact with infectious agents

- Hypodermic needles, syringes, scalpel blades, and broken glassware
- Discarded medical equipment that came in contact with infectious agents
- Dialysis wastes
- Biologicals made from living organisms and their products
- All other medical wastes that may pose a threat to human health or the environment

According to EPA, etiologic agents that are transported interstate pursuant to the requirements of DOT, DHHS, and other applicable shipping requirements are not to be treated as regulated medical waste. EPA's regulations on medical wastes have yet to be promulgated. When such regulations are promulgated, they will include a uniform manifest for tracking the medical waste from the generator to the ultimate disposal facility.

5.14   Inspection and Enforcement

Section 3007 of RCRA includes the inspection provisions for both the TSD facilities and the person who generates, transports, treats, stores, or disposes of hazardous waste. Under the statutory authority granted by RCRA and other environmental statutes, EPA and authorized state agency personnel may have access to the premises and to the related records. This includes the authority to copy records and to collect waste samples.

Section 3008 of RCRA authorizes EPA to bring appropriate enforcement actions against a statutory violation of subtitle C of RCRA or the regulations promulgated thereunder. These can include both civil penalties and criminal sanctions as well as issuance of administrative orders. A civil penalty can include a fine of up to $25,000 for each day of violation and can be imposed even in the absence of serving the violator with a compliance order. Criminal sanctions can include a fine of up to $50,000 for each day of violation or 2 years of imprisonment or both against persons who "knowingly" commit certain violations involving storage, transportation, treatment, and disposal of hazardous wastes. In addition, under the knowing-endangerment provision for placing another person in imminent danger of death or serious bodily injury, a convicted individual may be fined up to $250,000 and/or given up to 15 years of imprisonment, and an organizational defendant (viz., a corporation) can be fined up to $1 million.

RCRA's incorporation of the term *knowing* or *knowingly* as the mental culpability or scienter element required for criminal violation has been a highly debated legal issue. To date, two schools of thought have emerged since the statute is silent as to what such term implies. The conservative interpretation suggests that the government has to show that the defendant knew about the general nature of the waste that was treated, stored, or disposed of by the defendant. The liberal interpretation using a restrictive view requires that the government prove that the defendant knew each element of the offense and that he or she was specifically violating RCRA.

Apart from the regulatory enforcement, RCRA has a distinct and separate provision for citizen suits. This allows any person to bring a civil suit against an alleged violator of RCRA's statutory requirement or against the EPA administrator for failing to perform a nondiscretionary duty.

Last, under section 7003 of RCRA, EPA may bring suits against alleged violators to prevent an "imminent and substantial endangerment" to health or the environment. The RCRA Amendments of 1984 made it clear that any violation occurring prior to the enactment of RCRA is covered retroactively under the imminent-hazard provision. Hence, there cannot be a defense based on the constitutionality of retrospective application of RCRA provisions in such cases.

Recommended Reading

1. J. Gordon Arbuckle, M. E. Bosco, D. R. Case, E. P. Laws, J. C. Martin, M. L. Miller, R. V. Randle, R. G. Stoll, T. F. P. Sullivan, T. A. Vanderver, Jr., and P. A. J. Wilson, *Environmental Law Handbook,* 10th ed., Government Institute, Inc., Rockville, Md., 1989.

2. John-Mark Stensvaag, *Hazardous Waste Law and Practice,* vols. 1 and 2, Wiley, New York, 1990.

3. J. L. Leiter (ed.), *Underground Storage Tank Guide,* Thompson Publishing Group, Salisbury, Md., 1990.

4. "Hazardous Waste and Superfund 1991," American Bar Association, ABA satellite seminar, May 9, 1991.

5. Jennifer L. Machlin and Tammie R. Young, *Managing Environmental Risk—Real Estate Business and Transactions,* Clark Boardman Company, New York, 1990.

Important Cases

1. *Shell Oil v. EPA,* No. 80-1532 (D.C. Cir. Dec. 6, 1991).

2. *Hazardous Waste Treatment Council v. EPA,* 861 F. 2d 270 (D.C. Cir. 1988).

3. *American Mining Congress v. EPA,* 824 F. 2d 1177 (D.C. Cir. 1987).

4. *United States v. Vertac Chemical Corporation,* 489 F. Supp. 870 (E.D. Ark. 1980).

APPENDIX 5A   **Lists of Hazardous Wastes**

1. Hazardous wastes from nonspecific sources.

(a) The following solid wastes are listed hazardous wastes from nonspecific sources unless they are excluded under §§260.20 and 260.22 and listed in appendix IX.

Industry and EPA hazardous waste No.	Hazardous waste	Hazard code
Generic:		
F001.....................................	The following spent halogenated solvents used in degreasing: Tetrachloroethylene, trichloroethylene, methylene chloride, 1,1,1-trichloroethane, carbon tetrachloride, and chlorinated fluorocarbons; all spent solvent mixtures/blends used in degreasing containing, before use, a total of ten percent or more (by volume) of one or more of the above halogenated solvents or those solvents listed in F002, F004, and F005; and still bottoms from the recovery of these spent solvents and spent solvent mixtures.	(T)
F002.....................................	The following spent halogenated solvents: Tetrachloroethylene, methylene chloride, trichloroethylene, 1,1,1-trichloroethane, chlorobenzene, 1,1,2-trichloro-1,2,2-trifluoroethane, ortho-dichlorobenzene, trichlorofluoromethane, and 1,1,2-trichloroethane; all spent solvent mixtures/blends containing, before use, a total of ten percent or more (by volume) of one or more of the above halogenated solvents or those listed in F001, F004, or F005; and still bottoms from the recovery of these spent solvents and spent solvent mixtures.	(T)

Industry and EPA hazardous waste No.	Hazardous waste	Hazard code
F003	The following spent non-halogenated solvents: Xylene, acetone, ethyl acetate, ethyl benzene, ethyl ether, methyl isobutyl ketone, n-butyl ether, cyclohexanone, and methanol; all spent solvent mixtures/blends containing, before use, only the above spent non-halogenated solvents; and all spent solvent mixtures/blends containing, before use, one or more of the above non-halogenated solvents, and, a total of ten percent or more (by volume) of one or more of those solvents listed in F001, F002, F004, and F005; and still bottoms from the recovery of these spent solvents and spent solvent mixtures.	(I)*
F004	The following spent non-halogenated solvents: Cresols and cresylic acid, and nitrobenzene; all spent solvent mixtures/blends containing, before use, a total of ten percent or more (by volume) of one or more of the above non-halogenated solvents or those solvents listed in F001, F002, and F005; and still bottoms from the recovery of these spent solvents and spent solvent mixtures.	(T)
F005	The following spent non-halogenated solvents: Toluene, methyl ethyl ketone, carbon disulfide, isobutanol, pyridine, benzene, 2-ethoxyethanol, and 2-nitropropane; all spent solvent mixtures/blends containing, before use, a total of ten percent or more (by volume) of one or more of the above non-halogenated solvents or those solvents listed in F001, F002, or F004; and still bottoms from the recovery of these spent solvents and spent solvent mixtures.	(I,T)
F006	Wastewater treatment sludges from electroplating operations except from the following processes: (1) Sulfuric acid anodizing of aluminum; (2) tin plating on carbon steel; (3) zinc plating (segregated basis) on carbon steel; (4) aluminum or zinc-aluminum plating on carbon steel; (5) cleaning/stripping associated with tin, zinc and aluminum plating on carbon steel; and (6) chemical etching and milling of aluminum.	(T)
F007	Spent cyanide plating bath solutions from electroplating operations............................	(R, T)
F008	Plating bath residues from the bottom of plating baths from electroplating operations where cyanides are used in the process.	(R, T)
F009	Spent stripping and cleaning bath solutions from electroplating operations where cyanides are used in the process.	(R, T)
F010	Quenching bath residues from oil baths from metal heat treating operations where cyanides are used in the process.	(R, T)
F011	Spent cyanide solutions from salt bath pot cleaning from metal heat treating operations.	(R, T)
F012	Quenching waste water treatment sludges from metal heat treating operations where cyanides are used in the process.	(T)
F019	Wastewater treatment sludges from the chemical conversion coating of aluminum except from zirconium phosphating in aluminum can washing when such phosphating is an exclusive conversion coating process.	(T)
F020	Wastes (except wastewater and spent carbon from hydrogen chloride purification) from the production or manufacturing use (as a reactant, chemical intermediate, or component in a formulating process) of tri- or tetrachlorophenol, or of intermediates used to produce their pesticide derivatives. (This listing does not include wastes from the production of Hexachlorophene from highly purified 2,4,5-trichlorophenol.).	(H)
F021	Wastes (except wastewater and spent carbon from hydrogen chloride purification) from the production or manufacturing use (as a reactant, chemical intermediate, or component in a formulating process) of pentachlorophenol, or of intermediates used to produce its derivatives.	(H)
F022	Wastes (except wastewater and spent carbon from hydrogen chloride purification) from the manufacturing use (as a reactant, chemical intermediate, or component in a formulating process) of tetra-, penta-, or hexachlorobenzenes under alkaline conditions.	(H)
F023	Wastes (except wastewater and spent carbon from hydrogen chloride purification) from the production of materials on equipment previously used for the production or manufacturing use (as a reactant, chemical intermediate, or component in a formulating process) of tri- and tetrachlorophenols. (This listing does not include wastes from equipment used only for the production or use of Hexachlorophene from highly purified 2,4,5-trichlorophenol.).	(H)
F024	Process wastes, including but not limited to, distillation residues, heavy ends, tars, and reactor clean-out wastes, from the production of certain chlorinated aliphatic hydrocarbons by free radical catalyzed processes. These chlorinated aliphatic hydrocarbons are those having carbon chain lengths ranging from one to and including five, with varying amounts and positions of chlorine substitution. (This listing does not include wastewaters, wastewater treatment sludges, spent catalysts, and wastes listed in § 261.31 or § 261.32.).	(T)
F025	Condensed light ends, spent filters and filter aids, and spent desiccant wastes from the production of certain chlorinated aliphatic hydrocarbons, by free radical catalyzed processes. These chlorinated aliphatic hydrocarbons are those having carbon chain lengths ranging from one to and including five, with varying amounts and positions of chlorine substitution.	(T)

Industry and EPA hazardous waste No.	Hazardous waste	Hazard code
F026...................................	Wastes (except wastewater and spent carbon from hydrogen chloride purification) from the production of materials on equipment previously used for the manufacturing use (as a reactant, chemical intermediate, or component in a formulating process) of tetra-, penta-, or hexachlorobenzene under alkaline conditions.	(H)
F027...................................	Discarded unused formulations containing tri-, tetra-, or pentachlorophenol or discarded unused formulations containing compounds derived from these chlorophenols. (This listing does not include formulations containing Hexachlorophene sythesized from prepurified 2,4,5-trichlorophenol as the sole component.).	(H)
F028...................................	Residues resulting from the incineration or thermal treatment of soil contaminated with EPA Hazardous Waste Nos. F020, F021, F022, F023, F026, and F027.	(T)
F032 ¹	Wastewaters, process residuals, preservative drippage, and spent formulations from wood preserving processes generated at plants that currently use or have previously used chlorophenolic formulations (except potentially cross-contaminated wastes that have had the F032 waste code deleted in accordance with § 261.35 of this chapter and where the generator does not resume or initiate use of chlorophenolic formulations). This listing does not include K001 bottom sediment sludge from the treatment of wastewater from wood preserving processes that use creosote and/or pentachlorophenol. (NOTE: The listing of wastewaters that have not come into contact with process contaminants is stayed administratively. The listing for plants that have previously used chlorophenolic formulations is administratively stayed whenever these wastes are covered by the F034 or F035 listings. These stays will remain in effect until further administrative action is taken.).	(T)
F034 ¹	Wastewaters, process residuals, preservative drippage, and spent formulations from wood preserving process generated at plants that use creosote formulations. This listing does not include K001 bottom sediment sludge from the treatment of wastewater from wood preserving processes that use creosote and/or pentachlorophenol. (NOTE: The listing of wastewaters that have not come into contact with process contaminants is stayed administratively. The stay will remain in effect until further administrative action is taken.).	(T)
F035 ¹	Wastewaters, process residuals, preservative drippage, and spent formulations from wood preserving process generated at plants that use inorganic preservatives containing arsenic or chromium. This listing does not include K001 bottom sediment sludge from the treatment of wastewater from wood preserving processes that use creosote and/or pentachlorophenol. (NOTE: The listing of wastewaters that have not come into contact with process contaminants is stayed administratively. The stay will remain in effect until further administrative action is taken.).	(T)
F037...................................	Petroleum refinery primary oil/water/solids separation sludge—Any sludge generated from the gravitational separation of oil/water/solids during the storage or treatment of process wastewaters and oily cooling wastewaters from petroleum refineries. Such sludges include, but are not limited to, those generated in: oil/water/solids separators; tanks and impoundments; ditches and other conveyances; sumps; and stormwater units receiving dry weather flow. Sludge generated in stormwater units that do not receive dry weather flow, sludges generated from non-contact once-through cooling waters segregated for treatment from other process or oily cooling waters, sludges generated in aggressive biological treatment units as defined in § 261.31(b)(2) (including sludges generated in one or more additional units after wastewaters have been treated in aggressive biological treatment units) and K051 wastes are not included in this listing.	(T)
F038...................................	Petroleum refinery secondary (emulsified) oil/water/solids separation sludge—Any sludge and/or float generated from the physical and/or chemical separation of oil/water/solids in process wastewaters and oily cooling wastewaters from petroleum refineries. Such wastes include, but are not limited to, all sludges and floats generated in: induced air flotation (IAF) units, tanks and impoundments, and all sludges generated in DAF units. Sludges generated in stormwater units that do not receive dry weather flow, sludges generated from non-contact once-through cooling waters segregated for treatment from other process or oily cooling waters, sludges and floats generated in aggressive biological treatment units as defined in § 261.31(b)(2) (including sludges and floats generated in one or more additional units after wastewaters have been treated in aggressive biological treatment units) and F037, K048, and K051 wastes are not included in this listing.	(T)
F039...................................	Leachate (liquids that have percolated through land disposed wastes) resulting from the disposal of more than one restricted waste classified as hazardous under subpart D of this part. (Leachate resulting from the disposal of one or more of the following EPA Hazardous Wastes and no other Hazardous Wastes retains its EPA Hazardous Waste Number(s): F020, F021, F022, F026, F027, and/or F028.).	(T)

¹ The F032, F034, and F305 listings are administratively stayed with respect to the process area receiving drippage of these wastes provided persons desiring to continue operating notify EPA by August 6, 1991 of their intent to upgrade or install drip pads, and by November 6, 1991 provide evidence to EPA that they have adequate financing to pay for drip pad upgrades or installation, as provided in the administrative stay. The stay of the listings will remain in effect until February 6, 1992 for existing drip pads and until May 6, 1992 for new drip pads.
*(I,T) should be used to specify mixtures containing ignitable and toxic constituents.

2. Hazardous wastes from specific sources.
The following solid wastes are listed hazardous wastes from specific sources unless they are excluded under §§260.20 and 260.22 and listed in appendix IX.

Industry and EPA hazardous waste No.	Hazardous waste	Hazard code
Wood preservation: K001	Bottom sediment sludge from the treatment of wastewaters from wood preserving processes that use creosote and/or pentachlorophenol.	(T)
Inorganic pigments:		
K002	Wastewater treatment sludge from the production of chrome yellow and orange pigments.	(T)
K003	Wastewater treatment sludge from the production of molybdate orange pigments	(T)
K004	Wastewater treatment sludge from the production of zinc yellow pigments......................	(T)
K005	Wastewater treatment sludge from the production of chrome green pigments	(T)
K006	Wastewater treatment sludge from the production of chrome oxide green pigments (anhydrous and hydrated).	(T)
K007	Wastewater treatment sludge from the production of iron blue pigments	(T)
K008	Oven residue from the production of chrome oxide green pigments...............................	(T)
Organic chemicals:		
K009	Distillation bottoms from the production of acetaldehyde from ethylene	(T)
K010	Distillation side cuts from the production of acetaldehyde from ethylene........................	(T)
K011	Bottom stream from the wastewater stripper in the production of acrylonitrile..................	(R, T)
K013	Bottom stream from the acetonitrile column in the production of acrylonitrile..................	(R, T)
K014	Bottoms from the acetonitrile purification column in the production of acrylonitrile	(T)
K015	Still bottoms from the distillation of benzyl chloride..	(T)
K016	Heavy ends or distillation residues from the production of carbon tetrachloride.............	(T)
K017	Heavy ends (still bottoms) from the purification column in the production of epichlorohydrin.	(T)

Industry and EPA hazardous waste No.	Hazardous waste	Hazard code
K018	Heavy ends from the fractionation column in ethyl chloride production	(T)
K019	Heavy ends from the distillation of ethylene dichloride in ethylene dichloride production.	(T)
K020	Heavy ends from the distillation of vinyl chloride in vinyl chloride monomer production.	(T)
K021	Aqueous spent antimony catalyst waste from fluoromethanes production	(T)
K022	Distillation bottom tars from the production of phenol/acetone from cumene	(T)
K023	Distillation light ends from the production of phthalic anhydride from naphthalene	(T)
K024	Distillation bottoms from the production of phthalic anhydride from naphthalene	(T)
K025	Distillation bottoms from the production of nitrobenzene by the nitration of benzene	(T)
K026	Stripping still tails from the production of methy ethyl pyridines	(T)
K027	Centrifuge and distillation residues from toluene diisocyanate production	(R, T)
K028	Spent catalyst from the hydrochlorinator reactor in the production of 1,1,1-trichloroethane.	(T)
K029	Waste from the product steam stripper in the production of 1,1,1-trichloroethane	(T)
K030	Column bottoms or heavy ends from the combined production of trichloroethylene and perchloroethylene.	(T)
K083	Distillation bottoms from aniline production	(T)
K085	Distillation or fractionation column bottoms from the production of chlorobenzenes	(T)
K093	Distillation light ends from the production of phthalic anhydride from ortho-xylene	(T)
K094	Distillation bottoms from the production of phthalic anhydride from ortho-xylene	(T)
K095	Distillation bottoms from the production of 1,1,1-trichloroethane	(T)
K096	Heavy ends from the heavy ends column from the production of 1,1,1-trichloroethane.	(T)
K103	Process residues from aniline extraction from the production of aniline	(T)
K104	Combined wastewater streams generated from nitrobenzene/aniline production	(T)
K105	Separated aqueous stream from the reactor product washing step in the production of chlorobenzenes.	(T)
K107	Column bottoms from product separation from the production of 1,1-dimethylhydrazine (UDMH) from carboxylic acid hydrazines.	(C,T)
K108	Condensed column overheads from product separation and condensed reactor vent gases from the production of 1,1-dimethylhydrazine (UDMH) from carboxylic acid hydrazides.	(I,T)
K109	Spent filter cartridges from product purification from the production of 1,1-dimethylhydrazine (UDMH) from carboxylic acid hydrazides.	(T)
K110	Condensed column overheads from intermediate separation from the production of 1,1-dimethylhydrazine (UDMH) from carboxylic acid hydrazides.	(T)
K111	Product washwaters from the production of dinitrotoluene via nitration of toluene	(C,T)
K112	Reaction by-product water from the drying column in the production of toluenediamine via hydrogenation of dinitrotoluene.	(T)
K113	Condensed liquid light ends from the purification of toluenediamine in the production of toluenediamine via hydrogenation of dinitrotoluene.	(T)
K114	Vicinals from the purification of toluenediamine in the production of toluenediamine via hydrogenation of dinitrotoluene.	(T)
K115	Heavy ends from the purification of toluenediamine in the production of toluenediamine via hydrogenation of dinitrotoluene.	(T)
K116	Organic condensate from the solvent recovery column in the production of toluene diisocyanate via phosgenation of toluenediamine.	(T)
K117	Wastewater from the reactor vent gas scrubber in the production of ethylene dibromide via bromination of ethene.	(T)
K118	Spent adsorbent solids from purification of ethylene dibromide in the production of ethylene dibromide via bromination of ethene.	(T)
K136	Still bottoms from the purification of ethylene dibromide in the production of ethylene dibromide via bromination of ethene.	(T)
Inorganic chemicals:		
K071	Brine purification muds from the mercury cell process in chlorine production, where separately prepurified brine is not used.	(T)
K073	Chlorinated hydrocarbon waste from the purification step of the diaphragm cell process using graphite anodes in chlorine production.	(T)
K106	Wastewater treatment sludge from the mercury cell process in chlorine production	(T)
Pesticides:		
K031	By-product salts generated in the production of MSMA and cacodylic acid	(T)
K032	Wastewater treatment sludge from the production of chlordane	(T)
K033	Wastewater and scrub water from the chlorination of cyclopentadiene in the production of chlordane.	(T)
K034	Filter solids from the filtration of hexachlorocyclopentadiene in the production of chlordane.	(T)
K035	Wastewater treatment sludges generated in the production of creosote	(T)
K036	Still bottoms from toluene reclamation distillation in the production of disulfoton	(T)
K037	Wastewater treatment sludges from the production of disulfoton	(T)
K038	Wastewater from the washing and stripping of phorate production	(T)

Industry and EPA hazardous waste No.	Hazardous waste	Hazard code
K039	Filter cake from the filtration of diethylphosphorodithioic acid in the production of phorate.	(T)
K040	Wastewater treatment sludge from the production of phorate	(T)
K041	Wastewater treatment sludge from the production of toxaphene	(T)
K042	Heavy ends or distillation residues from the distillation of tetrachlorobenzene in the production of 2,4,5-T.	(T)
K043	2,6-Dichlorophenol waste from the production of 2,4-D	(T)
K097	Vacuum stripper discharge from the chlordane chlorinator in the production of chlordane.	(T)
K098	Untreated process wastewater from the production of toxaphene	(T)
K099	Untreated wastewater from the production of 2,4-D	(T)
K123	Process wastewater (including supernates, filtrates, and washwaters) from the production of ethylenebisdithiocarbamic acid and its salt.	(T)
K124	Reactor vent scrubber water from the production of ethylenebisdithiocarbamic acid and its salts.	(C, T)
K125	Filtration, evaporation, and centrifugation solids from the production of ethylenebisdithiocarbamic acid and its salts.	(T)
K126	Baghouse dust and floor sweepings in milling and packaging operations from the production or formulation of ethylenebisdithiocarbamic acid and its salts.	(T)
K131	Wastewater from the reactor and spent sulfuric acid from the acid dryer from the production of methyl bromide.	(C, T)
K132	Spent absorbent and wastewater separator solids from the production of methyl bromide.	(T)
Explosives:		
K044	Wastewater treatment sludges from the manufacturing and processing of explosives	(R)
K045	Spent carbon from the treatment of wastewater containing explosives	(R)
K046	Wastewater treatment sludges from the manufacturing, formulation and loading of lead-based initiating compounds.	(T)
K047	Pink/red water from TNT operations	(R)
Petroleum refining:		
K048	Dissolved air flotation (DAF) float from the petroleum refining industry	(T)
K049	Slop oil emulsion solids from the petroleum refining industry	(T)
K050	Heat exchanger bundle cleaning sludge from the petroleum refining industry	(T)
K051	API separator sludge from the petroleum refining industry	(T)
K052	Tank bottoms (leaded) from the petroleum refining industry	(T)
Iron and steel:		
K061	Emission control dust/sludge from the primary production of steel in electric furnaces.	(T)
K062	Spent pickle liquor generated by steel finishing operations of facilities within the iron and steel industry (SIC Codes 331 and 332).	(C,T)
Primary copper:		
K064	Acid plant blowdown slurry/sludge resulting from the thickening of blowdown slurry from primary copper production.	(T)
Primary lead:		
K065	Surface impoundment solids contained in and dredged from surface impoundments at primary lead smelting facilities.	(T)
Primary zinc:		
K066	Sludge from treatment of process wastewater and/or acid plant blowdown from primary zinc production.	(T)
Primary aluminum:		
K088	Spent potliners from primary aluminum reduction	(T)
Ferroalloys:		
K090	Emission control dust or sludge from ferrochromiumsilicon production	(T)
K091	Emission control dust or sludge from ferrochromium production	(T)
Secondary lead:		
K069	Emission control dust/sludge from secondary lead smelting. (NOTE: This listing is stayed administratively for sludge generated from secondary acid scrubber systems. The stay will remain in effect until further administrative action is taken. If EPA takes further action effecting this stay, EPA will publish a notice of the action in the **Federal Register**.	(T)
K100	Waste leaching solution from acid leaching of emission control dust/sludge from secondary lead smelting.	(T)
Veterinary pharmaceuticals:		
K084	Wastewater treatment sludges generated during the production of veterinary pharmaceuticals from arsenic or organo-arsenic compounds.	(T)
K101	Distillation tar residues from the distillation of aniline-based compounds in the production of veterinary pharmaceuticals from arsenic or organo-arsenic compounds.	(T)
K102	Residue from the use of activated carbon for decolorization in the production of veterinary pharmaceuticals from arsenic or organo-arsenic compounds.	(T)

Industry and EPA hazardous waste No.	Hazardous waste	Hazard code
Ink formulation:		
K086	Solvent washes and sludges, caustic washes and sludges, or water washes and sludges from cleaning tubs and equipment used in the formulation of ink from pigments, driers, soaps, and stabilizers containing chromium and lead.	(T)
Coking:		
K060	Ammonia still lime sludge from coking operations...	(T)
K087	Decanter tank tar sludge from coking operations...	(T)

3. Discarded commercial chemical products, off-specification species, container residues, and spill residues thereof.

a. Acute Hazardous Wastes

Haz-ardous waste No.	Chemical abstracts No.	Substance
P023	107–20–0	Acetaldehyde, chloro-
P002	591–08–2	Acetamide, N-(aminothioxomethyl)-
P057	640–19–7	Acetamide, 2-fluoro-
P058	62–74–8	Acetic acid, fluoro-, sodium salt
P002	591–08–2	1-Acetyl-2-thiourea
P003	107–02–8	Acrolein
P070	116–06–3	Aldicarb
P004	309–00–2	Aldrin
P005	107–18–6	Allyl alcohol
P006	20859–73–8	Aluminum phosphide (R,T)
P007	2763–96–4	5-(Aminomethyl)-3-isoxazolol
P008	504–24–5	4-Aminopyridine
P009	131–74–8	Ammonium picrate (R)
P119	7803–55–6	Ammonium vanadate
P099	506–61–6	Argentate(1-), bis(cyano-C)-, potassium
P010	7778–39–4	Arsenic acid H_3AsO_4
P012	1327–53–3	Arsenic oxide As_2O_3
P011	1303–28–2	Arsenic oxide As_2O_5
P011	1303–28–2	Arsenic pentoxide
P012	1327–53–3	Arsenic trioxide
P038	692–42–2	Arsine, diethyl-
P036	696–28–6	Arsonous dichloride, phenyl-
P054	151–56–4	Aziridine
P067	75–55–8	Aziridine, 2-methyl-
P013	542–62–1	Barium cyanide
P024	106–47–8	Benzenamine, 4-chloro-
P077	100–01–6	Benzenamine, 4-nitro-
P028	100–44–7	Benzene, (chloromethyl)-
P042	51–43–4	1,2-Benzenediol, 4-[1-hydroxy-2-(methylamino)ethyl]-, (R)-
P046	122–09–8	Benzeneethanamine, alpha,alpha-dimethyl-
P014	108–98–5	Benzenethiol
P001	[1] 81–81–2	2H-1-Benzopyran-2-one, 4-hydroxy-3-(3-oxo-1-phenylbutyl)-, & salts, when present at concentrations greater than 0.3%
P028	100–44–7	Benzyl chloride
P015	7440–41–7	Beryllium
P017	598–31–2	Bromoacetone
P018	357–57–3	Brucine
P045	39196–18–4	2-Butanone, 3,3-dimethyl-1-(methylthio)-, O-[methylamino)carbonyl] oxime
P021	592–01–8	Calcium cyanide
P021	592–01–8	Calcium cyanide Ca(CN)$_2$
P022	75–15–0	Carbon disulfide
P095	75–44–5	Carbonic dichloride
P023	107–20–0	Chloroacetaldehyde
P024	106–47–8	p-Chloroaniline
P026	5344–82–1	1-(o-Chlorophenyl)thiourea
P027	542–76–7	3-Chloropropionitrile
P029	544–92–3	Copper cyanide
P029	544–92–3	Copper cyanide Cu(CN)
P030	....	Cyanides (soluble cyanide salts), not otherwise specified
P031	460–19–5	Cyanogen
P033	506–77–4	Cyanogen chloride

Haz-ardous waste No.	Chemical abstracts No.	Substance
P033	506–77–4	Cyanogen chloride (CN)Cl
P034	131–89–5	2-Cyclohexyl-4,6-dinitrophenol
P016	542–88–1	Dichloromethyl ether
P036	696–28–6	Dichlorophenylarsine
P037	60–57–1	Dieldrin
P038	692–42–2	Diethylarsine
P041	311–45–5	Diethyl-p-nitrophenyl phosphate
P040	297–97–2	O,O-Diethyl O-pyrazinyl phosphorothioate
P043	55–91–4	Diisopropylfluorophosphate (DFP)
P004	309–00–2	1,4,5,8-Dimethanonaphthalene, 1,2,3,4,10,10-hexa- chloro-1,4,4a,5,8,8a,-hexahydro-, (1alpha,4alpha,4abeta,5alpha,8alpha,8abeta)-
P060	465–73–6	1,4,5,8-Dimethanonaphthalene, 1,2,3,4,10,10-hexa- chloro-1,4,4a,5,8,8a-hexahydro-, (1alpha,4alpha,4abeta,5beta,8beta,8abeta)-
P037	60–57–1	2,7:3,6-Dimethanonaphth[2,3-b]oxirene, 3,4,5,6,9,9-hexachloro-1a,2,2a,3,6,6a,7,7a-octahydro-, (1aalpha,2beta,2aalpha,3beta,6beta,6aalpha,7beta, 7aalpha)-
P051	[1] 72–20–8	2,7:3,6-Dimethanonaphth [2,3-b]oxirene, 3,4,5,6,9,9-hexachloro-1a,2,2a,3,6,6a,7,7a-octahydro-, (1aalpha,2beta,2abeta,3alpha,6alpha,6abeta,7beta, 7aalpha)-, & metabolites
P044	60–51–5	Dimethoate
P046	122–09–8	alpha,alpha-Dimethylphenethylamine
P047	[1] 534–52–1	4,6-Dinitro-o-cresol, & salts
P048	51–28–5	2,4-Dinitrophenol
P020	88–85–7	Dinoseb
P085	152–16–9	Diphosphoramide, octamethyl-
P111	107–49–3	Diphosphoric acid, tetraethyl ester
P039	298–04–4	Disulfoton
P049	541–53–7	Dithiobiuret
P050	115–29–7	Endosulfan
P088	145–73–3	Endothall
P051	72–20–8	Endrin
P051	72–20–8	Endrin, & metabolites
P042	51–43–4	Epinephrine
P031	460–19–5	Ethanedinitrile
P066	16752–77–5	Ethanimidothioic acid, N-[[(methylamino)carbonyl]oxy]-, methyl ester
P101	107–12–0	Ethyl cyanide
P054	151–56–4	Ethyleneimine
P097	52–85–7	Famphur
P056	7782–41–4	Fluorine
P057	640–19–7	Fluoroacetamide
P058	62–74–8	Fluoroacetic acid, sodium salt
P065	628–86–4	Fulminic acid, mercury(2+) salt (R,T)
P059	76–44–8	Heptachlor
P062	757–58–4	Hexaethyl tetraphosphate
P116	79–19–6	Hydrazinecarbothioamide
P068	60–34–4	Hydrazine, methyl-
P063	74–90–8	Hydrocyanic acid
P063	74–90–8	Hydrogen cyanide
P096	7803–51–2	Hydrogen phosphide
P060	465–73–6	Isodrin
P007	2763–96–4	3(2H)-Isoxazolone, 5-(aminomethyl)-
P092	62–38–4	Mercury, (acetato-O)phenyl-
P065	628–86–4	Mercury fulminate (R,T)
P082	62–75–9	Methanamine, N-methyl-N-nitroso-
P064	624–83–9	Methane, isocyanato-
P016	542–88–1	Methane, oxybis[chloro-
P112	509–14–8	Methane, tetranitro- (R)
P118	75–70–7	Methanethiol, trichloro-
P050	115–29–7	6,9-Methano-2,4,3-benzodioxathiepin, 6,7,8,9,10,10- hexachloro-1,5,5a,6,9,9a-hexahydro-, 3-oxide
P059	76–44–8	4,7-Methano-1H-indene, 1,4,5,6,7,8,8-heptachloro- 3a,4,7,7a-tetrahydro-
P066	16752–77–5	Methomyl
P068	60–34–4	Methyl hydrazine
P064	624–83–9	Methyl isocyanate
P069	75–86–5	2-Methyllactonitrile
P071	298–00–0	Methyl parathion
P072	86–88–4	alpha-Naphthylthiourea
P073	13463–39–3	Nickel carbonyl
P073	13463–39–3	Nickel carbonyl Ni(CO)₄, (T-4)-

Hazardous waste No.	Chemical abstracts No.	Substance
P074	557–19–7	Nickel cyanide
P074	557–19–7	Nickel cynaide Ni(CN)$_2$
P075	[1] 54–11–5	Nicotine,.& salts
P076	10102–43–9	Nitric oxide
P077	100–01–6	p-Nitroaniline
P078	10102–44–0	Nitrogen dioxide
P076	10102–43–9	Nitrogen oxide NO
P078	10102–44–0	Nitrogen oxide NO$_2$
P081	55–63–0	Nitroglycerine (R)
P082	62–75–9	N-Nitrosodimethylamine
P084	4549–40–0	N-Nitrosomethylvinylamine
P085	152–16–9	Octamethylpyrophosphoramide
P087	20816–12–0	Osmium oxide OsO$_4$, (T-4)-
P087	20816–12–0	Osmium tetroxide
P088	145–73–3	7-Oxabicyclo[2.2.1]heptane-2,3-dicarboxylic acid
P089	56–38–2	Parathion
P034	131–89–5	Phenol, 2-cyclohexyl-4,6-dinitro-
P048	51–28–5	Phenol, 2,4-dinitro-
P047	[1] 534–52–1	Phenol, 2-methyl-4,6-dinitro-, & salts
P020	88–85–7	Phenol, 2-(1-methylpropyl)-4,6-dinitro-
P009	131–74–8	Phenol, 2,4,6-trinitro-, ammonium salt (R)
P092	62–38–4	Phenylmercury acetate
P093	103–85–5	Phenylthiourea
P094	298–02–2	Phorate
P095	75–44–5	Phosgene
P096	7803–51–2	Phosphine
P041	311–45–5	Phosphoric acid, diethyl 4-nitrophenyl ester
P039	298–04–4	Phosphorodithioic acid, O,O-diethyl S-[2-(ethylthio)ethyl] ester
P094	298–02–2	Phosphorodithioic acid, O,O-diethyl S-[(ethylthio)methyl] ester
P044	60–51–5	Phosphorodithioic acid, O,O-dimethyl S-[2-(methylamino)-2-oxoethyl] ester
P043	55–91–4	Phosphorofluoridic acid, bis(1-methylethyl) ester
P089	56–38–2	Phosphorothioic acid, O,O-diethyl O-(4-nitrophenyl) ester
P040	297–97–2	Phosphorothioic acid, O,O-diethyl O-pyrazinyl ester
P097	52–85–7	Phosphorothioic acid, O-[4-[(dimethylamino)sulfonyl]phenyl] O,O-dimethyl ester
P071	298–00–0	Phosphorothioic acid, O,O,-dimethyl O-(4-nitrophenyl) ester
P110	78–00–2	Plumbane, tetraethyl-
P098	151–50–8	Potassium cyanide
P098	151–50–8	Potassium cyanide K(CN)
P099	506–61–6	Potassium silver cyanide
P070	116–06–3	Propanal, 2-methyl-2-(methylthio)-, O-[(methylamino)carbonyl]oxime
P101	107–12–0	Propanenitrile
P027	542–76–7	Propanenitrile, 3-chloro-
P069	75–86–5	Propanenitrile, 2-hydroxy-2-methyl-
P081	55–63–0	1,2,3-Propanetriol, trinitrate (R)
P017	598–31–2	2-Propanone, 1-bromo-
P102	107–19–7	Propargyl alcohol
P003	107–02–8	2-Propenal
P005	107–18–6	2-Propen-1-ol
P067	75–55–8	1,2-Propylenimine
P102	107–19–7	2-Propyn-1-ol
P008	504–24–5	4-Pyridinamine
P075	[1] 54–11–5	Pyridine, 3-(1-methyl-2-pyrrolidinyl)-, (S)-, & salts
P114	12039–52–0	Selenious acid, dithallium(1+) salt
P103	630–10–4	Selenourea
P104	506–64–9	Silver cyanide
P104	506–64–9	Silver cyanide Ag(CN)
P105	26628–22–8	Sodium azide
P106	143–33–9	Sodium cyanide
P106	143–33–9	Sodium cyanide Na(CN)
P108	[1] 57–24–9	Strychnidin-10-one, & salts
P018	357–57–3	Strychnidin-10-one, 2,3-dimethoxy-
P108	[1] 57–24–9	Strychnine, & salts
P115	7446–18–6	Sulfuric acid, dithallium(1+) salt
P109	3689–24–5	Tetraethyldithiopyrophosphate
P110	78–00–2	Tetraethyl lead
P111	107–49–3	Tetraethyl pyrophosphate

Haz-ardous waste No.	Chemical abstracts No.	Substance
P112	509–14–8	Tetranitromethane (R)
P062	757–58–4	Tetraphosphoric acid, hexaethyl ester
P113	1314–32–5	Thallic oxide
P113	1314–32–5	Thallium oxide Tl$_2$O$_3$
P114	12039–52–0	Thallium(I) selenite
P115	7446–18–6	Thallium(I) sulfate
P109	3689–24–5	Thiodiphosphoric acid, tetraethyl ester
P045	39196–18–4	Thiofanox
P049	541–53–7	Thioimidodicarbonic diamide [(H$_2$N)C(S)]$_2$NH
P014	108–98–5	Thiophenol
P116	79–19–6	Thiosemicarbazide
P026	5344–82–1	Thiourea, (2-chlorophenyl)-
P072	86–88–4	Thiourea, 1-naphthalenyl-
P093	103–85–5	Thiourea, phenyl-
P123	8001–35–2	Toxaphene
P118	75–70–7	Trichloromethanethiol
P119	7803–55–6	Vanadic acid, ammonium salt
P120	1314–62–1	Vanadium oxide V$_2$O$_5$
P120	1314–62–1	Vanadium pentoxide
P084	4549–40–0	Vinylamine, N-methyl-N-nitroso-
P001	[1] 81–81–2	Warfarin, & salts, when present at concentrations greater than 0.3%
P121	557–21–1	Zinc cyanide
P121	557–21–1	Zinc cyanide Zn(CN)$_2$
P122	1314–84–7	Zinc phosphide Zn$_3$P$_2$, when present at concentrations greater than 10% (R,T)

[1] CAS Number given for parent compound only.

b. Toxic Wastes (unless otherwise designated)

Haz-ardous waste No.	Chemical abstracts No.	Substance
U001	75–07–0	Acetaldehyde (I)
U034	75–87–6	Acetaldehyde, trichloro-
U187	62–44–2	Acetamide, N-(4-ethoxyphenyl)-
U005	53–96–3	Acetamide, N-9H-fluoren-2-yl-
U240	[1] 94–75–7	Acetic acid, (2,4-dichlorophenoxy)-, salts & esters
U112	141–78–6	Acetic acid ethyl ester (I)
U144	301–04–2	Acetic acid, lead(2+) salt
U214	563–68–8	Acetic acid, thallium(1+) salt
see F027	93–76–5	Acetic acid, (2,4,5-trichlorophenoxy)-
U002	67–64–1	Acetone (I)
U003	75–05–8	Acetonitrile (I,T)
U004	98–86–2	Acetophenone
U005	53–96–3	2-Acetylaminofluorene
U006	75–36–5	Acetyl chloride (C,R,T)
U007	79–06–1	Acrylamide
U008	79–10–7	Acrylic acid (I)
U009	107–13–1	Acrylonitrile
U011	61–82–5	Amitrole
U012	62–53–3	Aniline (I,T)
U136	75–60–5	Arsinic acid, dimethyl-
U014	492–80–8	Auramine

Haz-ardous waste No.	Chemical abstracts No.	Substance
U015	115-02-6	Azaserine
U010	50-07-7	Azirino[2',3':3,4]pyrrolo[1,2-a]indole-4,7-dione, 6-amino-8-[[(aminocarbonyl)oxy]methyl]-1,1a,2,8,8a,8b-hexahydro-8a-methoxy-5-methyl-, [1aS-(1aalpha, 8beta,8aalpha,8balpha)]-
U157	56-49-5	Benz[j]aceanthrylene, 1,2-dihydro-3-methyl-
U016	225-51-4	Benz[c]acridine
U017	98-87-3	Benzal chloride
U192	23950-58-5	Benzamide, 3,5-dichloro-N-(1,1-dimethyl-2-propynyl)-
U018	56-55-3	Benz[a]anthracene
U094	57-97-6	Benz[a]anthracene, 7,12-dimethyl-
U012	62-53-3	Benzenamine (I,T)
U014	492-80-8	Benzenamine, 4,4'-carbonimidoylbis[N,N-dimethyl-
U049	3165-93-3	Benzenamine, 4-chloro-2-methyl-, hydrochloride
U093	60-11-7	Benzenamine, N,N-dimethyl-4-(phenylazo)-
U328	95-53-4	Benzenamine, 2-methyl-
U353	106-49-0	Benzenamine, 4-methyl-
U158	101-14-4	Benzenamine, 4,4'-methylenebis[2-chloro-
U222	636-21-5	Benzenamine, 2-methyl-, hydrochloride
U181	99-55-8	Benzenamine, 2-methyl-5-nitro-
U019	71-43-2	Benzene (I,T)
U038	510-15-6	Benzeneacetic acid, 4-chloro-alpha-(4-chlorophenyl)-alpha-hydroxy-, ethyl ester
U030	101-55-3	Benzene, 1-bromo-4-phenoxy-
U035	305-03-3	Benzenebutanoic acid, 4-[bis(2-chloroethyl)amino]-
U037	108-90-7	Benzene, chloro-
U221	25376-45-8	Benzenediamine, ar-methyl-
U028	117-81-7	1,2-Benzenedicarboxylic acid, bis(2-ethylhexyl) ester
U069	84-74-2	1,2-Benzenedicarboxylic acid, dibutyl ester
U088	84-66-2	1,2-Benzenedicarboxylic acid, diethyl ester
U102	131-11-3	1,2-Benzenedicarboxylic acid, dimethyl ester
U107	117-84-0	1,2-Benzenedicarboxylic acid, dioctyl ester
U070	95-50-1	Benzene, 1,2-dichloro-
U071	541-73-1	Benzene, 1,3-dichloro-
U072	106-46-7	Benzene, 1,4-dichloro-
U060	72-54-8	Benzene, 1,1'-(2,2-dichloroethylidene)bis[4-chloro-
U017	98-87-3	Benzene, (dichloromethyl)-
U223	26471-62-5	Benzene, 1,3-diisocyanatomethyl- (R,T)
U239	1330-20-7	Benzene, dimethyl- (I,T)
U201	108-46-3	1,3-Benzenediol
U127	118-74-1	Benzene, hexachloro-
U056	110-82-7	Benzene, hexahydro- (I)
U220	108-88-3	Benzene, methyl-
U105	121-14-2	Benzene, 1-methyl-2,4-dinitro-
U106	606-20-2	Benzene, 2-methyl-1,3-dinitro-
U055	98-82-8	Benzene, (1-methylethyl)- (I)
U169	98-95-3	Benzene, nitro-
U183	608-93-5	Benzene, pentachloro-
U185	82-68-8	Benzene, pentachloronitro-
U020	98-09-9	Benzenesulfonic acid chloride (C,R)
U020	98-09-9	Benzenesulfonyl chloride (C,R)
U207	95-94-3	Benzene, 1,2,4,5-tetrachloro-
U061	50-29-3	Benzene, 1,1'-(2,2,2-trichloroethylidene)bis[4-chloro-
U247	72-43-5	Benzene, 1,1'-(2,2,2-trichloroethylidene)bis[4- methoxy-
U023	98-07-7	Benzene, (trichloromethyl)-
U234	99-35-4	Benzene, 1,3,5-trinitro-
U021	92-87-5	Benzidine
U202	¹81-07-2	1,2-Benzisothiazol-3(2H)-one, 1,1-dioxide, & salts
U203	94-59-7	1,3-Benzodioxole, 5-(2-propenyl)-
U141	120-58-1	1,3-Benzodioxole, 5-(1-propenyl)-
U090	94-58-6	1,3-Benzodioxole, 5-propyl-
U064	189-55-9	Benzo[rst]pentaphene
U248	¹81-81-2	2H-1-Benzopyran-2-one, 4-hydroxy-3-(3-oxo-1-phenyl-butyl)-, & salts, when present at concentrations of 0.3% or less
U022	50-32-8	Benzo[a]pyrene
U197	106-51-4	p-Benzoquinone
U023	98-07-7	Benzotrichloride (C,R,T)
U085	1464-53-5	2,2'-Bioxirane
U021	92-87-5	[1,1'-Biphenyl]-4,4'-diamine
U073	91-94-1	[1,1'-Biphenyl]-4,4'-diamine, 3,3'-dichloro-
U091	119-90-4	[1,1'-Biphenyl]-4,4'-diamine, 3,3'-dimethoxy-
U095	119-93-7	[1,1'-Biphenyl]-4,4'-diamine, 3,3'-dimethyl-
U225	75-25-2	Bromoform

Haz-ardous waste No.	Chemical abstracts No.	Substance
U030	101–55–3	4-Bromophenyl phenyl ether
U128	87–68–3	1,3-Butadiene, 1,1,2,3,4,4-hexachloro-
U172	924–16–3	1-Butanamine, N-butyl-N-nitroso-
U031	71–36–3	1-Butanol (I)
U159	78–93–3	2-Butanone (I,T)
U160	1338–23–4	2-Butanone, peroxide (R,T)
U053	4170–30–3	2-Butenal
U074	764–41–0	2-Butene, 1,4-dichloro- (I,T)
U143	303–34–4	2-Butenoic acid, 2-methyl-, 7-[[2,3-dihydroxy-2-(1-methoxyethyl)-3-methyl-1-oxobutoxy]methyl]-2,3,5,7a-tetrahydro-1H-pyrrolizin-1-yl ester, [1S-[1alpha(Z),7(2S*,3R*),7aalpha]]-
U031	71–36–3	n-Butyl alcohol (I)
U136	75–60–5	Cacodylic acid
U032	13765–19–0	Calcium chromate
U238	51–79–6	Carbamic acid, ethyl ester
U178	615–53–2	Carbamic acid, methylnitroso-, ethyl ester
U097	79–44–7	Carbamic chloride, dimethyl-
U114	[1] 111–54–6	Carbamodithioic acid, 1,2-ethanediylbis-, salts & esters
U062	2303–16–4	Carbamothioic acid, bis(1-methylethyl)-, S-(2,3-dichloro-2-propenyl) ester
U215	6533–73–9	Carbonic acid, dithallium(1+) salt
U033	353–50–4	Carbonic difluoride
U156	79–22–1	Carbonochloridic acid, methyl ester (I,T)
U033	353–50–4	Carbon oxyfluoride (R,T)
U211	56–23–5	Carbon tetrachloride
U034	75–87–6	Chloral
U035	305–03–3	Chlorambucil
U036	57–74–9	Chlordane, alpha & gamma isomers
U026	494–03–1	Chlornaphazin
U037	108–90–7	Chlorobenzene
U038	510–15–6	Chlorobenzilate
U039	59–50–7	p-Chloro-m-cresol
U042	110–75–8	2-Chloroethyl vinyl ether
U044	67–66–3	Chloroform
U046	107–30–2	Chloromethyl methyl ether
U047	91–58–7	beta-Chloronaphthalene
U048	95–57–8	o-Chlorophenol
U049	3165–93–3	4-Chloro-o-toluidine, hydrochloride
U032	13765–19–0	Chromic acid H_2CrO_4, calcium salt
U050	218–01–9	Chrysene
U051	....	Creosote
U052	1319–77–3	Cresol (Cresylic acid)
U053	4170–30–3	Crotonaldehyde
U055	98–82–8	Cumene (I)
U246	506–68–3	Cyanogen bromide (CN)Br
U197	106–51–4	2,5-Cyclohexadiene-1,4-dione
U056	110–82–7	Cyclohexane (I)
U129	58–89–9	Cyclohexane, 1,2,3,4,5,6-hexachloro-, (1alpha,2alpha,3beta,4alpha,5alpha,6beta)-
U057	108–94–1	Cyclohexanone (I)
U130	77–47–4	1,3-Cyclopentadiene, 1,2,3,4,5,5-hexachloro-
U058	50–18–0	Cyclophosphamide
U240	[1] 94–75–7	2,4-D, salts & esters
U059	20830–81–3	Daunomycin
U060	72–54–8	DDD
U061	50–29–3	DDT
U062	2303–16–4	Diallate
U063	53–70–3	Dibenz[a,h]anthracene
U064	189–55–9	Dibenzo[a,i]pyrene
U066	96–12–8	1,2-Dibromo-3-chloropropane
U069	84–74–2	Dibutyl phthalate
U070	95–50–1	o-Dichlorobenzene
U071	541–73–1	m-Dichlorobenzene
U072	106–46–7	p-Dichlorobenzene
U073	91–94–1	3,3'-Dichlorobenzidine
U074	764–41–0	1,4-Dichloro-2-butene (I,T)
U075	75–71–8	Dichlorodifluoromethane
U078	75–35–4	1,1-Dichloroethylene
U079	156–60–5	1,2-Dichloroethylene

Hazardous waste No.	Chemical abstracts No.	Substance
U025	111–44–4	Dichloroethyl ether
U027	108–60–1	Dichloroisopropyl ether
U024	111–91–1	Dichloromethoxy ethane
U081	120–83–2	2,4-Dichlorophenol
U082	87–65–0	2,6-Dichlorophenol
U084	542–75–6	1,3-Dichloropropene
U085	1464–53–5	1,2:3,4-Diepoxybutane (I,T)
U108	123–91–1	1,4-Diethyleneoxide
U028	117–81–7	Diethylhexyl phthalate
U086	1615–80–1	N,N'-Diethylhydrazine
U087	3288–58–2	O,O-Diethyl S-methyl dithiophosphate
U088	84–66–2	Diethyl phthalate
U089	56–53–1	Diethylstilbesterol
U090	94–58–6	Dihydrosafrole
U091	119–90–4	3,3'-Dimethoxybenzidine
U092	124–40–3	Dimethylamine (I)
U093	60–11–7	p-Dimethylaminoazobenzene
U094	57–97–6	7,12-Dimethylbenz[a]anthracene
U095	119–93–7	3,3'-Dimethylbenzidine
U096	80–15–9	alpha,alpha-Dimethylbenzylhydroperoxide (R)
U097	79–44–7	Dimethylcarbamoyl chloride
U098	57–14–7	1,1-Dimethylhydrazine
U099	540–73–8	1,2-Dimethylhydrazine
U101	105–67–9	2,4-Dimethylphenol
U102	131–11–3	Dimethyl phthalate
U103	77–78–1	Dimethyl sulfate
U105	121–14–2	2,4-Dinitrotoluene
U106	606–20–2	2,6-Dinitrotoluene
U107	117–84–0	Di-n-octyl phthalate
U108	123–91–1	1,4-Dioxane
U109	122–66–7	1,2-Diphenylhydrazine
U110	142–84–7	Dipropylamine (I)
U111	621–64–7	Di-n-propylnitrosamine
U041	106–89–8	Epichlorohydrin
U001	75–07–0	Ethanal (I)
U174	55–18–5	Ethanamine, N-ethyl-N-nitroso-
U155	91–80–5	1,2-Ethanediamine, N,N-dimethyl-N'-2-pyridinyl-N'-(2-thienylmethyl)-
U067	106–93–4	Ethane, 1,2-dibromo-
U076	75–34–3	Ethane, 1,1-dichloro-
U077	107–06–2	Ethane, 1,2-dichloro-
U131	67–72–1	Ethane, hexachloro-
U024	111–91–1	Ethane, 1,1'-[methylenebis(oxy)]bis[2-chloro-
U117	60–29–7	Ethane, 1,1'-oxybis-(I)
U025	111–44–4	Ethane, 1,1'-oxybis[2-chloro-
U184	76–01–7	Ethane, pentachloro-
U208	630–20–6	Ethane, 1,1,1,2-tetrachloro-
U209	79–34–5	Ethane, 1,1,2,2-tetrachloro-
U218	62–55–5	Ethanethioamide
U226	71–55–6	Ethane, 1,1,1-trichloro-
U227	79–00–5	Ethane, 1,1,2-trichloro-
U359	110–80–5	Ethanol, 2-ethoxy-
U173	1116–54–7	Ethanol, 2,2'-(nitrosoimino)bis-
U004	98–86–2	Ethanone, 1-phenyl-
U043	75–01–4	Ethene, chloro-
U042	110–75–8	Ethene, (2-chloroethoxy)-
U078	75–35–4	Ethene, 1,1-dichloro-
U079	156–60–5	Ethene, 1,2-dichloro-, (E)-
U210	127–18–4	Ethene, tetrachloro-
U228	79–01–6	Ethene, trichloro-
U112	141–78–6	Ethyl acetate (I)
U113	140–88–5	Ethyl acrylate (I)
U238	51–79–6	Ethyl carbamate (urethane)
U117	60–29–7	Ethyl ether (I)
U114	¹ 111–54–6	Ethylenebisdithiocarbamic acid, salts & esters
U067	106–93–4	Ethylene dibromide
U077	107–06–2	Ethylene dichloride
U359	110–80–5	Ethylene glycol monoethyl ether
U115	75–21–8	Ethylene oxide (I,T)
U116	96–45–7	Ethylenethiourea
U076	75–34–3	Ethylidene dichloride

Haz-ardous waste No.	Chemical abstracts No.	Substance
U118	97–63–2	Ethyl methacrylate
U119	62–50–0	Ethyl methanesulfonate
U120	206–44–0	Fluoranthene
U122	50–00–0	Formaldehyde
U123	64–18–6	Formic acid (C,T)
U124	110–00–9	Furan (I)
U125	98–01–1	2-Furancarboxaldehyde (I)
U147	108–31–6	2,5-Furandione
U213	109–99–9	Furan, tetrahydro-(I)
U125	98–01–1	Furfural (I)
U124	110–00–9	Furfuran (I)
U206	18883–66–4	Glucopyranose, 2-deoxy-2-(3-methyl-3-nitrosoureido)-, D-
U206	18883–66–4	D-Glucose, 2-deoxy-2-[[(methylnitrosoamino)-carbonyl]amino]-
U126	765–34–4	Glycidylaldehyde
U163	70–25–7	Guanidine, N-methyl-N′-nitro-N-nitroso-
U127	118–74–1	Hexachlorobenzene
U128	87–68–3	Hexachlorobutadiene
U130	77–47–4	Hexachlorocyclopentadiene
U131	67–72–1	Hexachloroethane
U132	70–30–4	Hexachlorophene
U243	1888–71–7	Hexachloropropene
U133	302–01–2	Hydrazine (R,T)
U086	1615–80–1	Hydrazine, 1,2-diethyl-
U098	57–14–7	Hydrazine, 1,1-dimethyl-
U099	540–73–8	Hydrazine, 1,2-dimethyl-
U109	122–66–7	Hydrazine, 1,2-diphenyl-
U134	7664–39–3	Hydrofluoric acid (C,T)
U134	7664–39–3	Hydrogen fluoride (C,T)
U135	7783–06–4	Hydrogen sulfide
U135	7783–06–4	Hydrogen sulfide H$_2$S
U096	80–15–9	Hydroperoxide, 1-methyl-1-phenylethyl- (R)
U116	96–45–7	2-Imidazolidinethione
U137	193–39–5	Indeno[1,2,3-cd]pyrene
U190	85–44–9	1,3-Isobenzofurandione
U140	78–83–1	Isobutyl alcohol (I,T)
U141	120–58–1	Isosafrole
U142	143–50–0	Kepone
U143	303–34–4	Lasiocarpine
U144	301–04–2	Lead acetate
U146	1335–32–6	Lead, bis(acetato-O)tetrahydroxytri-
U145	7446–27–7	Lead phosphate
U146	1335–32–6	Lead subacetate
U129	58–89–9	Lindane
U163	70–25–7	MNNG
U147	108–31–6	Maleic anhydride
U148	123–33–1	Maleic hydrazide
U149	109–77–3	Malononitrile
U150	148–82–3	Melphalan
U151	7439–97–6	Mercury
U152	126–98–7	Methacrylonitrile (I, T)
U092	124–40–3	Methanamine, N-methyl- (I)
U029	74–83–9	Methane, bromo-
U045	74–87–3	Methane, chloro- (I, T)
U046	107–30–2	Methane, chloromethoxy-
U068	74–95–3	Methane, dibromo-
U080	75–09–2	Methane, dichloro-
U075	75–71–8	Methane, dichlorodifluoro-
U138	74–88–4	Methane, iodo-
U119	62–50–0	Methanesulfonic acid, ethyl ester
U211	56–23–5	Methane, tetrachloro-
U153	74–93–1	Methanethiol (I, T)
U225	75–25–2	Methane, tribromo-
U044	67–66–3	Methane, trichloro-
U121	75–69–4	Methane, trichlorofluoro-
U036	57–74–9	4,7-Methano-1H-indene, 1,2,4,5,6,7,8,8-octachloro-2,3,3a,4,7,7a-hexahydro-
U154	67–56–1	Methanol (I)
U155	91–80–5	Methapyrilene
U142	143–50–0	1,3,4-Metheno-2H-cyclobuta[cd]pentalen-2-one, 1,1a,3,3a,4,5,5,5a,5b,6-decachlorooctahydro-
U247	72–43–5	Methoxychlor

Haz-ardous waste No.	Chemical abstracts No.	Substance
U154	67–56–1	Methyl alcohol (I)
U029	74–83–9	Methyl bromide
U186	504–60–9	1-Methylbutadiene (I)
U045	74–87–3	Methyl chloride (I,T)
U156	79–22–1	Methyl chlorocarbonate (I,T)
U226	71–55–6	Methyl chloroform
U157	56–49–5	3-Methylcholanthrene
U158	101–14–4	4,4'-Methylenebis(2-chloroaniline)
U068	74–95–3	Methylene bromide
U080	75–09–2	Methylene chloride
U159	78–93–3	Methyl ethyl ketone (MEK) (I,T)
U160	1338–23–4	Methyl ethyl ketone peroxide (R,T)
U138	74–88–4	Methyl iodide
U161	108–10–1	Methyl isobutyl ketone (I)
U162	80–62–6	Methyl methacrylate (I,T)
U161	108–10–1	4-Methyl-2-pentanone (I)
U164	56–04–2	Methylthiouracil
U010	50–07–7	Mitomycin C
U059	20830–81–3	5,12-Naphthacenedione, 8-acetyl-10-[(3-amino-2,3,6-trideoxy)-alpha-L-lyxo-hexopyranosyl)oxy]-7,8,9,10-tetrahydro-6,8,11-trihydroxy-1-methoxy-, (8S-cis)-
U167	134–32–7	1-Naphthalenamine
U168	91–59–8	2-Naphthalenamine
U026	494–03–1	Naphthalenamine, N,N'-bis(2-chloroethyl)-
U165	91–20–3	Naphthalene
U047	91–58–7	Naphthalene, 2-chloro-
U166	130–15–4	1,4-Naphthalenedione
U236	72–57–1	2,7-Naphthalenedisulfonic acid, 3,3'-[(3,3'-dimethyl[1,1'-biphenyl]-4,4'-diyl)bis(azo)bis[5-amino-4-hydroxy]-, tetrasodium salt
U166	130–15–4	1,4-Naphthoquinone
U167	134–32–7	alpha-Naphthylamine
U168	91–59–8	beta-Naphthylamine
U217	10102–45–1	Nitric acid, thallium(1+) salt
U169	98–95–3	Nitrobenzene (I,T)
U170	100–02–7	p-Nitrophenol
U171	79–46–9	2-Nitropropane (I,T)
U172	924–16–3	N-Nitrosodi-n-butylamine
U173	1116–54–7	N-Nitrosodiethanolamine
U174	55–18–5	N-Nitrosodiethylamine
U176	759–73–9	N-Nitroso-N-ethylurea
U177	684–93–5	N-Nitroso-N-methylurea
U178	615–53–2	N-Nitroso-N-methylurethane
U179	100–75–4	N-Nitrosopiperidine
U180	930–55–2	N-Nitrosopyrrolidine
U181	99–55–8	5-Nitro-o-toluidine
U193	1120–71–4	1,2-Oxathiolane, 2,2-dioxide
U058	50–18–0	2H-1,3,2-Oxazaphosphorin-2-amine, N,N-bis(2-chloroethyl)tetrahydro-, 2-oxide
U115	75–21–8	Oxirane (I,T)
U126	765–34–4	Oxiranecarboxyaldehyde
U041	106–89–8	Oxirane, (chloromethyl)-
U182	123–63–7	Paraldehyde
U183	608–93–5	Pentachlorobenzene
U184	76–01–7	Pentachloroethane
U185	82–68–8	Pentachloronitrobenzene (PCNB)
See F027	87–86–5	Pentachlorophenol
U161	108–10–1	Pentanol, 4-methyl-
U186	504–60–9	1,3-Pentadiene (I)
U187	62–44–2	Phenacetin
U188	108–95–2	Phenol
U048	95–57–8	Phenol, 2-chloro-
U039	59–50–7	Phenol, 4-chloro-3-methyl-
U081	120–83–2	Phenol, 2,4-dichloro-
U082	87–65–0	Phenol, 2,6-dichloro-
U089	56–53–1	Phenol, 4,4'-(1,2-diethyl-1,2-ethenediyl)bis-, (E)-
U101	105–67–9	Phenol, 2,4-dimethyl-
U052	1319–77–3	Phenol, methyl-
U132	70–30–4	Phenol, 2,2'-methylenebis[3,4,6-trichloro-
U170	100–02–7	Phenol, 4-nitro-

Haz-ardous waste No.	Chemical abstracts No.	Substance
See F027	87–86–5	Phenol, pentachloro-
See F027	58–90–2	Phenol, 2,3,4,6-tetrachloro-
See F027	95–95–4	Phenol, 2,4,5-trichloro-
See F027	88–06–2	Phenol, 2,4,6-trichloro-
U150	148–82–3	L-Phenylalanine, 4-[bis(2-chloroethyl)amino]-
U145	7446–27–7	Phosphoric acid, lead(2+) salt (2:3)
U087	3288–58–2	Phosphorodithioic acid, O,O-diethyl S-methyl ester
U189	1314–80–3	Phosphorus sulfide (R)
U190	85–44–9	Phthalic anhydride
U191	109–06–8	2-Picoline
U179	100–75–4	Piperidine, 1-nitroso-
U192	23950–58–5	Pronamide
U194	107–10–8	1-Propanamine (I,T)
U111	621–64–7	1-Propanamine, N-nitroso-N-propyl-
U110	142–84–7	1-Propanamine, N-propyl- (I)
U066	96–12–8	Propane, 1,2-dibromo-3-chloro-
U083	78–87–5	Propane, 1,2-dichloro-
U149	109–77–3	Propanedinitrile
U171	79–46–9	Propane, 2-nitro- (I,T)
U027	108–60–1	Propane, 2,2'-oxybis[2-chloro-
U193	1120–71–4	1,3-Propane sultone
See F027	93–72–1	Propanoic acid, 2-(2,4,5-trichlorophenoxy)-
U235	126–72–7	1-Propanol, 2,3-dibromo-, phosphate (3:1)
U140	78–83–1	1-Propanol, 2-methyl- (I,T)
U002	67–64–1	2-Propanone (I)
U007	79–06–1	2-Propenamide
U084	542–75–6	1-Propene, 1,3-dichloro-
U243	1888–71–7	1-Propene, 1,1,2,3,3,3-hexachloro-
U009	107–13–1	2-Propenenitrile
U152	126–98–7	2-Propenenitrile, 2-methyl- (I,T)
U008	79–10–7	2-Propenoic acid (I)
U113	140–88–5	2-Propenoic acid, ethyl ester (I)
U118	97–63–2	2-Propenoic acid, 2-methyl-, ethyl ester
U162	80–62–6	2-Propenoic acid, 2-methyl-, methyl ester (I,T)
U194	107–10–8	n-Propylamine (I,T)
U083	78–87–5	Propylene dichloride
U148	123–33–1	3,6-Pyridazinedione, 1,2-dihydro-
U196	110–86–1	Pyridine
U191	109–06–8	Pyridine, 2-methyl-
U237	66–75–1	2,4-(1H,3H)-Pyrimidinedione, 5-[bis(2-chloroethyl)amino]-
U164	56–04–2	4(1H)-Pyrimidinone, 2,3-dihydro-6-methyl-2-thioxo-
U180	930–55–2	Pyrrolidine, 1-nitroso-
U200	50–55–5	Reserpine
U201	108–46–3	Resorcinol
U202	¹ 81–07–2	Saccharin, & salts
U203	94–59–7	Safrole
U204	7783–00–8	Selenious acid
U204	7783–00–8	Selenium dioxide
U205	7488–56–4	Selenium sulfide
U205	7488–56–4	Selenium sulfide SeS$_2$ (R,T)
U015	115–02–6	L-Serine, diazoacetate (ester)
See F027	93–72–1	Silvex (2,4,5-TP)
U206	18883–66–4	Streptozotocin
U103	77–78–1	Sulfuric acid, dimethyl ester
U189	1314–80–3	Sulfur phosphide (R)
See F027	93–76–5	2,4,5-T
U207	95–94–3	1,2,4,5-Tetrachlorobenzene
U208	630–20–6	1,1,1,2-Tetrachloroethane
U209	79–34–5	1,1,2,2-Tetrachloroethane
U210	127–18–4	Tetrachloroethylene
See F027	58–90–2	2,3,4,6-Tetrachlorophenol

Hazardous waste No.	Chemical abstracts No.	Substance
U213	109–99–9	Tetrahydrofuran (I)
U214	563–68–8	Thallium(I) acetate
U215	6533–73–9	Thallium(I) carbonate
U216	7791–12–0	Thallium(I) chloride
U216	7791–12–0	Thallium chloride TlCl
U217	10102–45–1	Thallium(I) nitrate
U218	62–55–5	Thioacetamide
U153	74–93–1	Thiomethanol (I,T)
U244	137–26–8	Thioperoxydicarbonic diamide [(H$_2$N)C(S)]$_2$S$_2$, tetramethyl-
U219	62–56–6	Thiourea
U244	137–26–8	Thiram
U220	108–88–3	Toluene
U221	25376–45–8	Toluenediamine
U223	26471–62–5	Toluene diisocyanate (R,T)
U328	95–53–4	o-Toluidine
U353	106–49–0	p-Toluidine
U222	636–21–5	o-Toluidine hydrochloride
U011	61–82–5	1H-1,2,4-Triazol-3-amine
U227	79–00–5	1,1,2-Trichloroethane
U228	79–01–6	Trichloroethylene
U121	75–69–4	Trichloromonofluoromethane
See F027	95–95–4	2,4,5-Trichlorophenol
See F027	88–06–2	2,4,6-Trichlorophenol
U234	99–35–4	1,3,5-Trinitrobenzene (R,T)
U182	123–63–7	1,3,5-Trioxane, 2,4,6-trimethyl-
U235	126–72–7	Tris(2,3-dibromopropyl) phosphate
U236	72–57–1	Trypan blue
U237	66–75–1	Uracil mustard
U176	759–73–9	Urea, N-ethyl-N-nitroso-
U177	684–93–5	Urea, N-methyl-N-nitroso-
U043	75–01–4	Vinyl chloride
U248	[1] 81–81–2	Warfarin, & salts, when present at concentrations of 0.3% or less
U239	1330–20–7	Xylene (I)
U200	50–55–5	Yohimban-16-carboxylic acid, 11,17-dimethoxy-18-[(3,4,5-trimethoxybenzoyl)oxy]-, methyl ester, (3beta,16beta,17alpha,18beta,20alpha)-
U249	1314–84–7	Zinc phosphide Zn$_3$P$_2$, when present at concentrations of 10% or less

[1] CAS Number given for parent compound only.

APPENDIX 5B   Uniform Hazardous Waste Manifest

Please print or type.   (Form designed for use on elite (12-pitch) typewriter.)   Form. ..proved. OMB No. 2050-0039 Expires 9-30-91

UNIFORM HAZARDOUS WASTE MANIFEST	1. Generator's US EPA ID No		Manifest Document No.	2. Page 1 of	Information in the shaded areas is not required by Federal law.
3. Generator's Name and Mailing Address				A. State Manifest Document Number	
				B. State Generator's ID	
4. Generator's Phone ( )					
5. Transporter 1 Company Name	6.	US EPA ID Number		C. State Transporter's ID	
				D. Transporter's Phone	
7. Transporter 2 Company Name	8.	US EPA ID Number		E. State Transporter's ID	
				F. Transporter's Phone	
9. Designated Facility Name and Site Address	10.	US EPA ID Number		G. State Facility's ID	
				H. Facility's Phone	

	11. US DOT Description (Including Proper Shipping Name, Hazard Class, and ID Number)	12. Containers		13. Total Quantity	14. Unit Wt/Vol	I. Waste No.
		No.	Type			
G E N E R A T O R a.						
b.						
c.						
d.						
J. Additional Descriptions for Materials Listed Above			K. Handling Codes for Wastes Listed Above			

15. Special Handling Instructions and Additional Information

16. **GENERATOR'S CERTIFICATION:** I hereby declare that the contents of this consignment are fully and accurately described above by proper shipping name and are classified, packed, marked, and labeled, and are in all respects in proper condition for transport by highway according to applicable international and national government regulations.

If I am a large quantity generator, I certify that I have a program in place to reduce the volume and toxicity of waste generated to the degree I have determined to be economically practicable and that I have selected the practicable method of treatment, storage, or disposal currently available to me which minimizes the present and future threat to human health and the environment; OR, if I am a small quantity generator, I have made a good faith effort to minimize my waste generation and select the best waste management method that is available to me and that I can afford.

Printed/Typed Name	Signature	Month Day Year

T R A N S P O R T E R	17. Transporter 1 Acknowledgement of Receipt of Materials		
	Printed/Typed Name	Signature	Month Day Year
	18. Transporter 2 Acknowledgement of Receipt of Materials		
	Printed/Typed Name	Signature	Month Day Year

F A C I L I T Y	19. Discrepancy Indication Space		
	20. Facility Owner or Operator Certification of receipt of hazardous materials covered by this manifest except as noted in Item 19		
	Printed/Typed Name	Signature	Month Day Year

EPA Form 8700-22 (Rev. 9-88)   Previous editions are obsolete

Please print or type.   (Form designed for use on elite (12-pitch) typewriter.)   Form Approved OMB No. 2050-0039 Expires 9-30-91

UNIFORM HAZARDOUS WASTE MANIFEST (Continuation Sheet)	21. Generator's US EPA ID No.	Manifest Document No.	22. Page	Information in the shaded areas is not required by Federal law

23. Generator's Name

L. State Manifest Document Number

M. State Generator's ID

24. Transporter ____ Company Name	25. US EPA ID Number	N. State Transporter's ID
		O. Transporter's Phone
26. Transporter ____ Company Name	27. US EPA ID Number	P. State Transporter's ID
		Q. Transporter's Phone

28. US DOT Description (Including Proper Shipping Name, Hazard Class, and ID Number)	29. Containers		30. Total Quantity	31. Unit Wt./Vol	R. Waste No.
	No.	Type			
a.					
b.					
c.					
d.					
e.					
f.					
g.					
h.					
i.					

S. Additional Descriptions for Materials Listed Above

T. Handling Codes for Wastes Listed Above

32. Special Handling Instructions and Additional Information

33. Transporter ____ Acknowledgement of Receipt of Materials		Date
Printed/Typed Name	Signature	Month Day Year
34. Transporter ____ Acknowledgement of Receipt of Materials		Date
Printed/Typed Name	Signature	Month Day Year

35. Discrepancy Indication Space

EPA Form 8700-22A (Rev. 9-88) Previous edition is obsolete.

APPENDIX 5C   Notification for Underground Storage Tanks

Notification for Underground Storage Tanks

FORM APPROVED
OMB NO. 2050-0068
APPROVAL EXPIRES 9 30 91

EPA estimates public reporting burden for this form to average 30 minutes per response, including time for reviewing instructions, gathering and maintaining the data needed, and completing and reviewing the form. Send comments regarding this burden estimate to Chief, Information Policy Branch, PM-223, U.S. Environmental Protection Agency, 401 M St., S.W., Washington, D.C. 20460; and to the Office of Information and Regulatory Affairs, Office of Management and Budget, Washington, D.C. 20503, marked "Attention: Desk Officer for EPA."

STATE USE ONLY

I D Number

Date Received

GENERAL INFORMATION

Notification is required by Federal law for all underground tanks that have been used to store regulated substances since January 1, 1974, that are in the ground as of May 8, 1986, or that are brought into use after May 8, 1986. The information requested is required by Section 9002 of the Resource Conservation and Recovery Act, (RCRA), as amended.

The primary purpose of this notification program is to locate and evaluate underground tanks that store or have stored petroleum or hazardous substances. It is expected that the information you provide will be based on reasonably available records, or, in the absence of such records, your knowledge, belief, or recollection.

Who Must Notify? Section 9002 of RCRA, as amended, requires that, unless exempted, owners of underground tanks that store regulated substances must notify designated State or local agencies of the existence of their tanks. Owner means:
(a) in the case of an underground storage tank in use on November 8, 1984, or brought into use after that date, any person who owns an underground storage tank used for the storage, use, or dispensing of regulated substances, and
(b) in the case of any underground storage tank in use before November 8, 1984, but no longer in use on that date, any person who owned such tank immediately before the discontinuation of its use.

What Tanks Are Included? Underground storage tank is defined as any one or combination of tanks that (1) is used to contain an accumulation of "regulated substances," and (2) whose volume (including connected underground piping) is 10%, or more beneath the ground. Some examples are underground tanks storing: 1. gasoline, used oil, or diesel fuel, and 2. industrial solvents, pesticides, herbicides or fumigants.

What Tanks Are Excluded? Tanks removed from the ground are not subject to notification. Other tanks excluded from notification are:
1. farm or residential tanks of 1,100 gallons or less capacity used for storing motor fuel for noncommercial purposes.
2. tanks used for storing heating oil for consumptive use on the premises where stored.
3. septic tanks.

4. pipeline facilities (including gathering lines) regulated under the Natural Gas Pipeline Safety Act of 1968, or the Hazardous Liquid Pipeline Safety Act of 1979, or which is an intrastate pipeline facility regulated under State laws.
5. surface impoundments, pits, ponds, or lagoons.
6. storm water or waste water collection systems.
7. flow-through process tanks.
8. liquid traps or associated gathering lines directly related to oil or gas production and gathering operations.
9. storage tanks situated in an underground area (such as a basement, cellar, mineworking, drift, shaft, or tunnel) if the storage tank is situated upon or above the surface of the floor.

What Substances Are Covered? The notification requirements apply to underground storage tanks that contain regulated substances. This includes any substance defined as hazardous in section 101 (14) of the Comprehensive Environmental Response, Compensation and Liability Act of 1980 (CERCLA), with the exception of those substances regulated as hazardous waste under Subtitle C of RCRA. It also includes petroleum, e.g., crude oil or any fraction thereof which is liquid at standard conditions of temperature and pressure (60 degrees Fahrenheit and 14.7 pounds per square inch absolute).

Where To Notify? Completed notification forms should be sent to the address given at the top of this page.

When To Notify? 1. Owners of underground storage tanks in use or that have been taken out of operation after January 1, 1974, but still in the ground, must notify by May 8, 1986. 2. Owners who bring underground storage tanks into use after May 8, 1986, must notify within 30 days of bringing the tanks into use.

Penalties: Any owner who knowingly fails to notify or submits false information shall be subject to a civil penalty not to exceed $10,000 for each tank for which notification is not given or for which false information is submitted.

INSTRUCTIONS

Please type or print in ink all items except "signature" in Section V. This form must be completed for each location containing underground storage tanks. If more than 5 tanks are owned at this location, photocopy the reverse side, and staple continuation sheets to this form.

Indicate number of continuation sheets attached

I. OWNERSHIP OF TANK(S)

Owner Name (Corporation, Individual, Public Agency, or Other Entity)

Street Address

County

City   State   ZIP Code

Area Code   Phone Number

Type of Owner **(Mark all that apply ☒)**

☐ Current   ☐ State or Local Gov't   ☐ Private or Corporate
☐ Former   ☐ Federal Gov't (GSA facility I.D. no   ☐ Ownership uncertain

II. LOCATION OF TANK(S)

(If same as Section 1, mark box here ☐)

Facility Name or Company Site Identifier, as applicable

Street Address or State Road, as applicable

County

City (nearest)   State   ZIP Code

Indicate number of tanks at this location

Mark box here if tank(s) are located on land within an Indian reservation or on other Indian trust lands ☐

III. CONTACT PERSON AT TANK LOCATION

Name (If same as Section I, mark box here ☐)   Job Title   Area Code   Phone Number

IV. TYPE OF NOTIFICATION

☐ Mark box here only if this is an amended or subsequent notification for this location

V. CERTIFICATION (Read and sign after completing Section VI.)

I certify under penalty of law that I have personally examined and am familiar with the information submitted in this and all attached documents, and that based on my inquiry of those individuals immediately responsible for obtaining the information, I believe that the submitted information is true, accurate, and complete

Name and official title of owner or owner's authorized representative   Signature   Date Signed

CONTINUE ON REVERSE SIDE

EPA Form 7530-1 (Revised 9-88)   Page 1

Owner Name (from Section I) _____ Location (from Section II) _____ Page No. _____ of _____ Pages

VI DESCRIPTION OF UNDERGROUND STORAGE TANKS *(Complete for each tank at this location.)*					
Tank Identification No. (e.g., ABC-123), or Arbitrarily Assigned Sequential Number (e.g., 1,2,3...)	Tank No.	Tank No.	Tank No.	Tank No.	Tank No.
1. Status of Tank *(Mark all that apply ☒)* Currently in Use Temporarily Out of Use Permanently Out of Use Brought into Use after 5/8/86	☐☐☐☐	☐☐☐☐	☐☐☐☐	☐☐☐☐	☐☐☐☐
2. Estimated Age (Years)					
3. Estimated Total Capacity (Gallons)					
4. Material of Construction *(Mark one ☒)* Steel Concrete Fiberglass Reinforced Plastic Unknown Other, Please Specify	☐☐☐☐ ____	☐☐☐☐ ____	☐☐☐☐ ____	☐☐☐☐ ____	☐☐☐☐ ____
5. Internal Protection *(Mark all that apply ☒)* Cathodic Protection Interior Lining (e.g., epoxy resins) None Unknown Other, Please Specify	☐☐☐☐ ____	☐☐☐☐ ____	☐☐☐☐ ____	☐☐☐☐ ____	☐☐☐☐ ____
6. External Protection *(Mark all that apply ☒)* Cathodic Protection Painted (e.g., asphaltic) Fiberglass Reinforced Plastic Coated None Unknown Other, Please Specify	☐☐☐☐☐ ____	☐☐☐☐☐ ____	☐☐☐☐☐ ____	☐☐☐☐☐ ____	☐☐☐☐☐ ____
7. Piping *(Mark all that apply ☒)* Bare Steel Galvanized Steel Fiberglass Reinforced Plastic Cathodically Protected Unknown Other, Please Specify	☐☐☐☐☐ ____	☐☐☐☐☐ ____	☐☐☐☐☐ ____	☐☐☐☐☐ ____	☐☐☐☐☐ ____
8. Substance Currently or Last Stored in Greatest Quantity by Volume *(Mark all that apply ☒)* **a. Empty**	☐	☐	☐	☐	☐
b. Petroleum Diesel Kerosene Gasoline (including alcohol blends) Used Oil Other, Please Specify	☐☐☐☐	☐☐☐☐	☐☐☐☐	☐☐☐☐	☐☐☐☐
c. Hazardous Substance	☐	☐	☐	☐	☐
Please Indicate Name of Principal CERCLA Substance OR Chemical Abstract Service (CAS) No	____ ____	____ ____	____ ____	____ ____	____ ____
Mark box ☒ if tank stores a mixture of substances **d. Unknown**	☐☐	☐☐	☐☐	☐☐	☐☐
9. Additional Information (for tanks permanently taken out of service) **a.** Estimated date last used (mo/yr)	/	/	/	/	/
b. Estimated quantity of substance remaining (gal.)	____	____	____	____	____
c. Mark box ☒ if tank was filled with inert material (e.g., sand, concrete)	☐	☐	☐	☐	☐

Owner Name (from Section I) _____ Location (from Section II) _____ Page No. _____ of _____ Pages

VII CERTIFICATION OF COMPLIANCE (COMPLETE FOR ALL NEW TANKS AT THIS LOCATION)

10. Installation (mark all that apply):

☐ The installer has been certified by the tank and piping manufacturers

☐ The installer has been certified or licensed by the implementing agency

☐ The installation has been inspected and certified by a registered professional engineer

☐ The installation has been inspected and approved by the implementing agency

☐ All work listed on the manufacturer's installation checklists has been completed

☐ Another method was used as allowed by the implementing agency. Please specify:

_____

11. Release Detection (mark all that apply)

☐ Manual tank gauging

☐ Tank tightness testing with inventory controls

☐ Automatic tank gauging

☐ Vapor monitoring

☐ Ground-water monitoring

☐ Interstitial monitoring within a secondary barrier

☐ Interstitial monitoring within secondary containment

☐ Automatic line leak detectors

☐ Line tightness testing

☐ Another method allowed by the implementing agency. Please specify:

_____

12. Corrosion Protection (if applicable)

☐ As specified for coated steel tanks with cathodic protection

☐ As specified for coated steel piping with cathodic protection

☐ Another method allowed by the implementing agency. Please specify:

_____

13. I have financial responsibility in accordance with Subpart I. Please specify

Method _____

Insurer _____

Policy Number _____

14. OATH: I certify that the information concerning installation provided in Item 10 is true to the best of my belief and knowledge

Installer _____   _____

                          Name   Date

_____

                          Position

_____

                          Company

APPENDIX 5D   EPA's Final Approach for Existing Tank

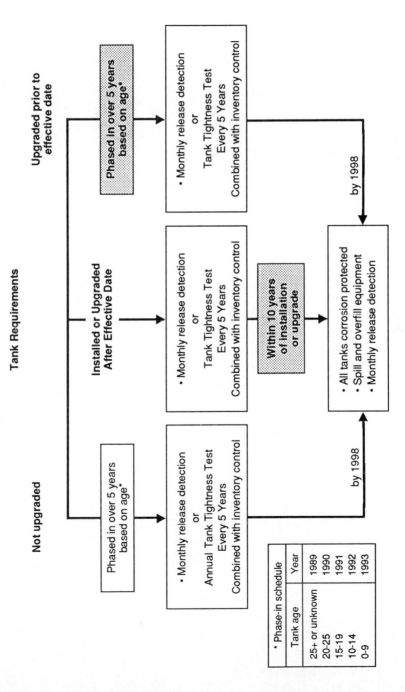

Tank Requirements

Not upgraded

Upgraded prior to
effective date

Installed or Upgraded
After Effective Date

Phased in over 5 years
based on age*

Phased in over 5 years
based on age*

- Monthly release detection
  or
  Annual Tank Tightness Test
  Every 5 Years
  Combined with inventory control

- Monthly release detection
  or
  Tank Tightness Test
  Every 5 Years
  Combined with inventory control

- Monthly release detection
  or
  Tank Tightness Test
  Every 5 Years
  Combined with inventory control

Within 10 years
of installation
or upgrade

- All tanks corrosion protected
- Spill and overfill equipment
- Monthly release detection

by 1998

by 1998

* Phase-in schedule

Tank age	Year
25+ or unknown	1989
20-25	1990
15-19	1991
10-14	1992
0-9	1993

Piping: **More stringent requirements for pressurized lines by 1990.**

SOURCE: *Federal Register*, Vol. 53, No. 185, September 23, 1988.

Hazardous Substances under CERCLA

6.1 Introduction

The Comprehensive Environmental Response, Compensation, and Liability Act (CERCLA), popularly known as the *Superfund,* was passed by the U.S. Congress and signed into law by President Carter on December 11, 1980. While the Resource Conservation and Recovery Act (RCRA) deals basically with the management of hazardous wastes that are generated, transferred, stored, treated, or disposed of, CERCLA provides a response action for the environmental releases of hazardous substances. Simplistically, RCRA deals with active or current generation, transportation, treatment, storage, and disposal operations of hazardous waste whereas CERCLA deals with inactive or past hazardous waste disposal sites from which there is an actual or threatened release of a hazardous substance.

A CERCLA-based response is statutorily triggered by an actual release or the threat of release of a "hazardous substance or pollutant or contaminant" into the environment.

The depth and breadth of the statutory provisions covered under CERCLA are extraordinary for an environmental law. The reach of CERCLA is not limited to a specific medium; rather, it covers all the environmental media—air, water (both surface water and groundwater), and land. In addition, all direct and indirect human activities that may cause release or the threat of release of a hazardous substance, pollutant, or contaminant can bring CERCLA's statutory provisions and the regulations promulgated thereunder into play. In this context, note that whenever a waste is designated as hazardous waste under RCRA, it automatically becomes a hazardous substance under CERCLA. Hence, CERCLA covers a much broader area than RCRA. Additionally, the original CERCLA has been amended by the Superfund Amendments and Reauthorization Act (SARA) of 1986 which has broadened the original scope of CERCLA considerably.

SARA also extended and expanded CERCLA's original 5-year program. However, the core responsibility of EPA under CERCLA still includes three major objectives:

- To identify and evaluate inactive hazardous waste disposal sites from where hazardous substances are being or may be released into the environment
- To impose or undertake cleanup requirements at those sites where there is a release or threatened release of a hazardous substance into the environment
- To evaluate and assess damages to natural resources and collect all response costs from the *potentially responsible parties* (PRPs)

This chapter discusses various issues involving CERCLA as amended by SARA. However, Title III of SARA is not included here. Rather, it is discussed in Chap. 7 because of the subject matter diversity.

6.2   Definitions

The applicable terms have been defined very broadly under CERCLA. The most important is *hazardous substance*. Section 101(14) of CERCLA defines a hazardous substance as one requiring special designation and consideration by EPA under its own statutory provision, the Clean Air Act (CAA), the Clean Water Act (CWA), the Toxic Substances Control Act (TSCA), and any hazardous waste as defined in RCRA.

Under CERCLA, the term *hazardous substance* has the underpinnings of these five federal statutes and it includes

- Any element, compound, mixture, solution, or substance designated under section 102 of CERCLA
- Any hazardous waste having the characteristics identified under or listed pursuant to section 3001 of the Solid Waste Disposal Act (i.e., RCRA) except those suspended by an act of Congress
- Any toxic pollutant listed under section 307(a) of the Federal Water Pollution Control Act (FWPCA)
- Any hazardous air pollutant listed under section 112 of the Clean Air Act
- Any imminently hazardous chemical substance or mixture with respect to which the EPA administrator has taken action pursuant to section 7 of the Toxic Substances Control Act

The term *hazardous substance* under CERCLA, however, does not include petroleum including "crude oil or any fraction thereof which is not otherwise specifically listed or designated" as a hazardous substance. Moreover, this term does not include natural gas, natural gas liquids, liquefied natural gas or synthetic gas usable as fuel, or a mixture of natural gas and synthetic gas.

In this context, note that certain volatile organic compounds, e.g., benzene, toluene, etc., which are listed individually as hazardous substances under CERCLA are indigenous to petroleum. In a July 1987 legal memorandum entitled "Scope of the CERCLA Petroleum Exclusion under Section 101(14) and 104(a)(2)," EPA asserted that the presence of those hazardous substances that are inherent in petroleum will not result in an automatic loss of the petroleum exclusion as recognized under CERCLA. However, this exclusion will be lost if the substance or material in question containing petroleum includes the listed substances which are not found in refined petroleum fractions or which exist at levels that exceed their natural levels in such fractions. In *City of Philadelphia v. Stepan Chemical Company et al.,* the court supported this position of EPA. In a 1991 case, *United States of America and the State of Washington v. The Western Processing Company, Inc., et al.,* a federal district court affirmed that CERCLA's petroleum exclusion applies to unrefined and refined gasoline even though certain indigenous components and additives added during the refining process are designated hazardous substances. But it held that waste oil may fall outside the petroleum exclusion since this exclusion does not extend to those hazardous substances that are added, either intentionally or unintentionally, to petroleum products.

Under section 101(33) of CERCLA, a *pollutant or contaminant* is any element, substance, compound, or mixture, including disease-causing agents, that is not designated or listed as a hazardous substance but which "will or may reasonably be anticipated to cause" adverse health effects including death, disease, behavioral abnormalities, cancer, genetic mutations, physiological malfunctions, or physical deformations in organisms or their offspring when such contaminant is released into the environment. This term does not include petroleum or crude oil constituents unless specifically listed or designated as a hazardous substance.

Section 101(22) defines *release* as follows:

> [A]ny spilling, leaking, pumping, pouring, emitting, emptying, discharging, injecting, escaping, leaching, dumping, or disposing into the environment (including the abandonment or discarding of barrels, containers, and other closed receptacles containing any hazardous substance or pollutant or contaminant), but excludes (A) any release which results in exposure to persons solely within a workplace, with respect to a claim which such persons may assert against the employer of such persons, (B) emissions from the engine exhaust of a motor vehicle, rolling stock, aircraft, vessel, or pipeline pumping station engine, (C) release of source, byproduct, or special nuclear material from a nuclear incident,...and (D) the normal application of fertilizer.

In addition to these four specific exclusions, certain types of releases defined as *federally permitted releases* will not be construed as a violated release. These include releases specifically allowed under federal permits issued to a facility owner or operator under certain environmental statutes, e.g., the Clean Air Act; the Clean Water Act; the Solid Waste Disposal Act (later amended as RCRA); the Marine Protection, Research, and Sanctuaries Act; the Safe Drinking Water Act; or other applicable state laws.

Section 101(9) of CERCLA defines *facility* as follows:

> "[F]acility" means (A) any building, structure, installation, equipment, pipe or pipeline (including any pipe into a sewer or publicly owned treatment works),

well, pit, pond, lagoon, impoundment, ditch, landfill, storage container, motor vehicle, rolling stock, or aircraft, or (B) any site or area where a hazardous substance has been deposited, stored, disposed of, or placed, or otherwise come to be
located; but does not include any common product in consumer use or any vessel.

In the context of CERCLA, any release of a hazardous substance from a facility into the environment must be judged by these defined "terms of art."
Under section 101(8), the term *environment* has a specific definition:

"[E]nvironment" means (A) the navigable waters, the waters of the contiguous
zone, and the ocean waters for which the natural resources are under the exclusive management authority of the United States under the Magnuson Fishery
Conservation and Management Act, and (B) any other surface water, ground water, drinking water supply, land surface or subsurface strata, or ambient air
within the United States or under the jurisdiction of the United States.

Thus, for the purpose of CERCLA, the term *environment* encompasses all
three media—air, water, and land—which makes the reach of CERCLA from
the ocean to the sky and anything in between.
Section (101)(20)(A) defines *owner or operator* as follows:

"[O]wner or operator" means (i) in the case of a vessel, any person owning, operating or chartering by demise, such vessel, (ii) in the case of an onshore facility
or an offshore facility, any person owning or operating such facility, and (iii) in
the case of any facility…any person who owned, operated, or otherwise controlled activities at such facility immediately beforehand. Such term does not include a person, who, without participating in the management of a vessel or facility, holds indicia of ownership primarily to protect his security interest in the
vessel or facility.

Hence, the term *owner or operator* has a flexible definition depending on
whether it is in relation to a vessel or an onshore or offshore facility. Moreover, mere security interest in such vessel or facility without an active participation in its management may not subject an owner or operator to the statutory requirements of CERCLA.
Since the purpose of CERCLA is to respond to and remove any actual or
threatened release of a hazardous substance, it is extremely important to understand the statutory definitions of such terms.
Section 101(25) of CERCLA defines *respond* or *response* to mean several activities which include *remove, removal, remedy,* and *remedial action.* Section
101(23) of CERCLA defines *remove* or *removal* as follows:

"[R]emove" or "removal" means the cleanup or removal of released hazardous
substances from the environment, such actions as may be necessary taken in the
event of the threat of release of hazardous substances into the environment, such
actions as may be necessary to monitor, assess, and evaluate the release or
threat of release of hazardous substances, the disposal of removed material, or
the taking of such other actions as may be necessary to prevent, minimize, or
mitigate damage to the public health or welfare or to the environment, which
may otherwise result from a release or threat of release…

The term also includes within its ambit other broad-range provisions such as security fencing or other measures limiting access, alternative water supplies, temporary evacuation and housing, and any emergency assistance under the Disaster Relief Act of 1974.

Section 101(24) of CERCLA incorporates the following:

> "[R]emedy" or "remedial action" means those actions consistent with permanent remedy taken instead of or in addition to removal actions in the event of a release or threatened release of a hazardous substance into the environment, to prevent or minimize the release of hazardous substances so that they do not migrate to cause substantial danger to present or future public health or welfare or the environment.

Hence, the broad distinction that can be made between a removal action and a remedial action under CERCLA is that the former entails an initial or temporary response action whereas the latter involves permanent measures. As a result, it is obvious that a remedial action, as opposed to a removal action, will typically be far more extensive as well as expensive. Also, a removal action may later be transformed to a remedial action if the significance of the release or potential release so warrants.

6.3   Purpose of CERCLA

As the name *Superfund* connotes, CERCLA is basically a federal funding program earmarked for necessary remediation and cleanup following the release of a hazardous substance into the environment. Section 104(a)(1) of CERCLA requires a removal and/or remedial action whenever (1) any hazardous substance is released or there is substantial threat of such release into the environment or (2) there is release or substantial threat of release into the environment of any pollutant or contaminant that may present an imminent and substantial threat to the public health or welfare.

The purpose of CERCLA is to provide a response mechanism for cleaning up any hazardous substance released into the environment, e.g., an accidental spill or release of a hazardous substance from an abandoned hazardous waste disposal facility. At present, there are two funding programs covered under CERCLA and its progeny, SARA: the Hazardous Substance Response Trust Fund (Response Trust Fund) and the Post-Closure Liability Trust Fund (Liability Trust Fund).

CERCLA achieves its goal of expedient cleanup programs through these two well-orchestrated funding mechanisms. The Response Trust Fund receives 87.5 percent of its revenue from petroleum and chemical feedstock taxes and the remaining 12.5 percent from a well-established appropriation program. For the Liability Trust Fund, a tax is imposed on qualified hazardous waste disposal facilities. In addition, all potentially responsible parties may be jointly and severally liable for all costs incurred during the cleanup and remedial actions. If a particular defendant is unavailable, the Response Trust Fund will absorb the costs of remediation of a released hazardous substance or substances, as the case may be. Section 111 of CERCLA provides details of how and when money in the funds will be available as well as their specific limitations and restrictions for such "orphan" sites.

A proper response mechanism under CERCLA may involve a removal action or a remedial action. A removal action involves a short-term, limited response to a necessary cleanup. It is thus more appropriate when the release is a small, simple, and short-term discharge of a hazardous substance into the environment. A remedial action is more appropriate when the release, accidental or intentional, requires a complex, large-scale, and long-term solution to achieve an acceptable cleanup or response action. While the semantics are important in grasping the seriousness and complexity of a release, the critical concern under CERCLA is the necessity of a prompt response action that is appropriate given the nature of an actual or a threatened release.

6.4 Hazardous Substance Management under CERCLA

EPA's regulations for the cleanup, control, and management of a release of a hazardous substance are set forth in 40 CFR part 300, subpart D. These incorporate a national oil and hazardous substances pollution contingency plan. In addition, table 302.4 in 40 CFR part 302 provides a list of regulated hazardous substances. This table has been reproduced in its entirety in Table 7.2.

The responsibilities of implementing various action plans under CERCLA are shared by several federal agencies and, in certain situations, by authorized state agencies. While EPA has the overall responsibility for plan implementation, several other agencies and departments are responsible for carrying out certain specific tasks. For example, all evacuations and relocations are handled by the Federal Emergency Management Agency (FEMA), while the U.S. Coast Guard provides appropriate response actions in coastal zones, the Great Lakes, ports, and harbors. The Department of Defense (DOD) deals with releases from its own vessels and facilities. Specific administrative functions are handled by the Transportation, Health and Human Services, Interior, Treasury, and Justice Departments and by other appropriate federal agencies and/or departments. This sharing of responsibilities has resulted in some specific interrelationships within several federal agencies and departments. In addition, it has contributed to certain overlaps between CERCLA and other federal statutes. This is well reflected in the statutory definition of a hazardous substance, discussed earlier, which has the underpinnings of five federal statutes.

Because of the complexities thus involved, any management of a hazardous substance by necessity includes several facets:

National priorities list

Section 105(8) of CERCLA requires EPA to develop certain criteria for determining priorities among releases or threatened releases for the purpose of taking long-term remedial evaluation and action. The importance of the *national priorities list* (NPL) is mostly stressed, however, in connection with the eligibility of an NPL site for Superfund-financed remedial action. As a result, inclusion on the NPL brings a certain amount of notoriety and national attention. In addition, EPA's regulations, set forth in 40 CFR §300.425, make only those releases and potential releases of hazardous substances from facilities listed in the NPL eligible for remedial action. The NPL, by definition, is not a final list. It can be shortened or expanded depending on additional findings about the inactive hazardous waste disposal facilities located anywhere within the United States.

Any inactive or abandoned hazardous waste disposal facility which reportedly poses a serious threat to human health or to the environment may be assessed by EPA based on EPA's own knowledge or on information received from other federal or state agencies as well as private citizens requesting its inclusion on the NPL. Nonetheless, EPA has a formal, structured program for the evaluation and listing of such a facility on the NPL. For this purpose, EPA uses a scoring method called the *hazard ranking system* (HRS); details of which can be found in 40 CFR part 300, appendix A. The HRS is the principal screening mechanism used by EPA to place sites on the NPL, and as a result, the HRS is only a measure of relative risk and not a yardstick for absolute risk measurement.

Hazard ranking system

On September 1, 1984, EPA published its original hazard ranking system, a scoring system for an objective evaluation of various inactive hazardous waste disposal facilities. This original HRS underwent several revisions, the most recent one on December 14, 1990. The revised HRS retains the original cutoff score and EPA's basic approach. However, it incorporates certain revisions based on the requirements of SARA.

Simplistically, the HRS measures the relative degree of risk to human health and to the environment due to the existence of an inactive hazardous waste disposal facility. To evaluate the hazardousness of such a facility, EPA first authorizes or undertakes a *preliminary assessment* (PA) or a *site inspection* (SI) to determine

- The quantity, concentration, and the toxicity of a release
- Any potential for migration, explosion, or other fire hazards
- All probable pathways for direct or indirect exposure to human beings and to the environment

For the purpose of appropriate HRS scoring, the site assessment process includes distinct specific tasks for PA and SI:

- For a PA, the focus should be on a visual inspection and collection of available local, state, and federal permitting data; site-specific information (i.e., topography, population, etc.); and historical industrial activity.
- For an SI, the data collected during the PA should be augmented by additional data collection, including sampling of appropriate environmental me-

dia and wastes, to determine the likelihood of a site's receiving a high enough HRS score for possible inclusion on the NPL.

At the conclusion of a facility evaluation, numerical scores are assigned to each of the factors evaluated, and the total is tabulated. A final score of 28.5 or higher in HRS scores ranging from 0 to 100 can theoretically lead to EPA's decision to list a facility on the NPL.

The latest final rule on HRS promulgated on December 14, 1990, is an attempt by EPA to make the HRS more accurate in assessing relative potential risks. As a result, the revised HRS, which became effective on March 14, 1991, is required to address those facilities which contain radioactive and mixed wastes. Some of the important provisions of the revised HRS include

- Evaluation of new exposure pathways or threats that may entail direct contact of people with contaminated soils and assessment of contamination of the aquatic food chain

- Expansion of toxicity evaluation, taking into consideration not only acute health effects but also carcinogenic and chronic noncarcinogenic effects

- Increase of the sensitive environments from just wetlands and endangered species to include other areas and items specifically designated by various federal and state agencies

- Evaluation of potential for air contamination and for contaminated groundwater that may enter any surface water

The recent EPA rule also incorporates several major changes in three specific areas: multiple pathways, groundwater migration pathway, and surface water migration pathway. For each pathway, EPA has incorporated new specific requirements. Note that under the revised HRS, higher scores will be assigned when people are actually exposed to contamination and when potentially exposed people and sensitive environments are closer to a site.

By incorporating these changes, EPA plans to meet its obligations for the revision of the hazard ranking system under section 125 of CERCLA which was added by SARA. Section 125 of CERCLA requires appropriate consideration of several site-specific characteristics. These include

- The quantity, toxicity, and concentrations of hazardous constituents present in the waste at the facility in comparison with those in other wastes

- The extent of, and potential for, a release of such hazardous constituents into the environment

- The degree of risk to human health and to the environment posed by such constituents

Removal and remedial actions

As stated earlier, section 101(23) of CERCLA defines *remove* or *removal* as the cleanup or removal of released hazardous substances from the particular medium (air, water, or soil) that may be contaminated due to such release. These response actions may also be necessary in the event of threat of release of a hazardous substance into the environment. These response actions may involve

- Monitoring, assessing, and evaluating the release or threat of release of hazardous substances
- Proper disposal of removed materials
- Undertaking such other actions as may be necessary to prevent, minimize, or mitigate damage to public health or welfare or to the environment which may otherwise result from a release or the threat of a release

The removal action may also include security fencing or other measures to limit access, provision of alternative water supplies, temporary evacuation, and housing of threatened individuals at a safe location.

A remedial action, by the statutory framework, is broader in scope than a removal action. It includes a much more comprehensive program than a removal action does. As discussed earlier, section 101(24) defines a *remedy* or *remedial action* as that action which is "consistent with permanent remedy taken" to prevent or eliminate the release of hazardous substances so that they do not migrate, causing substantial present or future damage to public health or welfare or to the environment. Remedial actions may include

- Storage, confinement, or perimeter protection by using dikes, trenches, ditches and clay cover, and other neutralization measures
- Cleanup of released hazardous substances or contaminated materials
- Recycling or reuse, diversion, destruction, or segregation of reactive wastes
- Dredging or excavation, repair or replacement of leaking containers, collection of leachate and runoff
- On-site treatment or incineration
- Provision of alternative water supplies
- Any monitoring reasonably required to ensure protection of public health, welfare, and the environment

Thus, a remedial action may include the costs of permanent relocation of residents, business, and community facilities in addition to the costs for permanent cleanup measures and appropriate monitoring. The options for the required cleanup and the costs of these options involve certain standardized procedures. They typically encompass

- Preparation of a work plan for the *remedial investigation/feasibility study* (RI/FS)
- Undertaking of the RI/FS as approved by EPA or an authorized state agency
- Selection of a remedial plan through a *record of decision* (ROD)

These procedures, adopted originally by EPA, are essential to determine the extent of cleanup necessary and to critically assess "how clean is clean" in the context of regulatory requirements. On July 9, 1991, EPA published its proposed rule for setting cleanup standards for groundwaters and surface waters, soil, and air that will ensure protection of human health and the environment. This proposed framework embraces dual standards. For carcinogens, the standard calls for a protection risk of 1 in 10,000. The standard for noncarcinogens is not restricted by such fixed limits.

Remedial investigation/feasibility study.   The remedial investigation/feasibility study (RI/FS) is the first step in any CERCLA action. EPA's guidance document on the subject of a removal or remedial action actually outlines a methodology that meets the CERCLA requirements. EPA's approach is said to be a dynamic and flexible process that can be tailored to specific circumstances by using a case-by-case analysis. In a simplified way, EPA's approach to a cleanup or response action is founded on a determination of how best to use the built-in flexibility. At the same time, EPA expects to conduct an appropriate and efficient RI/FS in a timely and cost-effective manner, using its flexible process.

Notwithstanding this built-in flexibility, an effective RI/FS study is centered on gathering sufficient information to support an informed risk management decision. It is important to select the most appropriate remedy from all conceivable options for an affected facility. As the name implies, the RI/FS process includes a two-prong approach. The first part, RI, is truly a site characterization process, and it may include

- Identification of the source, degree, and extent of contamination
- Pathways of possible migration of contaminants through various media (land, water, and air)
- Extent of potential exposure to humans and to the environment

The essential scope of a typical RI is to collect adequate data about a release episode in order to develop and evaluate various remedial options.

In the FS, however, the information put together in the RI is employed to consider and evaluate various technical or engineering options for the cleanup. However, even though the core of the FS is an evaluation of available engineering or construction options, it includes a detailed analysis of costs, benefits, engineering feasibility, and environmental impact of all such options.

Figure 6.1, developed by EPA, shows a step-by-step or phased-in approach to a typical remedial investigation based on RI/FS. Figure 6.2, also developed by EPA, provides important guidance with respect to analytical levels of data used during the RI/FS. It is fair to say that during a typical RI/FS of a hazardous waste site, proper understanding and delineation of environmental chemistry of chemicals and substances present at the site along with the basic physical characteristics of the site are very important.

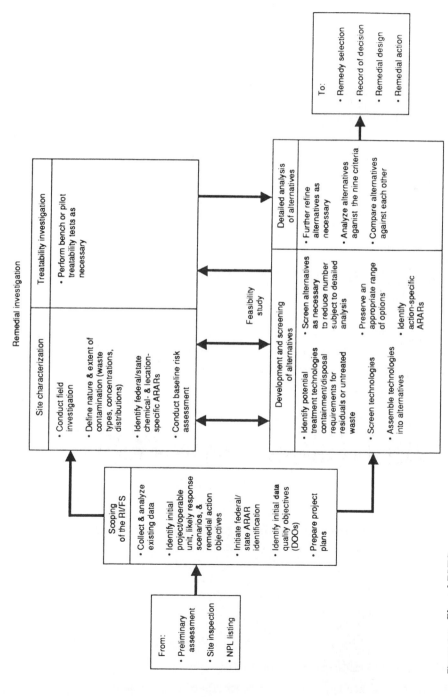

Figure 6.1 Phased RI/FS process.

Data uses	Analytical level	Type of analysis
Site characterization Monitoring during implementation	Level I	• Total organic/inorganic vapor detection using portable instruments • Field test kits
Site characterization Evaluation of alternatives Engineering design Monitoring during implementation	Level II	• Variety of organics by GC; inorganics by AA; XRF • Tentative ID; analyte-specific • Detection limits vary from low ppm to low ppb
Risk assessment PRP determination Site characterization Evaluation of alternatives Engineering design Monitoring during implementation	Level III	• Organics/inorganics using EPA procedures other than CLP can be analyte-specific • RCRA characteristic tests
Risk assessment PRP determination Evaluation of alternatives Engineering design	Level IV	• HSL organics/inorganics by GC/MS; AA; ICP • Low ppb detection limit
Risk assessment PRP determination	Level V	• Nonconventional parameters • Method-specific detection limits • Modification of existing methods • Appendix B parameters

Figure 6.2 Summary of analytical levels appropriate to data uses.

Note that the objective of the RI/FS process is not to eliminate all uncertainties, but to gather sufficient information for development and selection of the most appropriate remedial action for a specific facility. Such a selection, by its very nature, involves the balancing of various factors as well as the exercise of best professional judgment. In addition, the RI/FS process is not generally influenced by whether the selected remedy will be financed by federal or state government funds available under CERCLA or by a PRP for the release. However, it may be influenced by certain procedural matters, such as work plan preparation, reporting requirements, etc., with regard to the final outcome.

In October 1988, EPA issued an interim final guidance document for conducting an appropriate RI/FS study. It represents the methodology that the CERCLA program has established for characterizing the nature and extent of risks posed by uncontrolled hazardous waste facilities and for evaluating various potential remedies.

To select the proper remedy, it is *extremely important* to collect appropriate and adequate data. Such data may include several important factors:

■ The nature and extent of the facility

■ Types of waste deposited and methods of past disposal

- Quantity, nature, and fate of chemicals and chemical substances that may be present in disposed wastes

- Type and extent of environmental media (air, water, and soil) contaminated

- Hydrogeological characteristics of the facility, including surface water and groundwater flow rates and their directions

- Presence of any contaminant plume and its possible movement into or out of the facility

- Soil characteristics, thickness of strata, horizontal and vertical hydraulic conductivity of soils

- Presence of any aquifer or water-bearing stratum within the facility and its current physical and chemical nature

- All probable scenarios of contaminant exposure to humans, animals, flora, and fauna

- Appropriate remedial measures that are currently available and their relative merits and fallibilities based on site-specific considerations

Record of decision. Upon completion of the RI/FS, EPA's methodology calls for selection of the most appropriate *remedial action* (RA). This is done when EPA prepares and publishes a record of decision, which announces its tentative selection of the most appropriate remedy from the various alternatives considered and evaluated along with an explanation for such a selection. Under the new section 117 of CERCLA, EPA is required to provide an opportunity for public comments and a public hearing to further discuss its tentative selection of the remedial plan. Then EPA will issue a ROD, explaining the propriety of its final decision in view of the available options and the public comments made.

Design and construction. The issuance of the ROD by EPA sets the groundwork for development of a *remedial design* (RD). This includes technical specifications, drawings, engineering, and construction work for the remedial plan selected. At the design and engineering stage, appropriate engineering plans and drawings are prepared, technical specifications are drawn up, and construction activities are selected through the development of a project contract.

Upon selection of a general contractor and appropriate subcontractors, the construction activities are undertaken to implement the *remedial action* (RA) planned. Each of these phased processes must be approved by EPA prior to implementation. This is why it is extremely important to establish a somewhat objective concept of "how clean is clean" during the selection of the remedial plan and not to get shortchanged at a very late stage of the remedial process.

6.5   National Contingency Plan

The kingpin for response, cleanup, and remedial action is the National Oil and Hazardous Substances Pollution Contingency Plan. The *national contingency plan* (NCP), as it is popularly known, was originally developed under section 311 of the Clean Water Act, but its scope has been significantly broadened under CERCLA. The NCP is required by section 311(c)(2) of CWA and section 105 of CERCLA. It has been developed by EPA in coordination with members of the National Response Team (NRT) under Executive Order 12580. The NCP establishes "procedures and standards for responding to releases of hazardous substances, pollutants and contaminants."

The NCP applies to, and becomes effective when there are discharges of oil into or upon the navigable waters of the United States and adjoining shorelines; it also applies to releases of hazardous substances, pollutants, or contaminants which may present an imminent and substantial danger to public health or welfare.

In *United States v. Carolawn Company, Inc.*, the court agreed with EPA's position that a substance containing any amount of a listed hazardous substance will trigger CERCLA. Thus, any and all levels of release of hazardous substances, pollutants, and contaminants into land, water, and air may be addressed and included in the NCP. This is definitely an expansive interpretation of CERCLA, and it is hotly debated within the legal circle. A contrary decision was handed down by the court in *United States v. Ottati & Goss*, which held a minimum quantity requirement for triggering CERCLA. It appears, unless the U.S. Supreme Court renders a decision, that this no-minimum-requirement criterion, a conservative approach, may be worth considering. In a 1990 case, *City of New York v. Exxon Corp.*, the court held that CERCLA liability has no *de minimis* exemption if a waste or contaminant includes a hazardous substance. A similar decision was rendered in *B.F. Goodrich Co. v. Murtha* for municipal solid waste.

The importance of the NCP, however, is rooted in a cost recovery action. The cost recovery actions will be appropriate only when the cleanup activities are undertaken by the government or a third party. Bona fide conformance with the NCP, in turn, signifies the "CERCLA-quality cleanup." This implies a legitimacy for recovery of costs incurred in a removal or remedial action. Presently, EPA has adopted a "substantial compliance" test to meet the NCP consistency requirement for private cost recovery actions under section 107(a)(4)(B) of CERCLA. This EPA regulation is set forth in 40 CFR part 300, subpart H, and 40 CFR §300.700. By doing this, EPA has also exempted several other requirements completely.

A substantial-compliance test program must include certain remedy selection requirements. Any CERCLA compliance program, to be considered substantial, must

- Protect human health and the environment
- Provide permanent solutions and alternatives or resource recovery technologies to the maximum extent practicable
- Be cost-effective

In accordance with 40 CFR section 300.700(c), a private response action under section 107 of CERCLA must be consistent with the NCP for the purpose of cost recovery. Such an action will qualify if, as a whole, it is in substantial compliance with the CERCLA requirements and results in a CERCLA-quality cleanup. Any immaterial or insubstantial deviations from the regulatory provisions included in 40 CFR part 300 will not be automatically labeled as NCP-inconsistent. However, whenever possible, close conformance with the CERCLA requirements is appropriate to establish NCP consistency.

In a 1990 case, *General Electric Co. v. Litton Industrial Automation Systems, Inc.,* the Eighth U.S. Circuit Court of Appeals ruled that a private party, as in the case of the government, can recover its legal costs, under a CERCLA section 107 private cost recovery action. In view of this, the importance of NCP consistency cannot be stressed too much.

6.6   Applicable or Relevant and Appropriate Requirements

Section 121(d)(2) of CERCLA cites certain requirements for cleanup of affected facilities. The premise here is that such cleanups must attain the standard for *applicable or relevant and appropriate requirements* (ARARs) for the specific situation. The ARAR standard applies when any hazardous substance, pollutant, or contaminant remains on a facility. Under EPA's review process as set forth in section 121(c) of CERCLA, EPA is required to review an affected facility at least every 5 years when a remedial action results in leaving some hazardous substances, pollutants, or contaminants at the facility. Under the ARAR rule, even those facilities which meet all the requirements of section 121(d)(2) of CERCLA cleanup must still be reviewed every 5 years unless all hazardous substances, pollutants, and contaminants have been removed.

Section 121(d)(2)(A) of CERCLA provides the following cleanup guidelines to be used as specific ARAR standards for certain types of federal or state actions:

- Any standard, requirement, criteria, or limitation under any federal environmental law, including, but not limited to, the Toxic Substances Control Act, the Safe Drinking Water Act, the Clean Air Act, the Clean Water Act, the Marine Protection, Research and Sanctuaries Act, or the Solid Waste Disposal Act [RCRA]; or

- Any promulgated standard, requirement, criteria, or limitation under a state environmental or facility siting law that is more stringent than the federal standard, requirement, criteria, or limitation, including each such state standard, requirement, criteria or limitation contained in a program approved, authorized or delegated by the [EPA] Administrator under a statute...and that has been identified to the President by the state in a timely manner.

Thus, ARARs provide some ground rules for the required degree of cleanup as well as general guidance for "how clean is clean."

Section 121(d)(4) of CERCLA lists six exceptions to the basic ARAR requirements. These exceptions apply when

- The selected action is only a part of the total remedial action that will conform to the ARAR requirements when completed.

- Compliance with the ARAR requirements will present greater risk to human health and to the environment than other alternative options available.

- Compliance with the ARAR requirements is technically impracticable from an engineering perspective.

- An equivalent standard of performance by the selected remedial action is achieved through the use of another method or approach.

- A state fails to consistently apply the standard, requirement, criteria, or limitation in similar circumstances under the state requirement.

- Establishment of the ARAR standard will not provide a balance between the cleanup need for the facility in question and the needs for all other facilities requiring a cleanup under the government fund-financed program.

None of these exceptions, however, relieve or waive the basic requirement that all remedial plans be sufficient to protect human health and the environment. However, EPA's ARAR policy is not yet uniform nationwide. Even today, the ARAR policy and its underlying standards are subject to the vagaries of interpretation by individual regional offices of EPA and delegated state agencies.

For the purpose of judicial review, EPA is required to establish an administrative record upon which its remedy selection may be based and which may provide for public participation in the selection of such remedy.

6.7 Off-Site Disposal

Section 121(b)(1) of CERCLA emphasizes the undesirability of off-site transportation and disposal of hazardous substances, pollutants, and contaminants. In fact, such an option is designated as the "least favored alternative."

Section 121(d)(3) of CERCLA requires certain minimum qualifications for an off-site disposal facility to accept waste from an affected facility. It includes the following requirements:

- The off-site waste management facility which will accept such waste must not be releasing any hazardous waste or any hazardous constituent into the groundwater, surface water, or soil.

- All other units at the off-site waste management facility are being controlled against possible releases by an approved RCRA corrective-action program.

6.8 Liability of Potentially Responsible Parties

The basic theme of CERCLA liability is to "make the polluter pay." Generally, CERCLA imposes liability for environmental cleanup costs and other appropriate relief arising from the release of a hazardous substance from a facility into the environment. Such liability is imposed on four specific types of potentially responsible parties:

- Present owner(s) or operator(s) of the facility which is releasing or threatening to release any hazardous substance into the environment

- Past owner(s) and operator(s) of the facility at the time of disposal of any hazardous substance

- Transporter(s) of the hazardous substances to a self-chosen facility

- Generator(s) of the hazardous substances who arranged for the disposal or treatment or arranged with a transporter for disposal or treatment of such hazardous substances at the facility

Under the broad provisions of CERCLA, all such parties may be jointly and severally liable for site remedial costs. As a group, the PRPs may be responsible for all relevant costs incurred due to past improper disposal, starting with the preliminary investigation of the release, RI/FS, design, and construction costs as well as other administrative costs. In this context, note that this joint-and-several liability is not written in the statutory provision. However, the courts have generally accepted such liability based on the legislative history and the intent of the legislators behind the enactment of CERCLA.

It is generally agreed within the legal circle that CERCLA imposes strict liability on the owner or operator of a facility that releases hazardous substances for damage and cleanup costs, although the statute does not specifically suggest it. In addition, CERCLA contains important liability implications for several other groups. CERCLA liability may conceivably reach and include a host of individuals, groups, and entities, such as

- Real estate purchasers, landlords, and lessees

- Lenders or financial institutions that provided necessary money for procurement of a contaminated facility

- Debtors who have not filed petitions for bankruptcy

- Corporations which spun off assets, subsidiaries, or divisions

- Parent corporations acting as lenders or retaining direct management control over the activities of the subsidiaries

Thus, it is important to consider potential or hidden CERCLA liability in the event of a merger or acquisition. Without an express assumption of liability by the buyer, the seller may continue to remain liable. In certain states, the seller's failure in complying with the mandatory disclosure requirement of known or existing environmental contamination acts to safeguard the innocent buyer or landowner in spite of a purchase in an as-is condition. Furthermore, any broad contractual arrangement based on an as-is clause may not circumvent statutory liability unless such liabilities are explicitly assumed by the buyer. This is an issue currently debated nationwide since the first two sentences in section 107(e)(1) of CERCLA appear to contradict each other.

This has resulted in three distinct and separate opinions by the federal courts in general. One group of courts has held that the statutory liability under CERCLA cannot be privately contracted away by and between parties.

The second group, relying on the second sentence in section 107(e)(1), has ruled that private parties under the contract law can allot and shift CERCLA liability from and to each other and that such arrangement is not preempted or otherwise nullified by CERCLA. The third group, taking a somewhat centrist position, has held that private contractual agreements between liable parties are legally enforceable unless they absolve their liability to the government to whom they must remain jointly and severally liable irrespective of the contractual agreements.

As indicated earlier, there is a continuing debate within the legal circle over whether there must be a release of a minimum quantity of a hazardous substance to trigger CERCLA and the regulations promulgated thereunder. The debate is largely embedded in the textual language of section 101(22) of CERCLA which includes "any" type or physical form of release in the definition of a CERCLA release. But section 103 of CERCLA expressly states a "minimum" reportable quantity that suggests a threshold or *de minimis* quantity.

This apparent conflicting construction of statutory language can be resolved only by an appropriate congressional amendment. Until that happens, both plaintiffs and defendants in CERCLA actions will be under a cloud of doubt as to whether the statute is correctly interpreted. Thus far, the lower federal courts have not agreed to one single interpretation of this apparent conflict in the statutory construction of CERCLA provisions. Unfortunately, as indicated earlier, the U.S. Supreme Court is yet to rule on this state of confusion, thus making it difficult to reach a definitive conclusion based on existing statutory language.

Hence, the conservative approach here is to assume that for a defendant, the worst scenario is any release of a hazardous substance that may make her or him liable under CERCLA. For the plaintiff, the worst scenario is a release of a hazardous substance that must equal or exceed 1 pound or its reportable quantity as defined in section 102(b) of CERCLA to trigger CERCLA liability. But for both, the quantity of a released hazardous substance is immaterial if there is a determination that the actual or threatened release constitutes an "imminent and substantial endangerment" to public health or welfare or to the environment. In such a case, section 106 of CERCLA is automatically triggered, requiring emergency response action to thwart such dangers.

Defenses to CERCLA liability

CERCLA recognizes certain limited defenses against liabilities:

- An act of God
- An act of war
- An act or omission of a third party other than an employee or agent of the defendant charged with the cause of release or threatened release

In the last situation, it is important to establish that the owner or operator of the facility had not been negligent and that all suitable precautions were taken to prevent such release or a threatened release.

In *United States v. Kayser-Roth Corp.*, the court held that the mere fact that the parent corporation was not the owner is not sufficient to preclude it from liability that stemmed from a spill of trichloroethylene at a plant owned by a subsidiary that was dissolved shortly after the spill. The court held that the parent corporation was liable as an operator. The court here analyzed the disjunctive use of the words *owner* and *operator* in the statute and found that an active involvement in the activities of a subsidiary by a parent corporation makes it liable.

In a 1988 case entitled *United States v. Serafini,* the court drew a line with regard to the defense based on the "innocent-landowner" rule. The court held that defendants to a CERCLA liability action may not use the innocent-landowner defense if they fail to inspect the land prior to purchase. Hence, it is extremely important that one undertake, at the time of acquisition of land or property, an appropriate inquiry into the previous ownership and past uses of such land or property. This calls for a "due diligence review" based on "good commercial or customary practice" that includes all appropriate inquiry regarding the possible and probable past or present contamination of land or property. On June 6, 1989, EPA issued a guidance document entitled "Guidance on Landowner Liability," which provides the regulatory requirements of the innocent-landowner defense.

Statute of limitations

Section 112(d) of CERCLA deals with statutes of limitations for various cost recovery actions that may involve claims against the Superfund. Note that while the original statute specified no statute of limitations for government cost recovery action, section 112(d) of the 1986 CERCLA Amendments, otherwise known as SARA, requires that a claim for recovery of costs under section 111(a) of CERCLA, must be presented to the federal government within 6 years of completion of all response actions. This statute of limitations starts to toll only after completion of all removal and/or remedial actions and has no bearing as to the time of release of hazardous substances.

Section 112(d)(2) of CERCLA makes a distinction for statutes of limitations for claims against the Superfund for recovery of damages under section 107(a) of CERCLA. Such claims must be presented within 3 years after the latter of

- The date of the discovery of the loss and the related release
- The date on which the final EPA regulations for assessment of damages to natural resources are promulgated

There are also statutes of limitations for civil proceedings where claims are made against private parties. Section 113(g)(2) of CERCLA makes a distinction for the statutes of limitations for cost recovery actions under section 107 of CERCLA based on whether removal or remedial actions are involved. According to this statutory provision, an initial action for cost recovery must be commenced

- Within 3 years after the completion of a removal action [or 6 years after a determination to grant a waiver under section 104(c)(1)(C) of CERCLA for continued response action]

- Within 6 years after an initiation of physical on-site construction of the remedial action

Pursuant to section 113(g)(1), actions for natural resources damages must also be brought before the statutes of limitations run out. These statutes of limitations are the same as those for claims against the Superfund and as prescribed in section 112(d) of CERCLA.

In *United States v. Mottolo,* the court held that the statute of limitations as specified in section 112(d) of CERCLA applies only to claims of recovery against the government fund and not to section 107 of CERCLA private-party actions against the PRPs.

In this context, note that there is a commencement date for the clock to run. The term *federally required commencement date* means the date on which the plaintiff knew or reasonably should have known that the personal injury or property damages complained of were caused or contributed to by a hazardous substance, pollutant, or contaminant.

Thus, the statute of limitations for cost recovery of a remedial action is not strictly a fixed time period like other typical actions, but can stretch over several years depending on the stage of the response action and when the clock starts to run.

Before SARA, CERCLA did not specifically provide any statute of limitations in a private-party cost recovery action against the PRPs where no government funds were sought. It was generally held that the statute of limitations of 3 years applied only if one were seeking reimbursement from government funds.

6.9 Notification and Release Reporting Requirements

CERCLA requires immediate notification, but no later than 24 hours, to the National Response Center (1-800-424-8802) for any release of a hazardous substance that equals or exceeds its *reportable quantity* (RQ). Section 103 of CERCLA requires that any person in charge of a facility immediately notify the National Response Center (NRC) of the release of a hazardous substance in an amount equal to or greater than its RQ as soon as that person has such knowledge. Hence, both the knowledge of a release and the meeting of the specific RQ, taken together, trigger the notification requirements. However, release notification made to the NRC under 40 CFR part 300, subpart E, does not relieve the facility owner or operator from notification obligations under SARA Title III or applicable state laws. Such obligations are not mutually exclusive.

A "federally permitted" release does not activate the reporting requirements irrespective of the quantity and type of hazardous substances released. The CERCLA-based reporting requirements also exclude the application of registered pesticide products allowed under the Federal Insecticide, Fungicide, and Rodenticide Act (FIFRA) or to their handling and storage by an agricultural producer or farmer.

The reportable quantities of the hazardous substances are not the same for all substances. They vary depending on their respective degree of hazard. Generally, EPA provides three different quantities as the designated reportable quantity—1 pound, 10 pounds, and 1000 pounds—with the first two being more common.

Details of EPA's regulations covering release reporting requirements are included in 40 CFR part 302. These release reporting requirements are discussed in detail in Chap. 10.

6.10   Penalties and Sanctions

CERCLA imposes both civil and criminal penalties for failure to comply with the reporting of releases within the prescribed time. Even for those listed hazardous substances for which EPA has yet to establish specific RQs, reporting must be made using the statutorily established reportable quantity of 1 pound unless and until EPA's regulatory designation of RQ suggests otherwise. However, if a release is continuous as well as stable in quantity and rate, such a release need not be reported every hour or every day. A one-time-only notice will suffice in those circumstances as long as the release of the hazardous substances involved is unchanged. Unfortunately, under the current EPA regulatory framework, there is no legal definition of the term *continuous release*.

In addition to the release reporting obligation, section 103(c) of CERCLA requires a follow-up notification by the affected facilities to EPA. It states:

> Within one hundred and eighty days after December 11, 1980, any person who owns or operates, or who at the time of disposal owned or operated, or who accepted hazardous substances for transport and selected, a facility at which hazardous substances (as defined in section [101(14)(C)]) are or have been stored, treated, or disposed of shall, unless such facility has a permit issued under, or has been accorded interim status under, subtitle C of [RCRA], notify the [EPA] Administrator... of the existence of such facility, specifying the amount and type of any hazardous substance to be found there, and any known, suspected, or likely releases of such substances from such facility.

This imposes a tremendous burden on both present and past owners and operators of facilities that may contain hazardous substances or from which a release of hazardous substance has occurred or is suspected. Any knowing failure to comply with this provision may result in both civil and criminal penalties.

6.11   Risk Assessment

As discussed earlier, the core concern of a removal or remedial action plan is mitigation of any possible danger to human health or to the environment. Risk assessment is a tool that is routinely used to assess the degree of risk due to a release and subsequent exposure to environmental hazards by human beings. Risk assessment characterizes potential adverse health effects and thus provides an appropriate guide to the desired course and/or level of remedy.

EPA has not formally endorsed any specific format for risk assessment. However, an EPA insight has been offered through some of its publications, notably *Risk Assessment Guidelines* (1986) and *Superfund Public Health Evaluation Manual*. In addition, the National Research Council of the National Academy of Science recommends four distinct elements in risk assessment: hazard identification, dose-response assessment, exposure assessment, and risk characterization.

At the hazard identification stage, the question is whether the released agent or substance causes any adverse effect. This involves several considerations and findings based on

- Toxicological effects on the target organs and the nature of such effects due to exposure
- Toxicity mechanisms for both humans and animals
- Structure-activity relationships
- Metabolic and pharmacokinetic characteristics
- In vitro and in vivo tests
- Identification of any susceptible populations (e.g., children, women, older people)
- Evidence of an increase in the incidence of an adverse health condition based on human studies

The dose-response assessment provides a scientific relationship or corollary between the dose and the effect in humans as a result of the exposure. Such an assessment may include

- Data selection including determination of threshold versus nonthreshold response
- Evaluations of the results of bioassays to assess the critical dose causing a certain toxic effect
- Considerations of uncertainties involved in projecting human dose-response relationship using the results of animal bioassays while considering low dose levels, particularly for carcinogens
- Use of pharmacokinetic and toxicity mechanisms to develop an appropriate dose-response model

For exposure assessment, the critical question is, What exposure can currently be experienced or anticipated under a different set of conditions? A typical exposure assessment may deal with

- Type of exposure caused by a physical or chemical agent

- Amount of available agent at human exchange sites (e.g., skin, lungs, eyes) for a specified period

- Amount of total exposure due to all agents from all relevant media (air, water, soil) and food

- Separate determinations for exposure to a particular individual and group of the population under study

At the risk characterization stage, the basic inquiry is about the estimated incidence due to the adverse effect in a particular population. This is done by retrospectively estimating an individual's exposure to an agent. This may include

- Compilation of historical data on concentrations of chemicals in different media

- Use of information based on models involving potency, half-life, fate, mode of transportation, etc., along with data showing present levels of various chemicals

- Accounting of chemicals present in body fluid, parts, and organs

6.12 Cost Recovery under CERCLA

CERCLA includes cost recovery or reimbursement provisions for both a government cleanup and a private-party cleanup. Section 107(a)(4)(B) of CERCLA provides for a private-party cost recovery claim.

However, to recover the CERCLA response costs against a PRP, a private party must establish several elements:

- A release or a threatened release of a hazardous substance or substances existed.

- The release or threatened release was from a facility.

- The costs incurred and claims made are necessary response costs of a voluntary cleanup.

- The incurred costs being claimed are consistent with the NCP.

The statute, however, does not define a necessary response cost. Furthermore, CERCLA and its progeny, SARA, are silent with respect to the nature and amount of costs that may be recovered due to a voluntary response or cleanup action by a third party.

In the absence of statutory delineation of reimbursable response costs, the courts attempted to define what the costs should or can be. Generally, the courts have held that all costs directly related to removal or remedial actions including costs of investigation, planning, and administration are recoverable. There is, however, no clear indication of whether attorneys' fees are recoverable in a private-party cost recovery action. This is well exemplified by very contrasting decisions in several noted cases, including *United States v.*

Northeastern Pharmaceutical and Chemical Co. (NEPACCO), Pease and Curren Refining, Inc., v. Spectrolab, Inc., Regan v. Cherry Corp., and *General Electric Co. v. Litton Industrial Automations Systems, Inc.* There seems to be a disagreement among the courts based on their interpretations of response costs under section 107 of CERCLA and the legislative history. The Supreme Court has yet to rule on this controversy. However, for government cost recovery actions, there is no disagreement that the government can recover its attorneys' fees. This is based on section 104(b)(1) of CERCLA which specifically authorizes legal investigation by the President of the United States.

Additionally, the courts have generally held that indirect overhead costs associated with a government response or cleanup action are recoverable. Unfortunately, there is no such affirmative or negative determination by any federal court to date with respect to a private-party response action. From the stand point of consistency and uniformity, one may argue that any voluntary private response action, being a substitute for a government action, should be treated equally. As such, nondirect expenses may be recoverable as long as all the ground rules for cost recovery are met.

Two other statutory provisions that have been added to CERCLA by way of SARA should be carefully noted here. Section 113(f) of CERCLA provides for a right of contribution among the PRPs. It requires the courts to determine an equitable allocation of response costs among the culpable defendants. Additionally, section 122(a) of CERCLA authorizes the government to reach a settlement agreement with a PRP that may release the PRP from further liability. Section 122(b) of CERCLA allows a "mixed funding" program based on sharing of remedial costs from monies available from the Superfund and from the PRPs.

Additional discussion on cost recovery actions can be found in Chap. 16.

6.13   EPA Settlement Policies

For the purpose of government settlement of a CERCLA action, EPA may issue four different types of notice letters asking PRPs to settle:

- General notice letter
- RI/FS special notice letter
- RD/RA (remedial design/remedial action) special notice letter
- Removal-action notice letter

The last three notice letters are issued only to those PRPs against whom EPA has collected sufficient evidence when such evidence shows their potential liability under section 107 of CERCLA. The general notice letter, on the contrary, resembles a somewhat diabolic and cursory "shot in the dark" approach. As a result, it is routinely sent to all the readily identifiable PRPs without any hard and fast determination of their culpability.

The original statute as enacted in 1980 did not contain any language concerning notice letters or other forms of notification to the PRPs of a contaminated site from which a release or threat of release of a hazardous substance is apparent. The use of notice letters has evolved as a product of EPA's administrative operation. Nonetheless, the notice letters sent by EPA to the PRPs are founded in specific statutory authorizations under CERCLA.

Typically, once a contaminated site is listed on the NPL and thus becomes a so-called Superfund Site, EPA attempts to collect as much information as possible with regard to the identifiable PRPs. At this stage, it is still a shot in the dark and such preliminarily identified PRPs may not have any relationship to the activities that occurred at the site. At this preliminary stage, EPA mails out only a general notice letter to all such PRPs in order to gather additional information pertaining to the site and its past and current usage and activities pursuant to §104(e) and §107 of CERCLA.

Based on the information received through the responses to §104(e) notice letters, EPA may decide to send a second general notice letter or a special notice letter. This is done pursuant to §122 of CERCLA which contains settlement provisions. EPA usually mails the general notice letter when it prefers to initiate an informal negotiation with a PRP. On the other hand, EPA will mail the special notice letter when it plans to conduct a formal negotiation for settlement, thereby fully invoking its authority under §122(e) of CERCLA. In either case, the notice letters are mailed prior to the commencement of any RI/FS and RD/RA activity in order to allow the PRPs to come up with alternative plans, if they wish.

Section 122(e) notice letters are sent to the PRPs solely for the purpose of negotiating and facilitating early settlements for necessary response actions. This notice procedure is discretionary and EPA is not required to expedite remedial action working with the PRPs. Mailing of §122(e) notice letters, however, puts a moratorium on enforcement actions by EPA for 60 days. During this period, PRPs may submit a "good faith" offer to EPA which, if accepted, will extend the moratorium by another 30 days. EPA has the discretion to extend this moratorium even further if a settlement appears very likely.

In general, §122(e) notice letters are sent to all PRPs against whom EPA has sufficient evidence, on a preliminary basis, to establish their liability pursuant to §107 of CERCLA. However, by no means is this list of PRPs the final or the only list. EPA may revise (i.e., add or delete) its PRP list at any time and has the discretion to bring enforcement actions against some or all the PRPs that make its list.

IF EPA fails to reach a satisfactory settlement at the §122 stage, it may resort to issuance of a unilateral administrative order (UAO) to the PRPs pursuant to its authority under §106 of CERCLA. Section 106(a) of CERCLA is indeed EPA's ultimate abatement action tool. An enforcement action based on §106(a) of CERCLA should not be taken lightly, even if such an order is unilateral. Pursuant to §106(b) of CERCLA, a PRP that willfully violates, or fails or refuses to comply with the §106(a) order without "sufficient cause," may be subject to a fine of $25,000 per day for each day of violation. In addition, CERCLA's treble damage provision may allow EPA to obtain damages of up to 3 times the amount of its response costs. It should be emphasized that if there are settling PRPs, they may also bring a cost recovery action against the non-settling PRPs irrespective of EPA's action.

Over the years, EPA has issued numerous guidance notices for settlement of claims under CERCLA. These include some general guidelines for specific situations. They are useful as discretionary tools for settlement as opposed to commencing litigation. They include EPA's

- Basic settlement policy of February 1985
- 1986 guidelines under section 122 of CERCLA
- Guidelines for *de minimis* party settlements
- Guidelines on covenant not to sue
- Guidelines on notice letters, negotiations, and information exchanges
- Guidelines on mixed funding
- Guidelines for preparing preliminary nonbinding allocations of responsibility (NBAR)

In view of expected settlements, EPA has also developed standardized versions of consent decrees. On July 8, 1991, EPA published its *model consent decree* (MCD) for cleanup under CERCLA. During the early 1990s, the contents of several of the EPA consent decrees were approved by the courts, using the criteria of fairness, reasonableness, and jurisprudence. Hence, any settlement with EPA does not have to be a raw or one-sided deal. Historically, EPA's actions have been viewed by industries as a discouragement for settlement. However, the latest MCD appears to underscore a more realistic approach by EPA that encourages dialogue by the parties interested in reaching a settlement rather than use of a boilerplate language.

The settlement policies currently followed by EPA are discussed in detail in Chap. 15.

Recommended Reading

1. ABA Satellite Seminar, "Hazardous Waste and Superfund 1991," American Bar Association, May 9, 1991.

2. Christopher R. Schraff and Robert E. Steinberg, *RCRA and Superfund,* vol. 1, Shepard's/McGraw-Hill, Colorado Springs, 1991.

3. J. G. Arbuckle, M. E. Bosco, D. R. Case, E. P. Laws, J. C. Martin, M. L. Miller, R. V. Randle, R. G. Stoll, T. F. P. Sullivan, T. A. Vanderver, Jr., P. A. J. Wilson, *Environmental Law Handbook,* 10th ed., Government Institutes, Inc., Rockville, Md., 1989.

4. U.S. Environmental Protection Agency, *Guidance for Conducting Remedial Investigations and Feasibility Studies under CERCLA,* Interim Final, EPA/540/G-89/O04, October 1988.

Important Cases

1. *United States v. Kayser-Roth Corp.,* 910 F. 2d 24 (1st Cir. 1990)., *reh'g. denied* (Sept. 9, 1990).

2. *United States of America and the State of Washington v. The Western Processing Company, Inc., et al.,* 1991 WL 42117 (W.D. Wash. Mar. 22, 1991).

3. *United States v. Ottati & Goss,* 900 F. 2d 429 (1st Cir. 1990).

4. *General Electric Co. v. Litton Industrial Automation Systems, Inc.,* 715 F. Supp. 949 (W.D. Mo. 1989) *aff'd.* 920 F. 2d 1415 (8th Cir. 1990), *cert. denied* 59 U.S.L.W. 3652 (1991).

5. *B.F. Goodrich Co. v. Murtha,* 32 ERC 1487 (D. Conn. 1991).

6. *Purolator Products Corp. v. Allied Signal, Inc.,* 1991 WL 158978 (W.D.N.Y. Aug. 16, 1991).

7. *Pease and Curren Refining, Inc. v. Spectrolab, Inc.,* 744 F. Supp. 945 (C.D. Cal. 1990).

8. *City of New York v. Exxon Corp.,* 744 F. Supp. 474 (S.D.N.Y. 1990).

9. *Regan v. Cherry Corp.,* 706 F. Supp. 145 (D.R.I. 1989).

10. *City of Philadelphia v. Stepan Chemical Company, et al.,* Civ. No. 81-0851 (E.D. Pa., Dec. 13, 1988).

11. *United States v. Serafini,* 750 F. Supp. 168 (M.D. Pa. 1990).

12. *United States v. Carolawn Co., Inc.,* 698 F. Supp 616 (D.S.C. 1987).

13. *United States v. Mottolo,* 605 F. Supp. 898 (D.N.H. 1985).

14. *United States v. Northeastern Pharmaceutical and Chemical Co. (NEPACCO),* 579 F. Supp. 823 (W.D. Mo. 1984), *aff'd.* in part and *rev'd.* in part on other grounds, 810 F. 2d 26 (8th Cir. 1986), *cert. denied* 484 U.S. 848 (1987).

7

Emergency Planning and Right-to-Know under SARA Title III

7.1 Introduction

The Superfund Amendments and Reauthorization Act (SARA), enacted into law on October 17, 1986, is a progeny of the Comprehensive Environmental Response, Compensation, and Liability Act (CERCLA) of 1980. It addresses closed or inactive hazardous waste disposal sites if there is any actual or threatened environmental release of hazardous substances, pollutants, or contaminants. The most revolutionary part of SARA, however, is the emergency planning and community right-to-know provision covered under Title III.

Title III of SARA sets out to accomplish two goals: to encourage and support emergency planning efforts at the state and local levels and to provide the public and local governments with information concerning potential chemical hazards in their communities. Title III of SARA is basically comprised of four major categories: emergency planning, emergency notification, community right-to-know reporting requirements, and toxic chemical release reporting requirements.

From historical as well as legal perspectives, section 103(a) of CERCLA requires that a release into the environment of a hazardous substance in an amount equal to or greater than its reportable quantity (RQ), as established under section 102 of CERCLA, be reported to the National Response Center (NRC). Title III of SARA establishes a separate program that requires releases of hazardous substances as well as other "extremely hazardous substances" to be reported to designated state and local emergency planning authorities. Apart from this, section 104 of CERCLA provides the federal government with the authority to investigate certain releases. Specifically, SARA amends section 104 of CERCLA to

- Clarify EPA's investigating, record gathering, and access authorities
- Increase the flexibility of removal actions

■ Require that primary attention be paid to releases that pose a threat to human health

Title III of SARA includes a statutory agenda for planning and preparedness for any environmental release of hazardous substances. It reflects a congressional reaction to the tragedy in Bhopal, India, and thus has been aptly dubbed the "chemicals in the sunshine act." By enacting Title III of SARA, Congress essentially undertook emergency planning in the event of a disaster.

As discussed in Chap. 6, SARA generally emphasizes the importance of entering into negotiations and reaching settlements with potentially responsible parties (PRPs) through either direct undertaking or financing appropriate response actions. But Title III of SARA, also known as the Emergency Planning and Community Right-to-Know Act (EPCRA), creates a specific statutory approach to emergency planning. It provides a comprehensive program for reporting of releases of chemicals or chemical substances in all environmental media (air, water, and land).

EPCRA specifically requires notification to state and local emergency planning units regarding inventories and releases of hazardous chemical substances. EPCRA emphasizes the requirements for detailed disclosures by business entities about the use and on-site storage of potentially hazardous chemicals. It also requires appropriate state and local governments to make such information available to the public upon request.

Under EPCRA, three different groups of chemicals or chemical substances are identified:

■ Extremely hazardous substances (EHSs)

■ Hazardous chemicals

■ Toxic chemicals

Based on EPCRA's statutory scheme, the presence of any of these substances and chemicals at or above certain specified quantities may trigger certain reporting requirements.

There are three subtitles in EPCRA. Subtitle A includes the framework for emergency planning and environmental release notification. Subtitle B deals with the annual reporting requirements for two broad groups of chemicals and includes inventories of "hazardous chemicals" and routine disposal and accidental releases of "toxic chemicals" into the environment. Subtitle C contains general provisions of EPCRA and details both enforcement actions and penalties that may be imposed for various violations.

EPCRA addresses emergency planning and notification for various types of hazardous substances and chemicals. They can be categorized in four different groups:

■ CERCLA hazardous substances as listed pursuant to section 103(a) of CERCLA

■ Extremely hazardous substances as defined in EPCRA

- Hazardous chemicals as defined in EPCRA

- Toxic chemicals based on the standard industrial classification (SIC) codes 20 through 39 and as designated under section 313(b) of EPCRA

Each of these categories of hazardous substances and chemicals may be subject to emergency planning, emergency release notification, and other reporting requirements, as discussed later in this chapter.

7.2   Definitions

Pursuant to EPCRA, *hazardous chemicals* are those chemicals defined under 29 CFR §1910.1200(c). For such chemicals which can be physical or health hazards, the facility owner or operator is required to prepare or have available a current *material safety data sheet* (MSDS). The MSDSs must comply with the hazard communication regulations of the Occupational Safety and Health Administration (OSHA). The term *hazardous chemical,* however, excludes the following:

- Any food, food additive, color additive, drug, or cosmetic regulated by the Food and Drug Administration (FDA)

- Any substance present in a solid form in any manufactured item so as to prevent its exposure under normal conditions of use

- Any substance used for personal, family, or household purposes or present in the same form and concentration as a product packaged for public distribution

- Any substance used in a research laboratory or a hospital or other medical facility under the direct supervision of a technically qualified person

- Any substance that is used as a fertilizer or in routine agricultural operations

An *extremely hazardous substance* is a substance specifically listed pursuant to subtitle A of EPCRA. Details of the extremely hazardous substances including their listings are contained in 40 CFR part 355. Table 7.1 reproduces the list of extremely hazardous substances and their threshold planning quantities, provided by EPA in appendix A of 40 CFR part 355.

Toxic chemical means a chemical or chemical category listed in 40 CFR §372.65. The reporting requirement, as discussed later, for a toxic chemical is linked to a specific threshold quantity for that particular chemical.

The term *facility* includes all buildings, equipment, structures, and other stationary items which are located on a single site or on contiguous or adjacent sites and which are owned or operated by the same person.

The term *environment* includes air, water, and land and the interrelationship among and between them and all living things.

The term *reportable quantity* (RQ) means the reportable quantity established in Table 302.4 of 40 CFR part 302 for a CERCLA hazardous substance. This table has been reproduced in Table 7.2. For any substance not showing a specific RQ, the reportable quantity is 1 pound.

The *threshold planning quantity* (TPQ) means the quantity shown against a specific extremely hazardous substance as listed in appendices A and B of 40 CFR part 355. For an extremely hazardous substance that is solid, EPA provides two threshold planning quantities (for example, 500 and 10,000 pounds). The lower quantity will apply if the solid exists in powdered form and has a particle size of less than 100 microns (1 micron = 1×10^{-6} meter).

The term *mixture* means a heterogeneous association of substances in which individual substances retain their identities and can usually be separated from each other by mechanical means. If the reporting is based on each component of a mixture which is a hazardous chemical, the quantity must be determined from the concentration of the hazardous chemical, in weight percent (greater than 1.0 percent or 0.1 percent if carcinogenic) multiplied by the mass (in pounds) of the mixture. If the reporting is for the mixture itself, the total quantity of the mixture must be reported.

In the context of Title III of SARA, EPA's definition of *chemical* as set forth in 29 CPR section 1910.1200(c) includes any element, chemical compound, or mixture of elements and/or compounds.

TABLE 7.1   List of Extremely Hazardous Substances and Their Threshold Planning Quantities

[Alphabetical Order]

CAS No.	Chemical name	Notes	Reportable quantity* (pounds)	Threshold planning quantity (pounds)
75–86–5	Acetone Cyanohydrin		10	1,000
1752–30–3	Acetone Thiosemicarbazide	e	1	1,000/10,000
107–02–8	Acrolein		1	500
79–06–1	Acrylamide	d, l	5,000	1,000/10,000
107–13–1	Acrylonitrile	d, l	100	10,000
814–68–6	Acrylyl Chloride	e, h	1	100
111–69–3	Adiponitrile	e, l	1	1,000
116–06–3	Aldicarb	c	1	100/10,000
309–00–2	Aldrin	d	1	500/10,000
107–18–6	Allyl Alcohol		100	1,000
107–11–9	Allylamine	e	1	500
20859–73–8	Aluminum Phosphide	b	100	500
54–62–6	Aminopterin	e	1	500/10,000
78–53–5	Amiton	e	1	500
3734–97–2	Amiton Oxalate	e	1	100/10,000
7664–41–7	Ammonia	l	100	500
300–62–9	Amphetamine	e	1	1,000
62–53–3	Aniline	d, l	5,000	1,000
88–05–1	Aniline, 2,4,6-Trimethyl-	e	1	500
7783–70–2	Antimony Pentafluoride	e	1	500
1397–94–0	Antimycin A	c, e	1	1,000/10,000
86–88–4	ANTU		100	500/10,000
1303–28–2	Arsenic pentoxide	d	1	100/10,000
1327–53–3	Arsenous oxide	d,h	1	100/10,000
7784–34–1	Arsenous trichloride	d	1	500
7784–42–1	Arsine	e	1	100
2642–71–9	Azinphos-Ethyl	e	1	100/10,000
86–50–0	Azinphos-Methyl		1	10/10,000
98–87–3	Benzal Chloride	d	5,000	500
98–16–8	Benzenamine, 3-(Trifluoromethyl)-	e	1	500
100–14–1	Benzene, 1-(Chloromethyl)-4-Nitro-	e	1	500/10,000

TABLE 7.1   List of Extremely Hazardous Substances and Their Threshold Planning Quantities (Continued)

[Alphabetical Order]

CAS No.	Chemical name	Notes	Reportable quantity* (pounds)	Threshold planning quantity (pounds)
98–05–5	Benzenearsonic Acid	e	1	10/10,000
3615–21–2	Benzimidazole, 4,5-Dichloro-2-(Trifluoromethyl)-	e, g	1	500/10,000
98–07–7	Benzotrichloride	d	10	500
100–44–7	Benzyl Chloride	d	100	500
140–29–4	Benzyl Cyanide	e, h	1	500
15271–41–7	Bicyclo[2.2.1]Heptane-2-Carbonitrile, 5-Chloro-6-((((Methylamino)Carbonyl)Oxy)Imino)-, (1s-(1-alpha, 2-beta, 4-alpha, 5-alpha, 6E))-.	e	1	500/10,000
534–07–6	Bis(Chloromethyl) Ketone	e	1	10/10,000
4044–65–9	Bitoscanate	e	1	500/10,000
10294–34–5	Boron Trichloride	e	1	500
7637–07–2	Boron Trifluoride	e	1	500
353–42–4	Boron Trifluoride Compound With Methyl Ether (1:1)	e	1	1,000
28772–56–7	Bromadiolone	e	1	100/10,000
7726–95–6	Bromine	e. l	1	500
1306–19–0	Cadmium Oxide	e	1	100/10,000
2223–93–0	Cadmium Stearate	c, e	1	1,000/10,000
7778–44–1	Calcium arsenate	d	1	500/10,000
8001–35–2	Camphechlor	d	1	500/10,000
56–25–7	Cantharidin	e	1	100/10,000
51–83–2	Carbachol Chloride	e	1	500/10,000
26419–73–8	Carbamic Acid, Methyl-, 0-(((2,4-Dimethyl-1, 3-Dithiolan-2-yl)Methylene)Amino)-.	e	1	100/10,000
1563–66–2	Carbofuran		10	10/10,000
75–15–0	Carbon Disulfide	l	100	10,000
786–19–6	Carbophenothion	e	1	500
57–74–9	Chlordane	d	1	1,000
470–90–6	Chlorfenvinfos	e	1	500
7782–50–5	Chlorine		10	100
24934–91–6	Chlormephos	e	1	500
999–81–5	Chlormequat Chloride	e, h	1	100/10,000
79–11–8	Chloroacetic Acid	e	1	100/10,000
107–07–3	Chloroethanol	e	1	500
627–11–2	Chloroethyl Chloroformate	e	1	1,000
67–66–3	Chloroform	d,1	10	10,000
542–88–1	Chloromethyl ether	d,h	10	100
107–30–2	Chloromethyl methyl ether	e,d	10	100
3691–35–8	Chlorophacinone	e	1	100/10,000
1982–47–4	Chloroxuron	e	1	500/10,000
21923–23–9	Chlorthiophos	e, h	1	500
10025–73–7	Chromic Chloride	e	1	1/10,000
62207–76–5	Cobalt, ((2,2'-(1,2-Ethanediylbis (Nitrilomethylidyne))Bis(6-Fluorophenolato))(2-)-N,N',O,O')-,.	e	1	100/10,000
10210–68–1	Cobalt Carbonyl	e, h	1	10/10,000
64–86–8	Colchicine	e, h	1	10/10,000
56–72–4	Coumaphos		10	100/10,000
5836–29–3	Coumatetralyl	e	1	500/10,000
95–48–7	Cresol, o-	d	1,000	1,000/10,000
535–89–7	Crimidine	e	1	100/10,000
4170–30–3	Crotonaldehyde		100	1,000
123–73–9	Crotonaldehyde, (E)-		100	1,000
506–68–3	Cyanogen Bromide		1,000	500/10,000
506–78–5	Cyanogen Iodide	e	1	1,000/10,000
2636–26–2	Cyanophos	e	1	1,000
675–14–9	Cyanuric Fluoride	e	1	100
66–81–9	Cycloheximide	e	1	100/10,000
108–91–8	Cyclohexylamine	e, l	1	10,000
17702–41–9	Decaborane(14)	e	1	500/10,000
8065–48–3	Demeton	e	1	500
919–86–8	Demeton-S-Methyl	e	1	500
10311–84–9	Dialifor	e	1	100/10,000
19287–45–7	Diborane	e	1	100
111–44–4	Dichloroethyl ether	d	10	10,000
149–74–6	Dichloromethylphenylsilane	e	1	1,000
62–73–7	Dichlorvos		10	1,000
141–66–2	Dicrotophos	e	1	100

TABLE 7.1   List of Extremely Hazardous Substances and Their Threshold Planning Quantities (*Continued*)

[Alphabetical Order]

CAS No.	Chemical name	Notes	Reportable quantity* (pounds)	Threshold planning quantity (pounds)
1464–53–5	Diepoxybutane	d	10	500
814–49–3	Diethyl Chlorophospate	e, h	1	500
1642–54–2	Diethylcarbamazine Citrate	e	1	100/10,000
71–63–6	Digitoxin	c, e	1	100/10,000
2238–07–5	Diglycidyl Ether	e	1	1,000
20830–75–5	Digoxin	e, h	1	10/10,000
115–26–4	Dimefox	e	1	500
60–51–5	Dimethoate		10	500/10,000
2524–03–0	Dimethyl Phosphorochloridothioate	e	1	500
77–78–1	Dimethyl sulfate	d	100	500
75–78–5	Dimethyldichlorosilane	e, h	1	500
57–14–7	Dimethylhydrazine	d	10	1,000
99–98–9	Dimethyl-p-Phenylenediamine	e	1	10/10,000
644–64–4	Dimetilan	e	1	500/10,000
534–52–1	Dinitrocresol		10	10/10,000
88–85–7	Dinoseb		1,000	100/10,000
1420–07–1	Dinoterb	e	1	500/10,000
78–34–2	Dioxathion	e	1	500
82–66–6	Diphacinone	e	1	10/10,000
152–16–9	Diphosphoramide, Octamethyl-		100	100
298–04–4	Disulfoton		1	500
514–73–8	Dithiazanine Iodide	e	1	500/10,000
541–53–7	Dithiobiuret		100	100/10,000
316–42–7	Emetine, Dihydrochloride	e, h	1	1/10,000
115–29–7	Endosulfan		1	10/10,000
2778–04–3	Endothion	e	1	500/10,000
72–20–8	Endrin		1	500/10,000
106–89–8	Epichlorohydrin	d,1	100	1,000
2104–64–5	EPN	e	1	100/10,000
50–14–6	Ergocalciferol	c, e	1	1,000/10,000
379–79–3	Ergotamine Tartrate	e	1	500/10,000
1622–32–8	Ethanesulfonyl Chloride, 2-Chloro-	e	1	500
10140–87–1	Ethanol, 1.2-Dichloro-, Acetate	e	1	1,000
563–12–2	Ethion		10	1,000
13194–48–4	Ethoprophos	e	1	1,000
538–07–8	Ethylbis(2-Chloroethyl)Amine	e, h	1	500
371–62–0	Ethylene Fluorohydrin	c, e, h	1	10
75–21–8	Ethylene oxide	d,1	10	1,000
107–15–3	Ethylenediamine		5,000	10,000
151–56–4	Ethyleneimine	d	1	500
542–90–5	Ethylthiocyanate	e	1	10,000
22224–92–6	Fenamiphos	e	1	10/10,000
122–14–5	Fenitrothion	e	1	500
115–90–2	Fensulfothion	e, h	1	500
4301–50–2	Fluenetil	e	1	100/10,000
7782–41–4	Fluorine	k	10	500
640–19–7	Fluoroacetamide	j	100	100/10,000
144–49–0	Fluoroacetic Acid	e	1	10/10,000
359–06–8	Fluoroacetyl Chloride	c, e	1	10
51–21–8	Fluorouracil	e	1	500/10,000
944–22–9	Fonofos	e	1	500
50–00–0	Formaldehyde	d,1	100	500
107–16–4	Formaldehyde Cyanohydrin	e, h	1	1,000
23422–53–9	Formetanate Hydrochloride	e,h	1	500/10,000
2540–82–1	Formothion	e	1	100
17702–57–7	Formparanate	e	1	100/10,000
21548–32–3	Fosthietan	e	1	500
3878–19–1	Fuberidazole	e	1	100/10,000
110–00–9	Furan		100	500
13450–90–3	Gallium Trichloride	e	1	500/10,000
77–47–4	Hexachlorocyclopentadiene	d,h	10	100
4835–11–4	Hexamethylenediamine, N,N'-Dibutyl-	e	1	500
302–01–2	Hydrazine	d	1	1,000
74–90–8	Hydrocyanic Acid		10	100
7647–01–0	Hydrogen chloride (gas only)	e,1	5,000	500
7664–39–3	Hydrogen Fluoride		100	100

TABLE 7.1   List of Extremely Hazardous Substances and Their Threshold Planning Quantities (*Continued*)

[Alphabetical Order]

CAS No.	Chemical name	Notes	Reportable quantity* (pounds)	Threshold planning quantity (pounds)
7722-84-1	Hydrogen Peroxide (Conc >52%)	e, l	1	1,000
7783-07-5	Hydrogen Selenide	e	1	10
7783-06-4	Hydrogen Sulfide	l	100	500
123-31-9	Hydroquinone	l	1	500/10,000
13463-40-6	Iron, Pentacarbonyl-	e	1	100
297-78-9	Isobenzan	e	1	100/10,000
78-82-0	Isobutyronitrile	e, h	1	1,000
102-36-3	Isocyanic Acid, 3,4-Dichlorophenyl Ester	e	1	500/10,000
465-73-6	Isodrin		1	100/10,000
55-91-4	Isofluorphate	c	100	100
4098-71-9	Isophorone Diisocyanate	b, e	1	100
108-23-6	Isopropyl Chloroformate	e	1	1,000
119-38-0	Isopropylmethylpyrazolyl Dimethylcarbamate	e	1	500
78-97-7	Lactonitrile	e	1	1,000
21609-90-5	Leptophos	e	1	500/10,000
541-25-3	Lewisite	c, e, h	1	10
58-89-9	Lindane	d	1	1,000/10,000
7580-67-8	Lithium Hydride	b, e	1	100
109-77-3	Malononitrile		1,000	500/10,000
12108-13-3	Manganese, Tricarbonyl Methylcyclopentadienyl	e, h	1	100
51-75-2	Mechlorethamine	c, e	1	10
950-10-7	Mephosfolan	e	1	500
1600-27-7	Mercuric Acetate	e	1	500/10,000
7487-94-7	Mercuric Chloride	e	1	500/10,000
21908-53-2	Mercuric Oxide	e	1	500/10,000
10476-95-6	Methacrolein Diacetate	e	1	1,000
760-93-0	Methacrylic Anhydride	e	1	500
126-98-7	Methacrylonitrile	h	1,000	500
920-46-7	Methacryloyl Chloride	e	1	100
30674-80-7	Methacryloyloxyethyl Isocyanate	e, h	1	100
10265-92-6	Methamidophos	e	1	100/10,000
558-25-8	Methanesulfonyl Fluoride	e	1	1,000
950-37-8	Methidathion	e	1	500/10,000
2032-65-7	Methiocarb		10	500/10,000
16752-77-5	Methomyl	h	100	500/10,000
151-38-2	Methoxyethylmercuric Acetate	e	1	500/10,000
80-63-7	Methyl 2-Chloroacrylate	e	1	500
74-83-9	Methyl Bromide	l	1,000	1,000
79-22-1	Methyl Chloroformate	d, h	1,000	500
60-34-4	Methyl Hydrazine		10	500
624-83-9	Methyl Isocyanate	f	1	500
556-61-6	Methyl Isothiocyanate	b, e	1	500
74-93-1	Methyl Mercaptan	l	100	500
3735-23-7	Methyl Phenkapton	e	1	500
676-97-1	Methyl Phosphonic Dichloride	b, e	1	100
556-64-9	Methyl Thiocyanate	e	1	10,000
78-94-4	Methyl Vinyl Ketone	e	1	10
502-39-6	Methylmercuric Dicyanamide	e	1	500/10,000
75-79-6	Methyltrichlorosilane	e, h	1	500
1129-41-5	Metolcarb	e	1	100/10,000
7786-34-7	Mevinphos		10	500
315-18-4	Mexacarbate		1,000	500/10,000
50-07-7	Mitomycin C	d	10	500/10,000
6923-22-4	Monocrotophos	e	1	10/10,000
2763-96-4	Muscimol		1,000	500/10,000
505-60-2	Mustard Gas	e, h	1	500
13463-39-3	Nickel carbonyl	d	10	1
54-11-5	Nicotine	c	100	100
65-30-5	Nicotine sulfate	e	100	100/10,000
7697-37-2	Nitric Acid		1,000	1,000
10102-43-9	Nitric Oxide	c	10	100
98-95-3	Nitrobenzene	l	1,000	10,000
1122-60-7	Nitrocyclohexane	e	1	500
10102-44-0	Nitrogen Dioxide		10	100
62-75-9	Nitrosodimethylamine	d,h	10	1,000
991-42-4	Norbormide	e	1	100/10,000

TABLE 7.1 List of Extremely Hazardous Substances and Their Threshold Planning Quantities (Continued)

[Alphabetical Order]

CAS No.	Chemical name	Notes	Reportable quantity* (pounds)	Threshold planning quantity (pounds)
0	Organorhodium Complex (PMN-82-147)	e	1	10/10,000
630-60-4	Ouabain	c, e	1	100/10,000
23135-22-0	Oxamyl	e	1	100/10,000
78-71-7	Oxetane, 3,3-Bis(Chloromethyl)-	l	e	500
2497-07-6	Oxydisulfoton	e, h	1	500
10028-15-6	Ozone	e	1	100
1910-42-5	Paraquat	e	1	10/10,000
2074-50-2	Paraquat Methosulfate	e	1	10/10,000
56-38-2	Parathion	c,d	10	100
298-00-0	Parathion-Methyl	c	100	100/10,000
12002-03-8	Paris Green	d	1	500/10,000
19624-22-7	Pentaborane	e	1	500
2570-26-5	Pentadecylamine	e	1	100/10,000
79-21-0	Peracetic Acid	e	1	500
594-42-3	Perchloromethylmercaptan		100	500
108-95-2	Phenol		1,000	500/10,000
4418-66-0	Phenol, 2,2'-Thiobis(4-Chloro-6-Methyl)-	e	1	100/10,000
64-00-6	Phenol, 3-(1-Methylethyl)-, Methylcarbamate	e	1	500/10,000
58-36-6	Phenoxarsine, 10,10'-Oxydi-	e	1	500/10,000
696-28-6	Phenyl Dichloroarsine	d, h	1	500
59-88-1	Phenylhydrazine Hydrochloride	e	1	1,000/10,000
62-38-4	Phenylmercury Acetate		100	500/10,000
2097-19-0	Phenylsilatrane	e, h	1	100/10,000
103-85-5	Phenylthiourma		108	100/10,000
298-02-2	Phorate		10	10
4104-14-7	Phosacetim	e	1	100/10,000
947-02-4	Phosfolan	e	1	100/10,000
75-44-5	Phosgene	l	10	10
732-11-6	Phosmet	e	1	10/10,000
13171-21-6	Phosphamidon	e	1	100
7803-51-2	Phosphine		100	500
2703-13-1	Phosphonothioic Acid, Methyl-, O-Ethyl O-(4-(Methylthio)Phenyl) Ester.	e	1	500
50782-69-9	Phosphonothioic Acid, Methyl-, S-(2-(Bis(1-Methylethyl)Amino)Ethyl O-Ethyl Ester.	e	1	100
2665-30-7	Phosphonothioic Acid, Methyl-, O-(4-Nitrophenyl) O-Phenyl Ester	e	1	500
3254-63-5	Phosphoric Acid, Dimethyl 4-(Methylthio) Phenyl Ester	e	1	500
2587-90-8	Phosphorothioic Acid, O,O-Dimethyl-S-(2-Methylthio) Ethyl Ester	c, e, g	1	500
7723-14-0	Phosphorus	b, h	1	100
10025-87-3	Phosphorus Oxychloride	d	1,000	500
10026-13-8	Phosphorus Pentachloride	b, e	1	500
1314-56-3	Phosphorus Pentoxide	b, e	1	10
7719-12-2	Phosphorus Trichloride		1,000	1,000
57-47-6	Physostigmine	e	1	100/10,000
57-64-7	Physostigmine, Salicylate (1:1)	e	1	100/10,000
124-87-8	Picrotoxin	e	1	500/10,000
110-89-4	Piperidine	e	1	1,000
23505-41-1	Pirimifos-Ethyl	e	1	1,000
10124-50-2	Potassium arsenite	d	1	500/10,000
151-50-8	Potassium Cyanide	b	10	100
506-61-6	Potassium Silver Cyanide	b	1	500
2631-37-0	Promecarb	e, h	1	500/10,000
106-96-7	Propargyl Bromide	e	1	10
57-57-8	Propiolactone, Beta-	e	1	500
107-12-0	Propionitrile		10	500
542-76-7	Propionitrile, 3-Chloro-		1,000	1,000
70-69-9	Propiophenone, 4-Amino-	e, g	1	100/10,000
109-61-5	Propyl Chloroformate	e	1	500
75-56-9	Propylene Oxide	l	100	10,000
75-55-8	Propyleneimine	d	1	10,000
2275-18-5	Prothoate	e	1	100/10,000
129-00-0	Pyrene	c	5,000	1,000/10,000
140-76-1	Pyridine, 2-Methyl-5-Vinyl-	e	1	500
504-24-5	Pyridine, 4-Amino-	h	1,000	500/10,000
1124-33-0	Pyridine, 4-Nitro-, 1-Oxide	e	1	500/10,000
53558-25-1	Pyriminil	e, h	1	100/10,000

TABLE 7.1   List of Extremely Hazardous Substances and Their Threshold Planning Quantities (Continued)

[Alphabetical Order]

CAS No.	Chemical name	Notes	Reportable quantity* (pounds)	Threshold planning quantity (pounds)
14167-18-1	Salcomine	e	1	500/10,000
107-44-8	Sarin	e, h	1	10
7783-00-8	Selenious Acid		10	1,000/10,000
7791-23-3	Selenium Oxychloride	e	1	500
563-41-7	Semicarbazide Hydrochloride	e	1	1,000/10,000
3037-72-7	Silane, (4-Aminobutyl)Diethoxymethyl-	e	1	1,000
7631-89-2	Sodium Arsenate	d	1,000	1,000/10,000
7784-46-5	Sodium arsenite	d	1	500/10,000
26628-22-8	Sodium Azide (Na(N3))	b	1,000	500
124-65-2	Sodium Cacodylate	e	1	100/10,000
143-33-9	Sodium Cyanide (Na(CN))	b	10	100
62-74-8	Sodium Fluoroacetate		10	10/10,000
13410-01-0	Sodium Selenate	e	1	100/10,000
10102-18-8	Sodium Selenite	h	100	100/10,000
10102-20-2	Sodium Tellurite	e	1	500/10,000
900-95-8	Stannane, Acetoxytriphenyl-	e, g	1	500/10,000
57-24-9	Strychnine	c	10	100/10,000
60-41-3	Strychnine sulfate	e	10	100/10,000
3689-24-5	Sulfotep		100	500
3569-57-1	Sulfoxide, 3-Chloropropyl Octyl	e	1	500
7446-09-5	Sulfur Dioxide	e, l	1	500
7783-60-0	Sulfur Tetrafluoride	e	1	100
7446-11-9	Sulfur Trioxide	b, e	1	100
7664-93-9	Sulfuric Acid		1,000	1,000
77-81-6	Tabun	c, e, h	1	10
13494-80-9	Tellurium	e	1	500/10,000
7783-80-4	Tellurium Hexafluoride	e, k	1	100
107-49-3	TEPP		10	100
13071-79-9	Terbufos	e, h	1	100
78-00-2	Tetraethyllead	c, d	10	100
597-64-8	Tetraethyltin	c, e	1	100
75-74-1	Tetramethyllead	c, e, l	1	100
509-14-8	Tetranitromethane		10	500
10031-59-1	Thallium Sulfate	h	100	100/10,000
6533-73-9	Thallous Carbonate	c, h	100	100/10,000
7791-12-0	Thallous Chloride	c, h	100	100/10,000
2757-18-8	Thallous Malonate	c, e, h	1	100/10,000
7446-18-6	Thallous Sulfate		100	100/10,000
2231-57-4	Thiocarbazide	e	1	1,000/10,000
39196-18-4	Thiofanox		100	100/10,000
297-97-2	Thionazin		100	500
108-98-5	Thiophenol		100	500
79-19-6	Thiosemicarbazide		100	100/10,000
5344-82-1	Thiourea, (2-Chlorophenyl)-		100	100/10,000
614-78-8	Thiourea, (2-Methylphenyl)-	e	1	500/10,000
7550-45-0	Titanium Tetrachloride	e	1	100
584-84-9	Toluene 2,4-Diisocyanate		100	500
91-08-7	Toluene 2,6-Diisocyanate		100	100
110-57-6	Trans-1,4-Dichlorobutene	e	1	500
1031-47-6	Triamiphos	e	1	500/10,000
24017-47-8	Triazofos	e	1	500
76-02-8	Trichloroacety Chloride	e	1	500
115-21-9	Trichloroethylsilane	e, h	1	500
327-98-0	Trichloronate	e, k	1	500
98-13-5	Trichlorophenylsilane	e, h	1	500
1558-25-4	Trichloro(Chloromethyl)Silane	e	1	100
27137-85-5	Trichloro(Dichlorophenyl)Silane	e	1	500
998-30-1	Triethoxysilane	e	1	500
75-77-4	Trimethylchlorosilane	e	1	1,000
824-11-3	Trimethylolpropane Phosphite	e, h	1	100/10,000
1066-45-1	Trimethyltin Chloride	e	1	500/10,000
639-58-7	Triphenyltin Chloride	e	1	500/10,000
555-77-1	Tris(2-Chloroethyl)Amine	e, h	1	100
2001-95-8	Valinomycin	c, e	1	1,000/10,000
1314-62-1	Vanadium Pentoxide		1,000	100/10,000
108-05-4	Vinyl Acetate Monomer	d, l	5,000	1,000

TABLE 7.1   List of Extremely Hazardous Substances and Their Threshold Planning Quantities (*Continued*)

[Alphabetical Order]

CAS No.	Chemical name	Notes	Reportable quantity* (pounds)	Threshold planning quantity (pounds)
81–81–2	Warfarin		100	500/10,000
129–06–6	Warfarin sodium...	e,h	100	100/10,000
28347–13–9	Xylylene Dichloride..	e	1	100/10,000
58270–08–9	Zinc, Dichloro(4,4-Dimethyl-5((((Methylamino) Carbonyl)Oxy)Imino)Pentanenitrile)-,(T-4)-.	e	1	100/10,000
1314–84–7	Zinc Phosphide..	b	100	500

*Only the statutory or final RQ is shown. For more information, see 40 CFR Table 302.4
Notes:
a This chemical does not meet acute toxicity criteria. Its TPQ is set at 10,000 pounds.
b This material is a reactive solid. The TPQ does not default to 10,000 pounds for non-powder, non-molten, non-solution form.
c The calculated TPQ changed after technical review as described in the technical support document.
d Indicates that the RQ is subject to change when the assessment of potential carcinogenicity and/or other toxicity is completed.
e Statutory reportable quantity for purposes of notification under SARA sect 304(a)(2).
f The statutory 1 pound reportable quantity for methyl isocyanate may be adjusted in a future rulemaking action.
g New chemicals added that were not part of the original list of 402 substances.
h Revised TPQ based on new or re-evaluated toxicity data.
j TPQ is revised to its calculated value and does not change due to technical review as in proposed rule.
k The TPQ was revised after proposal due to calculation error.
l Chemicals on the original list that do not meet toxicity criteria but because of their high production volume and recognized toxicity are considered chemicals of concern ("Other chemicals").

Table 7.3 (see page 276) is a reproduction of EPA's list of toxic chemicals under Title III of SARA, otherwise known as the Title III "list of lists."

7.3   Planning and Implementation Schedule

Under the general provisions of EPCRA, several deadlines of emergency planning are set for the purpose of timely implementation:

- April 17, 1987: appointment of a state "emergency response commission" by the governor of each state

- May 17, 1987: determination whether the facilities are subject to Title III

- July 17, 1987: establishment of state emergency planning districts

- August 17, 1987: establishment of local emergency planning committees to prepare an emergency plan

- September 17, 1987: appointment of emergency coordinators

- October 17, 1987: provision to local and state committees and fire departments of the MSDSs or a list of on-site chemicals in each facility

- March 1, 1988: start of annual submittals of inventory and release forms to appropriate authorities

- October 17, 1988: preparation of an emergency plan containing certain minimum details by local emergency planning committees

- June 30, 1991: submission of a report by the comptroller general on toxic chemical release

- October 17, 1991: submission of a report by EPA on its mass balance study to determine the accuracy of toxic chemical release information

These deadlines for various implementation plans indicate that the emergency planning is fully effective at this time. As a result, every facility where any extremely hazardous substance is present in an amount equal to or greater than its TPQ must have an emergency plan to deal with the release of that EHS. In addition, such a facility is required to provide an emergency release notification in the event of a release. Pursuant to emergency release notification requirements, a facility will have to notify, as discussed below, under certain circumstances. To facilitate fulfillment of these obligations, a facility coordinator responsible for implementing the emergency plan must be designated by such facility.

Emergency release notification

EPA regulations for emergency release notification are set forth in 40 CFR section 355.40. This notification is required of a facility where

- A hazardous chemical is produced, used, or stored.

TABLE 7.2 List of Hazardous Substances and Reportable Quantities

[Note: All Comments/Notes Are Located at the End of This Table]

Hazardous Substance	CASRN	Regulatory Synonyms	Statutory RQ	Statutory Code †	Statutory RCRA Waste Number	Final RQ Category	Final RQ Pounds (Kg)
Acenaphthene	83329	....	1*	2		B	100 (45.4)
Acenaphthylene	208968		1*	2		D	5000 (2270)
Acetaldehyde	75070	Ethanal	1000	1,4	U001	C	1000 (454)
Acetaldehyde, chloro-	107200	Chloroacetaldehyde	1*	4	P023	C	1000 (454)
Acetaldehyde, trichloro-	75876	Chloral	1*	4	U034	D	5000 (2270)
Acetamide, N-(aminothioxomethyl)-	591082	1-Acetyl-2-thiourea	1*	4	P002	C	1000 (454)
Acetamide, N-(4-ethoxyphenyl)-	62442	Phenacetin	1*	4	U187	B	100 (45.4)
Acetamide, 2-fluoro-	640197	Fluoroacetamide	1*	4	P057	B	100 (45.4)
Acetamide, N-9H-fluoren-2-yl-	53963	2-Acetylaminofluorene	1*	4	U005	X	1 (0.454)
Acetic acid	64197		1000	1		D	5000 (2270)
Acetic acid (2,4-dichlorophenoxy)-	94757	2,4-D Acid; 2,4-D, salts and esters	100	1,4	U240	B	100 (45.4)
Acetic acid, lead(2+) salt	301042	Lead acetate	5000	1,4	U144	B	100 (45.4) #
Acetic acid, thallium(1+) salt	563688	Thallium(I) acetate	1*	4	U214	B	100 (45.4)
Acetic acid, (2,4,5-trichlorophenoxy)	93765	2,4,5-T	100	1,4	U232	C	1000 (454)
Acetic acid, ethyl ester	141786	Ethyl acetate	1*	4	U112	D	5000 (2270)
Acetic acid, fluoro-, sodium salt	62748	Fluoroacetic acid, sodium salt	1*	4	P058	A	10 (4.54)
Acetic anhydride	108247		1000	1		D	5000 (2270)
Acetone	67641	2-Propanone	1*	4	U002	D	5000 (2270)
Acetone cyanohydrin	75865	Propanenitrile, 2-hydroxy-2-methyl-2-Methyllactonitrile	10	1,4	P069	A	10 (4.54)
Acetonitrile	75058		1*	4	U003	D	5000 (2270)
Acetophenone	98862	Ethanone, 1-phenyl-	1*	4	U004	D	5000 (2270)
2-Acetylaminofluorene	53963	Acetamide, N-9H-fluoren-2-yl-	1*	4	U005	X	1 (0.454)
Acetyl bromide	506967		5000	1		D	5000 (2270)
Acetyl chloride	75365		5000	1,4	U006	D	5000 (2270)
1-Acetyl-2-thiourea	591082	Acetamide, N-(aminothioxomethyl)-	1*	4	P002	C	1000 (454)
Acrolein	107028	2-Propenal	1	1,2,4	P003	X	1 (0.454)
Acrylamide	79061	2-Propenamide	1*	4	U007	D	5000 (2270)
Acrylic acid	79107	2-Propenoic acid	1*	4	U008	D	5000 (2270)
Acrylonitrile	107131	2-Propenenitrile	100	1,2,4	U009	B	100 (45.4)
Adipic acid	124049		5000	1		D	5000 (2270)
Aldicarb	116063	Propanal, 2-methyl-2-(methylthio)-, O-[(methylamino)carbonyl]oxime	1*	4	P070	X	1 (0.454)
Aldrin	309002	1,4,5,8-Dimethanonaphthalene, 1,2,3,4,10,10-hexachloro-1,4,4a,5,8,8a-hexahydro-, (1alpha,4alpha,4abeta,5alpha,8alpha,8abeta)-	1	1,2,4	P004	X	1 (0.454)
Allyl alcohol	107186	2-Propen-1-ol	100	1,4	P005	B	100 (45.4)
Allyl chloride	107051		1000	1		C	1000 (454)

Name	CAS Number	Statutory RQ	Code	Waste No.	Category	Final RQ pounds (kg)
Aluminum phosphide	20859738	1*	4	P006	B	100 (45.4)
Aluminum sulfate	10043013	5000	4		D	5000 (2270)
5-(Aminomethyl)-3-isoxazolol [Muscimol 3(2H)-Isoxazolone, 5-(aminomethyl)-]	2763964	1*	4	P007	C	1000 (454)
4-Aminopyridine [4-Pyridinamine]	504245	1*	4	P008	C	1000 (454)
Amitrole [1H-1,2,4-Triazol-3-amine]	61825	1*	4	U011	A	10 (4.54)
Ammonia	7664417	100	1		B	100 (45.4)
Ammonium acetate	631618	5000	1		D	5000 (2270)
Ammonium benzoate	1863634	5000	1		D	5000 (2270)
Ammonium bicarbonate	1066337	5000	1		D	5000 (2270)
Ammonium bichromate	7789095	1000	1		A	10 (4.54)
Ammonium bifluoride	1341497	5000	1		B	100 (45.4)
Ammonium bisulfite	10192300	5000	1		D	5000 (2270)
Ammonium carbamate	1111780	5000	1		D	5000 (2270)
Ammonium carbonate	506876	5000	1		D	5000 (2270)
Ammonium chloride	12125029	5000	1		D	5000 (2270)
Ammonium chromate	7788989	1000	1		A	10 (4.54)
Ammonium citrate, dibasic	3012655	5000	1		D	5000 (2270)
Ammonium fluoborate	13826830	5000	1		D	5000 (2270)
Ammonium fluoride	12125018	5000	1		B	100 (45.4)
Ammonium hydroxide	1336216	1000	1		C	1000 (454)
Ammonium oxalate	6009707	5000	1		D	5000 (2270)
	5972736					
Ammonium picrate [Phenol, 2,4,6-trinitro-, ammonium salt]	14258492 131748	1*	4	P009	A	10 (4.54)
Ammonium silicofluoride	16919190	1000	1		C	1000 (454)
Ammonium sulfamate	77730060	5000	1		D	5000 (2270)
Ammonium sulfide	12135761	5000	1		B	100 (45.4)
Ammonium sulfite	10196040	5000	1		D	5000 (2270)
Ammonium tartrate	14307438 3164292	5000	1		D	5000 (2270)
Ammonium thiocyanate	1762954	5000	1		D	5000 (2270)
Ammonium vanadate [Vanadic acid, ammonium salt]	7803556	1*	4	P119	C	1000 (454)
Amyl acetate	628637	1000	1		D	5000 (2270)
iso-Amyl acetate	123922	1*				
sec-Amyl acetate	626380	1*				
tert-Amyl acetate	625161	1*				
Aniline [Benzenamine]	62533	1000	1,4	U012	D	5000 (2270)
Anthracene	120127	1*	2		D	5000 (2270)
Antimony ††	7440360	1*	2		D	5000 (2270)
ANTIMONY AND COMPOUNDS	N.A.	1*	2			**
Antimony pentachloride	7647189	1000	1		C	1000 (454)
Antimony potassium tartrate	28300745	1000	1		B	100 (45.4)
Antimony tribromide	7789619	1000	1		C	1000 (454)
Antimony trichloride	10025919	1000	1		C	1000 (454)
Antimony trifluoride	7783564	1000	1		C	1000 (454)
Antimony trioxide	1309644	1000	1		C	1000 (454)
Argentate(1-), bis(cyano-C)-, potassium [Potassium silver cyanide]	506616	5000	4	P099	X	1000 (454)
Aroclor 1016 POLYCHLORINATED BIPHENYLS (PCBs)	12674112	10	1,2		X	1 (0.454)
Aroclor 1221 POLYCHLORINATED BIPHENYLS (PCBs)	11104282	10	1,2		X	1 (0.454)
Aroclor 1232 POLYCHLORINATED BIPHENYLS (PCBs)	11141165	10	1,2		X	1 (0.454)

TABLE 7.2 List of Hazardous Substances and Reportable Quantities (Continued)

[Note: All Comments/Notes Are Located at the End of This Table]

Hazardous Substance	CASRN	Regulatory Synonyms	Statutory			Final RQ	
			RQ	Code †	RCRA Waste Number	Category	Pounds (Kg)
Aroclor 1242	53469219	POLYCHLORINATED BIPHENYLS (PCBs)	10	1,2		X	1 (0.454)
Aroclor 1248	12672296	POLYCHLORINATED BIPHENYLS (PCBs)	10	1,2		X	1 (0.454)
Aroclor 1254	11097691	POLYCHLORINATED BIPHENYLS (PCBs)	10	1,2		X	1 (0.454)
Aroclor 1260	11096825	POLYCHLORINATED BIPHENYLS (PCBs)	10	1,2		X	1 (0.454)
Arsenic ††	7440382		1*	2,3		X	1 (0.454)
Arsenic acid	1327522	Arsenic acid H3AsO4	1*	4	P010	X	1 (0.454)
	7778394						
Arsenic acid H3AsO4	1327522	Arsenic acid	1*	4	P010	X	1 (0.454)
	7778394						
ARSENIC AND COMPOUNDS	N.A.		1*	2			**
Arsenic disulfide	1303328		5000	1		X	1 (0.454)
Arsenic oxide As2O3	1327533	Arsenic trioxide	5000	1,4	P012	X	1 (0.454)
Arsenic oxide As2O5	1303282	Arsenic pentoxide	5000	1,4	P011	X	1 (0.454)
Arsenic pentoxide	1303282	Arsenic oxide As2O5	5000	1,4	P011	X	1 (0.454)
Arsenic trichloride	7784341		5000	1		X	1 (0.454)
Arsenic trioxide	1327533	Arsenic oxide As2O3	5000	1,4	P012	X	1 (0.454)
Arsenic trisulfide	1303339		5000	1		X	1 (0.454)
Arsine, diethyl-	692422	Diethylarsine	1*	4	P038	X	1 (0.454)
Arsinic acid, dimethyl-	75605	Cacodylic acid	1*	4	U136	X	1 (0.454)
Arsonous dichloride, phenyl-	696286	Dichlorophenylarsine	1*	4	P036	X	1 (0.454)
Asbestos †††	1332214		1*	2,3		A	100 (45.4)
Auramine	492808	Benzenamine, 4,4'-carbonimidoylbis (N,N-dimethyl-	1*	4	U014	B	1 (0.454)
Azaserine	115026	L-Serine, diazoacetate (ester)	1*	4	U015	X	1 (0.454)
Aziridine, 2-methyl-	151564	1,2-Propylenimine	1*	4	P054	X	1 (0.454)
Azirino[2',3':3,4]pyrrolo[1,2-a]indole-4,7-dione,6-amino-8-[[(aminocarbonyl)oxy]methyl]-1,1a,2,8,8a,8b-hexahydro-8a-methoxy-5-methyl-,[1aS-(1aalpha,8beta,8aalpha,8balpha)]-	50077	Mitomycin C	1*	4	U010	A	10 (4.54)
Barium cyanide	542621		10	1,4	P013	A	10 (4.54)
Benz[j]aceanthrylene, 1,2-dihydro-3-methyl-	56495	3-Methylcholanthrene	1*	4	U157	A	10 (4.54)
Benz[c]acridine	225514		1*	4	U016	B	100 (45.4)
Benzal chloride	98873	Benzene, dichloromethyl-	1*	4	U017	D	5000 (2270)
Benzamide, 3,5-dichloro-N-(1,1-dimethyl-2-propynyl)-	23950585	Pronamide	1*	4	U192	D	5000 (2270)
Benz[a]anthracene	56553	1,2-Benzanthracene	1*	2,4	U018	A	10 (4.54)
1,2-Benzanthracene	56553	Benz[a]anthracene	1*	2,4	U018	A	10 (4.54)
Benz[a]anthracene, 7,12-dimethyl-	57976	7,12-Dimethylbenz[a]anthracene	1*	4	U094	X	1 (0.454)
Benzenamine	62533	Aniline	1000	1,4	U012	D	5000 (2270)

Substance	CAS No.	Synonym			Code		RQ
Benzenamine, 4,4'-carbonimidoylbis (N,N-dimethyl-	492808	Auramine	1*	4	U014	B	100 (45.4)
Benzenamine, 4-chloro-	106478	p-Chloroaniline	1*	4	P024	C	1000 (454)
Benzenamine, 4-chloro-2-methyl-, hydrochloride	3165933	4-Chloro-o-toluidine, hydrochloride	1*	4	U049	B	100 (45.4)
Benzenamine, N,N-dimethyl-4-(phenylazo)-	60117	p-Dimethylaminoazobenzene	1*	4	U093	A	10 (4.54)
Benzenamine, 2-methyl-	95534	o-Toluidine	1*	4	U328	B	100 (45.4)
Benzenamine, 4-methyl-	106490	p-Toluidine	1*	4	U353	B	100 (45.4)
Benzenamine, 4,4'-methylenebis(2-chloro-	101144	4,4'-Methylenebis(2-chloroaniline)	1*	4	U158	A	10 (4.54)
Benzenamine, 2-methyl- hydrochloride	636215	o-Toluidine hydrochloride	1*	4	U222	B	100 (45.4)
Benzenamine, 2-methyl-5-nitro-	99558	5-Nitro-o-toluidine	1*	4	U181	B	100 (45.4)
Benzenamine, 4-nitro-	100016	p-Nitroaniline	1000	4	P077	D	5000 (2270)
Benzene	71432	Benzene	1*	1,2,3,4	U019	A	10 (4.54)
Benzeneacetic acid, 4-chloro-alpha-(4-chloro-phenyl)-alpha-hydroxy-, ethyl ester	510156	Chlorobenzilate	1*	4	U038	A	10 (4.54)
Benzene, 1-bromo-4-phenoxy-	101553	4-Bromophenyl phenyl ether	1*	2,4	U030	B	100 (45.4)
Benzenebutanoic acid, 4-[bis(2-chloroethyl)amino]-	305033	Chlorambucil	1*	4	U035	A	10 (4.54)
Benzene, chloro-	108907	Chlorobenzene	100	1,2,4	U037	B	100 (45.4)
Benzene, chloromethyl-	100447	Benzyl chloride	100	1,4	P028	B	100 (45.4)
Benzenediamin, ar-methyl-	95807	Toluenediamine	1*	4	U221	A	10 (4.54)
	496720						
	823405						
1,2-Benzenedicarboxylic acid, dioctyl ester	117840	Di-n-octyl phthalate	1*	2,4	U107	D	5000 (2270)
1,2-Benzenedicarboxylic acid, [bis(2-ethylhexyl)]-ester.	117817	Bis (2-ethylhexyl)phthalate / Diethylhexyl phthalate	1*	2,4	U028	B	100 (45.4)
1,2-Benzenedicarboxylic acid, dibutyl ester.	84742	Di-n-butyl phthalate / Dbutyl phthalate / n-Butyl phthalate	100	1,2,4	U069	A	10 (4.54)
1,2-Benzenedicarboxylic acid, diethyl ester.	84662	Diethyl phthalate	1*	2,4	U088	C	1000 (454)
1,2-Benzenedicarboxylic acid, dimethyl ester	131113	Dimethyl phthalate	1*	2,4	U102	D	5000 (2270)
Benzene, 1,2-dichloro-	95501	o-Dichlorobenzene / 1,2-Dichlorobenzene	100	1,2,4	U070	B	100 (45.4)
Benzene, 1,3-dichloro-	541731	m-Dichlorobenzene / 1,3-Dichlorobenzene	1*	2,4	U071	B	100 (45.4)
Benzene, 1,4-dichloro-	106467	p-Dichlorobenzene / 1,4-Dichlorobenzene	100	1,2,4	U072	B	100 (45.4)
Benzene, 1,1'-(2,2-dichloroethylidene)bis[4-chloro-	72548	DDD / TDE / 4,4' DDD	1	1,2,4	U060	X	1 (0.454)
Benzene, dichloromethyl-	98873	Benzal chloride	1*	4	U017	D	5000 (2270)
Benzene, 1,3-diisocyanatomethyl-	584849	Toluene diisocyanate	1*	4	U223	B	100 (45.4)
	91087						
	26471625						
Benzene, dimethyl-	1330207	Xylene (mixed)	1000	1,4	U239	C	1000 (454)
m-Benzene, dimethyl	108383	m-Xylene					
o-Benzene, dimethyl	95476	o-Xylene					
p-Benzene, dimethyl	106423	p-Xylene					
1,3-Benzenediol	108463	Resorcinol	1000	1,4	U201	D	5000 (2270)
1,2-Benzenediol,4-[1-hydroxy-2-(methylamino)ethyl]-	51434	Epinephrine	1*	4	P042	C	1000 (454)
Benzeneethanamine, alpha,alpha-dimethyl-	122098	alpha,alpha-Dimethylphenethylamine	1*	4	P046	D	5000 (2270)
Benzene, hexachloro-	118741	Hexachlorobenzene	1*	2,4	U127	A	10 (4.54)
Benzene, hexahydro-	110827	Cyclohexane	1000	1,4	U056	C	1000 (454)

TABLE 7.2 List of Hazardous Substances and Reportable Quantities (*Continued*)

[Note: All Comments/Notes Are Located at the End of This Table]

Hazardous Substance	CASRN	Regulatory Synonyms	Statutory			Final RQ	
			RQ	Code †	RCRA Waste Number	Category	Pounds (Kg)
Benzene, hydroxy-	108952	Phenol	1000	1,2,4	U188	C	1000 (454)
Benzene, methyl-	108883	Toluene	1000	1,2,4	U220	C	1000 (454)
Benzene, 2-methyl-1,3-dinitro-	606202	2,6-Dinitrotoluene	1000	1,2,4	U106	B	100 (45.4)
Benzene, 1-methyl-2,4-dinitro-	121142	2,4-Dinitrotoluene	1000	1,2,4	U105	A	10 (4.54)
Benzene, 1-methylethyl-	98828	Cumene	1*	4	U055	D	5000 (2270)
Benzene, nitro-	98953	Nitrobenzene	1000	1,2,4	U169	C	1000 (454)
Benzene, pentachloro-	608935	Pentachlorobenzene	1*	4	U183	A	10 (4.54)
Benzene, pentachloronitro-	82688	Pentachloronitrobenzene (PCNB)	1*	4	U185	B	100 (45.4)
Benzenesulfonic acid chloride	98099	Benzenesulfonyl chloride	1*	4	U020	B	100 (45.4)
Benzenesulfonyl chloride	98099	Benzenesulfonic acid chloride	1*	4	U020	B	100 (45.4)
Benzene, 1,2,4,5-tetrachloro-	95943	1,2,4,5-Tetrachlorobenzene	1*	4	U207	D	5000 (2270)
Benzenethiol	108985	Thiophenol	1*	4	P014	B	100 (45.4)
Benzene, 1,1'-(2,2,2-tri-chloroethylidene)bis[4-chloro-	50293	DDT	1	1,2,4	U061	X	1 (0.454)
Benzene, 1,1'-(2,2,2-trichloroethylidene) bis[4-methoxy-	72435	4,4'DDT Methoxychlor	1	1,4	U247	X	1 (0.454)
Benzene, (trichloromethyl)-	98077	Benzotrichloride	1*	4	U023	A	10 (4.54)
Benzene, 1,3,5-trinitro-	99354	1,3,5-Trinitrobenzene	1*	4	U234	A	10 (4.54)
Benzidine	92875	(1,1'-Biphenyl)-4,4'diamine	1*	2,4	U021	X	1 (0.454)
1,2-Benzisothiazol-3(2H)-one, 1,1-dioxide	81072	Saccharin and salts	1*	4	U202	B	100 (45.4)
Benzo[a]anthracene	56553	1,2-Benzanthracene	1*	2,4	U018	A	10 (4.54)
Benzo[b]fluoranthene	205992		1*	2		X	1 (0.454)
Benzo[k]fluoranthene	207089		1*	2		D	5000 (2270)
Benzo[j,k]fluorene	206440	Fluoranthene	1*	2,4	U120	B	100 (45.4)
1,3-Benzodioxole, 5-)1-propenyl)-	120581	Isosafrole	1*	4	U141	B	100 (45.4)
1,3-Benzodioxole, 5-(2-propenyl)-	94597	Safrole	1*	4	U203	B	100 (45.4)
1,3-Benzodioxole, 5-propyl-	94586	Dihydrosafrole	1*	4	U090	A	10 (4.54)
Benzoic acid	65850		5000	1		D	5000 (2270)
Benzonitrile	100470		1000	1		D	5000 (2270)
Benzo [rst]pentaphene	189559	Dibenz[a,i]pyrene	1*	4	U064	A	10 (4.54)
Benzo[ghi]perylene	191242		1*	2		D	5000 (2270)
2H-1-Benzopyran-2-one, 4-hydroxy-3-(3-oxo-1-phenyl-butyl)-, & salts, when present at concentrations greater than 0.3%	81812	Warfarin, & salts, when present at concentrations greater than 0.3%	1*	4	P001	B	100 (45.4)
3,4-Benzopyrene	50328	Benzo[a]pyrene	1*	2,4	U022	X	1 (0.454)
3,4-Benzopryene	50328	Benzo[a]pyrene	1*	2,4	U022	X	1 (0.454)
p-Benzoquinone	106514	2,5-Cyclohexadiene-1,4-dione	1*	4	U197	A	10 (4.54)
Benzotrichloride	98077	Benzene, (trichloromethyl)-	1*	4	U023	A	10 (4.54)
Benzoyl chloride	98884		1000	1		C	1000 (454)

Hazardous substance	CAS No.	Synonym			RCRA waste No.		RQ in pounds (kilograms)
1,2-Benzphenanthrene	218019	Chrysene	1*	2,4	U050	B	100 (45.4)
Benzyl chloride	100447	Benzene, chloromethyl-	100	1,4	P028	B	100 (45.4)
Beryllium ††	7440417	Beryllium dust ††	1*	2,3,4	P015	A	10 (4.54)**
BERYLLIUM AND COMPOUNDS	N.A.		1*	2			
Beryllium chloride	7787475		5000	1		X	1 (0.454)
Beryllium dust ††	7440417	Beryllium ††	1*	2,3,4	P015	A	10 (4.54)
Beryllium fluoride	7787497		5000	1		X	1 (0.454)
Beryllium nitrate	13597994 7787555		5000	1		X	1 (0.454)
alpha—BHC	319846		1*	2		A	10 (4.54)
beta—BHC	319857		1*	2		X	1 (0.454)
delta—BHC	319868		1*	2		X	1 (0.454)
gamma—BHC	58899	Cyclohexane, 1,2,3,4,5,6-hexachloro-,(1alpha, 2alpha,3beta,4alpha,5alpha,6beta)- Hexachlorocyclohexane (gamma isomer) Lindane	1	1,2,4	U129	X	1 (0.454)
2,2'-Bioxirane	1464535	1,2:3,4-Diepoxybutane	1*	4	U085	A	10 (4.54)
(1,1'-Biphenyl)-4,4'diamine	92875	Benzidine	1*	2,4	U021	X	1 (0.454)
[1,1'-Biphenyl]-4,4'diamine,3,3'dichloro-	91941	3,3'-Dichlorobenzidine	1*	2,4	U073	B	1 (0.454)
[1,1'-Biphenyl]-4,4'diamine,3,3'dimethoxy-	119904	3,3'-Dimethoxybenzidine	1*	4	U091	A	100 (45.4)
[1,1'Biphenyl]-4,4'-diamine,3,3'-dimethyl-	119937	3,3'-Dimethylbenzidine	1*	4	U095	A	10 (4.54)
Bis (2-chloroethyl) ether.	111444	Dichloroethyl ether / Ethane,1,1'-oxybis[2-chloro-	1*	2,4	U025	C	10 (4.54)
Bis(2-chloroethoxy) methane	111911	Dichloromethoxy ethane / Ethane, 1,1'-[methylenebis(oxy)]bis[2-chloro-	1*	2,4	U024	B	1000 (454)
Bis (2-ethylhexyl)phthalate	117817	Diethylhexyl phthalate / 1,2-Benzenedicarboxylic acid, [bis(2-ethylhexyl)] ester	1*	2,4	U028	C	100 (45.4)
Bromoacetone	598312	2-Propanone, 1-bromo-	1*	4	P017	B	1000 (454)
Bromoform	75252	Methane, tribromo-	1*	2,4	U225	B	100 (45.4)
4-Bromophenyl phenyl ether	101553	Benzene, 1-bromo-4-phenoxy-	1*	2,4	U030	B	100 (45.4)
Brucine	357573	Strychnidin-10-one, 2,3-dimethoxy-	1*	4	P018	X	100 (45.4)
1,3-Butadiene, 1,1,2,3,4,4-hexachloro-	87683	Hexachlorobutadiene	1*	2,4	U128	A	1 (0.454)
1-Butanamine, N-butyl-N-nitroso-	924163	N-Nitrosodi-n-butylamine	1*	4	U172	D	10 (4.54)
1-Butanol	71363	n-Butyl alcohol	1*	4	U031	D	5000 (2270)
2-Butanone	78933	Methyl ethyl ketone (MEK)	1*	4	U159	A	5000 (2270)
2-Butanone peroxide	1338234	Methyl ethyl ketone peroxide	1*	4	U160	B	10 (4.54)
2-Butanone, 3,3-dimethyl-1-(methylthio)-, O[(methylamino)carbonyl] oxime.	39196184	Thiofanox	1*	4	P045	B	100 (45.4)
2-Butenal	123739 4170303	Crotonaldehyde	100	1,4	U053	B	100 (45.4)
2-Butene, 1,4-dichloro-	764410	1,4-Dichloro-2-butene	1*	4	U074	X	1 (0.454)
2-Butenoic acid, 2-methyl-, 7[[2,3-dihydroxy-2-(1-methoxyethyl)-3-methyl-1-oxobutoxy]methyl]-2,3,5,7a-tetrahydro-1H-pyrrolizin-1-yl ester, [1S-[1alpha(Z),7(2S*,3R*),7aalpha]]-	303344	Lasiocarpine	1*	4	U143	A	10 (4.54)
Butyl acetate	123864		5000	1		D	5000 (2270)
iso-Butyl acetate	110190						
sec-Butyl acetate	105464						
tert-Butyl acetate	540885						
n-Butyl alcohol	71363	1-Butanol	1*	4	U031	D	5000 (2270)

TABLE 7.2 List of Hazardous Substances and Reportable Quantities *(Continued)*

[Note: All Comments/Notes Are Located at the End of This Table]

Hazardous Substance	CASRN	Regulatory Synonyms	Statutory			Final RQ	
			RQ	Code †	RCRA Waste Number	Category	Pounds (Kg)
Butylamine	109739		1000	1		C	1000 (454)
iso-Butylamine	78819						
sec-Butylamine	513495						
	13952846						
tert-Butylamine	75649						
Butyl benzyl phthalate	85687		1*	2		B	100 (45.4)
n-Butyl phthalate	84742	Di-n-butyl phthalate	100	1,2,4	U069	A	10 (4.54)
		Dibutyl phthalate					
		1,2-Benzenedicarboxylic acid, dibutyl ester					
Butyric acid	107926		5000	1		D	5000 (2270)
iso-Butyric acid	79312						
Cacodylic acid	75605	Arsinic acid, dimethyl-	1*	4	U136	X	1 (0.454)
Cadmium ††	7440439		1*	2		A	10 (4.54)
Cadmium acetate	543908		100	1		A	10 (4.54)
CADMIUM AND COMPOUNDS	N.A.		1*	2			**
Cadmium bromide	7789426		100	1		A	10 (4.54)
Cadmium chloride	10108642		100	1		A	10 (4.54)
Calcium arsenate	7778441		1000	1		X	1 (0.454)
Calcium arsenite	52740166		1000	1		X	1 (0.454)
Calcium carbide	75207		5000	1		A	10 (4.54)
Calcium chromate	13765190	Chromic acid H2CrO4, calcium salt	1000	1,4	U032	A	10 (4.54)
Calcium cyanide	592018	Calcium cyanide Ca(CN)2	1*	1,4	P021	A	10 (4.54)
Calcium cyanide Ca(CN)2	592018	Calcium cyanide	10	1,4	P021	A	10 (4.54)
Calcium dodecylbenzenesulfonate	26264062		1000	1		C	1000 (454)
Calcium hypochlorite	7778543		100	1		A	10 (4.54)
Camphene, octachloro-	8001352	Toxaphene	1	1,2,4	P123	X	1 (0.454)
Captan	133062		10	1		A	10 (4.54)
Carbamic acid, ethyl ester	51796	Ethyl carbamate (urethane)	1*	4	U238	B	100 (45.4)
Carbamic acid, methylnitroso-, ethyl ester	615532	N-Nitroso-N-methylurethane	1*	4	U178	X	1 (0.454)
Carbamic chloride, dimethyl-	79447	Dimethylcarbamoyl chloride	1*	4	U097	X	1 (0.454)
Carbamodithioic acid, 1,2-ethanediylbis, salts & esters	111546	Ethylenebisdithiocarbamic acid, salts & esters	1*	4	U114	D	5000 (2270)
Carbamothioic acid, bis(1-methylethyl)-, S-(2,3-dichloro-2-propenyl) ester	2303164	Diallate	1*	4	U062	B	100 (45.4)
Carbaryl	63252		100	1		B	100 (45.4)
Carbofuran	1563662		10	1		A	10 (4.54)
Carbon disulfide	75150		5000	1,4	P022	B	100 (45.4)
Carbon oxyfluoride	353504	Carbonic difluoride	1*	4	U033	C	1000 (454)
Carbon tetrachloride	56235	Methane, tetrachloro-	5000	1,2,4	U211	A	10 (4.54)
Carbonic acid, dithallium(1+) salt	6533739	Thallium(I) carbonate	1*	4	U215	B	100 (45.4)

Name	CAS No.	Synonyms			Code		RQ
Carbonic dichloride	75445	Phosgene	5000	1,4	P095	A	10 (4.54)
Carbonic difluoride	353504	Carbon oxyfluoride	1*	4	U033	C	1000 (454)
Carbonochloridic acid, methyl ester	79221	Methyl chlorocarbonate; Methyl chloroformate	1*	4	U156	C	1000 (454)
Chloral	75876	Acetaldehyde, trichloro-	1*	4	U034	D	5000 (2270)
Chlorambucil	305033	Benzenebutanoic acid, 4-[bis(2-chloroethyl)amino]-	1*	4	U035	A	10 (4.54)
Chlordane	57749	Chlordane, alpha & gamma isomers; Chlordane, technical; 4,7-Methano-1H-indene, 1,2,4,5,6,7,8,8-octachloro-2,3,3a,4,7,7a-hexahydro-	1	1,2,4	U036	X	1 (0.454)
CHLORDANE (TECHNICAL MIXTURE AND METABOLITES)	N.A.		1*	2			**
Chlordane, alpha & gamma isomers	57749	Chlordane, technical; 4,7-Methano-1H-indene, 1,2,4,5,6,7,8,8-octachloro-2,3,3a,4,7,7a-hexahydro-	1	1,2,4	U036	X	1 (0.454)
Chlordane, technical	57749	Chlordane; Chlordane, alpha & gamma isomers; 4,7-Methano-1H-indene, 1,2,4,5,6,7,8,8-octachloro-2,3,3a,4,7,7a-hexahydro-	1	1,2,4	U036	X	1 (0.454)
CHLORINATED BENZENES	N.A.		1*	2			**
CHLORINATED ETHANES	N.A.		1*	2			**
CHLORINATED NAPHTHALENE	N.A.		1*	2			**
CHLORINATED PHENOLS	N.A.		1*	1			**
Chlorine	7782505		10	1		A	10 (4.54)
Chlornaphazine	494031	Naphthalenamine, N,N'-bis(2-chloroethyl)-	1*	4	U026	B	100 (45.4)
Chloroacetaldehyde	107200	Acetaldehyde, chloro-	1*	4	P023	C	1000 (454)
CHLOROALKYL ETHERS	N.A.		1*	2			**
p-Chloroaniline	106478	Benzenamine, 4-chloro-	1*	4	P024	C	1000 (454)
Chlorobenzene	108907	Benzene, chloro-	100	1,2,4	U037	B	100 (45.4)
Chlorobenzilate	510156	Benzeneacetic acid, 4-chloro-alpha-(4-chlorophenyl)-alpha-hydroxy-, ethyl ester	1*	4	U038	A	10 (4.54)
4-Chloro-m-cresol	59507	p-Chloro-m-cresol; Phenol, 4-chloro-3-methyl-	1*	2,4	U039	D	5000 (2270)
p-Chloro-m-cresol	59507	Phenol, 4-chloro-3-methyl-; 4-Chloro-m-cresol	1*	2,4	U039	D	5000 (2270)
Chlorodibromomethane	124481		1*	2		B	100 (45.4)
Chloroethane	75003		1*	2		B	100 (45.4)
2-Chloroethyl vinyl ether	110758	Ethene, 2-chloroethoxy-	1*	2,4	U042	C	1000 (454)
Chloroform	67663	Methane, trichloro-	5000	1,2,4	U044	A	10 (4.54)
Chloromethyl methyl ether	107302	Methane, chloromethoxy-	1*	4	U046	A	10 (4.54)
beta-Chloronaphthalene	91587	Naphthalene, 2-chloro-; 2-Chloronaphthalene	1*	2,4	U047	D	5000 (2270)
2-Chloronaphthalene	91587	beta-Chloronaphthalene; Naphthalene, 2-chloro-	1*	2,4	U047	D	5000 (2270)
2-Chlorophenol	95578	o-Chlorophenol; Phenol, 2-chloro-	1*	2,4	U048	B	100 (45.4)
o-Chlorophenol	95578	Phenol, 2-chloro-; 2-Chlorophenol	1*	2,4	U048	B	100 (45.4)
4-Chlorophenyl phenyl ether	7005723		1*	2		D	5000 (2270)

TABLE 7.2 List of Hazardous Substances and Reportable Quantities (Continued)

[Note: All Comments/Notes Are Located at the End of This Table]

Hazardous Substance	CASRN	Regulatory Synonyms	Statutory			Final RQ	
			RQ	Code †	RCRA Waste Number	Category	Pounds (Kg)
1-(o-Chlorophenyl)thiourea	5344821	Thiourea, (2-chlorophenyl)-	1*	4	P026	B	100 (45.4)
3-Chloropropionitrile	542767	Propanenitrile, 3-chloro-	1*	4	P027	C	1000 (454)
Chlorosulfonic acid	7790945		1000	4		C	1000 (454)
4-Chloro-o-toluidine, hydrochloride	3165933	Benzenamine, 4-chloro-2-methyl-, hydrochloride	1*	4	U049	B	100 (45.4)
Chlorpyrifos	2921882		1	1		X	1 (0.454)
Chromic acetate	1066304		1000	1		C	1000 (454)
Chromic acid	11115745		1000	1		A	10 (4.54)
Chromic acid H2CrO4, calcium salt	7738945	Calcium chromate	1000	1,4	U032	A	10 (4.54)
Chromic sulfate	13765190		1000	1		C	1000 (454)
Chromium ††	10101538		1*	2		D	5000 (2270)
CHROMIUM AND COMPOUNDS	7440473		1*	2			**
Chromous chloride	10049055		1000	1		C	1000 (454)
Chrysene	218019	1,2-Benzphenanthrene	1*	2,4	U050	B	100 (45.4)
Cobaltous bromide	7789437		1000	1		C	1000 (454)
Cobaltous formate	544183		1000	1		C	1000 (454)
Cobaltous sulfamate	14017415		1000	1		C	1000 (454)
Coke Oven Emissions	N.A.		1*	3		X	1 (0.454)
Copper cyanide CuCN	544923	Copper cyanide	1*	4	P029	A	10 (4.54)
Copper ††	7440508		1*	2		D	5000 (2270)
COPPER AND COMPOUNDS	N.A.		1*	2			**
Copper cyanide	544923	Copper cyanide CuCN	1*	4	P029	A	10 (4.54)
Coumaphos	56724		10	1		A	10 (4.54)
Creosote	8001589		1*	1	U051	X	1 (0.454)
Cresol(s)	1319773	Cresylic acid; Phenol, methyl-	1000	1,4	U052	C	1000 (454)
m-Cresol	108394	m-Cresylic acid					
o-Cresol	95487	o-Cresylic acid					
p-Cresol	106445	p-Cresylic acid					
Cresylic acid	1319773	Cresol(s); Phenol, methyl-	1000	1,4	U052	C	1000 (454)
m-Cresol	108394	m-Cresol					
o-Cresol	95487	o-Cresol					
p-Cresol	106445	p-Cresol					
Crotonaldehyde	123739 4170303	2-Butenal	100	1,4	U053	B	100 (45.4)
Cumene	98828	Benzene, 1-methylethyl-	1*	4	U055	D	5000 (2270)
Cupric acetate	142712		100	1		B	100 (45.4)
Cupric acetoarsenite	12002038		100	1		X	1 (0.454)
Cupric chloride	7447394		10	1		A	10 (4.54)

Substance	CAS No.	Statutory (lbs)	Category	RCRA No.	Code	Final RQ (lbs (kg))
Cupric nitrate	3251238	100	1		B	100 (45.4)
Cupric oxalate	5893663	100	1		B	100 (45.4)
Cupric sulfate	7758987	10	1		A	10 (4.54)
Cupric sulfate, ammoniated	10380297	100	1		B	100 (45.4)
Cupric tartrate	815827	100*	2		B	100 (45.4)**
CYANIDES	N.A.					
Cyanides (soluble salts and complexes) not otherwise specified	57125	1*	4	P030	A	10 (4.54)
Cyanogen	460195	1*	4	P031	B	100 (45.4)
Ethanedinitrile						
Cyanogen bromide (CN)Br	506683	1*	4	U246	C	1000 (454)
Cyanogen bromide (CN)Br	506683	1*	4	U246	C	1000 (454)
Cyanogen chloride (CN)Cl	506774	10	1,4	P033	A	10 (4.54)
Cyanogen chloride	506774	10	1,4	P033	A	10 (4.54)
2,5-Cyclohexadiene-1,4-dione	106514	1*	4	U197	A	10 (4.54)
p-Benzoquinone						
Cyclohexane	110827	1000	1,4	U056	C	1000 (454)
Cyclohexane, 1,2,3,4,5,6-hexachloro-, (1alpha,2alpha,3beta,4alpha,5alpha,6beta)-	58899	1	1,2,4	U129	X	1 (0.454)
Benzene, hexahydro-						
gamma—BHC						
Hexachlorocyclohexane (gamma isomer)						
Lindane						
Cyclohexanone	108941	1*	4	U057	D	5000 (2270)
2-Cyclohexyl-4,6-dinitrophenol	131895	1*	4	P034	B	100 (45.4)
Phenol, 2-cyclohexyl-4,6-dinitro-						
1,3-Cyclopentadiene, 1,2,3,4,5,5-hexachloro-	77474	1	1,2,4	U130	A	10 (4.54)
Hexachlorocyclopentadiene						
Cyclophosphamide	50180	1*	4	U058	A	10 (4.54)
2H-1,3,2-Oxazaphosphorin-2-amine, N,N-bis(2-chloroethyl)tetrahydro-, 2-oxide						
2,4-D Acid	94757	100	1,4	U240	B	100 (45.4)
Acetic acid (2,4-dichlorophenoxy)-2,4-D, salts and esters						
2,4-D Ester	94111	100	1		B	100 (45.4)
	94791					
	94804					
	1320189					
	1928387					
	1928616					
	1929733					
	2971382					
	25168267					
	53467111					
2,4-D, salts and esters	94757	100	1,4	U240	B	100 (45.4)
Daunomycin	20830813	1*	4	U059	A	10 (4.54)
5,12-Naphthacenedione, 8-acetyl-10-[3-amino-2,3,6- trideoxy-alpha-L-lyxo-hexo- pyranosyl)oxyl]-7,8,9,10- tetrahydro-6,8,11-trihydroxy-1-methoxy-, (8S-cis)-						
DDD	72548	1	1,2,4	U060	X	1 (0.454)
Benzene, 1,1'-(2,2-dichloroethylidene)bis[4-chloro-TDE						
4,4' DDD	72548	1	1,2,4	U060	X	1 (0.454)
Benzene, 1,1'-(2,2-dichloroethylidene)bis[4-chloro-DDD / TDE						
DDE	72559	1*	2		X	1 (0.454)
4,4' DDE	72559	1*	2		X	1 (0.454)
DDE						
DDT	50293	1	1,2,4	U061	X	1 (0.454)
Benzene, 1,1'-(2,2,2-trichloroethylidene)bis[4-chloro-4,4'DDT						

TABLE 7.2 List of Hazardous Substances and Reportable Quantities (Continued)

[Note: All Comments/Notes Are Located at the End of This Table]

Hazardous Substance	CASRN	Regulatory Synonyms	Statutory			Final RQ	
			RQ	Code †	RCRA Waste Number	Category	Pounds (Kg)
4,4'DDT	50293	Benzene, 1,1'-(2,2,2-trichloroethylidene)bis[4-chloro- DDT	1	1,2,4	U061	X	1 (0.454)
DDT AND METABOLITES	N.A.		1*				**
Diallate	2303164	Carbamothioic acid, bis(1-methylethyl)-, S-(2,3-dich-loro-2-propenyl) ester	1*	4	U062	B	100 (45.4)
Diazinon	333415		1	1			1 (0.454)
Dibenz[a,h]anthracene	53703	Dibenzo[a,h]anthracene 1,2:5,6-Dibenzanthracene	1*	2,4	U063	X	1 (0.454)
1,2:5,6-Dibenzanthracene	53703	Dibenz[a,h]anthracene Dibenzo[a,h]anthracene	1*	2,4	U063	X	1 (0.454)
Dibenzo[a,h]anthracene	53703	Dibenz[a,h]anthracene 1,2:5,6-Dibenzanthracene	1*	2,4	U063	X	1 (0.454)
Dibenz[a,l]pyrene	189559	Benzo[rst]pentaphene	1*	4	U064	A	10 (4.54)
1,2-Dibromo-3-chloropropane	96128	Propane, 1,2-dibromo-3-chloro-	1*	4	U066	X	1 (0.454)
Dibutyl phthalate	84742	n-Butyl phthalate 1,2-Benzenedicarboxylic acid, dibutyl ester	100	1,2,4	U069	A	10 (4.54)
Di-n-butyl phthalate	84742	Dibutyl phthalate n-Butyl phthalate 1,2-Benzenedicarboxylic acid, dibutyl ester	100	1,2,4	U069	A	10 (4.54)
Dicamba	1918009		1000	1		C	1000 (454)
Dichlobenil	1194656		1000	1		B	100 (45.4)
Dichlone	117806		1	1		X	1 (0.454)
Dichlorobenzene	25321226		100	1		B	100 (45.4)
1,2-Dichlorobenzene	95501	Benzene, 1,2-dichloro- o-Dichlorobenzene	100	1,2,4	U070	B	100 (45.4)
1,3-Dichlorobenzene	541731	Benzene, 1,3-dichloro m-Dichlorobenzene	1*	2,4	U071	B	100 (45.4)
1,4-Dichlorobenzene	106467	Benzene, 1,4-dichloro p-Dichlorobenzene	100	1,2,4	U072	B	100 (45.4)
m-Dichlorobenzene	541731	Benzene, 1,3-dichloro 1,3-Dichlorobenzene	1*	2,4	U071	B	100 (45.4)
o-Dichlorobenzene	95501	Benzene, 1,2-dichloro 1,2-Dichlorobenzene	100	1,2,4	U070	B	100 (45.4)
p-Dichlorobenzene	106467	Benzene,1,4-dichloro 1,4-Dichlorobenzene	100	1,2,4	U072	B	100 (45.4)
DICHLOROBENZIDINE	N.A.		1*	2			**
3,3'-Dichlorobenzidine	91941	[1,1'-Biphenyl]-4,4'diamine,3,3'dichloro-	1*	2,4	U073	X	1 (0.454)
Dichlorobromomethane	75274		1*	2		D	5000 (2270)
1,4-Dichloro-2-butene	764410	2-Butene, 1,4-dichloro-	1*	2	U074	X	1 (0.454)
Dichlorodifluoromethane	75718	Methane, dichlorodifluoro-	1*	4	U075	D	5000 (2270)
1,1-Dichloroethane	75343	Ethane, 1,1-dichloro- Ethylidene dichloride	1*	2,4	U076	C	1000 (454)
1,2-Dichloroethane	107062	Ethane, 1,2-dichloro- Ethylene dichloride	5000	1,2,4	U077	B	100 (45.4)

Hazardous Substance	Regulatory Synonyms	CAS No.	Statutory RQ	Code	RCRA Waste No.	Category	Final RQ Pounds (Kg)
1,1-Dichloroethylene	Ethene, 1,1-dichloro-; Vinylidene chloride	75354	5000	1,2,4	U078	B	100 (45.4)
1,2-Dichloroethylene	Ethene, 1,2-dichloro- (E)	156605	1*	2,4	U079	C	1000 (454)
Dichloroethyl ether	Bis (2-chloroethyl) ether	111444	1*	2,4	U025	A	10 (4.54)
Dichloroisopropyl ether	Propane, 2,2'-oxybis[2-chloro-	108601	1*	2,4	U027	C	1000 (454)
Dichloromethoxy ethane	Bis(2-chloroethoxy) methane; Ethane, 1,1'-[methylenebis(oxy)]bis(2-chloro-	111911	1*	2,4	U024	C	1000 (454)
Dichloromethyl ether	Methane, oxybis(chloro-	542881	1*	4	P016	A	10 (4.54)
2,4-Dichlorophenol	Phenol, 2,4-dichloro-	120832	1*	2,4	U081	B	100 (45.4)
2,6-Dichlorophenol	Phenol, 2,6-dichloro-	87650	1*	4	U082	B	100 (45.4)
Dichlorophenylarsine	Arsonous dichloride, phenyl-	696286	1*	4	P036	X	1 (0.454)
Dichloropropane		26638197	5000	1		C	1000 (454)
1,1-Dichloropropane		78999					
1,3-Dichloropropane		142289					
1,2-Dichloropropane	Propane, 1,2-dichloro-; Propylene dichloride	78875	5000	1,2,4	U083	C	1000 (454)
Dichloropropane—Dichloropropene (mixture)		8003198	5000	1		B	100 (45.4)
Dichloropropene		26952238	5000	1		B	100 (45.4)
2,3-Dichloropropene		78886					
1,3-Dichloropropene	1-Propene, 1,3-dichloro-	542756	5000	1,2,4	U084	B	100 (45.4)
2,2-Dichloropropionic acid		75990	5000	1		D	5000 (2270)
Dichlorvos		62737	10	1		A	10 (4.54)
Dicofol		115322	5000	4		X	10 (4.54)
Dieldrin	2,7:3,6-Dimethanonaphth[2,3-b]oxirene, 3,4,5,6,9,9-hexachloro-1a,2,2a,3,6,6a,7,7a-octahydro-, (1aalpha,2beta,2aalpha,3beta,6beta,6aalpha,7beta,7aalpha)-	60571	1	1,2,4	P037	A	1 (0.454)
1,2,3,4-Diepoxybutane	2,2'-Bioxirane	1464535	1*	4	U085	B	10 (4.54)
Diethylamine		109897	1000	1		X	100 (454.4)
Diethylarsine	Arsine, diethyl-	692422	1*	4	P038	B	1 (0.454)
1,4-Diethylenedioxide	1,4-Dioxane	123911	1*	4	U108	B	100 (45.4)
Diethylhexyl phthalate	Bis (2-ethylhexyl)phthalate; 1,2-Benzenedicarboxylic acid, [bis(2-ethylhexyl)] ester	117817	1*	2,4	U028	B	100 (45.4)
N,N-Diethylhydrazine	Hydrazine, 1,2-diethyl-	1615801	1*	4	U086	A	10 (4.54)
O,O-Diethyl S-methyl dithiophosphate	Phosphorodithioic acid, O,O-diethyl S-methyl ester	3288582	1*	4	U087	D	5000 (2270)
Diethyl-p-nitrophenyl phosphate	Phosphoric acid, diethyl 4-nitrophenyl ester	311455	1*	4	P041	B	100 (45.4)
Diethyl phthalate	1,2-Benzenedicarboxylic acid, diethyl ester	84662	1*	4	U088	C	1000 (454)
O,O-Diethyl O-pyrazinyl phosphorothioate	Phosphorothioic acid, O,O-diethyl O-pyrazinyl ester	297972	1*	4	P040	B	100 (45.4)
Diethylstilbestrol	Phenol, 4,4'-(1,2-diethyl-1,2-ethenediyl)bis-, (E)	56531	1*	4	U089	X	1 (0.454)
Dihydrosafrole	1,3-Benzodioxole, 5-propyl-	94586	1*	4	U090	A	10 (4.54)
Diisopropylfluorophosphate	Phosphorofluoridic acid, bis(1-methylethyl) ester	55914	1*	4	P043	B	100 (45.4)
Aldrin	1,4,5,8-Dimethanonaphthalene, 1,2,3,4,10,10-hexachloro-1,4,4a,5,8,8a-hexahydro-, (1alpha,4alpha,4abeta,5alpha,8alpha,8abeta)-	309002	1	1,2,4	P004	X	1 (0.454)
Isodrin	1,2,3,4,10,10-hexachloro-1,4,4a,5,8,8a-hexahydro-, (1alpha,4alpha,4abeta,5alpha,8beta,8abeta)-1,4,5,8-Dimethanonaphthalene,	465736	1*	4	P060	X	1 (0.454)

TABLE 7.2 List of Hazardous Substances and Reportable Quantities (Continued)

[Note: All Comments/Notes Are Located at the End of This Table]

Hazardous Substance	CASRN	Regulatory Synonyms	Statutory RQ	Statutory Code †	Statutory RCRA Waste Number	Final RQ Category	Final RQ Pounds (Kg)
8abeta)-2,7:3,6-Dimethanonaphth[2,3-b]oxirene, 3,4,5,6,9,9-hexachloro-1a,2,2a,3,6,6a,7,7a-octahydro-, (1aalpha,2beta,2aalpha,3beta,6beta,6aalpha,7beta,7aalpha)-2,7:3,6-	60571	Dieldrin	1	1,2,4	P037	X	1 (0.454)
Dimethanonaphth[2,3-b]oxirene, 3,4,5,6,9,9-hex-achloro-1a,2,2a,3,6,6a,7,7a-octa-hydro-, (1aalpha,2beta,2abeta,3alpha,6alpha,6abeta,7beta,7aalpha)-Dimethanoate	72208	Endrin Endrin, & metabolites	1	1,2,4	P051	X	1 (0.454)
Phosphorodithioic acid, O,O-dimethyl S-[2(methyla-mino)-2-oxoethyl] ester	60515		1*	4	P044	A	10 (4.54)
3,3'-Dimethoxybenzidine	119904	[1,1'-Biphenyl]-4,4'diamine,3,3'dimethoxy-	1*	4	U091	B	100 (45.4)
Dimethylamine	124403	Methanamine, N-methyl-	1000	1,4	U092	C	1000 (454)
p-Dimethylaminoazobenzene	60117	Benzenamine, N,N-dimethyl-4-(phenylazo-)	1*	4	U093	X	10 (4.54)
7,12-Dimethylbenz[a]anthracene	57976	Benz[a]anthracene, 7,12-dimethyl-	1*	4	U094	A	1 (0.454)
3,3'-Dimethylbenzidine	119937	[1,1'Biphenyl]-4,4'-diamine,3,3'-dimethyl-	1*	4	U095	A	10 (4.54)
alpha,alpha-Dimethylbenzylhydroperoxide	80159	Hydroperoxide, 1-methyl-1-phenylethyl-	1*	4	U096	X	10 (4.54)
Dimethylcarbamoyl chloride	79447	Carbamic chloride, dimethyl-	1*	4	U097	A	1 (0.454)
1,1-Dimethylhydrazine	57147	Hydrazine, 1,1-dimethyl-	1*	4	U098	X	10 (4.54)
1,2-Dimethylhydrazine	540738	Hydrazine, 1,2-dimethyl-	1*	4	U099	A	1 (0.454)
alpha,alpha-Dimethylphenethylamine	122098	Benzeneethanamine, alpha,alpha-dimethyl-	1*	4	P046	D	5000 (2270)
2,4-Dimethylphenol	105679	Phenol, 2,4-dimethyl-	1*	2,4	U101	B	100 (45.4)
Dimethyl phthalate	131113	1,2-Benzenedicarboxylic acid, dimethyl ester	1*	2,4	U102	D	5000 (2270)
Dimethyl sulfate	77781	Sulfuric acid, dimethyl ester	1*	1	U103	B	100 (45.4)
Dinitrobenzene (mixed)	25154545		1000	1		B	100 (45.4)
m-Dinitrobenzene	99650						
o-Dinitrobenzene	528290						
p-Dinitrobenzene	100254						
4,6-Dinitro-o-cresol and salts	534521	Phenol, 2-methyl-4,6-dinitro-	1*	2,4	P047	A	10 (4.54)
Dinitrophenol	25550587		1000	1		A	10 (4.54)
2,5-Dinitrophenol	329715						
2,6-Dinitrophenol	573568						
2,4-Dinitrophenol	51285	Phenol, 2,4-dinitro-	1000	1,2,4	P048	A	10 (4.54)
Dinitrotoluene	25321146		1000	1,2		A	10 (4.54)
3,4-Dinitrotoluene	610399						
2,4-Dinitrotoluene	121142	Benzene, 1-methyl-2,4-dinitro-	1000	1,2,4	U105	A	10 (4.54)
2,6-Dinitrotoluene	606202	Benzene, 2-methyl-1,3-dinitro-	1000	1,2,4	U106	B	100 (45.4)
Dinoseb	88857	Phenol, 2-(1-methylpropyl)-4,6-dinitro	1*	4	P020	C	1000 (454)
Di-n-octyl phthalate	117840	1,2-Benzenedicarboxylic acid, dioctyl ester	1*	2,4	U107	D	5000 (2270)
1,4-Dioxane	123911	1,4-Diethyleneoxide	1*	4	U108	B	100 (45.4) **
DIPHENYLHYDRAZINE	N.A.			2			
1,2-Diphenylhydrazine	122667	Hydrazine, 1,2-diphenyl-	1*	2,4	U109	A	10 (4.54)

Chemical	Synonym	CAS No.			Code		Final RQ pounds (kilograms)
Diphosphoramide, octamethyl-	Octamethylpyrophosphoramide	152169	1*	4	P085	B	100 (45.4)
Diphosphoric acid, tetraethyl ester	Tetraethyl pyrophosphate	107493	100	1,4	P111	A	10 (4.54)
Dipropylamine	1-Propanamine, N-propyl-	142847	1*	4	U110	D	5000 (2270)
Di-n-propylnitrosamine	1-Propanamine, N-nitroso-N-propyl-	62647	1*	2,4	U111	A	10 (4.54)
Diquat		85007 2764729	1000	1		C	1000 (454)
Disulfoton	Phosphorodithioic acid, o,o-diethyl S-[2-(ethylthio)ethyl]ester	298044	1	1,4	P039	X	1 (0.454)
Dithiobiuret	Thioimidodicarbonic diamide [(H2N)C(S)]2NH	541537	1*	4	P049	B	100 (45.4)
Diuron		330541	100	·		B	100 (45.4)
Dodecylbenzenesulfonic acid		27176870	1000	1		C	1000 (454)
Endosulfan	6,9-Methano-2,4,3-benzodioxathiepin, 6,7,8,9,10,10-hexachloro-1,5,5a,6,9,9a- hexahydro-, 3-oxide	115297	1	1,2,4	P050	X	1 (0.454)
alpha - Endosulfan		959988	1*	2		X	1 (0.454)
beta - Endosulfan		33213659	1*	2		X	1 (0.454)
ENDOSULFAN AND METABOLITES		N.A.	1*	2			**
Endosulfan sulfate		1031078	1*	2		X	1 (0.454)
Endothall	7-Oxabicyclo[2.2.1]heptane-2,3-dicarboxylic acid	145733	1*	4	P088	C	1000 (454)
Endrin	Endrin, & metabolites 2,7:3,6-Dimethanonaphth[2,3-b]oxirene, 3,4,5,6,9,9 -hexachloro-1a,2,2a,3, 6,6a,7,7a-octa-hydro-, (1aalpha, 2beta,2abeta,3alpha,6alpha, 6abeta,7beta, 7aalpha)-	72208	1	1,2,4	P051	X	1 (0.454)
Endrin aldehyde		7421934	1*	2		X	1 (0.454)
ENDRIN AND METABOLITES		N.A.	1*	2			**
Endrin, & metabolites	Endrin 2,7:3,6-Dimethanonaphth[2,3-b]oxirene, 3,4,5,6,9,9-hexachloro-1a,2,2a,3, 6,6a,7,7a-octa-hydro-, (1aalpha, 2beta,2abeta,3alpha,6alpha, 6abeta,7beta, 7aalpha)-	72208	1	1,2,4	P051	X	1 (0.454)
Epichlorohydrin	Oxirane, (chloromethyl)-	106898	1000	1,4	U041	B	100 (45.4)
Epinephrine	1,2-Benzenediol,4-[1-hydroxy-2-(methylamino)ethyl]-	51434	1*	4	P042	C	1000 (454)
Ethanal	Acetaldehyde	75070	1000	1,4	U001	C	1000 (454)
Ethanamine, N-ethyl-N-nitroso-	N-Nitrosodiethylamine	55185	1*	4	U174	X	1 (0.454)
1,2-Ethanediamine, N,N-dimethyl-N'-2-pyridinyl-N'-(2-thienylmethyl)-	Methapyrilene	91805	1*	4	U155	D	5000 (2270)
Ethane, 1,2-dibromo-	Ethylene dibromide	106934	1000	1,4	U067	X	1 (0.454)
Ethane, 1,1-dichloro-	Ethylidene dichloride	75343	1*	2,4	U076	C	1000 (454)
Ethane, 1,2-dichloro-	1,1-Dichloroethane 1,2-Dichloroethane	107062	5000	1,2,4	U077	B	100 (45.4)
Ethanedinitrile	Cyanogen	460195	1*	4	P031	B	100 (45.4)
Ethane, hexachloro-	Hexachloroethane	67721	1*	2,4	U131	B	100 (45.4)
Ethane, 1,1'-[methylenebis(oxy)]bis(2-chloro-	Bis(2-chloroethoxy) methane	111911	1*	2,4	U024	C	1000 (454)
Ethane, 1,1'-oxybis-	Dichloromethoxy ethane Ethyl ether	60297	1*	4	U117	B	100 (45.4)

TABLE 7.2 List of Hazardous Substances and Reportable Quantities (*Continued*)

[Note: All Comments/Notes Are Located at the End of This Table]

Hazardous Substance	CASRN	Regulatory Synonyms	Statutory RQ	Statutory Code †	Statutory RCRA Waste Number	Final RQ Category	Final RQ Pounds (Kg)
Ethane, 1,1'-oxybis[2-chloro-	111444	Bis (2-chloroethyl) ether / Dichloroethyl ether	1*	2,4	U025	A	10 (4.54)
Ethane, pentachloro-	76017	Pentachloroethane	1*	4	U184	A	10 (4.54)
Ethane, 1,1,1,2-tetrachloro-	630206	1,1,1,2-Tetrachloroethane	1*	4	U208	B	100 (45.4)
Ethane, 1,1,2,2-tetrachloro-	79345	1,1,2,2-Tetrachloroethane	1*	2,4	U209	B	100 (45.4)
Ethanethioamide	62555	Thioacetamide	1*	4	U218	A	10 (4.54)
Ethane, 1,1,1-trichloro-	71556	Methyl chloroform / 1,1,1-Trichloroethane	1*	2,4	U226	C	1000 (454)
Ethane, 1,1,2-trichloro-	79005	1,1,2-Trichloroethane	1*	2,4	U227	B	100 (45.4)
Ethanimidothioic acid, N-[[(methyl-amino)carbonyl]oxy]-, methyl ester	16752775	Methomyl	1*	4	P066	B	100 (45.4)
Ethanol, 2-ethoxy-	110805	Ethylene glycol monoethyl ether	1*	4	U359	C	1000 (454)
Ethanol, 2,2'-(nitrosoimino)bis-	1116547	N-Nitrosodiethanolamine	1*	4	U173	X	1 (0.454)
Ethanone, 1-phenyl-	98862	Acetophenone	1*	4	U004	D	5000 (2270)
Ethene, chloro-	75014	Vinyl chloride	1*	2,3,4	U043	X	1 (0.454)
Ethene, 2-chloroethoxy-	110758	2-Chloroethyl vinyl ether	1*	2,4	U042	C	1000 (454)
Ethene, 1,1-dichloro-	75354	Vinylidene chloride / 1,1-Dichloroethylene	5000	1,2,4	U078	B	100 (45.4)
Ethene, 1,2-dichloro- (E)	156605	1,2-Dichloroethylene	1*	2,4	U079	C	1000 (454)
Ethene, tetrachloro-	127184	Perchloroethylene / Tetrachloroethene / Tetrachloroethylene	1*	2,4	U210	B	100 (45.4)
Ethene, trichloro-	79016	Trichloroethene / Trichloroethylene	1000	1,2,4	U228	B	100 (45.4)
Ethion	563122		10	1		A	10 (4.54)
Ethyl acetate	141786	Acetic acid, ethyl ester	1*	4	U112	D	5000 (2270)
Ethyl acrylate	140885	2-Propenoic acid, ethyl ester	1*	4	U113	C	1000 (454)
Ethylbenzene	100414		1000	1,2		C	1000 (454)
Ethyl carbamate (urethane)	51796	Carbamic acid, ethyl ester	1*	4	U238	B	100 (45.4)
Ethyl cyanide	107120	Propanenitrile	1*	4	P101	A	10 (4.54)
Ethylenebisdithiocarbamic acid, salts & esters	111546	Carbamodithioic acid, 1,2-ethanediylbis, salts & esters	1*	4	U114	D	5000 (2270)
Ethylenediamine	107153		1000	1		D	5000 (2270)
Ethylenediamine-tetraacetic acid (EDTA)	60004		5000	1		D	5000 (2270)
Ethylene dibromide	106934	Ethane, 1,2-dibromo-	1000	1,4	U067	X	1 (0.454)
Ethylene dichloride	107062	Ethane, 1,2-dichloro-	5000	1,2,4	U077	B	100 (45.4)
Ethylene glycol monoethyl ether	110805	Ethanol, 2-ethoxy-	1*	4	U359	C	1000 (454)
Ethylene oxide	75218	Oxirane	1*	4	U115	A	10 (4.54)
Ethylenethiourea	96457	2-Imidazolidinethione	1*	4	U116	A	10 (4.54)

Hazardous substance (synonym)	CAS No.	Cat.	Code		RQ	RQ lbs (kg)
Ethylenimine (Aziridine)	151564	4	P054	X	1*	1 (0.454)
Ethyl ether (Ethane, 1,1'-oxybis-)	60297	4	U117	B	1*	100 (45.4)
Ethylidene dichloride (Ethane, 1,1-dichloro-; 1,1-Dichloroethane)	75343	2,4	U076	C	1*	1000 (454)
Ethyl methacrylate (2-Propenoic acid, 2-methyl-, ethyl ester)	97632	4	U118	C	1*	1000 (454)
Ethyl methanesulfonate (Methanesulfonic acid, ethyl ester)	62500	4	U119	X	1*	1 (0.454)
Famphur (Phosphorothioic acid, O,[4-[(di-methylamino) sulfonyl] phenyl] O,O-dimethyl ester)	52857	4	P097	C	1*	1000 (454)
Ferric ammonium citrate	1185575	1		C	1000	1000 (454)
Ferric ammonium oxalate	2944674 / 55488874	1		C	1000	1000 (454)
Ferric chloride	7705080	1		C	1000	1000 (454)
Ferric fluoride	7783508	1		B	100	100 (45.4)
Ferric nitrate	10421484	1		C	1000	1000 (454)
Ferric sulfate	10028225	1		C	1000	1000 (454)
Ferrous ammonium sulfate	10045893	1		C	1000	1000 (454)
Ferrous chloride	7758943	1		B	100	100 (45.4)
Ferrous sulfate	7720787 / 7782630	1		C	1000	1000 (454)
Fluoranthene (Benzo[j,k]fluorene)	206440	2,4	U120	B	1*	100 (45.4)
Fluorene	86737	2		D	1*	5000 (2270)
Fluorine	7782414	4	P056	A	1*	10 (4.54)
Fluoroacetamide, 2-fluoro- (Acetamide, 2-fluoro-)	640197	4	P057	B	1*	100 (45.4)
Fluoroacetic acid, sodium salt (Acetic acid, fluoro-, sodium salt)	62748	4	P058	A	1*	10 (4.54)
Formaldehyde	50000	1,4	U122	B	1000	100 (45.4)
Formic acid	64186	1,4	U123	D	5000	5000 (2270)
Fulminic acid, mercury(2+)salt (Mercury fulminate)	628864	4	P065	A	1*	10 (4.54)
Fumaric acid	110178	4	U124	D	5000	5000 (2270)
Furan, tetrahydro-	110009	4	U213	B	1*	100 (45.4)
2-Furancarboxaldehyde (Furfural)	109999	4	U125	C	1*	1000 (454)
2,5-Furandione (Maleic anhydride)	98011	1,4	U147	D	1000	5000 (2270)
Furfural (2-Furancarboxaldehyde)	108316	1,4	U125	D	5000	5000 (2270)
Furfuran (Furan)	98011 / 110009	1,4	U124	D	1000	5000 (2270)
Glucopyranose, 2-deoxy-2-(3-methyl-3-nitrosoureido)- (Streptozotocin)	18883664	4	U206	B	1*	100 (45.4)
D-Glucose, 2-deoxy-2-[[(methylnitrosoamino)carbonyl]amino]- (Streptozotocin)	18883664	4	U206	X	1*	1 (0.454)
Glycidylaldehyde (Oxiranecarboxyaldehyde)	765344	4	U126	A	1*	10 (4.54)
Guanidine, N-methyl-N'-nitro-N-nitroso- (MNNG)	70257	4	U163	A	1*	10 (4.54)
Guthion	86500	1		X	1	1 (0.454)
HALOETHERS	N.A.	2			1*	**
HALOMETHANES	N.A.	2			1*	
Heptachlor (4,7-Methano-1H-indene, 1,4,5,6,7,8,8-heptachloro-3a,4,7,7a-tetrahydro-)	76448	1,2,4	P059	X	1	1 (0.454)
HEPTACHLOR AND METABOLITES	N.A.	2			1*	**
Heptachlor epoxide	1024573	2		X	1*	1 (0.454)
Hexachlorobenzene	118741	2,4	U127	A	1*	1 (0.454)
Hexachlorobutadiene (1,3-Butadiene, 1,1,2,3,4,4-hexachloro-)	87683	2,4	U128	X	1*	10 (4.54)

TABLE 7.2 List of Hazardous Substances and Reportable Quantities (*Continued*)

[Note: All Comments/Notes Are Located at the End of This Table]

Hazardous Substance	CASRN	Regulatory Synonyms	Statutory RQ	Statutory Code†	Statutory RCRA Waste Number	Final RQ Category	Final RQ Pounds (Kg)
HEXACHLOROCYCLOHEXANE (all isomers)	608731		1*	2			**
Hexachlorocyclohexane (gamma isomer)	58899	Cyclohexane, 1,2,3,4,5,6-hexachloro-, (1alpha,2alpha,3beta,4alpha,5alpha, 6beta)- gamma-BHC Lindane	1	1,2,4	U129	X	1 (0.454)
Hexachlorocyclopentadiene	77474	1,3-Cyclopentadiene,1,2,3,4,5,5-hexachloro-	1	1,2,4	U130	A	10 (4.54)
Hexachloroethane	67721	Ethane, hexachloro-	1*	2,4	U131	B	100 (45.4)
Hexachlorophene	70304	Phenol, 2,2'-methylenebis[3,4,6-trichloro-	1*	4	U132	B	100 (45.4)
Hexachloropropene	1888717	1-Propene, 1,1,2,3,3,3-hexachloro-	1*	4	U243	C	1000 (454)
Hexaethyl tetraphosphate	757584	Tetraphosphoric acid, hexaethyl ester	1*	4	P062	B	100 (45.4)
Hydrazine	302012		1*	4	U133	X	1 (0.454)
Hydrazine, 1,2-diethyl-	1615801	N,N'-Diethylhydrazine	1*	4	U086	A	10 (4.54)
Hydrazine, 1,1-dimethyl-	57147	1,1-Dimethylhydrazine	1*	4	U098	A	10 (4.54)
Hydrazine, 1,2-dimethyl-	540738	1,2-Dimethylhydrazine	1*	4	U099	X	1 (0.454)
Hydrazine, 1,2-diphenyl-	122667	1,2-Diphenylhydrazine	1*	2,4	U109	A	10 (4.54)
Hydrazine, methyl-	60344	Methyl hydrazine	1*	4	P068	A	10 (4.54)
Hydrazinecarbothioamide	79196	Thiosemicarbazide	1*	4	P116	B	100 (45.4)
Hydrochloric acid	7647010		5000	1		D	5000 (2270)
Hydrocyanic acid	74908	Hydrogen cyanide	10	1,4	P063	A	10 (4.54)
Hydrofluoric acid	7664393	Hydrogen fluoride	5000	1,4	U134	B	100 (45.4)
Hydrogen chloride	7647010	Hydrochloric acid	5000	1		D	5000 (2270)
Hydrogen cyanide	74908	Hydrocyanic acid	10	1,4	P063	A	10 (4.54)
Hydrogen fluoride	7664393	Hydrofluoric acid	5000	1,4	U134	B	100 (45.4)
Hydrogen sulfide	7783064	Hydrogen sulfide H2S	100	1,4	U135	B	100 (45.4)
Hydrogen sulfide H2S	7783064	Hydrogen sulfide	100	1,4	U135	B	100 (45.4)
Hydroperoxide, 1-methyl-1-phenylethyl-	80159	alpha,alpha-Dimethylbenzylhydroperoxide	1*	4	U096	A	10 (4.54)
2-Imidazolidinethione	96457	Ethylenethiourea	1*	4	U116	A	10 (4.54)
Indeno(1,2,3-cd)pyrene	193395	1,10-(1,2-Phenylene)pyrene	1*	2,4	U137	B	100 (45.4)
1,3-Isobenzofurandione	85449	Phthalic anhydride	1*	4	U190	D	5000 (2270)
Isobutyl alcohol	78831	1-Propanol, 2-methyl-	1*	4	U140	D	5000 (2270)
Isodrin	465736	1,4,5,8-Dimethanonaphthalene, 1,2,3,4,10,10-hexachloro-1,4,4a,5,8,8a-hexahydro, (1alpha,4alpha,4abeta,5beta, 8beta,8abeta)-	1*	4	P060	X	1 (0.454)
Isophorone	78591		1*	2		D	5000 (2270)
Isoprene	78795		1000	1		B	100 (45.4)
Isopropanolamine dodecylbenzenesulfonate	42504461		1000	1		C	1000 (454)
Isosafrole	120581	1,3-Benzodioxole, 5-(1-propenyl)-	1*	4	U141	B	100 (45.4)
3(2H)-Isoxazolone, 5-(aminomethyl)-	2763964	Muscimol 5-(Aminomethyl)-3-isoxazolol	1*	4	P007	C	1000 (454)

Name	CAS No.		Cat.	Waste No.		Final RQ Pounds (Kg)
Kepone	143500	1	1,4	U142	X	1 (0.454)
Lasiocarpine — 1,3,4-Metheno-2H-cyclobuta[cd]pentalen-2-one, 1,1a,3,3a,4,5,5,5a,5b,6-decachlorooctahydro- (Kepone); 2-Butenoic acid, 2-methyl-, 7[[2,3-dihydroxy-2-(1-methoxyethyl)-3-methyl-1-oxobutoxy]methyl]-2,3,5,7a-tetrahydro-1H-pyrrolizin-1-yl ester, [1S-[1alpha(Z), 7(2S*,3R*),7aalpha]]-	303344	1*	4	U143	A	10 (4.54)
Lead ††	7439921		2			#
Lead acetate — Acetic acid, lead(2+) salt	301042	5000	1,4	U144		#
LEAD AND COMPOUNDS	N.A.	1*	2			**
Lead arsenate	7784409, 7645252, 10102484	5000	1		X	1 (0.454)
Lead, bis(acetato-O)tetrahydroxytri	1335326	1*	4	U146	B	100 (45.4)
Lead chloride	7758954	5000	1		B	100 (45.4)
Lead fluoborate	13814965	5000	1		B	100 (45.4)
Lead fluoride	7783462	1000	1		B	100 (45.4)
Lead iodide	10101630	5000	1		B	100 (45.4)
Lead nitrate	10099748	5000	1		B	100 (45.4)
Lead phosphate — Phosphoric acid, lead(2+) salt (2:3)	7446277, 7428480, 1072351, 52652592, 56189094	1*	1	U145	D	5000# (2270)
Lead subacetate — Lead, bis(acetato-O)tetrahydroxytri	1335326	5000	4	U146	B	100 (45.4)
Lead sulfate	15739807, 7446142	1*	1		B	100 (45.4)
Lead sulfide	1314870	5000	1		D	5000# (2270)
Lead thiocyanate	592870	5000	1		B	100 (45.4)
Lindane — Cyclohexane, 1,2,3,4,5,6-hexachloro-, (1alpha,2alpha,3beta,4alpha,5alpha,6beta)-; gamma-BHC; Hexachlorocyclohexane (gamma isomer)	58899	1	1,2,4	U129	X	1 (0.454)
Lithium chromate	14307358	1000	1		A	10 (4.54)
Malathion	121755	10	1		B	100 (45.4)
Maleic acid	110167	5000	1		D	5000 (2270)
Maleic anhydride — 2,5-Furandione	108316	1*	1,4	U147	D	5000 (2270)
Maleic hydrazide — 3,6-Pyridazinedione, 1,2-dihydro-	123331	1*	4	U148	C	5000 (2270)
Malononitrile — Propanedinitrile	109773	1*	4	U149	X	1000 (454)
Melphalan — L-Phenylalanine, 4-[bis(2-chloroethyl) amino]-	148823	100	4	U150	X	1 (0.454)
Mercaptodimethur	2032657	1	1		A	10 (4.54)
Mercuric cyanide	592041	10	1		X	1 (0.454)
Mercuric nitrate	10045940	10	1		A	10 (4.54)
Mercuric sulfate	7783359	10	1		A	10 (4.54)
Mercuric thiocyanate	592858	10	1		A	10 (4.54)
Mercurous nitrate	10415755	1*	1		A	10 (4.54)
Mercury	7782867, 7439976	1*	2,3,4	U151	X	1 (0.454)
MERCURY AND COMPOUNDS	N.A.	1*	2			**
Mercury, (acetate-O)phenyl- — Phenylmercury acetate	62384	1*	4	P092	B	100 (45.4)
Mercury fulminate — Fulminic acid, mercury(2+)salt	628864	1*	4	P065	A	10 (4.54)

TABLE 7.2 List of Hazardous Substances and Reportable Quantities (Continued)

[Note: All Comments/Notes Are Located at the End of This Table]

Hazardous Substance	CASRN	Regulatory Synonyms	Statutory RQ	Statutory Code †	RCRA Waste Number	Final RQ Category	Final RQ Pounds (Kg)
Methacrylonitrile	126987	2-Propenenitrile, 2-methyl-	1*	4	U152	C	1000 (454)
Methanamine, N-methyl-	124403	Dimethylamine	1000	1,4	U092	C	1000 (454)
Methanamine, N-methyl-N-nitroso-	62759	N-Nitrosodimethylamine	1*	2,4	P082	A	10 (4.54)
Methane, bromo-	74839	Methyl bromide	1*	2,4	U029	C	1000 (454)
Methane, chloro-	74873	Methyl chloride	1*	2,4	U045	B	100 (45.4)
Methane, chloromethoxy-	107302	Chloromethyl methyl ether	1*	4	U046	A	10 (4.54)
Methane, dibromo-	74953	Methylene bromide	1*	4	U068	C	1000 (454)
Methane, dichloro-	75092	Methylene chloride	1*	2,4	U080	C	1000 (454)
Methane, dichlorodifluoro-	75718	Dichlorodifluoromethane	1*	4	U075	D	5000 (2270)
Methane, iodo-	74884	Methyl iodide	1*	4	U138	B	100 (45.4)
Methane, isocyanato-	624839	Methyl isocyanate	1*	4	P064		# #
Methane, oxybis(chloro-	542881	Dichloromethyl ether	1*	4	P016	A	10 (4.54)
Methanesulfenyl chloride, trichloro-	594423	Trichloromethanesulfenyl chloride	1*	4	P118	B	100 (45.4)
Methanesulfonic acid, ethyl ester	62500	Ethyl methanesulfonate	1*	4	U119	X	1 (0.454)
Methane, tetrachloro-	56235	Carbon tetrachloride	5000	1,2,4	U211	A	10 (4.54)
Methane, tetranitro-	509148	Tetranitromethane	1*	4	P112	A	10 (4.54)
Methane, tribromo-	75252	Bromoform	1*	2,4	U225	B	100 (45.4)
Methane, trichloro-	67663	Chloroform	5000	1,2,4	U044	A	10 (4.54)
Methane, trichlorofluoro-	75694	Trichloromonofluoromethane	1*	4	U121	D	5000 (2270)
Methanethiol	74931	Methylmercaptan	100	1,4	U153	B	100 (45.4)
6,9-Methano-2,4,3-benzodioxathiepin, 6,7,8,9,10,10-hexachloro-1,5,5a,6,9,9a- hexahydro-, 3-oxide	115297	Endosulfan	1	1,2,4	P050	X	1 (0.454)
1,3,4-Metheno-2H-cyclobuta[cd]pentalen-2-one, 1,1a,3,3a,4,5,5,5a,5b,6-decachlorooctahydro-	143500	Kepone	1	1,4	U142	X	1 (0.454)
4,7-Methano-1H-indene, 1,4,5,6,7,8,8-heptachloro-3a,4,7,7a-tetrahydro-	76448	Heptachlor	1	1,2,4	P059	X	1 (0.454)
4,7-Methano-1H-indene, 1,2,4,5,6,7,8,8-octachloro-2,3,3a,4,7,7a-hexahydro-	57749	Chlordane, alpha & gamma isomers; Chlordane, technical	1	1,2,4	U036	X	1 (0.454)
Methanol	67561	Methyl alcohol	1*	4	U154	D	5000 (2270)
Methapyrilene	91805	1,2-Ethanediamine, N,N-dimethyl-N'-2-pyridinyl-N'-(2-thienylmethyl)-	1*	4	U155	D	5000 (2270)
Methomyl	16752775	Ethanimidothioic acid, N-[[(methyl-amino)carbonyl]oxy]-, methyl ester	1*	4	P066	B	100 (45.4)
Methoxychlor	72435	Benzene, 1,1'-(2,2,2-trichloroethylidene) bis[4-methoxy-	1	1,4	U247	X	1 (0.454)
Methyl alcohol	67561	Methanol	1*	4	U154	D	5000 (2270)
Methyl bromide	74839	Methane, bromo-	1*	2,4	U029	C	1000 (454)
1-Methylbutadiene	504609	1,3-Pentadiene	1*	4	U186	B	100 (45.4)

Hazardous Substance	CASRN	Regulatory Synonyms	Statutory RQ	Code	Waste No.	RQ Category	Final RQ lb (kg)
Methyl chloride	74873	Methane, chloro-	1*	2,4	U045	B	100 (45.4)
Methyl chlorocarbonate	79221	Carbonochloridic acid, methyl ester; Methyl chloroformate	1*	4	U156	C	1000 (454)
Methyl chloroform	71556	Ethane, 1,1,1-trichloro-; 1,1,1-Trichloroethane	1*	2,4	U226	C	1000 (454)
Methyl chloroformate	79221	Carbonochloridic acid, methyl ester; Methyl chlorocarbonate	1*	4	U156	C	1000 (454)
3-Methylcholanthrene	56495	Benz[j]aceanthrylene, 1,2-dihydro-3-methyl-	1*	4	U157	A	10 (4.54)
4,4'-Methylenebis(2-chloroaniline)	101144	Benzenamine, 4,4'-methylenebis(2-chloro-	1*	4	U158	A	10 (4.54)
Methylene bromide	74953	Methane, dibromo-	1*	4	U068	C	1000 (454)
Methylene chloride	75092	Methane, dichloro-	1*	2,4	U080	C	1000 (454)
Methyl ethyl ketone (MEK)	78933	2-Butanone	1*	4	U159	D	5000 (2270)
Methyl ethyl ketone peroxide	1338234	2-Butanone peroxide	1*	4	U160	A	10 (4.54)
Methyl hydrazine	60344	Hydrazine, methyl-	1*	4	P068	A	10 (4.54)
Methyl iodide	74884	Methane, iodo-	1*	4	U138	B	100 (45.4)
Methyl isobutyl ketone	108101	4-Methyl-2-pentanone	1*	4	U161	D	5000 (2270) # #
Methyl isocyanate	624839	Methane, isocyanato-	1*	4	P064	A	10 (4.54)
2-Methyllactonitrile	75865	Acetone cyanohydrin; Propanenitrile, 2-hydroxy-2-methyl-	10	1,4	P069	A	10 (4.54)
Methylmercaptan	74931	Methanethiol; Thiomethanol	100	1,4	U153	B	100 (45.4)
Methyl methacrylate	80626	2-Propenoic acid, 2-methyl-, methyl ester	5000	1,4	U162	C	1000 (454)
Methyl parathion	298000	Phosphorothioic acid, O,O-dimethyl O-(4-nitro-phenyl) ester	100	1,4	P071	B	100 (45.4)
4-Methyl-2-pentanone	108101	Methyl isobutyl ketone	1*	4	U161	D	5000 (2270)
Methylthiouracil	56042	4(1H)-Pyrimidinone, 2,3-dihydro-6-methyl-2-thioxo-	1*	4	U164	A	10 (4.54)
Mevinphos	7786347		1	1		A	10 (4.54)
Mexacarbate	315184		1000	4		C	1000 (454)
Mitomycin C	50077	Azirino[2',3':3,4]pyrrolo[1,2-a]indole-4,7-dione,6-amino-8-[[(aminocarbonyl)oxy]methyl]-1,1a,2,8,8a,8b-hexahydro-8a-methoxy-5-methyl-,[1aS-(1aalpha, 8beta, 8aalpha, 8balpha)]-	1*	4	U010	A	10 (4.54)
MNNG	70257	Guanidine, N-methyl-N'-nitro-N-nitroso-	1*	4	U163	A	10 (4.54)
Monoethylamine	75047		1000	1		B	100 (45.4)
Monomethylamine	74895		1000	1		B	100 (45.4)
Multi Source Leachate			1*	4	F039	X	1 (0.454)
Muscimol	2763964	3(2H)-Isoxazolone, 5-(aminomethyl)-; 5-(Aminomethyl)-3-isoxazolol	1*	4	P007	C	1000 (454)
Naled	300765		10	1		A	10 (4.54)
5,12-Naphthacenedione, 8-acetyl-10-[3-amino-2,3,6-trideoxy-alpha-L-lyxo-hexopyranosyl)oxy]-7,8,9,10-tetrahydro-6,8,11-trihydroxy-1-methoxy-, (8S-cis)-	20830813	Daunomycin	1*	4	U059	A	10 (4.54)
1-Naphthalenamine	134327	alpha-Naphthylamine	1*	4	U167	B	100 (45.4)
2-Naphthalenamine	91598	beta-Naphthylamine	1*	4	U168	A	10 (4.54)
Naphthalenamine, N,N'-bis(2-chloroethyl)-	494031	Chlornaphazine	1*	1,2,4	U026	B	100 (45.4)
Naphthalene	91203		5000	2,4	U165	B	100 (45.4)
Naphthalene, 2-chloro-	91587	beta-Chloronaphthalene; 2-Chloronaphthalene	1*	2,4	U047	D	5000 (2270)
1,4-Naphthalenedione	130154	1,4-Naphthoquinone	1*	4	U166	D	5000 (2270)

TABLE 7.2 List of Hazardous Substances and Reportable Quantities (*Continued*)

[Note: All Comments/Notes Are Located at the End of This Table]

Hazardous Substance	CASRN	Regulatory Synonyms	Statutory		RCRA Waste Number	Final RQ	
			RQ	Code †		Category	Pounds (Kg)
2,7-Naphthalenedisulfonic acid, 3,3'-[(3,3'-dimethyl-(1,1'-biphenyl)-4,4'-diyl)-bis(azo)]bis(5-amino-4-hydroxy)-tetrasodium salt.	72571	Trypan blue	1*	4	U236	A	10 (4.54)
Naphthenic acid	1338245		100			A	100 (45.4)
1,4-Naphthoquinone	130154	1,4-Naphthalenedione	1*	1	U166	B	100 (45.4)
alpha-Naphthylamine	134327	1-Naphthalenamine	1*	4	U167	D	5000 (2270)
beta-Naphthylamine	91598	2-Naphthalenamine	1*	4	U168	B	100 (45.4)
alpha-Naphthylthiourea	86884	Thiourea, 1-naphthalenyl-	1*	4	P072	A	10 (4.54)
Nickel ††	7440020		1*	2		B	100 (45.4)
Nickel ammonium sulfate	15699180		5000	1		B	100 (45.4)
NICKEL AND COMPOUNDS	N.A.		1*	2		B	100 (45.4)**
Nickel carbonyl	13463393	Nickel carbonyl Ni(CO)4, (T-4)-	1*	4	P073	A	10 (4.54)
Nickel carbonyl Ni(CO)4, (T-4)-	13463393	Nickel carbonyl	1*	4	P073	A	10 (4.54)
Nickel chloride	7718549 37211055		5000	1		B	100 (45.4)
Nickel cyanide	557197	Nickel cyanide Ni(CN)2	1*	4	P074	A	10 (4.54)
Nickel cyanide Ni(CN)2	557197	Nickel cyanide	1*	4	P074	A	10 (4.54)
Nickel hydroxide	12054487		1000	1		A	10 (4.54)
Nickel nitrate	14216752		5000	1		B	100 (45.4)
Nickel sulfate	7786814		5000	1		B	100 (45.4)
Nicotine, & salts	54115	Pyridine, 3-(1-methyl-2-pyrrolidinyl)-, (S)-	1*	1	P075	B	100 (45.4)
Nitric acid	7697372		1000	1		C	1000 (454)
Nitric acid, thallium (1+) salt	10102451	Thallium (I) nitrate	1*	4	U217	B	100 (45.4)
Nitric oxide	10102439	Nitrogen oxide NO	1*	4	P076	A	10 (4.54)
p-Nitroaniline	100016	Benzenamine, 4-nitro-	1*	4	P077	D	5000 (2270)
Nitrobenzene	98953	Benzene, nitro-	1000	1,2,4	U169	C	1000 (454)
Nitrogen dioxide	10102440	Nitrogen oxide NO2	1000	1,4	P078	A	10 (4.54)
Nitrogen oxide NO	10102439	Nitric oxide	1*	4	P076	A	10 (4.54)
Nitrogen oxide NO2	10102440	Nitrogen dioxide	1000	1,4	P078	A	10 (4.54)
Nitroglycerine	55630	1,2,3-Propanetriol, trinitrate-	1*	4	P081	A	10 (4.54)
Nitrophenol (mixed)	25154556		1000	1		B	100 (45.4)
m-Nitrophenol	554847					B	100 (45.4)
o-Nitrophenol	88755	2-Nitrophenol				B	100 (45.4)
p-Nitrophenol	100027	Phenol, 4-nitro- 4-Nitrophenol				B	100 (45.4)
o-Nitrophenol	88755	2-Nitrophenol	1000	1,2		B	100 (45.4)
p-Nitrophenol	100027	Phenol, 4-nitro- 4-Nitrophenol	1000	1,2,4	U170	B	100 (45.4)
2-Nitrophenol	88755	o-Nitrophenol	1000	1,2		B	100 (45.4)

TABLE 7.2 List of Hazardous Substances and Reportable Quantities (*Continued*)

[Note: All Comments/Notes Are Located at the End of This Table]

Hazardous Substance	CASRN	Regulatory Synonyms	Statutory RQ	Statutory Code †	Statutory RCRA Waste Number	Final RQ Category	Final RQ Pounds (Kg)
Phenol, 4,4'-(1,2-diethyl-1,2-ethenediyl)bis-, (E)	56531	Diethylstilbestrol	1*	4	U089	X	1 (0.454)
Phenol, 2,4-dimethyl-	105679	2,4-Dimethylphenol	1*	2,4	U101	B	100(45.4)
Phenol, 2,4-dinitro-	51285	2,4-Dinitrophenol	1000	1,2,4	P048	A	10 (4.54)
Phenol, methyl-	1319773	Cresol(s) Cresylic acid	1000	1,4	U052	C	1000 (454)
m-Cresol	108394	m-Cresylic acid					
o-Cresol	95487	o-Cresylic acid					
p-Cresol	106445	p-Cresylic acid					
Phenol, 2-methyl-4,6-dinitro-	534521	4,6-Dinitro-o-cresol and salts	1*	2,4	P047	A	10 (4.54)
Phenol, 2,2'-methylenebis[3,4,6-trichloro-	70304	Hexachlorophene	1*	4	U132	B	100 (45.4)
Phenol, 2-(1-methylpropyl)-4,6-dinitro-	88857	Dinoseb	1*	4	P020	C	1000 (454)
Phenol, 4-nitro-	100027	p-Nitrophenol / 4-Nitrophenol	1000	1,2,4	U170	B	100 (45.4)
Phenol, pentachloro-	87865	Pentachlorophenol	10	1,2,4	U242	A	10 (4.54)
Phenol, 2,3,4,6-tetrachloro-	58902	2,3,4,6-Tetrachlorophenol	1*		U212	A	10 (4.54)
Phenol, 2,4,5-trichloro-	95954	2,4,5-Trichlorophenol	10	1,4	U230	A	10 (4.54)
Phenol, 2,4,6-trichloro-	88062	2,4,6-Trichlorophenol	10	1,2,4	U231	A	10 (4.54)
Phenol, 2,4,6-trinitro-, ammonium salt	131748	Ammonium picrate	1*	4	P009	A	10 (4.54)
L-Phenylalanine, 4-[bis(2-chloroethyl) amino]	148823	Melphalan	1*	4	U150	X	1 (0.454)
1,10-(1,2-Phenylene)pyrene	193395	Indeno(1,2,3-cd)pyrene	1*	2,4	U137	B	100 (45.4)
Phenylmercury acetate	62384	Mercury, (acetato-O)phenyl-	1*	4	P092	B	100 (45.4)
Phenylthiourea	103855	Thiourea, phenyl-	1*	4	P093	B	100 (45.4)
Phorate	298022	Phosphorodithioic acid, O,O-diethyl S-(ethylthio), methyl ester	1*	4	P094	A	10 (4.54)
Phosgene	75445	Carbonic dichloride	5000	1,4	P095	A	10 (4.54)
Phosphine	7803512		1*	4	P096	B	100 (45.4)
Phosphoric acid	7664382		5000	1		D	5000 (2270)
Phosphoric acid, diethyl 4-nitrophenyl ester	311455	Diethyl-p-nitrophenyl phosphate	1*	1	P041	B	100 (45.4)
Phosphoric acid, lead(2+) salt (2:3)	7446277	Lead phosphate	1*	4	U145	X	1 (0.454) #
Phosphorodithioic acid, O,O-diethyl S-[2-(ethylthio)ethyl]ester	298044	Disulfoton	1	1,4	P039	X	1 (0.454)
Phosphorodithioic acid, O,O-diethyl S-(ethylthio), methyl ester	298022	Phorate	1*	4	P094	A	10 (4.54)
Phosphorodithioic acid, O,O-diethyl S-methyl ester	3288582	O,O-Diethyl S-methyl dithiophosphate	1*	4	U087	D	5000 (2270)
Phosphorodithioic acid, O,O-dimethyl S-[2(methyla-mino)-2-oxoethyl] ester	60515	Dimethoate	1*	4	P044	A	10 (4.54)
Phosphorofluoridic acid, bis(1-methylethyl) ester	55914	Diisopropylfluorophosphate	1*	4	P043	B	100 (45.4)
Phosphorothioic acid, O,O-diethyl O-(4-nitrophenyl) ester	56382	Parathion	1	1,4	P089	A	10 (4.54)
Phosphorothioic acid, O,[4-[(dimethylamino) sulfonyl]phenyl]O,O-dimethyl ester	52857	Famphur	1*	4	P097	C	1000 (454)

Name	CAS No.	Synonym					Quantity
Phosphorothioic acid, O,O-dimethyl O-(4-nitrophenyl) ester	298000	Methyl parathion	100	1,4	P071	B	100 (45.4)
Phosphorothioic acid, O,O-diethyl O-pyrazinyl ester	297972	O,O-Diethyl O-pyrazinyl phosphorothioate	1*	4	P040	B	100 (45.4)
Phosphorus	7723140			1		X	1 (0.454)
Phosphorus oxychloride	10025873		5000	1		C	1000 (454)
Phosphorus pentasulfide	1314803	Phosphorus sulfide Sulfur phosphide	100	1,4	U189	B	100 (45.4)
Phosphorus sulfide	1314803	Phosphorus pentasulfide Sulfur phosphide	100	1,4	U189	B	100 (45.4)
Phosphorus trichloride	7719122		5000	1		C	1000 (454)
PHTHALATE ESTERS	N.A.		1*	2			**
Phthalic anhydride	85449	1,3-Isobenzofurandione	1*	4	U190	D	5000 (2270)
2-Picoline	109068	Pyridine, 2-methyl-	1*	4	U191	D	5000 (2270)
Piperidine, 1-nitroso-	100754	N-Nitrosopiperidine	1*	4	U179	A	10 (4.54)
Plumbane, tetraethyl-	78002	Tetraethyl lead	100	1,4	P110	A	10 (4.54)
POLYCHLORINATED BIPHENYLS (PCBs)	1336363		10	1,2		X	1 (0.454)
Arocor 1016	12674112	POLYCHLORINATED BIPHENYLS (PCBs)					
Arocor 1221	11104282	POLYCHLORINATED BIPHENYLS (PCBs)					
Arocor 1232	11141165	POLYCHLORINATED BIPHENYLS (PCBs)					
Arocor 1242	53469219	POLYCHLORINATED BIPHENYLS (PCBs)					
Arocor 1248	12672296	POLYCHLORINATED BIPHENYLS (PCBs)					
Arocor 1254	11097691	POLYCHLORINATED BIPHENYLS (PCBs)					
Arocor 1260	11096825	POLYCHLORINATED BIPHENYLS (PCBs)					
POLYNUCLEAR AROMATIC HYDROCARBONS	N.A.		1*	2			**
Potassium arsenate	7784410		1000	1		X	1 (0.454)
Potassium arsenite	10124502		1000	1		X	1 (0.454)
Potassium bichromate	7778509		1000	1		A	10 (4.54)
Potassium chromate	7789006		1000	1		A	10 (4.54)
Potassium cyanide	151508	Potassium cyanide K (CN)	10	1,4	P098	A	10 (4.54)
Potassium cyanide K(CN)	151508	Potassium cyanide	10	1,4	P098	A	10 (4.54)
Potassium hydroxide	1310583		1000	1		C	1000 (454)
Potassium permanganate	7722647		100	1		B	100 (45.4)
Potassium silver cyanide	506616	Argentate (1-), bis(cyano-C), potassium	1*	4	P099	X	1 (0.454)
Pronamide	23950585	Benzamide, 3,5-dichloro-N-(1,1-dimethyl-2-propynyl)-	1*	4	U192	D	5000 (2270)
Propanal, 2-methyl-2-(methylthio)-, O-[(methylamino)carbonyl]oxime	116063	Aldicarb	1*	4	P070	X	1 (0.454)
1-Propanamine	107108	n-Propylamine	1*	4	U194	D	5000 (2270)
1-Propanamine, N-propyl-	142847	Dipropylamine	1*	4	U110	D	5000 (2270)
1-Propanamine, N-nitroso-N-propyl-	621647	Di-n-propylnitrosamine	1*	2,4	U111	A	10 (4.54)
Propane, 1,2-dibromo-3-chloro-	96128	1,2-Dibromo-3-chloropropane	1*	4	U066	X	1 (0.454)
Propane, 2-nitro-	79469	2-Nitropropane	1*	4	U171	A	10 (4.54)
1,3-Propane sultone	1120714	1,2-Oxathiolane, 2,2-dioxide	1*	4	U193	A	10 (4.54)
Propane, 1,2-dichloro-	78875	Propylene dichloride	5000	1,2,4	U083	C	1000 (454)
Propanedinitrile	109773	Malononitrile	1*	4	U149	C	1000 (454)
Propanenitrile	107120	Ethyl cyanide	1*	4	P101	A	10 (4.54)
Propanenitrile, 3-chloro-	542767	3-Chloropropionitrile	1*	4	P027	C	1000 (454)
Propanenitrile, 2-hydroxy-2-methyl-	75865	Acetone cyanohydrin 2-Methyllactonitrile	10	1,4	P069	A	10 (4.54)
Propane, 2,2'-oxybis[2-chloro-	108601	Dichloroisopropyl ether	1*	2,4	U027	C	1000 (454)
1,2,3-Propanetriol, trinitrate-	55630	Nitroglycerine	1*	4	P081	A	10 (4.54)
1-Propanol, 2,3-dibromo-, phosphate (3:1)	126727	Tris(2,3-dibromopropyl) phosphate	1*	4	U235	A	10 (4.54)

TABLE 7.2 List of Hazardous Substances and Reportable Quantities (*Continued*)

[Note: All Comments/Notes Are Located at the End of This Table]

Hazardous Substance	CASRN	Regulatory Synonyms	Statutory			Final RQ	
			RQ	Code †	RCRA Waste Number	Category	Pounds (Kg)
1-Propanol, 2-methyl-	78831	Isobutyl alcohol	1*	4	U140	D	5000 (2270)
2-Propanone	67641	Acetone	1*	4	U002	D	5000 (2270)
2-Propanone, 1-bromo-	598312	Bromoacetone	1*	4	P017	C	1000 (454)
Propargite	2312358		10	1		A	10 (4.54)
Propargyl alcohol	107197	2-Propyn-1-ol	1*	4	P102	C	1000 (454)
2-Propenal	107028	Acrolein	1	1,2,4	P003	X	1 (0.454)
2-Propenamide	79061	Acrylamide	1*	4	U007	D	5000 (2270)
1-Propene, 1,1,2,3,3,3-hexachloro-	1888717	Hexachloropropene	1*	4	U243	D	1000 (454)
1-Propene, 1,3-dichloro-	542756	1,3-Dichloropropene	5000	1,2,4	U084	B	100 (45.4)
2-Propenenitrile	107131	Acrylonitrile	100	1,2,4	U009	B	100 (45.4)
2-Propenenitrile, 2-methyl-	126987	Methacrylonitrile	1*	4	U152	C	1000 (454)
2-Propenoic acid	79107	Acrylic acid	1*	4	U008	D	5000 (2270)
2-Propenoic acid, ethyl ester	140885	Ethyl acrylate	1*	4	U113	C	1000 (454)
2-Propenoic acid, 2-methyl-, ethyl ester	97632	Ethyl methacrylate	1*	4	U118	C	1000 (454)
2-Propenoic acid, 2-methyl-, methyl ester	80626	Methyl methacrylate	5000	1,4	U162	C	1000 (454)
2-Propen-1-ol	107186	Allyl alcohol	100	1,4	P005	B	100 (45.4)
Propionic acid	79094		5000	1		D	5000 (2270)
Propionic acid, 2-(2,4,5-trichlorophenoxy)-	93721	Silvex (2,4,5-TP) / 2,4,5-TP acid	100	1,4	U233	B	100 (45.4)
Propionic anhydride	123626		5000	1		D	5000 (2270)
n-Propylamine	107108	1-Propanamine	1*	4	U194	D	5000 (2270)
Propylene dichloride	78875	Propane, 1,2-dichloro- / 1,2-Dichloropropane	5000	1,2,4	U083	C	1000 (454)
Propylene oxide	75569		5000	1		B	100 (45.4)
1,2-Propylenimine	75558	Aziridine, 2-methyl-	1*	4	P067	X	1 (0.454)
2-Propyn-1-ol	107197	Propargyl alcohol	1*	4	P102	C	1000 (454)
Pyrene	129000		1*	2		D	5000 (2270)
Pyrethrins	121299 / 121211 / 8003347		1000	1		X	1 (0.545)
3,6-Pyridazinedione, 1,2-dihydro-	123331	Maleic hydrazide	1*	4	U148	D	5000 (2270)
4-Pyridinamine	504245	4-Aminopyridine	1*	4	P008	C	1000 (454)
Pyridine	110861		1*	4	U196	C	1000 (454)
Pyridine, 2-methyl-	109068	2-Picoline	1*	4	U191	D	5000 (2270)
Pyridine, 3-(1-methyl-2-pyrrolidinyl)-, (S)-	54115	Nicotine, & salts	1*	4	P075	B	100 (45.4)
2,4-(1H,3H)-Pyrimidinedione, 5-[bis(2-chloroethyl)amino]-	66751	Uracil mustard	1*	4	U237	A	10 (4.54)
4(1H)-Pyrimidinone, 2,3-dihydro-6-methyl-2-thioxo-	56042	Methylthiouracil	1*	4	U164	A	10 (4.54)
Pyrrolidine, 1-nitroso-	930552	N-Nitrosopyrrolidine	1*	4	U180	X	1 (0.454)
Quinoline	91225		1000	1		D	5000 (2270)

Chemical	CAS Number	Statutory RQ	Code	RCRA Waste No.	Category	Final RQ lb (kg)
RADIONUCLIDES	N.A.	1*	3		D	5000 (2270) §
Reserpine — Yohimban-16-carboxylic acid, 11,17-dimethoxy-18-[(3,4,5-trimethoxybenzoyl)oxy]-, methyl ester (3beta,16beta,17alpha,18beta,20alpha)-	50555	1*	4	U200	D	5000 (2270)
Resorcinol — 1,3-Benzenediol	108463	1000		U201	B	5000 (2270)
Saccharin and salts — 1,2-Benzisothiazol-3(2H)-one, 1,1-dioxide	81072	1*	1,4	U202	B	100 (45.4)
Safrole — 1,3-Benzodioxole, 5-(2-propenyl)-	94597	1*	4	U203	A	100 (45.4)
Selenious acid	7783008	1*	4	U204	C	10 (4.54)
Selenious acid, dithallium (1+) salt — Thallium selenite	12039520	1*	4	P114	B	1000 (454)
Selenium ††	7782492	1*	2			100 (45.4) **
SELENIUM AND COMPOUNDS	N.A.		2			
Selenium dioxide	7446084	1000	1,4	U204	A	10 (4.54)
Selenium oxide — Selenium dioxide	7446084	1000	1,4	U204	A	10 (4.54)
Selenium sulfide — Selenium sulfide SeS2	7488564	1*	4	U205	A	10 (4.54)
Selenium sulfide SeS2 — Selenium sulfide	7488564	1*	4	U205	A	10 (4.54)
Selenourea	630104	1*	4	P103	C	1000 (454)
L-Serine, diazoacetate (ester) — Azaserine	115026	1*	4	U015	X	1 (0.454)
SILVER ††	7440224	1*	2		C	1000 (454)
SILVER AND COMPOUNDS	N.A.		2			
Silver cyanide — Silver cyanide Ag (CN)	506649	1*	4	P104	X	1 (0.454)
Silver cyanide Ag (CN) — Silver cyanide	506649	1*	4	P104	X	1 (0.454)
Silver nitrate	7761888	1	1		X	1 (0.454) **
Silvex (2,4,5-TP) — Propionic acid, 2-(2,4,5-trichlorophenoxy)- / 2,4,5-TP acid	93721	100	1,4	U233	B	100 (45.4)
Sodium	7440235	1000	1		A	10 (4.54)
Sodium arsenate	7631892	1000	1		A	1 (0.454)
Sodium arsenite	7784465	1000	1		X	1 (0.454)
Sodium azide	26628228	1*	4	P105	C	1000 (454)
Sodium bichromate	10588019	1000	1		A	10 (4.54)
Sodium bifluoride	1333831	5000	1		B	100 (45.4)
Sodium bisulfite	7631905	5000	1		D	5000 (2270)
Sodium chromate	7775113	1000	1		A	10 (4.54)
Sodium cyanide	143339	10	1,4	P106	A	10 (4.54)
Sodium cyanide Na (CN) — Sodium cyanide	143339	10	1,4	P106	A	10 (4.54)
Sodium dodecylbenzenesulfonate	25155300	1000	1		C	1000 (454)
Sodium fluoride	7681494	5000	1		C	1000 (454)
Sodium hydrosulfide	16721805	5000	1		D	5000 (2270)
Sodium hydroxide	1310732	1000	1		C	1000 (454)
Sodium hypochlorite	7681529 / 10022705	100	1		B	100 (45.4)
Sodium methylate	124414	1000	1		C	1000 (454)
Sodium nitrite	7632000	100	1		B	100 (45.4)
Sodium phosphate, dibasic	7558794 / 10039324 / 10140655	5000	1		D	5000 (2270)

TABLE 7.2 List of Hazardous Substances and Reportable Quantities (*Continued*)

[Note: All Comments/Notes Are Located at the End of This Table]

Hazardous Substance	CASRN	Regulatory Synonyms	Statutory RQ	Statutory Code †	RCRA Waste Number	Final RQ Category	Final RQ Pounds (Kg)
Sodium phosphate, tribasic	7601549 7758294 7785844 10101890 10124568 10361894		5000	1		D	5000 (2270)
Sodium selenite	10102188 7782823		1000	1		B	100 (45.4)
Streptozotocin	18883664	D-Glucose, 2-deoxy-2-[[(methylnitrosoamino)-carbonyl]amino]-; Glucopyranose, 2-deoxy-2-(3-methyl-3-nitrosour-eido)-	1*	4	U206	X	1 (0.454)
Strontium chromate	7789062		1000	1		A	10 (4.54)
Strychnidin-10-one	57249		10	1,4	P108	A	10 (4.54)
Strychnidin-10-one, 2,3-dimethoxy-	357573	Brucine	1*	4	P018	B	100 (45.4)
Strychnine, & salts	57249	Strychnidin-10-one	10	1,4	P108	A	10 (4.54)
Styrene	100425		1000	1		C	1000 (454)
Sulfur monochloride	12771083		1000	1		C	1000 (454)
Sulfur phosphide	1314803	Phosphorus pentasulfide; Phosphorus sulfide	100	1,4	U189	B	100 (45.4)
Sulfuric acid	7664939 8014957		1000	1		C	1000 (454)
Sulfuric acid, dithallium (1+) salt	7446186 10031591	Thallium (I) sulfate	1000	1,4	P115	B	100 (45.4)
Sulfuric acid, dimethyl ester	77781	Dimethyl sulfate	1*	4	U103	B	100 (45.4)
2,4,5-T acid	93765	Acetic acid, (2,4,5-trichlorophenoxy); 2,4,5-T	100	1,4	U232	C	1000 (454)
2,4,5-T amines	2008460 1319728 3813147 6369966 6369977		100	1		D	5000 (2270)
2,4,5-T esters	93798 1928478 2545597 25168154 61792072		100	1		C	1000 (454)
2,4,5-T salts	13560991	Acetic acid, (2,4,5-trichlorophenoxy); 2,4,5-T acid	100	1		C	1000 (454)
2,4,5-T	93765		100	1,4	U232	C	1000 (454)

Name	CAS No.	Synonym / Description	RQ	Code	Waste No.		Final RQ
TDE	72548	Benzene, 1,1'-(2,2-dichloroethylidene)bis[4-chloro- DDD 4,4' DDD	1*	1,2,4	U060	X	1 (0.454)
1,2,4-Tetrachlorobenzene	95943	Benzene, 1,2,4,5-tetrachloro-	1*	4	U207	D	5000 (2270)
2,3,7,8-Tetrachlorodibenzo-p-dioxin (TCDD)	1746016		1*	2		X	1 (0.454)
1,1,1,2-Tetrachloroethane	630206	Ethane, 1,1,1,2-tetrachloro-	1*	4	U208	B	100 (45.4)
1,1,2,2-Tetrachloroethane	79345	Ethane, 1,1,2,2-tetrachloro-	1*	2,4	U209	B	100 (45.4)
Tetrachloroethene	127184	Ethene, tetrachloro- Perchloroethylene Tetrachloroethylene	1*	2,4	U210	B	100 (45.4)
Tetrachloroethylene	127184	Ethene, tetrachloro- Perchloroethylene Tetrachloroethene	1*	2,4	U210	B	100 (45.4)
2,3,4,6-Tetrachlorophenol	58902	Phenol, 2,3,4,6-tetrachloro-	1*	4	U212	A	10 (4.54)
Tetraethyl lead	78002	Plumbane, tetraethyl-	100	1,4	P110	A	10 (4.54)
Tetraethyl pyrophosphate	107493	Diphosphoric acid, tetraethyl ester	100	1,4	P111	A	10 (4.54)
Tetraethyldithiopyrophosphate	3689245	Thiodiphosphoric acid, tetraethyl ester	1*	4	P109	B	100 (45.4)
Tetrahydrofuran	109999	Furan, tetrahydro-	1*	4	U213	C	1000 (454)
Tetranitromethane	509148	Methane, tetranitro-	1*	4	P112	A	10 (4.54)
Tetraphosphoric acid, hexaethyl ester	757584	Hexaethyl tetraphosphate	1*	4	P062	B	100 (45.4)
Thallic oxide	1314325	Thallium oxide Tl2O3	1*	4	P113	B	100 (45.4)
Thallium ††	7440280		1*	2		C	1000 (454) **
Thallium and compounds	N.A.		1*	2			
Thallium (I) acetate	563688	Acetic acid, thallium(1+) salt	1*	4	U214	B	100 (45.4)
Thallium (I) carbonate	6533739	Carbonic acid, dithallium(1+) salt	1*	4	U215	B	100 (45.4)
Thallium (I) chloride	7791120	Thallium chloride TlCl	1*	4	U216	B	100 (45.4)
Thallium chloride TlCl	7791120	Thallium(I) chloride	1*	4	U216	B	100 (45.4)
Thallium (I) nitrate	10102451	Nitric acid, thallium (1+) salt	1*	4	U217	B	100 (45.4)
Thallium oxide Tl2O3	1314325	Thallic oxide	1*	4	P113	B	100 (45.4)
Thallium selenite	12039520	Selenious acid, dithallium(1+) salt	1*	4	P114	C	1000 (454)
Thallium (I) sulfate	7446186	Sulfuric acid, dithallium(1+) salt	1000	1,4	P115	B	100 (45.4)
	10031591						
Thioacetamide	62555	Ethanethioamide	1*	4	U218	A	10 (4.54)
Thiodiphosphoric acid, tetraethyl ester	3689245	Tetraethyldithiopyrophosphate	1*	4	P109	B	100 (45.4)
Thiofanox	39196184	2-Butanone, 3,3-dimethyl-1-(methylthio)-, O[(methylamino)carbonyl) oxime	1*	4	P045	B	100 (45.4)
Thioimidodicarbonic diamide [(H2N)C(S)] 2NH	541537	Dithiobiuret	1*	4	P049	B	100 (45.4)
Thiomethanol	74931	Methanethiol Methylmercaptan	100	1,4	U153	B	100 (45.4)
Thioperoxydicarbonic diamide [(H2N)C(S)] 2S2, tetramethyl-	137268	Thiram	1*	4	U244	A	10 (4.54)
Thiophenol	108985	Benzenethiol	1*	4	P014	B	100 (45.4)
Thiosemicarbazide	79196	Hydrazinecarbothioamide	1*	4	P116	B	100 (45.4)
Thiourea	62566		1*	4	U219	A	10 (4.54)
Thiourea, (2-chlorophenyl)-	5344821	1-(o-Chlorophenyl)thiourea	1*	4	P026	B	100 (45.4)
Thiourea, 1-naphthalenyl-	86884	alpha-Naphthylthiourea	1*	4	P072	B	100 (45.4)
Thiourea, phenyl-	103855	Phenylthiourea	1*	4	P093	B	100 (45.4)
Thiram	137268	Thioperoxydicarbonic diamide [(H2N)C(S)] 2S2, tetramethyl-	1*	4	U244	A	10 (4.54)
Toluene	108883	Benzene, methyl-	1000	1,2,4	U220	C	1000 (454)

TABLE 7.2 List of Hazardous Substances and Reportable Quantities (*Continued*)

[Note: All Comments/Notes Are Located at the End of This Table]

Hazardous Substance	CASRN	Regulatory Synonyms	Statutory			Final RQ	
			RQ	Code †	RCRA Waste Number	Category	Pounds (Kg)
Toluenediamine	95807; 496720; 823405; 25376458	Benzenediamine, ar-methyl-	1*	4	U221	A	10 (4.54)
Toluene diisocyanate	584849; 91087; 26471625	Benzene, 1,3-diisocyanatomethyl-	1*	4	U223	B	100 (45.4)
o-Toluidine	95534	Benzenamine, 2-methyl-	1*	4	U328	B	100 (45.4)
p-Toluidine	106490	Benzenamine, 4-methyl-	1*	4	U353	B	100 (45.4)
o-Toluidine hydrochloride	636215	Benzenamine, 2-methyl-, hydrochloride	1*	4	U222	B	100 (45.4)
Toxaphene	8001352	Camphene, octachloro-	1*	1,2,4,	P123	X	1 (0.454)
2,4,5-TP acid	93721	Propionic acid, 2-(2,4,5-trichlorophenoxy)- Silvex (2,4,5-TP)	100	1,4	U233	B	100 (45.4)
2,4,5-TP esters	32534955		100	1		B	100 (45.4)
1H-1,2,4-Triazol-3-amine	61825	Amitrole	1*	4	U011	A	10 (4.54)
Trichlorfon	52686		1000	1		B	100 (45.4)
1,2,4-Trichlorobenzene	120821		1*	2		B	100 (45.4)
1,1,1-Trichloroethane	71556	Ethane, 1,1,1-trichloro- Methyl chloroform	1*	2,4	U226	C	1000 (454)
1,1,2-Trichloroethane	79005	Ethane, 1,1,2-trichloro-	1*	2,4	U227	B	100 (45.4)
Trichloroethene	79016	Ethene, trichloro- Trichloroethylene	1000	1,2,4	U228	B	100 (45.4)
Trichloroethylene	79016	Ethene, trichloro- Trichloroethene	1000	1,2,4	U228	B	100 (45.4)
Trichloromethanesulfenyl chloride	594423	Methanesulfenyl chloride, trichloro-	1*	4	P118	B	100 (45.4)
Trichloromonofluoromethane	75694	Methane, trichlorofluoro-	1*	4	U121	D	5000 (2270)
Trichlorophenol	25167822; 15950660; 933788; 933755		10	1		A	10 (4.54)
2,3,4-Trichlorophenol							
2,3,5-Trichlorophenol							
2,3,6-Trichlorophenol							
2,4,5-Trichlorophenol	95954	Phenol, 2,4,5-trichloro-	10*	1,4	U230	A	10 (4.54)
2,4,6-Trichlorophenol	88062	Phenol, 2,4,6-trichloro-	10*	1,2,4	U231	A	10 (4.54)
3,4,5-Trichlorophenol	609198						
2,4,5-Trichlorophenol	95954	Phenol, 2,4,5-trichloro-	10*	1,4	U230	A	10 (4.54)
2,4,6-Trichlorophenol	88062	Phenol, 2,4,6-trichloro-	10	1,2,4	U231	A	10 (4.54)
Triethanolamine dodecylbenzenesulfonate	27323417		1000	1		C	1000 (454)
Triethylamine	121448		5000	1		D	5000 (2270)
Trimethylamine	75503		1000	1		D	1000 (454)
1,3,5-Trinitrobenzene	99354	Benzene, 1,3,5-trinitro-	1*	1	U234	B	100 (45.4)
1,3,5-Trioxane, 2,4,6-trimethyl-	123637	Paraldehyde	1*	4	U182	A	10 (4.54)
						C	1000 (454)

Substance	CAS No.		Statutory RQ	Footnotes	Code	Category	Final RQ pounds (kg)
Tris(2,3-dibromopropyl) phosphate	126727	N.A.	1*	4	U235	A	10 (4.54)
Trypan blue	72571	N.A.	1*	4	U236	A	10 (4.54)
Unlisted Hazardous Wastes Characteristic of Corrosivity		N.A.	1*	4	D002	B	100 (45.4)
Unlisted Hazardous Wastes Characteristics:		N.A.					
Characteristic of Toxicity:							
Arsenic (D004)		N.A.	*1	4	D004	X	1 (0.454)
Barium (D005)		N.A.	*1	4	D005	C	1,000 (454)
Benzene (D018)		N.A.	1000	1, 2, 3, 4	D018	A	10 (4.54)
Cadmium (D006)		N.A.	*1	4	D006	A	10 (4.54)
Carbon tetrachloride (D019)		N.A.	5,000	1, 2, 4	D019	A	10 (4.54)
Chlordane (D020)		N.A.	1	1, 2, 4	D020	X	1 (0.454)
Chlorobenzene (D021)		N.A.	100	1, 2, 4	D021	B	100 (45.4)
Chloroform (D022)		N.A.	5,000	1, 2, 4	D022	A	10 (4.54)
Chromium (D007)		N.A.	*1	4	D007	A	10 (4.54)
o-Cresol (D023)		N.A.	1,000	4	D023	C	1,000 (454)
m-Cresol (D024)		N.A.	1,000	1, 4	D024	C	1,000 (454)
p-Cresol (D025)		N.A.	1,000	1, 4	D025	C	1,000 (454)
Cresol (D026)		N.A.	1,000	1, 4	D026	C	1,000 (454)
2,4-D (D016)		N.A.	100	1, 4	D016	B	100 (45.4)
1,4-Dichlorobenzene (D027)		N.A.	100	1, 2, 4	D027	B	100 (45.4)
1,2-Dichloroethane (D028)		N.A.	5,000	1, 2, 4	D028	B	100 (45.4)
1,1-Dichloroethylene (D029)		N.A.	5,000	1, 2, 4	D029	B	100 (45.4)
2,4-Dinitrotoluene (D030)		N.A.	1,000	1, 2, 4	D030	A	10 (4.54)
Endrin (D012)		N.A.	*1	1, 4	D012	X	1 (0.454)
Heptachlor (and epoxide) (D031)		N.A.	1	1, 2, 4	D031	X	1 (0.454)
Hexachlorobenzene (D032)		N.A.	*1	1, 2, 4	D032	A	10 (4.54)
Hexachlorobutadiene (D033)		N.A.	*1	2, 4	D033	X	1 (0.454)
Hexachloroethane (D034)		N.A.	*1	2, 4	D034	B	100 (45.4)
Lead (D008)		N.A.	*1	4	D008		(#)
Lindane (D013)		N.A.	*1	1, 4	D013	X	1 (0.454)
Mercury (D009)		N.A.	*1	4	D009	X	1 (0.454)
Methoxychlor (D014)		N.A.	1	1, 4	D014	X	1 (0.454)
Methyl ethyl ketone (D035)		N.A.	*1	4	D035	D	5,000 (2270)
Nitrobenzene (D036)		N.A.	1,000	1, 2, 4	D036	C	1,000 (454)
Pentachlorophenol (D037)		N.A.	10	1, 2, 4	D037	A	10 (4.54)
Pyridine (D038)		N.A.	*1	4	D038	C	1,000 (454)
Selenium (D010)		N.A.	*1	4	D010	A	10 (4.54)
Silver (D011)		N.A.	*1	4	D011	X	1 (0.454)
Tetrachloroethylene (D039)		N.A.	*1	4	D039	B	100 (45.4)
Toxaphene (D015)		N.A.	*1	2, 4	D015	X	1 (0.454)
Trichloroethylene (D040)		N.A.	1000	1, 4	D040	B	100 (45.4)
2,4,5-Trichlorophenol (D041)		N.A.	10	1, 2, 4	D041	A	10 (4.54)
2,4,6-Trichlorophenol (D042)		N.A.	10	1, 4	D042	A	10 (4.54)
2,4,5-TP (D017)		N.A.	100	1, 2, 4	D017	B	100 (45.4)
Vinyl chloride (D043)		N.A.	*1	1, 4	D043	X	1 (0.454)
Unlisted Hazardous Wastes Characteristic of Ignitability		N.A.	1*	2, 3, 4	D001	B	100 (45.4)

1-Propanol, 2,3-dibromo-, phosphate [(3:1)]
2,7-Naphthalenedisulfonic acid, 3,3'-3,3'-dimethyl-(1,1'-biphenyl)-4,4'-diyl)-bis(azo)]bis(5-amino-4-hydroxy)-tetrasodium salt

TABLE 7.2 List of Hazardous Substances and Reportable Quantities (Continued)

[Note: All Comments/Notes Are Located at the End of This Table]

Hazardous Substance	CASRN	Regulatory Synonyms	Statutory				Final RQ
			RQ	Code †	RCRA Waste Number	Category	Pounds (Kg)
Unlisted Hazardous Wastes Characteristic of Reactivity.	N.A.		1*	4	D003	B	100 (45.4)
Uracil mustard	66751	2,4-(1H,3H)-Pyrimidinedione, 5-[bis(2-chloroethyl)amino]-	1*	4	U237	A	10 (4.54)
Uranyl acetate	541093		5000	1		B	100 (45.4)
Uranyl nitrate	10102064 36478769		5000	1		B	100 (45.4)
Urea, N-ethyl-N-nitroso-	759739	N-Nitroso-N-ethylurea	1*	4	U176	X	1 (0.454)
Urea, N-methyl-N-nitroso-	684935	N-Nitroso-N-methylurea	1*	4	U177	X	1 (0.454)
Vanadic acid, ammonium salt	7803556	Ammonium vanadate	1*	4	P119	C	1000 (454)
Vanadium oxide V205	1314621	Vanadium pentoxide	1000	1,4	P120	C	1000 (454)
Vanadium pentoxide	1314621	Vanadium oxide V205	1000	1,4	P120	C	1000 (454)
Vanadyl sulfate	27774136		1000	1		C	1000 (454)
Vinyl chloride	75014	Ethene, chloro-	1*	2,3,4	U043	X	1 (0.454)
Vinyl acetate	108054	Vinyl acetate monomer	1000	1		D	5000 (2270)
Vinyl acetate monomer	108054	Vinyl acetate	1000	1		D	5000 (2270)
Vinylamine, N-methyl-N-nitroso-	4549400	N-Nitrosomethylvinylamine	1*	4	P084	A	10 (4.54)
Vinylidene chloride	75354	Ethene, 1,1-dichloro- 1,1-Dichloroethylene	5000	1,2,4	U078	B	100 (45.4)
Warfarin, & salts, when present at concentrations greater than 0.3%.	81812	2H-1-Benzopyran-2-one, 4-hydroxy-3-(3-oxo-1-phenyl-butyl)-, & salts, when present at concentrations greater than 0.3%	1*	4	P001	B	100 (45.4)
Xylene (mixed)	1330207	Benzene, dimethyl	1000	1,4	U239	C	1000 (454)
m-Benzene, dimethyl	108383	m-Xylene					
o-Benzene, dimethyl	95476	o-Xylene					
p-Benzene, dimethyl	106423	p-Xylene					
Xylenol	1300716		1000	1	U200	C	1000 (454)
Yohimban-16-carboxylic acid,11,17-dimethoxy-18-[(3,4,5-trimethoxybenzoyl)oxy]-, methyl ester (3beta,16beta,17alpha, 18beta, 20alpha)-	50555	Reserpine	1*	4		D	5000 (2270)
Zinc ††	7440666		1*	2		C	1000 (454)
ZINC AND COMPOUNDS	N.A.		1*	2			**
Zinc acetate	557346		1000	1		C	1000 (454)
Zinc ammonium chloride	52628258 14639975 14639986		5000	1		C	1000 (454)
Zinc borate	1332076		1000	1		C	1000 (454)
Zinc bromide	7699458		5000	1		C	1000 (454)
Zinc carbonate	3486359		1000	1		C	1000 (454)

Name	CAS No.					
Zinc chloride	7646857	5000	1		C	1000 (454)
Zinc cyanide	557211	10	1,4	P121	A	10 (4.54)
Zinc cyanide Zn(CN)2	557211	10	1,4	P121	A	10 (4.54)
Zinc fluoride	7783495	1000	1		C	1000 (454)
Zinc formate	557415	1000	1		C	1000 (454)
Zinc hydrosulfite	7779864	1000	1		C	1000 (454)
Zinc nitrate	7779886	5000	1		C	1000 (454)
Zinc phenosulfonate	127822	5000	1		D	5000 (2270)
Zinc phosphide	1314847	1000	1,4	P122	B	100 (45.4)
Zinc phosphide Zn3P2, when present at concentrations greater than 10%.	1314847	1000	1,4	P122	B	100 (45.4)
Zinc silicofluoride	16871719	5000	1		D	5000 (2270)
Zinc sulfate	7733020	1000	1		C	1000 (454)
Zirconium nitrate	13746899	5000	1		D	5000 (2270)
Zirconium potassium fluoride	16923958	5000	1		D	1000 (454)
Zirconium sulfate	14644612	5000	1		D	5000 (2270)
Zirconium tetrachloride	10026116	5000	1		D	5000 (2270)
F001		1*	4	F001	A	10 (4.54)

The following spent halogenated solvents used in degreasing; all spent solvent mixtures/blends used in degreasing containing, before use, a total of ten percent or more (by volume) of one or more of the above halogenated solvents or those solvents listed in F002, F004, and F005; and still bottoms from the recovery of these spent solvents and spent solvent mixtures.

Name	CAS No.					
(a) Tetrachloroethylene	127184	1*	2,4	U210	B	100 (45.4)
(b) Trichloroethylene	79016	1000	1,2,4	U228	B	100 (45.4)
(c) Methylene chloride	75092	1*	2,4	U080	C	1000 (454)
(d) 1,1,1-Trichloroethane	71556	1*	2,4	U226	C	1000 (454)
(e) Carbon tetrachloride	56235	5000	1,2,4	U211	A	10 (4.54)
(f) Chlorinated fluorocarbons	N.A.				D	5000 (2270)
F002		1*	4	F002	A	10 (4.54)

The following spent halogenated solvents; all spent solvent mixtures/blends containing, before use, a total of ten percent or more (by volume) of one or more of the above halogenated solvents or those listed in F001, F004, or F005; and still bottoms from the recovery of these spent solvents and spent solvent mixtures.

Name	CAS No.					
(a) Tetrachloroethylene	127184	1*	2,4	U210	B	100 (45.4)
(b) Methylene chloride	75092	1*	2,4	U080	C	1000 (454)
(c) Trichloroethylene	79016	1000	1,2,4	U228	B	100 (45.4)
(d) 1,1,1-Trichloroethane	71556	1*	2,4	U226	C	1000 (454)
(e) Chlorobenzene	108907	100	1,2,4	U037	B	100 (45.4)
(f) 1,1,2-Trichloro-1,2,2-trifluoroethane	76131				D	5000 (2270)
(g) o-Dichlorobenzene	95501	100	1,2,4	U070	B	100 (45.4)
(h) Trichlorofluoromethane	75694	1*	4	U121	D	5000 (2270)

TABLE 7.2 List of Hazardous Substances and Reportable Quantities (*Continued*)

[Note: All Comments/Notes Are Located at the End of This Table]

Hazardous Substance	CASRN	Regulatory Synonyms	Statutory RQ	Statutory Code †	Statutory RCRA Waste Number	Final RQ Category	Final RQ Pounds (Kg)
(i) 1,1,2-Trichloroethane............	79005		1*	2,4	U227	B	100 (45.4)
F003............			1*	4	F003	B	100 (45.4)
The following spent non-halogenated solvents and the still bottoms from the recovery of these solvents:							
(a) Xylene	1330207					C	1000 (454)
(b) Acetone	67641					D	5000 (2270)
(c) Ethyl acetate	141786					D	5000 (2270)
(d) Ethylbenzene	100414					C	1000 (454)
(e) Ethyl ether	60297					B	100 (45.4)
(f) Methyl isobutyl ketone	108101					D	5000 (2270)
(g) n-Butyl alcohol	71363					D	5000 (2270)
(h) Cyclohexanone	108941					D	5000 (2270)
(i) Methanol	67561					D	5000 (2270)
F004............			1*	4	F004	C	1000 (454)
The following spent non-halogenated solvents and the still bottoms from the recovery of these solvents:							
(a) Cresols/Cresylic acid	1319773		1000	1,4	U052	C	1000 (454)
(b) Nitrobenzene	98953		1000	1,2,4	U169	C	1000 (454)
F005............			1*	4	F005	B	100 (45.4)
The following spent non-halogenated solvents and the still bottoms from the recovery of these solvents:							
(a) Toluene............	108883		1000	1,2,4	U220	C	1000 (454)
(b) Methyl ethyl ketone............	78933		1*	4	U159	D	5000 (2270)
(c) Carbon disulfide............	75150		5000	1,4	P022	B	100 (45.4)
(d) Isobutanol............	78831		1*	4	U140	D	5000 (2270)
(e) Pyridine............	110861		1*	4	U196	C	1000 (454)
F006............			1*	4	F006	A	10 (4.54)

Wastewater treatment sludges from electroplating operations except from the following processes: (1) sulfuric acid anodizing of aluminum, (2) tin plating on carbon steel, (3) zinc plating (segregated basis) on carbon steel, (4) aluminum or zinc-aluminum plating on carbon steel, (5) cleaning/stripping associated with tin, zinc and aluminum plating on carbon steel, and (6) chemical etching and milling of aluminum.						
F007	1*	4	F007	A	10 (4.54)
Spent cyanide plating bath solutions from electroplating operations.	1*	4	F008	A	10 (4.54)
F008					
Plating bath residues from the bottom of plating baths from electroplating operations where cyanides are used in the process.	1*	4	F009	A	10 (4.54)
F009					
Spent stripping and cleaning bath solutions from electroplating operations where cyanides are used in the process.	1*	4	F010	A	10 (4.54)
F010					
Quenching bath residues from oil baths from metal heat treating operations where cyanides are used in the process.	1*	4	F011	A	10 (4.54)
F011					
Spent cyanide solution from salt bath pot cleaning from metal heat treating operations.	1*	4	F012	A	10 (4.54)
F012					
Quenching wastewater treatment sludges from metal heat treating operations where cyanides are used in the process.	1	4	F019	A	10 (4.54)
F019					
Wastewater treatment sludges from the chemical conversion coating of aluminum except from zirconium phosphating in aluminum can washing when such phosphating is an exclusive conversion coating process	1*	4	F020	X	1 (0.454)
F020					

TABLE 7.2 List of Hazardous Substances and Reportable Quantities (*Continued*)

[Note: All Comments/Notes Are Located at the End of This Table]

Hazardous Substance	CASRN	Regulatory Synonyms	Statutory			Final RQ	
			RQ	Code †	RCRA Waste Number	Category	Pounds (Kg)
Wastes (except wastewater and spent carbon from hydrogen chloride purification) from the production or manufacturing use (as a reactant, chemical intermediate, or component in a formulating process) of tri-or-tetrachlorophenol, or of intermediates used to produce their pesticide derivatives. (This listing does not include wastes from the production of hexachlorophene from highly purified 2,4,5-trichlorophenol.) F021.......			1*	4	F021	X	1 (0.454)
Wastes (except wastewater and spent carbon from hydrogen chloride purification) from the production or manufacturing use (as a reactant, chemical intermediate, or component in a formulating process) of pentachlorophenol, or of intermediates used to produce its derivatives. F022.......			1*	4	F022	X	1 (0.454)
Wastes (except wastewater and spent carbon from hydrogen chloride purification) from the manufacturing use (as a reactant, chemical intermediate, or component in a formulating process) of tetra-, penta-, or hexachlorobenzenes under alkaline conditions. F023.......			1*	4	F023	X	1 (0.454)
Wastes (except wastewater and spent carbon from hydrogen chloride purification) from the production of materials on equipment previously used for the production or manufacturing use (as a reactant, chemical intermediate, or component in a formulating process) of tri- and tetrachlorophenols. (This listing does not include wastes from equipment used only for the production or use of hexa-chlorophene from highly purified 2,4,5-tri-chlorophenol.) F024.......			1*	4	F024	X	1 (0.454)

Hazardous waste					
F025 Wastes, including but not limited to distillation residues, heavy ends, tars, and reactor cleanout wastes, from the production of chlorinated aliphatic hydrocarbons, having carbon content from one to five, utilizing free radical catalyzed processes. (This listing does not include light ends, spent filters and filter aids, spent dessicants(sic), wastewater, wastewater treatment sludges, spent catalysts, and wastes listed in Section 261.32.)	4	1*	X	F025	##1 (0.454)
F026 Condensed light ends, spent filters and filter aids, and spent desiccant wastes from the production of certain chlorinated aliphatic hydrocarbons, by free radical catalyzed processes. These chlorinated aliphatic hydrocarbons are those having carbon chain lengths ranging from one to and including five, with varying amounts and positions of chlorine substitution.	4	1*	X	F026	1 (0.454)
F027 Wastes (except wastewater and spent carbon from hydrogen chloride purification) from the production of materials on equipment previously used for the manufacturing use (as a reactant, chemical intermediate, or component in a formulating process) of tetra-, penta-, or hexachlorobenzene under alkaline conditions.	4	1*	X	F027	1 (0.454)
F028 Discarded unused formulations containing tri-, tetra-, or pentachlorophenol or discarded unused formulations containing compounds derived from these chlorophenols. (This listing does not include formulations containing hexachlorophene synthesized from prepurified 2,4,5-tri-chlorophenol as the sole component.)	4	1*	X	F028	1 (0.454)
F032 Residues resulting from the incineration or thermal treatment of soil contaminated with EPA Hazardous Waste Nos. F020, F021, F022, F023, F026, and F027.	4	1*	X	F032	1(0.454)

TABLE 7.2 List of Hazardous Substances and Reportable Quantities (*Continued*)

[Note: All Comments/Notes Are Located at the End of This Table]

| Hazardous Substance | CASRN | Regulatory Synonyms | Statutory | | | Final RQ | |
			RQ	Code †	RCRA Waste Number	Category	Pounds (Kg)
Wastewaters, process residuals, preservative drippage, and spent formulations from wood preserving processes generated at plants that currently use or have previously used chlorophenolic formulations (except wastes from processes that have had the F032 waste code deleted in accordance with § 261.35 and do not resume or initiate use of chlorophenolic formulations). This listing does not include K001 bottom sediment sludge from the treatment of wastewater from wood preserving processes that use creosote and/or pentachlorophenol. F034................			1*	4	F034	X	1(0.454)
Wastewaters, process residuals, preservative drippage, and spent formulations from wood preserving processes generated at plants that use creosote formulations. This listing does not include K001 bottom sediment sludge from the treatment of wastewater from wood preserving processes that use creosote and/or pentachlorophenol. F035................			1*	4	F035	X	1(0.454)
Wastewaters, process residuals, preservative drippage, and spent formulations from wood preserving processes generated at plants that use inorganic preservatives containing arsenic or chromium. This listing does not include K001 bottom sediment sludge from the treatment of wastewater from wood preserving processes that use creosote and/or pentachlorophenol. F037................			1*	4	F037	X	1 (0.454)

264

F037 Petroleum refinery primary oil/water/solids separation sludge—Any sludge generated from the gravitational separation of oil/water/solids during the storage or treatment of process wastewaters and oily cooling wastewaters from petroleum refineries. Such sludges include, but are not limited to, those generated in: oil/water/solids separators; tanks and impoundments; ditches and other conveyances; sumps; and stormwater units receiving dry weather flow. Sludge generated in stormwater units that do not receive dry weather flow, sludges generated from non-contact once-through cooling waters segregated for treatment from other process or oily cooling waters, sludges generated in aggressive biological treatment units as defined in § 261.31(b)(2) (including sludges generated in one or more additional units after wastewaters have been treated in aggressive biological treatment units) and K051 wastes are not included in this listing.

F038 ....

	1*	4	F038	X	1 (0.454)

F038 Petroleum refinery secondary (emulsified) oil/water/solids separation sludge—Any sludge and/or float generated from the physical and/or chemical separation of oil/water/solids in process wastewaters and oily cooling wastewaters from petroleum refineries. Such wastes include, but are not limited to, all sludges and floats generated in: induced air flotation (IAF) units, tanks and impoundments, and all sludges generated in DAF units. Sludges generated in stormwater units that do not receive dry weather flow, sludges generated from once-through non-contact cooling waters segregated for treatment from other process or oil cooling wastes, sludges and floats generated in aggressive biological treatment units as defined in § 261.31(b)(2) (including sludges and floats generated in one or more additional units after wastewaters have been treated in aggressive biological treatment units) and F037, K048, and K051 wastes are not included in this listing.

K001 ....

	1*	4	K001	X	1 (0.454)

TABLE 7.2 List of Hazardous Substances and Reportable Quantities (*Continued*)

[Note: All Comments/Notes Are Located at the End of This Table]

| Hazardous Substance | CASRN | Regulatory Synonyms | Statutory | | | Final RQ | | |
			RQ	Code †	RCRA Waste Number	Category	Pounds (Kg)
Bottom sediment sludge from the treatment of wastewaters from wood preserving processes that use creosote and/or pentachlorophenol.							
K002.............. Wastewater treatment sludge from the production of chrome yellow and orange pigments.			1*	4	K002		#
K003.............. Wastewater treatment sludge from the production of molybdate orange pigments.			1*	4	K003		#
K004.............. Wastewater treatment sludge from the production of zinc yellow pigments.			1*	4	K004	A	10 (4.54)
K005.............. Wastewater treatment sludge from the production of chrome green pigments.			1*	4	K005		#
K006.............. Wastewater treatment sludge from the production of chrome oxide green pigments (anhydrous and hydrated).			1*	4	K006	A	10 (4.54)
K007.............. Wastewater treatment sludge from the production of iron blue pigments.			1*	4	K007	A	10 (4.54)
K008.............. Oven residue from the production of chrome oxide green pigments.			1*	4	K008	A	10 (4.54)
K009.............. Distillation bottoms from the production of acetaldehyde from ethylene.			1*	4	K009	A	10 (4.54)
K010.............. Distillation side cuts from the production of acetaldehyde from ethylene.			1*	4	K010	A	10 (4.54)
K011.............. Bottom stream from the wastewater stripper in the production of acrylonitrile.			1*	4	K011	A	10 (4.54)
K013..............			1*	4	K013	A	10 (4.54)

Description		Code				Quantity
Bottom stream from the acetonitrile column in the production of acrylonitrile.		K014	D	4	1*	5000 (2270)
Bottoms from the acetonitrile purification column in the production of acrylonitrile.		K015	A	4	1*	10 (4.54)
Still bottoms from the distillation of benzyl chloride.		K016	X	4	1*	1 (0.454)
Heavy ends or distillation residues from the production of carbon tetrachloride.		K017	A	4	1*	10 (4.54)
Heavy ends (still bottoms) from the purification column in the production of epi-chlorohydrin.		K018	X	4	1*	1 (0.454)
Heavy ends from the fractionation column in ethyl chloride production.		K019	X	4	1*	1 (0.454)
Heavy ends from the distillation of ethylene dichloride in ethylene dichloride production.		K020	X	4	1*	1 (0.454)
Heavy ends from the distillation of vinyl chloride in vinyl chloride monomer production.		K021	A	4	1*	10 (4.54)
Aqueous spent antimony catalyst waste from fluoromethanes production.		K022	X	4	1*	1 (0.454)
Distillation bottom tars from the production of phenol/acetone from cumene.		K023	D	4	1*	5000 (2270)
Distillation light ends from the production of phthalic anhydride from naphthalene.		K024	D	4	1*	5000 (2270)
Distillation bottoms from the production of phthalic anhydride from naphthalene.		K025	A	4	1*	10 (4.54)
Distillation bottoms from the production of nitrobenzene by the nitration of benzene.		K026	C	4	1*	1000 (454)
Stripping still tails from the production of methyl ethyl pyridines.		K027	A	4	1*	10 (4.54)
Centrifuge and distillation residues from toluene diisocyanate production.		K028	X	4	1*	1 (0.454)

TABLE 7.2 List of Hazardous Substances and Reportable Quantities (*Continued*)

[Note: All Comments/Notes Are Located at the End of This Table]

Hazardous Substance	CASRN	Regulatory Synonyms	Statutory RQ	Statutory Code †	RCRA Waste Number	Final RQ Category	Final RQ Pounds (Kg)
Spent catalyst from the hydrochlorinator reactor in the production of 1,1,1-trichloroethane. **K029**			1*	4	K029	X	1 (0.454)
Waste from the product steam stripper in the production of 1,1,1-trichloroethane. K030			1*	4	K030	X	1 (0.454)
Column bottoms or heavy ends from the combined production of trichloroethylene and perchloroethylene. K031			1*	4	K031	X	1 (0.454)
By-product salts generated in the production of MSMA and cacodylic acid. K032			1*	4	K032	A	10 (4.54)
Wastewater treatment sludge from the production of chlordane. K033			1*	4	K033	A	10 (4.54)
Wastewater and scrub water from the chlorination of cyclopentadiene in the production of chlordane. K034			1*	4	K034	A	10 (4.54)
Filter solids from the filtration of hexachlorocyclopentadiene in the production of chlordane. K035			1*	4	K035	X	1 (0.454)
Wastewater treatment sludges generated in the production of creosote. K036			1*	4	K036	X	1 (0.454)
Still bottoms from toluene reclamation distillation in the production of disulfoton. K037			1*	4	K037	X	1 (0.454)
Wastewater treatment sludges from the production of disulfoton. K038			1*	4	K038	A	10 (4.54)

268

Description			Code		Quantity
Wastewater from the washing and stripping of phorate production. K039	1*	4	K039	A	10 (4.54)
Filter cake from the filtration of diethylphosphorodithioic acid in the production of phorate. K040	1*	4	K040	A	10 (4.54)
Wastewater treatment sludge from the production of phorate. K041	1*	4	K041	X	1 (0.454)
Wastewater treatment sludge from the production of toxaphene. K042	1*	4	K042	A	10 (4.54)
Heavy ends or distillation residues from the distillation of tetrachlorobenzene in the production of 2,4,5-T. K043	1*	4	K043	A	10 (4.54)
2,6-Dichlorophenol waste from the production of 2,4-D. K044	1*	4	K044	A	10 (4.54)
Wastewater treatment sludges from the manufacturing and processing of explosives. K045	1*	4	K045	A	10 (4.54)
Spent carbon from the treatment of wastewater containing explosives. K046	1*	4	K046	B	100 (45.4)
Wastewater treatment sludges from the manufacturing, formulation and loading of lead-based initiating compounds. K047	1*	4	K047	A	10 (4.54)
Pink/red water from TNT operations. K048	1*	4	K048		#
Dissolved air flotation (DAF) float from the petroleum refining industry. K049	1*	4	K049		#
Slop oil emulsion solids from the petroleum refining industry. K050	1*	4	K050	A	10 (4.54)
Heat exchanger bundle cleaning sludge from the petroleum refining industry. K051	1*	4	K051		#

TABLE 7.2 List of Hazardous Substances and Reportable Quantities (*Continued*)

[Note: All Comments/Notes Are Located at the End of This Table]

Hazardous Substance	CASRN	Regulatory Synonyms	Statutory			Final RQ		
			RQ	Code †	RCRA Waste Number	Category	Pounds (Kg)	
API separator sludge from the petroleum refining industry. K052			1*	4	K052	A	10 (4.54)	
Tank bottoms (leaded) from the petroleum refining industry. K060			1*	4	K060	X	1 (0.454)	
Ammonia still lime sludge from coking operations. K061			1*	4	K061		#	
Emission control dust/sludge from the primary production of steel in electric furnaces. K062			1*	4	K062		#	
Spent pickle liquor generated by steel finishing operations of facilities within the iron and steel industry (SIC Codes 331 and 332). K064			1*	4	K064		##	
Acid plant blowdown slurry/sludge resulting from thickening of blowdown slurry from primary copper production. K065			1*	4	K065		##	
Surface impoundment solids contained in and dredged from surface impoundments at primary lead smelting facilities. K066			1*	4	K066		##	
Sludge from treatment of process wastewater and/or acid plant blowdown from primary zinc production. K069			1*	4	K069		#	
Emission control dust/sludge from secondary lead smelting. K071			1*	4	K071	X	1 (0.454)	
Brine purification muds from the mercury cell process in chlorine production, where separately prepurified brine is not used. K073			1*	4	K073	A	10 (4.54)	

Hazardous waste description			Hazardous waste No.	Hazard code	RQ in pounds (kilograms)
Chlorinated hydrocarbon waste from the purification step of the diaphragm cell process using graphite anodes in chlorine production.	4	1*	K083	B	100 (45.4)
Distillation bottoms from aniline extraction.	4	1*	K084	X	1 (0.454)
Wastewater treatment sludges generated during the production of veterinary pharmaceuticals from arsenic or organo-arsenic compounds.	4	1*	K085	A	10 (4.54)
Distillation or fractionation column bottoms from the production of chlorobenzenes.	4	1*	K086		#
Solvent washes and sludges, caustic washes and sludges, or water washes and sludges from cleaning tubs and equipment used in the formulation of ink from pigments, driers, soaps, and stabilizers containing chromium and lead.	4	1*	K087	B	100 (45.4)
Decanter tank tar sludge from coking operations.	4	1*	K088		
Spent potliners from primary aluminum reduction.	4	1*	K090		
Emission control dust or sludge from ferrochromiumsilicon production.	4	1	K091		
Emission control dust or sludge from ferrochromium production.	4	1*	K093	D	5000 (2270)
Distillation light ends from the production of phthalic anhydride from ortho-xylene.	4	1*	K094	D	5000 (2270)
Distillation bottoms from the production of phthalic anhydride from ortho-xylene.	4	1*	K095	B	100 (45.4)
Distillation bottoms from the production of 1,1,1-trichloroethane.	4	1*	K096	B	100 (45.4)
Heavy ends from the heavy ends column from the production of 1,1,1-trichloroethane.	4	1*	K097	X	1 (0.454)

271

TABLE 7.2 List of Hazardous Substances and Reportable Quantities (*Continued*)

[Note: All Comments/Notes Are Located at the End of This Table]

Hazardous Substance	CASRN	Statutory			Final RQ	
		RQ	Code †	RCRA Waste Number	Category	Pounds (Kg)
Vacuum stripper discharge from the chlordane chlorinator in the production of chlordane. K098		1*	4	K098	X	1 (0.454)
Untreated process wastewater from the production of toxaphene. K099		1*	4	K099	A	10 (4.54)
Untreated wastewater from the production of 2,4-D. K100		1*	4	K100		#
Waste leaching solution from acid leaching of emission control dust/sludge from secondary lead smelting. K101		1*	4	K101	X	1 (0.454)
Distillation tar residues from the distillation of aniline-based compounds in the production of veterinary pharmaceuticals from arsenic or organo-arsenic compounds. K102		1*	4	K102	X	1 (0.454)
Residue from the use of activated carbon for decolorization in the production of veterinary pharmaceuticals from arsenic or organo-arsenic compounds. K103		1*	4	K103	B	100 (45.4)
Process residues from aniline extraction from the production of aniline. K104		1*	4	K104	A	10 (4.54)
Combined wastewater streams generated from nitrobenzene/aniline production. K105		1*	4	K105	A	10 (4.54)
Separated aqueous stream from the reactor product washing step in the production of chlorobenzenes. K106		1*	4	K106	X	1 (0.454)

272

Wastewater treatment sludge from the mercury cell process in chlorine production.				
K107........... Column bottoms from product separation from the production of 1,1-dimethylhydrazine (UDMH) from carboxylic acid hydrazines.	10	4	K107	X	10 (4.54)
K108........... Condensed column overheads from product separation and condensed reactor vent gases from the production of 1,1-dimethylhydrazine (UDMH) from carboxylic acid hydrazides.	10	4	K108	X	10 (4.54)
K109........... Spent filter cartridges from product purification from the production of 1,1- dimethylhydrazine (UDMH) from carboxylic acid hydrazides.	10	4	K109	X	10 (4.54)
K110........... Condensed column overheads from intermediate separation from the production of 1,1-dimethyl-hydrazine (UDMH) from carboxylic acid hydrazides.	10	4	K110	X	10 (4.54)
K111........... Product washwaters from the production of dinitro-toluene via nitration of toluene.	1*	4	K111	A	10 (4.54)
K112........... Reaction by-product water from the drying column in the production of toluenediamine via hydrogenation of dinitrotoluene.	1*	4	K112	A	10 (4.54)
K113........... Condensed liquid light ends from the purification of toluenediamine in the production of toluenedia-mine via hydrogenation of dinitrotoluene.	1*	4	K113	A	10 (4.54)
K114........... Vicinals from the purification of toluenediamine in the production of toluenediamine via hydrogena-tion of dinitrotoluene.	1*	4	K114	A	10 (4.54)
K115........... Heavy ends from the purification of toluenediamine in the production of toluenediamine via hydrogen-ation of dinitrotoluene.	1*	4	K115	A	10 (4.54)
K116........... Organic condensate from the solvent recovery column in the production of toluene diisocyanate via phosgenation of toluenediamine.	1*	4	K116	A	10 (4.54)
K117...........	1*	4	K117	X	1 (0.454)

TABLE 7.2 List of Hazardous Substances and Reportable Quantities (*Continued*)

[Note: All Comments/Notes Are Located at the End of This Table]

Hazardous Substance	CASRN	Regulatory Synonyms	Statutory			Final RQ	
			RQ	Code †	RCRA Waste Number	Category	Pounds (Kg)
Wastewater from the reaction vent gas scrubber in the production of ethylene bromide via bromination of ethene. K118......			1*	4	K118	X	1 (0.454)
Spent absorbent solids from purification of ethylene dibromide in the production of ethylene dibromide. K123......			1*	4	K123	A	10 (4.54)
Process wastewater (including supernates, filtrates, and washwaters) from the production of ethylenebisdithiocarbamic acid and its salts. K124......			1*	4	K124	A	10 (4.54)
Reactor vent scrubber water from the production of ethylenebisdithiocarbamic acid and its salts. K125......			1*	4	K125	A	10 (4.54)
Filtration, evaporation, and centrifugation solids from the production of ethylenebisdithiocarbamic acid and its salts. K126......			1*	4	K126	A	10 (4.54)
Baghouse dust and floor sweepings in milling and packaging operations from the production or formulation of ethylenebisdithiocarbamic acid and its salts. K131......			100	4	K131	X	100 (45.4)
Wastewater from the reactor and spent sulfuric acid from the acid dryer in the production of methyl bromide. K132......			1000	4	K132	X	1000 (454)
Spent absorbent and wastewater solids from the production of methyl bromide. K136......			1*	4	K136	X	1 (0.454)
Still bottoms from the purification of ethylene dibromide in the production of ethylene dibromide via bromination of ethene.							

† Indicates the statutory source as defined by 1, 2, 3, and 4 below.

†† No reporting of releases of this hazardous substance is required if the diameter of the pieces of the solid metal released is equal to or exceeds 100 micrometers (0.004 inches).
††† The RQ for asbestos is limited to friable forms only.
1—indicates that the statutory source for designation of this hazardous substance under CERCLA is CWA Section 311(b)(4).
2—indicates that the statutory source for designation of this hazardous substance under CERCLA is CWA Section 307(a).
3—indicates that the statutory source for designation of this hazardous substance under CERCLA is CAA Section 112.
4—indicates that the statutory source for designation of this hazardous substance under CERCLA is RCRA Section 3001.
1*—indicates that the 1-pound RQ is a CERCLA statutory RQ.
# indicates that the RQ is subject to change when the assessment of potential carcinogenicity is completed.
The Agency may adjust the statutory RQ for this hazardous substance in a future rulemaking; until then the statutory RQ applies.
§—The adjusted RQs for radionuclides may be found in Appendix B to this table.
*.—indicates that no RQ is being assigned to the generic or broad class.

TABLE 7.3 Specific Toxic Chemical Listings

Chemical name	CAS No.	Effective date
Acetaldehyde	75–07–0	01/01/87
Acetamide	60–35–5	01/01/87
Acetone	67–64–1	01/01/87
Acetonitrile	75–05–8	01/01/87
2–Acetylaminofluorene	53–96–3	01/01/87
Acrolein	107–02–8	01/01/87
Acrylamide	79–06–1	01/01/87
Acrylic acid	79–10–7	01/01/87
Acrylonitrile	107–13–1	01/01/87
Aldrin[1,4:5,8-Dimethanonaphthalene,1,2,3,4,10,10-hexachloro-1,4,4a,5,8,8a-hexahydro-(1.alpha.,4.alpha.,4a.beta.,5.alpha.,8.alpha., 8a.beta.)-]	309–00–2	01/01/87
Allyl alcohol	107–18–6	1/01/90
Allyl chloride	107–05–1	01/01/87
Aluminum (fume or dust)	7429–90–5	01/01/87
Aluminum oxide (fibrous forms)	1344–28–1	01/01/87
2-Aminoanthraquinone	117–79–3	01/01/87
4-Aminoazobenzene	60–09–3	01/01/87
4-Aminobiphenyl	92–67–1	01/01/87
1-Amino-2-methylanthraquinone	82–28–0	01/01/87
Ammonia	7664–41–7	01/01/87
Ammonium nitrate (solution)	6484–52–2	01/01/87
Ammonium sulfate (solution)	7783–20–2	01/01/87
Aniline	62–53–3	01/01/87
o-Anisidine	90–04–0	01/01/87
p-Anisidine	104–94–9	01/01/87
o-Anisidine hydrochloride	134–29–2	01/01/87
Anthracene	120–12–7	01/01/87
Antimony	7440–36–0	01/01/87
Arsenic	7440–38–2	01/01/87
Asbestos (friable)	1332–21–4	01/01/87
Barium	7440–39–3	01/01/87
Benzal chloride	98–87–3	01/01/87
Benzamide	55–21–0	01/01/87
Benzene	71–43–2	01/01/87
Benzidine	92–87–5	01/01/87
Benzoic trichloride (Benzotrichloride)	98–07–7	01/01/87
Benzoyl chloride	98–88–4	01/01/87
Benzoyl peroxide	94–36–0	01/01/87
Benzyl chloride	100–44–7	01/01/87
Beryllium	7440–41–7	01/01/87
Biphenyl	92–52–4	01/01/87
Bis(2-chloroethyl) ether	111–44–4	01/01/87
Bis(chloromethyl) ether	542–88–1	01/01/87
Bis(2-chloro-1-methylethyl) ether	108–60–1	01/01/87
Bis(2-ethylhexyl) adipate	103–23–1	01/01/87
Bromochlorodifluoromethane (Halon 1211)	353–59–3	7/8/90
Bromoform (Tribromomethane)	75–25–2	01/01/87
Bromomethane (Methyl bromide)	74–83–9	01/01/87
Bromotrifluoromethane (Halon 1301)	75–63–8	7/8/90
1,3-Butadiene	106–99–0	01/01/87
Butyl acrylate	141–32–2	01/01/87
n-Butyl alcohol	71–36–3	01/01/87
sec-Butyl alcohol	78–92–2	01/01/87
tert-Butyl alcohol	75–65–0	01/01/87
Butyl benzyl phthalate	85–68–7	01/01/87

TABLE 7.3   Specific Toxic Chemical Listings (*Continued*)

Chemical name	CAS No.	Effective date
1,2-Butylene oxide	106–88–7	01/01/87
Butyraldehyde	123–72–8	01/01/87
C.I. Acid Green 3	4680–78–8	01/01/87
C.I. Basic Green 4	569–64–2	01/01/87
C.I. Basic Red 1	989–38–8	01/01/87
C.I. Direct Black 38	1937–37–7	01/01/87
C.I. Direct Blue 6	2602–46–2	01/01/87
C.I. Direct Brown 95	16071–86–6	01/01/87
C.I. Disperse Yellow 3	2832–40–8	01/01/87
C.I. Food Red 5	3761–53–3	01/01/87
C.I. Food Red 15	81–88–9	01/01/87
C.I. Solvent Orange 7	3118–97–6	01/01/87
C.I. Solvent Yellow 3	97–56–3	01/01/87
C.I. Solvent Yellow 14	842–07–9	01/01/87
C.I. Solvent Yellow 34 (Aurimine)	492–80–8	01/01/87
C.I. Vat Yellow 4	128–66–5	01/01/87
Cadmium	7440–43–9	01/01/87
Calcium cyanamide	156–62–7	01/01/87
Captan[1H-Isoindole-1,3(2H)-dione,3a,4,7,7a-tetrahydro-2-[(trichloromethyl)thio]-]	133–06–2	01/01/87
Carbaryl [1-Naphthalenol, methylcarbamate]	63–25–2	01/01/87
Carbon disulfide	75–15–0	01/01/87
Carbon tetrachloride	56–23–5	01/01/87
Carbonyl sulfide	463–58–1	01/01/87
Catechol	120–80–9	01/01/87
Chloramben [Benzoic acid,3-amino-2,5-dichloro-]	133–90–4	01/01/87
Chlordane [4,7-Methanoindan,1,2,4,5,6,7,8,8-octachloro-2,3,3a,4,7,7a-hexahydro-]	57–74–9	01/01/87
Chlorine	7782–50–5	01/01/87
Chlorine dioxide	10049–04–4	01/01/87
Chloroacetic acid	79–11–8	01/01/87
2-Chloroacetophenone	532–27–4	01/01/87
Chlorobenzene	108–90–7	01/01/87
Chlorobenzilate [Benzeneacetic acid, 4-chloro-.alpha.-(4-.chlorophenyl)-.alpha.-hydroxy-, ethyl ester]	510–15–6	01/01/87
Chloroethane (Ethyl chloride)	75–00–3	01/01/87
Chloroform	67–66–3	01/01/87
Chloromethane (Methyl chloride)	74–87–3	01/01/87
Chloromethyl methyl ether	107–30–2	01/01/87
Chloroprene	126–99–8	01/01/87
Chlorothalonil [1,3-Benzenedicarbonitrile,2,4,5,6-tetrachloro-]	1897–45–6	01/01/87
Chromium	7440–47–3	01/01/87
Cobalt	7440–48–4	01/01/87
Copper	7440–50–8	01/01/87
Creosote	8001–58–9	1/01/90
p-Cresidine	120–71–8	01/01/87
Cresol (mixed isomers)	1319–77–3	01/01/87
m-Cresol	108–39–4	01/01/87
o-Cresol	95–48–7	01/01/87
p-Cresol	106–44–5	01/01/87
Cumene	98–82–8	01/01/87
Cumene hydroperoxide	80–15–9	01/01/87
Cupferron[Benzeneamine, N-hydroxy-N-nitroso, ammonium salt]	135–20–6	01/01/87
Cyclohexane	110–82–7	01/01/87
2,4-D [Acetic acid, (2,4-dichlorophenoxy)-]	94–75–7	01/01/87
Decabromodiphenyl oxide	1163–19–5	01/01/87
Diallate [Carbamothioic acid, bis(1-methylethyl)-, S-(2,3-dichloro-2-propenyl) ester]	2303–16–4	01/01/87
2,4-Diaminoanisole	615–05–4	01/01/87
2,4-Diaminoanisole sulfate	39156–41–7	01/01/87
4,4′-Diaminodiphenyl ether	101–80–4	01/01/87
Diaminotoluene (mixed isomers)	25376–45–8	01/01/87
2,4-Diaminotoluene	95–80–7	01/01/87
Diazomethane	334–88–3	01/01/87
Dibenzofuran	132–64–9	01/01/87
1,2-Dibromo-3-chloropropane (DBCP)	96–12–8	01/01/87
1,2-Dibromoethane (Ethylene dibromide)	106–93–4	01/01/87
Dibromotetrafluoroethane (Halon 2402)	124–73–2	7/8/90
Dibutyl phthalate	84–74–2	01/01/87
Dichlorobenzene (mixed isomers)	25321–22–6	01/01/87
1,2-Dichlorobenzene	95–50–1	01/01/87
1,3-Dichlorobenzene	541–73–1	01/01/87
1,4-Dichlorobenzene	106–46–7	01/01/87
3,3′-Dichlorobenzidine	91–94–1	01/01/87
Dichlorobromomethane	75–27–4	01/01/87
Dichlorodifluoromethane (CFC–12)	75–71–8	7/8/90

TABLE 7.3   Specific Toxic Chemical Listings *(Continued)*

Chemical name	CAS No.	Effective date
1,2-Dichloroethane (Ethylene dichloride)	107–06–2	01/01/87
1,2-Dichlorethylene	540–59–0	01/01/87
Dichloromethane (Methylene chloride)	75–09–2	01/01/87
2,4-Dichlorophenol	120–83–2	01/01/87
1,2-Dichloropropane	78–87–5	01/01/87
2,3-Dichloropropene	78–88–6	1/01/90
1,3-Dichloropropylene	542–75–6	01/01/87
Dichlorotetrafluoroethane (CFC–114)	76–14–2	7/8/90
Dichlorvos [Phosphoric acid, 2,2-dichloroethenyl dimethyl ester]	62–73–7	01/01/87
Dicofol [Benzenemethanol,4-chloro-.alpha.-(4-chlorophenyl)-.alpha.-(trichloromethyl)-]	115–32–2	01/01/87
Diepoxybutane	1464–53–5	01/01/87
Diethanolamine	111–42–2	01/01/87
Di-(2-ethylhexyl) phthalate (DEHP)	177–81–7	01/01/87
Diethyl phthalate	84–66–2	01/01/87
Diethyl sulfate	64–67–5	01/01/87
3,3´-Dimethoxybenzidine	119–90–4	01/01/87
4-Dimethylaminoazobenzene	60–11–7	01/01/87
3,3´-Dimethylbenzidine (o-Tolidine)	119–93–7	01/01/87
Dimethylcarbamyl chloride	79–44–7	01/01/87
1,1-Dimethyl hydrazine	57–14–7	01/01/87
2,4-Dimethylphenol	105–67–9	01/01/87
Dimethyl phthalate	131–11–3	01/01/87
Dimethyl sulfate	77–78–1	01/01/87
m-Dinitrobenzene	99–65–0	1/01/90
o-Dinitrobenzene	528–29–0	1/01/90
p-Dinitrobenzene	100–25–4	1/01/90
4,6-Dinitro-o-cresol	534–52–1	01/01/87
2,4-Dinitrophenol	51–28–5	01/01/87
2,4-Dinitrotoluene	121–14–2	01/01/87
2,6-Dinitrotoluene	606–20–2	01/01/87
Dinitrotoluene (mixed isomers)	25321–14–6	1/01/90
n-Dioctyl phthalate	117–84–0	01/01/87
1,4-Dioxane	123–91–1	01/01/87
1,2-Diphenylhydrazine (Hydrazobenzene)	122–66–7	01/01/87
Epichlorohydrin	106–89–8	01/01/87
2-Ethoxyethanol	110–80–5	01/01/87
Ethyl acrylate	140–88–5	01/01/87
Ethylbenzene	100–41–4	01/01/87
Ethyl chloroformate	541–41–3	01/01/87
Ethylene	74–85–1	01/01/87
Ethylene glycol	107–21–1	01/01/87
Ethyleneimine(Aziridine)	151–56–4	01/01/87
Ethylene oxide	75–21–8	01/01/87
Ethylene thiourea	96–45–7	01/01/87
Fluometuron [Urea, N,N-dimethyl-N´-[3-(trifluoromethyl)phenyl]-]	2164–17–2	01/01/87
Formaldehyde	50–00–0	01/01/87
Freon 113 [Ethane, 1,1,2-trichloro-1,2,2-trifluoro-]	76–13–1	01/01/87
Heptachlor[1,4,5,6,7,8,8-Heptachloro-3a,4,7,7a-tetrahydro-4,7-methano-1H-indene]	76–44–8	01/01/87
Hexachlorobenzene	118–74–1	01/01/87
Hexachloro-1,3-butadiene	87–68–3	01/01/87
Hexachlorocyclopentadiene	77–47–4	01/01/87
Hexachloroethane	67–72–1	01/01/87
Hexachloronaphthalene	1335–87–1	01/01/87
Hexamethylphosphoramide	680–31–9	01/01/87
Hydrazine	302–01–2	01/01/87
Hydrazine sulfate	10034–93–2	01/01/87
Hydrochloric acid	7647–01–0	01/01/87
Hydrogen cyanide	74–90–8	01/01/87
Hydrogen fluoride	7664–39–3	01/01/87
Hydroquinone	123–31–9	01/01/87
Isobutyraldehyde	78–84–2	01/01/87
Isopropyl alcohol (Only persons who manufacture by the strong acid process are subject, no supplier notifiction.)	67–63–0	01/01/87
4,4´-Isopropylidenediphenol	80–05–7	01/01/87
Isosafrole	120–58–1	1/01/90
Lead	7439–92–1	01/01/87
Lindane [Cyclohexane, 1,2,3,4,5,6-hexachloro-(1.alpha.,2.alpha.,3.beta.,4.alpha.,5.alpha.,6.beta.)-]	58–89–9	01/01/87
Maleic anhydride	108–31–6	01/01/87
Maneb [Carbamodithioic acid, 1,2-ethanediylbis-, manganese complex]	12427–38–2	01/01/87
Manganese	7439–96–5	01/01/87
Mercury	7439–97–6	01/01/87
Methanol	67–56–1	01/01/87

TABLE 7.3 Specific Toxic Chemical Listings (*Continued*)

Chemical name	CAS No.	Effective date
Methoxychlor [Benzene, 1,1'-(2,2,2-trichloroethylidene)bis[4-methoxy-]	72-43-5	01/01/87
2-Methoxyethanol	109-86-4	01/01/87
Methyl acrylate	96-33-3	01/01/87
Methyl *tert*-butyl ether	1634-04-4	01/01/87
4,4'-Methylenebis(2-chloroaniline) (MBOCA)	101-14-4	01/01/87
4,4'-Methylenebis(*N,N*-dimethyl) benzenamine	101-61-1	01/01/87
Methylenebis(phenylisocyanate) (MBI)	101-68-8	01/01/87
Methylene bromide	74-95-3	01/01/87
4,4'-Methylenedianiline	101-77-9	01/01/87
Methyl ethyl ketone	78-93-3	01/01/87
Methyl hydrazine	60-34-4	01/01/87
Methyl iodide	74-88-4	01/01/87
Methyl isobutyl ketone	108-10-1	01/01/87
Methyl isocyanate	624-83-9	01/01/87
Methyl methacrylate	80-62-6	01/01/87
Michler's ketone	90-94-8	01/01/87
Molybdenum trioxide	1313-27-5	01/01/87
(Mono)chloropentafluoroethane (CFC-115)	76-15-3	7/8/90
Mustard gas [Ethane, 1,1'-thiobis[2-chloro-]	505-60-2	01/01/87
Naphthalene	91-20-3	01/01/87
alpha-Naphthylamine	134-32-7	01/01/87
beta-Naphthylamine	91-59-8	01/01/87
Nickel	7440-02-0	01/01/87
Nitric acid	7697-37-2	01/01/87
Nitrilotriacetic acid	139-13-9	01/01/87
5-Nitro-*o*-anisidine	99-59-2	01/01/87
Nitrobenzene	98-95-3	01/01/87
4-Nitrobiphenyl	92-93-3	01/01/87
Nitrofen [Benzene, 2,4-dichloro-1-(4-nitrophenoxy)-]	1836-75-5	01/01/87
Nitrogen mustard [2-Chloro-N-(2-chloroethyl)-N-methylethanamine]	51-75-2	01/01/87
Nitroglycerin	55-63-0	01/01/87
2-Nitrophenol	88-75-5	01/01/87
4-Nitrophenol	100-02-7	01/01/87
2-Nitropropane	79-46-9	01/01/87
p-Nitrosodiphenylamine	156-10-5	01/01/87
N,N-Dimethylaniline	121-69-7	01/01/87
N-Nitrosodi-*n*-butylamine	924-16-3	01/01/87
N-Nitrosodiethylamine	55-18-5	01/01/87
N-Nitrosodimethylamine	62-75-9	01/01/87
N-Nitrosodiphenylamine	86-30-6	01/01/87
N-Nitrosodi-*n*-propylamine	621-64-7	01/01/87
N-Nitrosomethylvinylamine	4549-40-0	01/01/87
N-Nitrosomorpholine	59-89-2	01/01/87
N-Nitroso-*N*-ethylurea	759-73-9	01/01/87
N-Nitroso-*N*-methylurea	684-93-5	01/01/87
N-Nitrosonornicotine	16543-55-8	01/01/87
N-Nitrosopiperidine	100-75-4	01/01/87
Octachloronaphthalene	2234-13-1	01/01/87
Osmium tetroxide	20816-12-0	01/01/87
Parathion [Phosphorothioic acid, O,O-diethyl-O-(4-nitrophenyl) ester]	56-38-2	01/01/87
Pentachlorophenol (PCP)	87-86-5	01/01/87
Peracetic acid	79-21-0	01/01/87
Phenol	108-95-2	01/01/87
p-Phenylenediamine	106-50-3	01/01/87
2-Phenylphenol	90-43-7	01/01/87
Phosgene	75-44-5	01/01/87
Phosphoric acid	7664-38-2	01/01/87
Phosphorus (yellow or white)	7723-14-0	01/01/87
Phthalic anhydride	85-44-9	01/01/87
Picric acid	88-89-1	01/01/87
Polychlorinated biphenyls (PCBs)	1336-36-3	01/01/87
Propane sultone	1120-71-4	01/01/87
beta-Propiolactone	57-57-8	01/01/87
Propionaldehyde	123-38-6	01/01/87
Propoxur [Phenol, 2-(1-methylethoxy)-, methylcarbamate]	114-26-1	01/01/87
Propylene (Propene)	115-07-1	01/01/87
Propyleneimine	75-55-8	01/01/87
Propylene oxide	75-56-9	01/01/87
Pyridine	110-86-1	01/01/87
Quinoline	91-22-5	01/01/87
Quinone	106-51-4	01/01/87
Quintozene [Pentachloronitrobenzene]	82-68-8	01/01/87

TABLE 7.3   Specific Toxic Chemical Listings (Continued)

Chemical name	CAS No.	Effective date
Saccharin (only persons who manufacture are subject, no supplier notification) [1,2-Benzisothiazol-3(2H)-one,1,1-dioxide]	81–07–2	01/01/87
Safrole	94–59–7	01/01/87
Selenium	7782–49–2	01/01/87
Silver	7440–22–4	01/01/87
Styrene	100–42–5	01/01/87
Styrene oxide	96–09–3	01/01/87
Sulfuric acid	7664–93–9	01/01/87
1,1,2,2-Tetrachloroethane	79–34–5	01/01/87
Tetrachloroethylene (Perchloroethylene)	127–18–4	01/01/87
Tetrachlorvinphos [Phosphoric acid, 2-chloro-1-(2,4,5-trichlorophenyl)ethenyl dimethyl ester]	961–11–5	01/01/87
Thallium	7440–28–0	01/01/87
Thioacetamide	62–55–5	01/01/87
4,4'-Thiodianiline	139–65–1	01/01/87
Thiourea	62–56–6	01/01/87
Thorium dioxide	1314–20–1	01/01/87
Titanium tetrachloride	7550–45–0	01/01/87
Toluene	108–88–3	01/01/87
Toluene-2,4-diisocyanate	584–84–9	01/01/87
Toluene-2,6-diisocyanate	91–08–7	01/01/87
Toluenediisocyanate (mixed isomers)	26471–62–5	1/01/90
o-Toluidine	95–53–4	01/01/87
o-Toluidine hydrochloride	636–21–5	01/01/87
Toxaphene	8001–35–2	01/01/87
Triaziquone [2,5-Cyclohexadiene-1,4-dione,2,3,5-tris(1-aziridinyl)-]	68–76–8	01/01/87
Trichlorfon [Phosphonic acid, (2,2,2-trichloro-1-hydroxyethyl)-, dimethyl ester]	52–68–6	01/01/87
1,2,4-Trichlorobenzene	120–82–1	01/01/87
1,1,1-Trichloroethane (Methyl chloroform)	71–55–6	01/01/87
1,1,2-Trichloroethane	79–00–5	01/01/87
Trichloroethylene	79–01–6	01/01/87
Trichlorofluoromethane (CFC–11)	75–69–4	7/8/90
2,4,5-Trichlorophenol	95–95–4	01/01/87
2,4,6-Trichlorophenol	88–06–2	01/01/87
Trifluralin [Benzeneamine, 2,6-dinitro-N,N-dipropyl-4-(trifluoromethyl)-1]	1582–09–8	01/01/87
1,2,4-Trimethylbenzene	95–63–6	01/01/87
Tris(2,3-dibromopropyl) phosphate	126–72–7	01/01/87
Urethane (Ethyl carbamate)	51–79–6	01/01/87
Vanadium (fume or dust)	7440–62–2	01/01/87
Vinyl acetate	108–05–4	01/01/87
Vinyl bromide	593–60–2	01/01/87
Vinyl chloride	75–01–4	01/01/87
Vinylidene chloride	75–35–4	01/01/87
Xylene (mixed isomers)	1330–20–7	01/01/87
m-Xylene	108–38–3	01/01/87
o-Xylene	95–47–6	01/01/87
p-Xylene	106–42–3	01/01/87
2,6-Xylidine	87–62–7	01/01/87
Zinc (fume or dust)	7440–66–6	01/01/87
Zineb [Carbamodithioic acid, 1,2-ethanediylbis-, zinc complex]	12122–67–7	01/01/87

- There is a release of a reportable quantity of an EHS or a CERCLA-designated hazardous substance.

The following types of releases are excluded from the emergency release notification under EPCRA:

- Any release which results in exposure to persons solely within the boundaries of the facility
- Any release which is a "federally permitted release" covered under an appropriate permit program as defined in section 101(10) of CERCLA
- Any release which is "continuous," as defined in section 103(f) of CERCLA, and stable in quantity and rate and which does not constitute "statistically significant increases"
- Any release of a pesticide product that is exempted from section 103(e) of CERCLA
- Any release that does not meet the definition of release under section 101(22) of CERCLA
- Any release of a radionuclide which occurs (1) naturally in soil from the land, (2) naturally from the disturbance of land, except mining, (3) from the dumping of coal and coal ash at utilities and industrial facilities, and (4) from coal and coal ash piles at utilities and industrial facilities.

The owner or operator of a facility is subject to this notification requirement. Such an owner or operator must notify the community emergency coordinator for the local emergency planning committee immediately in the event of an emergency release of an EHS or a CERCLA hazardous substance.

An emergency release notice must include certain information. At a minimum, it should include the following items to the extent they are known at the time of the notice:

- Chemical name or identity of any substance involved in the release
- An indication whether the substance is an EHS
- An estimate of the quantity released
- The time and duration of the release
- The medium (or media) into which the release occurred
- Any known or anticipated acute or chronic health risks, including any advice for necessary medical attention for exposed individuals
- Proper precautions recommended as a result of the release, including any necessity for temporary evacuation and restriction of certain activities
- The name(s) and telephone number(s) of the person(s) to be contacted for further information

A verbal notice of an emergency release should be followed by a written follow-up notice by the owner or operator of the affected facility as soon as practicable. The written notice should set forth and update or revise the verbal notice as more information becomes available. Also such a written notice should include

- Any action taken to respond to and contain the release
- Any known or anticipated acute or chronic health risks in connection with the release
- Any advice regarding appropriate medical attention and care for the exposed and susceptible individuals

If the release is transportation-related, the owner or operator of the facility may meet the emergency release notification requirements by providing the above information to a 911 operator in the affected area. In the event there is no 911 emergency telephone number, such information should be given to any available telephone operator in that area.

Routine release notification

Every facility that (1) employs ten or more full-time employees, and (2) manufactures, imports, processes, or otherwise uses any "toxic chemical" listed under section 313(c) of EPCRA in an amount greater than the threshold quantity must submit annual reports detailing routine releases from such facility into the environment.

In this context, note that the statutory definition of the term *manufacture* under section 313(b)(1)(C) of EPCRA means to "produce, prepare, import, or compound a toxic chemical." However, in EPA's regulations, this term, as set forth in 40 CFR §372.3, has been broadened significantly and covers a host of activities beyond this statutory definition:

> Manufacture also applies to a toxic chemical that is produced coincidentally during the manufacture, processing, use or disposal of another chemical or mixture of chemicals, including a toxic chemical that is separated from that other chemical or mixture of chemicals as a byproduct, and a toxic chemical that remains in that other chemical or other mixture of chemicals as an impurity.

When a toxic chemical is present in another chemical or in a mixture of chemicals as an impurity, a *de minimis* rule by EPA is to be applied for determining its threshold quantity. Based on this rule, if the toxic chemical present as an impurity is no greater than 1 percent (or 0.1 percent for a carcinogenic toxic chemical), such a chemical is not included for the purpose of reporting the toxic chemical release.

The EPA regulations provide the necessary guidelines for this reporting requirement. Section 313(b) of EPCRA specifically designates those manufacturing facilities that fall within the standard industrial classification (SIC)

TABLE 7.4   SIC Groups Subject to 40 CFR Part 313 Notification Requirements

SIC code	Industry group
20	Food
21	Tobacco
22	Textiles
23	Apparel
24	Lumber and wood
25	Furniture
26	Paper
27	Printing and publishing
28	Chemicals
29	Petroleum and coal
30	Rubber and plastics
31	Leather
32	Stone, clay, and glass
33	Primary metals
34	Fabricated metals
35	Machinery (excluding electric)
36	Electric and electronic equipment
37	Transportation equipment
38	Instruments
39	Miscellaneous manufacturing

codes 20 through 39, thus exempting other manufacturing operations from this annual reporting requirement. These toxic chemicals are listed in section 313(c) of CERCLA and codified in 40 CFR part 372. Table 7.4 provides a list of the SIC groups subject to 40 CFR part 372.

The designated threshold quantities of toxic chemicals that trigger this notification are as follows:

- 10,000 pounds or more of a toxic chemical if such quantities are used at a facility per year
- 25,000 pounds or more of a toxic chemical if such quantities are manufactured or processed at a facility per year

Note that the EPA administrator is empowered by EPCRA to establish a threshold quantity for a toxic chemical which may be different from the above quantities. This is obviously an area where revisions are to be expected based on the known health effects of any toxic chemical. As a result, it is important to be cognizant of the latest applicable designated threshold quantities.

7.4   Community Right to Know

Under the statutory scheme, the reports filed by an appropriate facility under EPCRA are available to any interested person. This reflects one of the major objectives of this federal statute—to make the citizens of a community aware of the names and quantities of chemicals released annually by a facility located in that community. Pursuant to this, every facility that falls within the scope of EPCRA must prepare or make available MSDSs for all the hazardous chemicals. EPA's regulations, set forth in 40 CFR section 370.20, also allow alternative and supplemental reportings.

Besides the reporting obligation through the availability of the MSDS, the affected facility is required to provide an annual chemical inventory report. This inventory reporting may only be accomplished by using specific EPA forms developed for chemical inventory reporting. These reporting requirements are discussed further below.

MSDS reporting

The owner or operator of a facility subject to OSHA's hazard communication (Hazcom) standard for hazardous chemicals must submit an MSDS for each such chemical present in the facility on the basis of a minimum threshold level. At present, there is no minimum threshold quantity for reporting under the MSDS program. This implies that an MSDS for each hazardous chemical present at the facility must be provided irrespective of the quantity of such a chemical.

The MSDS must be developed on the basis of 29 CFR §1910.1200(g). Section 311 of EPCRA requires that the owners or operators of the affected facilities provide MSDSs or a list of MSDS chemicals to the state emergency response commission, the local emergency planning committee, and the local fire department.

Pursuant to the EPA regulations as set forth in 40 CFR part 370, the owner or operator may use an alternative reporting in lieu of the submission of an MSDS for each hazardous chemical. For such an alternative reporting, the following items must be provided:

- A list of the hazardous chemicals for which the MSDS is required, grouped by their hazard category
- The chemical or common name of each hazardous chemical as provided on the MSDS
- The hazardous component of each hazardous chemical as provided on the MSDS except for mixtures

Apart from this routine MSDS submittal, the owner or operator of a facility may have to provide a revised MSDS to the state commission, local committee, and the fire department with jurisdiction over the facility. This is required upon discovery of any significant new information concerning the hazardous chemical for which the original MSDS has been submitted. This revised MSDS must be submitted within 3 months of such a discovery.

Inventory reporting

Besides the MSDS reporting, the owner or operator of a facility subject to EPCRA is required to submit a chemical inventory form to the state commission, the local committee, and the fire department with appropriate jurisdiction. This inventory reporting must be done on an annual basis.

There are two inventory forms that can be used for reporting tier 1 and tier 2 information. The inventory form containing tier 1 information may be used for those hazardous chemicals that exceed the threshold levels. Such a form must be submitted on or before March 1 of each year and cover the information for the preceding calendar year. Unlike tier 1, submittal of a tier 2 information form is not mandatory. A tier 2 form is only to be submitted within 30 days of the receipt of such a request in writing from an appropriate authority having jurisdiction over the facility. Alternatively, the facility owner or operator may submit a tier 2 form in lieu of a tier 1 form.

Appendix 7A contains reproduced copies of tier 1 and tier 2 inventory reporting forms developed by EPA.

7.5 Toxic Chemical Release Reporting

Toxic chemical release reporting falls within the general scope of routine release reporting discussed earlier. Section 313 of EPCRA requires the reporting of environmental releases of certain toxic chemicals. These chemicals are so designated because of their anticipated causation of adverse effects on human health or the environment. Such a designation is made using "reasonable available data."

As indicated under routine releases, the toxic chemical release reporting can be triggered when the affected facility

- Employs ten or more full-time employees
- Is classified under SIC codes 20 through 39
- Manufactures, processes, or otherwise uses a listed toxic chemical in quantities greater than the reporting threshold levels during the calendar year for which a release form was required pursuant to section 313 of EPCRA

It has also been indicated earlier that the threshold level for all the years starting in 1989 has been 25,000 pounds of the chemical manufactured or processed for the year. For chemicals that are otherwise used at a facility, the threshold level is 10,000 pounds for the applicable calendar year.

The toxic chemical release reporting must be in writing. In addition, it should include

- A statement that the mixture or trade name product contains a toxic chemical or chemicals subject to the reporting requirements of section 313 of EPCRA
- The name of each toxic chemical and the associated Chemical Abstracts Service (CAS) registry number of each such chemical, if available
- The percentage by weight of each toxic chemical in the mixture or in the trade name product

A written notice for a mixture or trade name product containing a listed toxic chemical should be provided to each new recipient at the time of the first shipment of each such mixture or trade name product. For all other recipients, such notice is to be provided in each calendar year beginning January 1, 1989. In addition, if an MSDS is required to be prepared and distributed for the mixture or trade name product for hazard communication under 29 CFR §1910.1200, the notice must be attached to or otherwise incorporated into such MSDS.

Any addition to or deletion from the list of toxic chemicals provided in Table 7.3 can be made by EPA under the statutory grant of authority. Other parties may also petition the EPA administrator for these changes or revisions. The factors considered for listing or delisting a chemical are

- Causation or capability of causing significant adverse or acute human health effects at concentrations reasonably likely to exist beyond the boundaries of a facility as a result of continuous or frequently recurring releases

- Knowledge pertaining to causation of cancers, teratogenic effects, serious or irreversible reproductive dysfunctions, neurological disorders, heritable genetic mutations, or other chronic health effects

- Causation of or capability of causing toxicity, persistence in the environment, or tendency to bioaccumulate in the environment

Under the general provision of EPCRA, the EPA administrator can modify the reporting thresholds of toxic chemicals as well as the frequency of such reportings. For this reason, section 313 of EPCRA requires EPA to develop and maintain a computerized database for a national toxic chemical inventory.

When more than one threshold level applies to the activities at a facility, the owner or operator of the facility must report all activities involving these chemicals which exceed the applicable threshold levels.

If a toxic chemical is listed based on a specific physical form or classification (e.g., fume or dust, solution, or friable nature), or a color, the reporting requirement is related to that specific form or color only. Owners or operators of facilities that solely manufacture, process, or use a toxic chemical in a different form or color from what is specified are not required to submit a report for that chemical.

Any toxic chemical reporting by the owner or operator of the appropriate facility must be made on the prescribed EPA form R (i.e., form 9350-1) in accordance with the instructions provided by EPA. This completed EPA form must be submitted to EPA as well as to the state where the facility is located. On September 21, 1991, EPA proposed new rules for reporting on form R under section 313 of EPCRA. If these rules are adopted, a facility will have to provide certain data for each toxic chemical at the time of submission of form R:

- The amount of toxic chemical that will be released into the environment during the year of submittal, percentage change from such release in the preceding year, and estimates for the following 2 years

- The amount of toxic chemical that will be recycled during the year, the percentage change from the preceding year, and the estimates for the following 2 years

- The amount of toxic chemical that will be treated during the year and the percentage change from the preceding year

- The types of recycling and source reduction practices that are used at the facility and how they have been developed and pursued

- Any change in production plans compared to the preceding year

- Any amount of toxic chemical that may have been released accidentally

Specific exemptions from chemical release reporting are provided in 40 CFR part 372. These are available when

- A toxic chemical is present in a *de minimis* concentration of the mixture (i.e., below 1 percent of the mixture or 0.1 percent of the mixture if the toxic chemical is a carcinogen) or the trade name product

- A toxic chemical is present in an article at the facility

- A toxic chemical is used at a facility for certain specific purposes (e.g., as a structural component of the facility, as a routine janitorial or grounds maintenance product, personal use by employees, in process or noncontact cooling water, and in the maintenance of motor vehicles)

- A toxic chemical is manufactured, processed, or used in a laboratory of the facility under the supervision of a technically qualified individual

In addition to the above, certain owners of leased property and certain operators of establishments on leased property are not subject to the reporting requirements with respect to toxic chemical release.

In the above context, the terms *mixture* and *trade name product* include

- Any article defined in 40 CFR §372.3

- Food, drug, cosmetic, alcoholic beverage, tobacco, or tobacco product packaged for distribution to the general public

- Any consumer product covered under the Consumer Product Safety Act and packaged for distribution to the general public

7.6 Enforcement Actions and Citizen Suits

Subtitle C of EPCRA provides details of enforcement actions that can be applied for statutory violations. It identifies the causes for civil actions that can be initiated by the state and local governments as well as the citizens through the filing of "citizen suits."

Under the statutory scheme, there are specific civil penalties for violation of emergency planning as established by EPCRA. The EPA administrator may order a facility owner or operator to comply with emergency planning requirements. Any person who violates or fails to obey such an order is liable for a civil penalty of not more than $25,000 for each day of such violation.

Additionally, there are civil, administrative, and criminal penalties for violations of emergency notification requirements under EPCRA. The administrative penalties can fall within class 1 or class 2 penalties as set forth in section 325(b) of EPCRA. Under class 1 administrative penalty, a civil penalty of not more than $25,000 per violation may be assessed. In a class 2 administrative penalty, a civil penalty of not more than $25,000 per day for the first violation and up to $75,000 per day for each day of second or subsequent violation may be assessed.

In addition, any person who knowingly and willfully fails to provide notice pursuant to section 304 of EPCRA can be fined not more than $25,000 or imprisoned for not more than 2 years or both. In the case of second or subsequent conviction, the penalty can be no more than $50,000 or imprisonment not longer than 5 years or both.

In a case of first impression, a federal district court held in 1991 that citizens can sue companies for their failures to file timely reports as required under EPCRA. The court in *Atlantic States Legal Foundation, Inc., v. Buffalo Envelope Company* also held that EPCRA confers federal jurisdiction over citizen suits for past violations. According to this court, such citizen suits may be maintained even when the violations are corrected after the filing of the notice of the suit, but before the filing of the suit itself. This holding is inapposite to an earlier U.S. Supreme Court's ruling in *Gwaltney of Smithfield v. Chesapeake Bay Foundation, Inc.* In that case, the Supreme Court held that citizens cannot sue polluters for "wholly past violations" under the Clean Water Act. The *Atlantic States* Court distinguished the decision by relying on the plain meaning of the operating language of EPCRA. It ruled that unlike section 505 of the Clean Water Act, section 326 of EPCRA does not contain "pervasive use of the present tense."

7.7 Trade Secrets

Section 322 of EPCRA authorizes withholding of certain information required in tier 2 form as trade secrets. However, any person who plans to withhold such information is required to establish certain verifiable claims. Any factual assertion of a claim of trade secret must include

- An explanation of the reason why such information is claimed to be a trade secret using specified trade secret factors and showing why such factors apply
- Submission of the information withheld as trade secrets separately from the other information submitted

Any person that withholds the specific chemical identity under the pretext of trade secrets is, however, required to provide the generic class or category of the hazardous chemical, extremely hazardous substance, or toxic chemical, as the case may be. Appendix 7B reproduces EPA's form which must be completed to substantiate any claims of trade secrecy under EPCRA.

Recommended Reading

1. Neil Orloff and Susan Sakai, *Community Right-to-Know Handbook,* Environmental Law Series, Clark Boardman Co. Ltd., New York, 1988.

2. J. Gordon Arbuckle, M. E. Bosco, D. R. Case, E. P. Laws, J. C. Martin, M. L. Miller, R. V. Randle, R. G. Stoll, T. F. P. Sullivan, T. A. Vanderver, Jr., and P. A. J. Wilson, *Environmental Law Handbook,* 10th ed., Government Institutes, Inc., Rockville, Md., 1989.

3. J. Gordon Arbuckle, Timothy A. Vanderver, Jr., and Paul A. J. Wilson, *Emergency Planning and Community Right-to-Know Act Handbook,* Government Institutes, Inc., Rockville, Md., 1989.

4. Kenneth B. Clansky, *Suspect Chemicals Sourcebook,* 6th ed., Roytech Publications, Inc., Burlingame, Calif., 1987.

Important Cases

1. *Gwaltney of Smithfield Ltd. v. Chesapeake Bay Foundation, Inc.,* 484 U.S. 49 (1987).

2. *Atlantic States Legal Foundation, Inc. v. Buffalo Envelope Company,* No. CIV-90-1110S (W.D.N.Y. Sept. 9, 1991).

APPENDIX 7A   Tier 1 and Tier 2 Forms

Page _____ of _____ pages
Form Approved OMB No. 2050-0072

Revised June 1990

Tier One

EMERGENCY AND HAZARDOUS CHEMICAL INVENTORY

Aggregate Information by Hazard Type

FOR OFFICIAL USE ONLY

ID #

Date Received

Important: Read instructions before completing form

Reporting Period   From January 1 to December 31, 19_____

Facility Identification

Name _____

Street _____

City _____ County _____ State _____ Zip _____

SIC Code ☐☐☐☐   Dun & Brad Number ☐☐-☐☐☐-☐☐☐☐

Emergency Contacts

Name _____

Title _____

Phone ( )

24 Hour Phone ( )

Name _____

Title _____

Phone ( )

24 Hour Phone ( )

Owner/Operator

Name _____

Mail Address _____

Phone ( )

☐ Check if information below is identical to the information submitted last year.

☐ Check if site plan is attached

	Hazard Type	Max Amount*	Average Daily Amount*	Number of Days On-Site	General Location
Physical Hazards	Fire	☐☐	☐☐	☐☐☐	
	Sudden Release of Pressure	☐☐	☐☐	☐☐☐	
	Reactivity	☐☐	☐☐	☐☐☐	
Health Hazards	Immediate (acute)	☐☐	☐☐	☐☐☐	
	Delayed (Chronic)	☐☐	☐☐	☐☐☐	

Certification   *(Read and sign after completing all sections)*

I certify under penalty of law that I have personally examined and am familiar with the information submitted in pages one through _____, and that based on my inquiry of those individuals responsible for obtaining the information, I believe that the submitted information is true, accurate and complete

_____
Name and official title of owner/operator OR owner/operator's authorized representative

_____
Signature   Date signed

* **Reporting Ranges**

Range Code	Weight Range in Pounds From...	To...
01	0	99
02	100	999
03	1000	9,999
04	10,000	99,999
05	100,000	999,999
06	1,000,000	9,999,999
07	10,000,000	49,999,999
08	50,000,000	99,999,999
09	100,000,000	499,999,999
10	500,000,000	999,999,999
11	1 billion	higher than 1 billion

Revised June 1990

Page _____ of _____ pages
Form Approved OMB No. 2050-0072

Tier Two
EMERGENCY AND HAZARDOUS CHEMICAL INVENTORY

Specific Information by Chemical

Facility Identification

Name _____
Street _____
City _____ County _____ State _____ Zip _____
SIC Code ▢▢▢▢   Dun & Brad Number ▢▢-▢▢▢-▢▢▢▢

FOR OFFICIAL USE ONLY
ID # _____
Date Received _____

Owner/Operator Name

Name _____ Phone ( )
Mail Address _____

Emergency Contact

Name _____ Title _____
Phone ( ) 24 Hr. Phone ( )

Name _____ Title _____
Phone ( ) 24 Hr. Phone ( )

Important: Read all instructions before completing form | **Reporting Period** From January 1 to December 31, 19___ | ▢ Check if information below is identical to the information submitted last year.

Chemical Description	**Physical and Health Hazards** (check all that apply)	**Inventory**	Container Type / Temperature / Pressure	**Storage Codes and Locations (Non-Confidential)** *Storage Locations*	Optional
CAS ▢▢▢▢▢▢▢▢▢ Trade Secret ▢ Chem. Name _____ Check all that apply: ▢ Pure ▢ Mix ▢ Solid ▢ Liquid ▢ Gas ▢ EHS EHS Name _____	▢ Fire ▢ Sudden Release of Pressure ▢ Reactivity ▢ Immediate (acute) ▢ Delayed (chronic)	▢▢ Max. Daily Amount (code) ▢▢ Avg. Daily Amount (code) ▢▢ No. of Days On-site (days)		_____ _____ _____ _____ _____	▢
CAS ▢▢▢▢▢▢▢▢▢ Trade Secret ▢ Chem. Name _____ Check all that apply: ▢ Pure ▢ Mix ▢ Solid ▢ Liquid ▢ Gas ▢ EHS EHS Name _____	▢ Fire ▢ Sudden Release of Pressure ▢ Reactivity ▢ Immediate (acute) ▢ Delayed (chronic)	▢▢ Max. Daily Amount (code) ▢▢ Avg. Daily Amount (code) ▢▢ No. of Days On-site (days)		_____ _____ _____ _____ _____	▢
CAS ▢▢▢▢▢▢▢▢▢ Trade Secret ▢ Chem. Name _____ Check all that apply: ▢ Pure ▢ Mix ▢ Solid ▢ Liquid ▢ Gas ▢ EHS EHS Name _____	▢ Fire ▢ Sudden Release of Pressure ▢ Reactivity ▢ Immediate (acute) ▢ Delayed (chronic)	▢▢ Max. Daily Amount (code) ▢▢ Avg. Daily Amount (code) ▢▢ No. of Days On-site (days)		_____ _____ _____ _____ _____	▢

Certification   *(Read and sign after completing all sections)*

I certify under penalty of law that I have personally examined and am familiar with the information submitted in pages one through _____, and that based on my inquiry of those individuals responsible for obtaining the information, I believe that the submitted information is true, accurate, and complete.

Name and official title of owner/operator OR owner/operator's authorized representative _____ Signature _____ Date signed _____

Optional Attachments
▢ I have attached a site plan
▢ I have attached a list of site coordinate abbreviations
▢ I have attached a description of dikes and other safeguard measures

Page _____ of _____ pages
Form Approved OMB No 2050-0072

Tier Two

EMERGENCY AND HAZARDOUS CHEMICAL INVENTORY

Specific Information by Chemical

Facility Identification

Name _____
Street _____
City _____ County _____ State _____ Zip _____

SIC Code [ ][ ][ ][ ]   Dun & Brad Number [ ][ ]-[ ][ ][ ]-[ ][ ][ ][ ]

FOR OFFICIAL USE ONLY

ID # _____
Date Received _____

Owner/Operator Name

Name _____   Phone ( )
Mail Address _____

Emergency Contact

Name _____   Title _____
Phone ( )   24 Hr. Phone ( )

Name _____   Title _____
Phone ( )   24 Hr. Phone ( )

| *Important: Read all instructions before completing form* | **Reporting Period** From January 1 to December 31, 19____ | ☐ Check if information below is identical to the information submitted last year. |

Confidential Location Information Sheet

Container Type **Temperature** **Pressure**

Storage Codes and Locations (Confidential)

Storage Locations

Optional

CAS # [ ][ ][ ][ ][ ][ ] [ ][ ][ ] [ ]   Chem. Name _____

☐

CAS # [ ][ ][ ][ ][ ][ ] [ ][ ][ ] [ ]   Chem. Name _____

☐

CAS # [ ][ ][ ][ ][ ][ ] [ ][ ][ ] [ ]   Chem. Name _____

☐

Certification   *(Read and sign after completing all sections)*

I certify under penalty of law that I have personally examined and am familiar with the information submitted in pages one through _____, and that based on my inquiry of those individuals responsible for obtaining the information, I believe that the submitted information is true, accurate, and complete.

Name and official title of owner/operator OR owner/operator's authorized representative _____

Signature _____   Date signed _____

Optional Attachments

☐ I have attached a site plan
☐ I have attached a list of site coordinate abbreviations
☐ I have attached a description of dikes and other safeguard measures

APPENDIX 7B

	United States Environmental Protection Agency Washington, DC 20460	
EPA	**Substantiation To Accompany Claims of Trade Secrecy Under the Emergency Planning and Community Right-To-Know Act of 1986**	Form Approved OMB No. 2050-0078 Approval expires 10-31-90

Paperwork Reduction Act Notice

Public reporting burden for this collection of information is estimated to vary from 27.7 hours to 33.2 hours per response, with an average of 28.8 hours per response, including time for reviewing instructions, searching existing data sources, gathering and maintaining the data needed, and completing and reviewing the collection of information. Send comments regarding the burden estimate or any other aspect of this collection of information, including suggestions for reducing this burden, to Chief, Information Policy Branch, PM-223, U.S. Environmental Protection Agency, 401 M Street, SW, Washington, DC 20460; and to the Office of Information and Regulatory Affairs, Office of Management and Budget, Washington, DC 20503.

Part 1. Substantiation Category

1.1 Title III Reporting Section (check only one)

☐ 303   ☐ 311   ☐ 312   ☐ 313

1.2 Reporting Year   19 _____

1.3 Indicate Whether This Form Is (check only one)

1.3a. ☐ **Sanitized**

(answer 1.3.1a below)

1.3.1a. Generic Class or Category

1.3b. ☐ **Unsanitized**

(answer 1.3.1b. and 1.3.2b. below)

1.3.1b. CAS Number

☐☐☐☐☐☐ – ☐☐ – ☐

1.3.2b. Specific Chemical Identity

Part 2. Facility Identification Information

2.1 Name

2.2 Street Address

2.3 City, State, and ZIP Code

2.4 Dun and Bradstreet Number

☐☐ – ☐☐ – ☐☐☐

EPA Form 9510-1 (7-88)   Page 1 of 5

Part 3. Responses to Substantiation Questions

3.1   Describe the specific measures you have taken to safeguard the confidentiality of the chemical identity claimed as trade secret, and indicate whether these measures will continue in the future.

3.2   Have you disclosed the information claimed as trade secret to any other person (other than a member of a local emergency planning committee, officer or employee of the United States or a State or local government, or your employee) who is not bound by a confidentiality agreement to refrain from disclosing this trade secret information to others?

☐ Yes   ☐ No

3.3   List all local, State, and Federal government entities to which you have disclosed the specific chemical identity. For each, indicate whether you asserted a confidentiality claim for the chemical identity and whether the government entity denied that claim.

Government Entity	Confidentiality Claim Asserted		Confidentiality Claim Denied	
	Yes	No	Yes	No

3.4   In order to show the validity of a trade secrecy claim, you must identify your specific use of the chemical claimed as trade secret and explain why it is a secret of interest to competitors. Therefore:

(i)   Describe the specific use of the chemical claimed as trade secret, identifying the product or process in which it is used. (If you use the chemical other than as a component of a product or in a manufacturing process, identify the activity where the chemical is used.)

(ii)   Has your company or facility identity been linked to the specific chemical identity claimed as trade secret in a patent, or in publications or other information sources available to the public or your competitors (of which you are aware)?

☐ Yes   ☐ No

If so, explain why this knowledge does not eliminate the justification for trade secrecy.

(iii) If this use of the chemical claimed as trade secret is unknown outside your company, explain how your competitors could deduce this use from disclosure of the chemical identity together with other information on the Title III submittal form.

3.4 (iv) Explain why your use of the chemical claimed as trade secret would be valuable information to your competitors.

3.5   Indicate the nature of the harm to your competitive position that would likely result from disclosure of the specific chemical identity, and indicate why such harm would be substantial.

3.6 (i) To what extent is the chemical claimed as trade secret available to the public or your competitors in products, articles, or environmental releases?

3.6 (ii)   Describe the factors which influence the cost of determining the identity of the chemical claimed as trade secret by chemical analysis of the product, article, or waste which contains the chemical (e.g., whether the chemical is in pure form or is mixed with other substances).

Part 4. Certification (Read and sign after completing all sections)

I certify under penalty of law that I have personally examined the information submitted in this and all attached documents. Based on my inquiry of those individuals responsible for obtaining the information, I certify that the submitted information is true, accurate, and complete, and that those portions of the substantiation claimed as confidential would, if disclosed, reveal the chemical identity being claimed as a trade secret, or would reveal other confidential business or trade secret information. I acknowledge that I may be asked by the Environmental Protection Agency to provide further detailed factual substantiation relating to this claim of trade secrecy, and certify to the best of my knowledge and belief that such information is available. I understand that if it is determined by the Administrator of EPA that this trade secret claim is frivolous, EPA may assess a penalty of up to $25,000 per claim.

I acknowledge that any knowingly false or misleading statement may be punishable by fine or imprisonment or both under applicable law.

4.1 Name and official title of owner or operator or senior management official

4.2 Signature *(All signatures must be original)*	4.3 Date Signed

Pesticides under FIFRA

8.1 Introduction

Persons associated with the agricultural sector such as farmers have well recognized the benefits of various pesticides in the form of insecticides, fungicides, and rodenticides. However, the question of safety regarding their use and disposal has become a matter of serious concern to the public over the last few decades. Indeed, these days, to balance the hazards and the benefits associated with pesticides is sometimes difficult. This conflict is more pronounced when the debate is over crop infestation, uncontrolled growth of weeds, and spoilage of crops and foods while in storage. However, of equal concern is the public fear which is exacerbated by the discovery of pesticides in groundwater and surface water, milk, and various processed and unprocessed foods.

Historically, chemical pesticides have been regulated by the federal government under the Federal Insecticide Act of 1910. However, compared to recent environmental laws, this statute had minimal impact partly because pesticides were not in common use in the early part of this century. The pesticide use pattern was drastically changed during World War II when the use of pesticides, mostly for the purpose of weed and pest control, suddenly increased. As a result of this unprecedented growth in the manufacture and use of pesticides, Congress became concerned, and thus the first comprehensive Federal Insecticide, Fungicide, and Rodenticide Act (FIFRA) was enacted on June 25, 1947.

This original act has been amended several times, notably in 1972, 1975, 1978, 1980, 1988, and 1990. FIFRA as amended through November 28, 1990, is the current federal legislation. However, the basic framework of the 1972 statute has not been significantly altered. The current statute, as amended, does not provide individual definitions of insecticides, fungicides, and rodenticides. However, because of their interrelationships and common properties, they are statutorily addressed as pesticides.

Also note that as the federal statute underwent several amendments, so did the authorized agency entrusted with regulating the provisions of the statute. In 1947, all pesticides subject to the earlier pesticide statute and distributed in interstate commerce were required to be registered with the U.S. Department of Agriculture (USDA). With the inception of the Environmental Protection Agency (EPA) in 1970, EPA became the federal agency responsible for pesticide regulation. As a result, EPA is now empowered both to promulgate and enforce pesticide regulations under FIFRA and to influence the regulations promulgated by other federal agencies concerning pesticides and their residues, e.g., the Food and Drug Administration (FDA), USDA, and the Department of Health and Human Services (DHHS). EPA's regulations promulgated under FIFRA are set forth in 40 CFR part 152.

8.2   Definitions

Section 2 of FIFRA provides definitions of various terms which are part of the statutory language. Some of these definitions are worth understanding since the word *pesticide* in the statute may include almost anything which is intended to be used to control, repel, or kill various lower forms of life. By this broad coverage, a pesticide can be a fungicide, rodenticide, or insecticide.

Under the statute, the term *pesticide* means (1) any substance or mixture of substances intended to prevent, destroy, repel, or mitigate any pest and (2) any substance or mixture of substances intended for use as a plant regulator, defoliant, or desiccant. The term *pesticide* does not include any article that is a "new animal drug" under a specific provision of the Federal Food, Drug, and Cosmetic Act (FFDCA) or an animal feed bearing or containing a new animal drug.

The term *defoliant* means any substance or mixture of substances intended to cause the leaves or foliage to drop from a plant, with or without causing abscission.

The term *desiccant* means any substance or mixture of substances intended for artificially accelerating the drying of plant tissue.

The term *insect* means any of the numerous small invertebrate animals generally having the body more or less segmented and for the most part belonging to the class of insects, comprising six-legged, usually winged forms. It includes beetles, bees, flies, and other allied anthropods, such as spiders, mites, ticks, centipedes, and wood lice.

The term *pest* means any insect, rodent, nematode, fungus, weed, or any other form of terrestrial or aquatic plant or animal life. It includes viruses, bacteria, and other microorganisms on or in living humans and animals which the EPA administrator declares to be a pest.

The term *fungus* means any non-chlorophyll-bearing thallophyte or plant of a lower order than mosses and liverworts. It includes rust, smut, mildew, mold, yeast, and certain types of bacteria that are not generally associated with living beings or processed food, beverages, or pharmaceuticals.

Pursuant to FIFRA, the term "active ingredient" may mean different things, depending on whether it is related to a pesticide, plant regulator, defoliant, or desiccant. In the case of a pesticide, it means an ingredient which will prevent, destroy, repel, or mitigate any pest. By the statutory definition, an ingredient which is not active is an "inert ingredient."

8.3   FIFRA Framework

The 1988 FIFRA Amendments require EPA to complete a pesticide safety review program within 9 years. However, the basic regulatory framework of FIFRA since the 1972 statute remains unchanged. The major thrust is still on the prevention of unreasonable adverse impacts on humans, plants, animals, and the environment due to the use of pesticides.

The basic FIFRA framework includes four major considerations:

- Evaluation of risks posed by pesticides by using a system of registration and reregistration
- Classification and certification of pesticides for specific uses
- Suspension, cancellation, or restriction of pesticides that pose a risk
- Enforcement of regulations by means of inspections, labeling, notices, and involvement by state regulatory authorities

Unlike other major environmental laws, FIFRA requires EPA to take a balanced approach to regulating a pesticide. Thus, EPA must consider the actual and potential risks posed by a pesticide and must weigh economic, social, health, and other environmental benefits. It is conceivable, at some future date, under this overall concept of balancing of risks and benefits, that a pesticide may be allowed in certain specific situations while being banned from general or widespread use. Such a restricted use, if and when legally mandated, will receive an ongoing scrutiny from EPA and other agencies about possible future expansion or constriction or outright ban of such regulated use.

8.4   Pesticide Registration and Reregistration

FIFRA and the regulations promulgated thereunder require that all pesticides and pesticide products with certain exceptions first be registered with EPA prior to their importation or distribution in commerce. As set forth in 40 CFR §152.15, no one may distribute or sell any pesticide product that is not registered with EPA unless it is exempted under 40 CFR §§152.20, 152.25, and 152.30. These exempted classes of pesticides include certain biological control agents, human drugs, treated articles or substances, pheromones, and pheromone traps, preservatives for biological specimens, vitamin hormone products and foods. Also exempted from registration are those pesticides which are transferred between registered establishments or distributed or sold for export or under an experimental use permit.

This registration requirement applies both to newly formulated pesticide products and to any new combinations or mixtures of pesticides that were individually registered earlier. In addition, any new brand of pesticides whose composition is similar to existing, registered pesticide products must be registered with EPA. Furthermore, if an already registered product finds a new use, it must again be registered for the intended new use. EPA can refuse to register any pesticide that it determines to be unduly hazardous or harmful.

Section 3 of FIFRA provides the registration requirements for pesticides. Whenever any person wants to manufacture, formulate, import, or distribute in commerce a pesticide or a pesticide product, an application for registration must be submitted to EPA prior to such activities. Registration of a pesticide is generally approved by EPA if several conditions are satisfied:

- The composition of the pesticide is in conformity with its claims
- The pesticide labeling and other submitted data comply with the requirements of FIFRA
- The intended use of the pesticide will not result in any unreasonable adverse effects on the environment
- The pesticide will not generally cause unreasonable adverse effects on the environment if used in accordance with prescribed and commonly recognized practice.

Any registration thus obtained is issued only for specific uses of that pesticide. The registration itself specifies the crops and insects to which the registered pesticide may be applied. A new registration must be obtained if any additional uses are intended beyond what is specified in the existing registration. Such uses must be supported by appropriate research data with respect to their efficacy and safety before a new registration will be considered by EPA. Also, all registrations are good for a 5-year period unless they are renewed for additional periods.

Section 3(g) of FIFRA requires that all currently registered pesticides be reregistered. Reregistration of existing pesticides is allowed if test data pass the regulatory criterion of not causing "unreasonable adverse effects on the environment" through their intended use. These data may be derived from studies involving toxicity, mutagenicity, teratogenicity, carcinogenicity, and other health, safety, and environmental impacts that may be caused by an exposure to a pesticide.

For existing pesticides that were originally registered by USDA, reregistration involves new and considerably more stringent health and safety test data. All such pesticides must meet the current standards for registration irrespective of the fact that they have been registered earlier and that the intended uses of such pesticides or pesticide products are well known to both the regulators and the general public. The 1988 FIFRA Amendments, however, allowed EPA to streamline the reregistration procedure by grouping pesticides by their active ingredients and registering them on a generic basis.

In this context, note that the registration document available from EPA provides the regulatory policy for pesticide products that may contain a regulated chemical for which a standard has been developed. This document also includes labeling, use, and safety criteria which will apply to products containing that chemical or a specific active ingredient.

8.5 Suspension and Cancellation

Section 6 of FIFRA authorizes EPA to restrict, suspend, or cancel registration of a pesticide if it constitutes an "imminent hazard" to public health or the environment or if there is a suspicion of a "substantial question of safety" regarding the pesticide. Of all these measures, restriction is the least prohibitive because it does not suspend or cancel an existing registration. It simply imposes a burden by restricting the use or distribution in commerce of that pesticide. Both suspension and cancellation carry a stricter prohibition when a pesticide poses an unreasonable risk to humans or to the environment. As a result, cancellation and suspension orders are typically published in the Federal Register.

Section 6(g) of FIFRA, which was added in 1988, requires notification to EPA and appropriate state and local officials regarding the quantity and location of any suspended or canceled pesticides that are in the possession of any producer, exporter, registrant, applicant, commercial applicator, permit holder for an experimental use, or any person who distributes, sells, or possesses such pesticides. On March 28, 1991, EPA published its proposed policy for notification of the quantity and location of any suspended or canceled pesticides that any such person possesses. EPA also recommended a standard format to be used by all respondents. This standard format is reproduced in App. 8A.

Suspension

A suspension is generally ordered when a pesticide causes an imminent hazard to humans or the environment. Grounds for suspension of registration of a pesticide are set forth in 40 CFR §152.150. Pursuant to section 6(c)(1) of FIFRA, EPA may suspend registration if it is determined to be necessary by EPA to prevent an imminent hazard. The statute authorizes EPA to impose two types of suspension: ordinary suspension and emergency suspension. In either case, a suspension order by EPA creates an immediate ban on the manufacture, import, and distribution in commerce of a pesticide.

Ordinary suspension is imposed when EPA is considering a cancellation or a change in classification regarding a particular pesticide. The word *ordinary* should not be taken lightly, for such suspension may relate to an emergency procedure prior to the possible or eventual cancellation of the pesticide.

Procedurally, an ordinary suspension resembles judicial proceedings involving preliminary injunction. In almost all suspension situations, EPA must inform the registrant about its intent to cancel the pesticide registration. As a result, a suspension order may be effective, often for years, before EPA may cancel the registration of a pesticide.

The registrant also may disagree with EPA's assessment and can ask for an administrative hearing for proper adjudication. An expedited hearing may be requested within 5 days of receipt of the notice from the EPA administrator. If no such hearing is requested, the suspension order becomes effective immediately, and that order is not subject to judicial review.

The emergency suspension, by comparison, reflects a much harsher ban that takes effect immediately. In an emergency suspension, the registrant is not given any notice or the opportunity to be heard at an expedited hearing before the suspension becomes effective. Moreover, the emergency suspension puts an immediate stop to further production, use, and distribution in commerce of the pesticide. However, this action can be taken only when an emergency is determined to exist by the EPA administrator. The registrant's only recourse is a judicial review to determine the propriety of this emergency order based on a substantial likelihood of serious, immediate harm. In *Dow Chemical v. Blum,* the court held that the EPA administrator must consider the following five factors before invoking an emergency suspension:

- The seriousness of the threatened harm
- The immediacy of the threatened harm
- The probability of occurrence of the threatened harm
- The public benefits derived from the continued use of the pesticide while it is being suspended
- The nature and extent of information available to the EPA administrator at the time of the suspension decision

Also, like any other agency action, an emergency suspension order may be thrown out by a court which determines that such order is arbitrary and capricious or an abuse of EPA's discretionary power under the law.

Cancellation

Grounds for cancellation of registration of a pesticide product are set forth in 40 CFR §152.148. Cancellation can be either voluntary or nonvoluntary. In a voluntary cancellation, a registrant requests that her or his pesticide registration be canceled or amended so as to delete one or more uses. However, prior to acting on such a request, the EPA administrator will have to publish a notice of the receipt of the request in the Federal Register. A nonvoluntary cancellation is initiated by the regulatory agency.

All cancellations result in a total and final recall of a pesticide from production, use, and distribution in commerce. It is indeed the end of the process in every sense of the term. However, the cancellation process is a lengthy process and takes considerable time, often years. The cancellation review process is initiated when a pesticide raises a "substantial question of safety" with regard to human beings or to the environment.

Until a cancellation order is formally issued by EPA, the pesticide in question may be manufactured, used, or distributed in commerce. If a cancellation order is issued, it becomes final unless a challenge to EPA's order is formally made within 30 days of its issuance. Such a challenge and the final outcome of the cancellation order may involve a public hearing or a scientific review or both. Thereafter, the administrative law judge recommends her or his decision to the EPA administrator or the authorized representative of EPA who in turn decides the final outcome. If cancellation is sustained, the pesticide will be banned for any further use or distribution in commerce.

As indicated earlier, section 6(g) of FIFRA requires certain persons to notify EPA and state and local officials of the quantity and location of any suspended or canceled pesticides in their possession. Generally, EPA will require the submission of such information only if the cancellation or suspension action is the result of adverse consequences on human health or environmental concern. EPA has also developed a standard format which the respondents should use for notification purposes.

Special reviews

EPA regulations as set forth in 40 CFR part 154 describe the criteria and the process of a special review involving cancellation of a pesticide. Procedurally, when EPA determines that a pesticide has exceeded certain established hazard criteria, it notifies the affected registrants of that product. These registrants can challenge the agency's decision based on unreasonable risk within 30 days. After 30 days, EPA has to announce whether a special review will be initiated to resolve the dispute.

Any eventual cancellation after a special review may ban all uses or some uses of the pesticide in question. Also, a special review may result in continued use of the product with certain restrictions. As indicated earlier, the ultimate cancellation decision is made after both the benefits and the risks from the use of the pesticide have been considered.

8.6 Reregistration Acceleration

The revised section 4 of the 1988 Amendments to FIFRA mandates five phases of the reregistration program to be undertaken by EPA. The purpose is to accelerate the reregistration of pesticides that were registered prior to November 1, 1984. These phases are as follows:

Phase 1. During this phase, EPA must identify, categorize, and publish lists of all active ingredients in the pesticides that will be reregistered. Four lists (A, B, C, and D) are to be published in the Federal Register by EPA. These listings are based on the statutory priority which places emphasis on the use (i.e., postharvest residues) of pesticides in foods or feedlots and those pesticides which may leach into groundwater.

Phase 2. This is reflective of responses by registrants interested in seeking reregistration. Within 90 days of the publication of lists B, C, and D in the Federal Register, registrants of pesticides containing the listed active ingredients must inform EPA whether they will seek reregistration. If so, they must also inform whether they will share the cost of generating data regarding the safety of the uses of such pesticides. EPA may allow any reasonable time, up to a maximum of 48 months, for the submittal of such data.

Phase 3. During this phase, each registrant committed to seeking reregistration of an active ingredient as listed will be given 1 year to provide specific information to EPA. Such information may include existing data as well as additional data that must be generated from studies and by following EPA's guidelines for testing.

Phase 4. During this phase, EPA will review the submitted information under phases 2 and 3 and determine whether the proposed testing by the registrants seeking reregistration will meet EPA's requirements. A list of the registrants against whom there are still outstanding data requirements for one or more active ingredients will be published in the Federal Register by EPA.

Phase 5. During this phase, EPA must review the submitted data and other information provided by the registrants and determine within 1 year whether the pesticides containing the listed active ingredients may be continued and thus reregistered. EPA may also require additional product specific data from the registrants before making such decisions.

Table 8-1 reproduces list A of chemical substances, for which pesticide registration standards have been issued based on EPA's notice of February 22, 1989. This reflects the phase 1 requirement.

Section 3(c)(3) of FIFRA includes a "fast-track" provision which requires EPA to carry out an expeditious review of an application for registration of a pesticide product. This will determine whether the pesticide product, if registered as proposed, will be (1) identical or substantially "similar in composition and labeling" to that of a registered product or (2) different in such a manner that will not significantly increase the risks of "unreasonable adverse effects" on human beings or the environment.

The 1988 amendments to FIFRA require EPA to undertake an expeditious review of generic or "me-too" applications. Under this fast-track provision of section 3(c)(3) of the amended statute, EPA is required to expedite its review process of a me-too application. This includes amendments to existing registrations that do not necessitate any scientific review of data. Pursuant to this, EPA must notify the registrant within 45 days of data submission whether the application is complete. EPA has 90 days after the receipt of a complete application to act on and grant or deny registration. If the application is denied, EPA is required to notify the registrant in writing and to provide the specific reason for its denial.

8.7   Experimental Use

Section 5 of FIFRA contains provisions for experimental-use permits. Any person may apply for an experimental-use permit for a pesticide. The EPA administrator is required to either grant or deny the permit upon completion of the review. This review must be completed within 120 days after receipt of the application and all required supporting data.

In the event of a denial, the EPA administrator is required to notify the applicant about the denial along with reasons for the denial. The applicant may

TABLE 8.1 List A under Phase 1

Registration standard	Date	Case	CAS No.	Chemical/Common name
Acephate	87	0042	30560-19-1	Acephate
ADBAC	85	0350	63449-41-2	Alkyl dimethyl benzyl ammonium chloride
			8001-54-5	Alkyl dimethyl ethylbenzyl ammonium chloride
			87175-02-8	Alkyl dimethyl dodecylbenzyl ammonium chloride
			61789-68-2	Alkyl bis(2-hydroxyethyl) benzyl ammonium chloride
			68424-87-3	Dialkyl methyl benzyl ammonium chloride
			1733-96-6	Alkyl dimethyl 1-napthylmethyl ammonium chloride
			1330-85-4	Trimethyl dodecylbenzyl ammonium chloride
			3734-33-6	((2,6-Xylylcarbamoyl)methyl diethyl ammonium benzoate
			121-54-0	2-(2-(p-(Diisobutyl)phenoxy)ethoxy)ethyl dimethyl benzyl ammonium chloride
			25155-18-4	2-(2-(p-(Diisobutyl)cresoxy)ethoxy)ethyl dimethyl benzyl ammonium chloride
Alachlor	84	0063	15972-60-8	Alachlor
Aldicarb	84	0140	116-06-3	Aldicarb
Aldrin	86	0172	309-00-2	Aldrin
Aliette	86	0646	39148-24-8	Fosetyl-Al
Allethrin stereoisomers	88	0437	584-79-2	Allethrin
				Allethrin Coil
			28057-48-9	d-trans-Allethrin
			28434-00-6	S-Bioallethrin
			42534-61-2	d-cis-trans-Allethrin
Al & Mg Phosphide	86	0025	20859-73-8	Aluminum phosphide
		0645	12057-74-8	Magnesium phosphide
4-Aminopyridine	80	0015	504-24-5	4-Aminopyridine
Amitraz	87	0234	33089-61-1	Amitraz
Amitrole	84	0095	61-82-5	Amitrole
Ammonium sulfamate	81	0016	7773-06-0	Ammonium sulfamate
Anilazine	83	0114	101-05-3	Anilazine
Arsenic acid	86	0389	7778-39-4	Arsenic acid
Aspon	80	0019	3244-90-4	Aspon
Asulam	88	0265	3337-71-1	Asulam
			2302-17-2	Sodium asulam
Atrazine	83	0062	1912-24-9	Atrazine
Azinphos-Methyl	86	0235	86-50-0	Azinphos-methyl
BT	88	0247	68038-71-1	Bacillus thuringiensis (Berliner) var. israelensis
			68038-71-1	Bacillus thuringiensis (Berliner) var. kurstaki
			68038-71-1	Bacillus thuringiensis (Berliner) var. aizswai
			68038-71-1	Bacillus thuringiensis (Berliner) var. tenebrionis
			68038-71-1	Bacillus thuringiensis (Berliner) var. san diego
Barium metaborate	83	0632	13701-59-2	Barium metaborate
			7775-19-1	Sodium metaborate

TABLE 8.1 List A under Phase 1 (*Continued*)

Registration standard	Date	Case	CAS No.	Chemical/Common name
Bendiocarb	87	0409	22781-23-3	Bendiocarb
Benomyl	87	0119	17804-35-2 52316-55-9	Benomyl Methyl (2-benzimidazole)carbamate phosphate
Bentazon	85	0182	50723-80-3	Sodium bentazon
Bifenox	81	0013	42576-02-3	Bifenox
BKLFI-2	81	0074		Iodine complex of POE monoester of 5- (and 6-)carboxy-4-hexyl-2-cyclohexene-1-octanoic acid
Boric Acid	83	0024	10043-35-3 1303-86-2 1303-96-4 12280-03-4	Boric acid Boric oxide Borax Disodium octaborate tetrahydrate
Bromacil	82	0041	314-40-9 53404-19-6 69484-12-4	Bromacil Lithium bromacil Sodium bromacil Bromacil, dimethylamine salt
Brominated Salicylanilide	85	0347	87-12-7 87-10-5 2577-72-2	4',5-Dibromosalicylanilide 3,4',5-Tribromosalicylanilide 3,5-Dibromosalicylanilide
Butoxycarboxim	81	0077	34681-23-7	Butoxycarboxim
Butylate	83	0071	2008-41-5	Butylate
Captafol	84	0116	2939-80-2	Captafol
Captan	86	0120	133-06-2	Captan
Carbaryl	84	0080	63-25-2	Carbaryl
Carbofuran	84	0101	1563-66-2	Carbofuran
Carbophenothion	84	0108	786-19-6	Carbophenothion
Carboxin	81	0012	5234-68-4 5259-88-1	Carboxin Oxycarboxin
Chloramben	81	0086	1076-46-6 7286-84-2 1954-81-0	Ammonium chloramben Chloramben, methyl ester Sodium chloramben
Chlordane	86	0173	57-74-9	Chlordane
Chlordimeform HCl	85	0142	6164-98-3 19750-95-9	Chlordimeform Chlordimeform hydrochloride
Chlorinated Isocyanurates	87	0569	2782-57-2 108-80-5 2244-21-5 2893-78-9 87-90-1 30622-37-8 51580-86-0	Dichloroisocyanuric acid Cyanuric acid Potassium dichloroisocyanurate Sodium dichloroisocyanurate Trichloroisocyanuric acid Penta-s-triazinetrione Sodium dichloroisocyanurate dihydrate
Chlorobenzilate	83	0072	510-15-6	Chlorobenzilate
Chloroneb	80	0007	2675-77-6	Chloroneb
Chloropicrin	82	0040	76-06-2	Chloropicrin
Chlorothalonil	88	0097	1897-45-6	Chlorothalonil
Chlorpropham	87	0271	101-21-3	Chlorpropham

TABLE 8.1 List A under Phase 1 (*Continued*)

Registration standard	Date	Case	CAS No.	Chemical/Common name
Chlorpyrifos	84	0100	2921-88-2	Chlorpyrifos
Chlorsulfuron	82	0631	64902-72-3	2- Chlorsulfuron
Chromated Arsenicals	86	0132	1303-28-2	Arsenic pentacide
			1327-53-3	Arsenic trioxide
			13464-38-5	Sodium arsenate
			13464-42-1	Sodium pyroarsenate
			7738-94-5	Chromic acid
			7778-50-9	Potassium dichromate
			10588-01-9	Sodium dichromate
Coal Tar/Creosote	88	0139	8052-42-4	Asphalt
			8052-42-4	Bitumen
			8007-45-2	Coal tar
			8007-45-2	Tar
			8002-29-7	Coal tar neutral oils
			8021-39-4	Wood creosote
			70321-79-8	Creosote oil (Derived from any source)
			8001-58-9	Coal tar creosote
Copper Compounds: Grp II	87	0649	1332-40-7	Basic copper chloride
			16828-95-8	Copper in the form of an ammonia complex
			33113-08-5	Copper ammonium carbonate
			1184-64-1	Copper carbonate
			20427-59-2	Copper hydroxide
			1332-40-7	Copper oxychloride
			8012-69-9	Copper oxychloride sulfate
			10402-15-0	Chelates of copper citrate
			527-09-3	Chelates of copper gluconate
			814-91-5	Copper oxalate
			10125-13-0	Copper chloride dihydrate
			3251-23-8	Copper nitrate
Copper Sulfate	86	0636	1344-73-6	Basic copper sulfate
			7758-99-8	Copper sulfate, pentahydrate
			10257-54-2	Copper sulfate, monohydrate
			7758-98-7	Copper sulfate (anhydrous)
Coumaphos	81	0018	56-72-4	Coumaphos
Cryolite	88	0087	15036-52-3	Cryolite
Cyanazine	84	0066	21725-46-2	Cyanazine
Cycloheximide	82	0038	66-81-9	Cycloheximide
Cyhexatin	85	0237	13121-70-5	Cyhexatin
Dalapon	87	0274	75-99-0	Dalapon
			127-20-8	Sodium dalapon
			29110-22-3	Magnesium dalapon
Daminozide	84	0032	1596-84-5	Daminozide
DCNA	83	0113	99-30-9	Dichloran
DCPA	88	0270	1861-32-1	Chlorthal-dimethyl
DDVP	87	0310	62-73-7	Dichlorvos
Deet	85	0002	134-62-3	N,N-Diethyl-meta-toluamide and other isomers
Demeton	85	0143	8065-48-3	Demeton
Dialifor	81	0010	10311-84-9	Dialifor
Diallate	83	0098	2303-16-4	Diallate

TABLE 8.1   List A under Phase 1 (*Continued*)

Registration standard	Date	Case	CAS No.	Chemical/Common name
Diazinon	88	0238	333-41-5	Diazinon
Dicamba	83	0065	1918-00-9	Dicamba
			2300-66-5	Dimethylamine dicamba
			25059-78-3	Diethanolamine dicamba
			53404-28-7	Monoethanolamine dicamba
			1982-69-0	Sodium dicamba
				Dicamba, diglycoamine salt
				Dicamba, isopropylamine salt
Dichlobenil	87	0263	1194-65-6	Dichlobenil
Dichlone	81	0008	117-00-6	Dichlone
Dicofol	83	0021	115-32-2	1,1-Bis(chlorophenyl)-2,2,2-trichloroethanol
Dicrotophos	82	0145	141-66-2	Dicrotophos
Difenzoquat	88	0223	43222-48-6	Difenzoquat methyl sulfate
Diflubenzuron	85	0144	35367-38-5	Diflubenzuron
Dimethoate	83	0088	60-15-5	Dimethoate
Dioxathion	83	0089	78-34-2	Dioxathion
Diphenamid	87	0269	957-51-7	Diphenamid
Dipropetryn	85	0224	4147-51-7	Dipropetryn
Diquat Dibromide	86	0288	85-00-7	Diquat dibromide
Disulfoton	85	0102	298-04-4	Disulfoton
Diuron	83	0046	330-54-1	Diuron
Dodine	87	0161	2439-10-3	Dodine
			19727-17-4	N-Dodecylguanidine terephthalate
			13590-97-1	Dodecylguanidine hydrochloride
Endosulfan	82	0014	115-29-7	Endosulfan
EPN	87	0147	2104-64-5	O-Ethyl O(p-nitrophenyl) phenylphosphonothioate
EPTC	83	0064	759-94-4	S-Ethyl dipropylthiocarbamate
Ethephon	88	0382	16672-87-0	Ethepon
Ethion	82	0090	563-12-2	Ethion
Ethoprop	88	0106	13194-46-4	Ethoprop
Ethoxyquin	81	0003	91-53-2	Ethoxyquin
Ethyl-Parathion	86	0155	56-38-2	Parathion
Fenaminosulf	83	0126	140-56-7	Fenaminosulf
Fenamiphos	87	0333	22224-92-6	Fenamiphos
Fenitrothion	87	0445	122-14-5	Fenitrothion
Fensulfothion	83	0107	115-90-2	Fensulfothion
Fenthion	88	0290	55-38-9	Fenthion
Fluchloralin	85	0189	33245-39-5	Fluchloralin
Fluometuron	85	0049	2164-17-2	Fluometuron
Folpet	87	0630	133-07-3	Folpet
Fonofos	84	0105	944-22-9	Fonofos

TABLE 8.1   List A under Phase 1 (*Continued*)

Registration standard	Date	Case	CAS No.	Chemical/Common name
Formaldehyde	88	0556	50-00-0 30525-89-4	Formaldehyde Paraformaldelhyde
Formetanate HCl	83	0091	23422-53-9	Formetanate hydrochloride
Fumarin	80	0004	117-52-2 34490-93-2	Coumafuryl Sodium fumarin
Glyphosate	86	0178	38641-94-0 34494-03-6	Isopropylamine glyphosate Sodium glyphosate
Heliothis NPV	84	0151		Polyhedral inclusion bodies of Heliothis nuclear polyhedrosis virus
Heptachlor	86	0175	76-44-8	Heptachlor
Hexazinone	88	0266	51235-04-2	Hexazinone
Isopropalin	81	0005	33820-53-0	Isopropalin
Lindane	85	0315	58-89-9	Lindane
Linuron	84	0047	330-55-2	Linuron
Malathion	88	0248	121-75-5	Malathion
Maleic Hydrazide	88	0381	123-33-1 28382-15-2 5716-15-4	1,2-Dihydro-3,6-pyridazinedione Potassium 1,2-dihydro-3,6-pyridazinedione Diethanolamine 1,2-dihydro-3,6-pyridazinedione
Mancozeb	87	0643	8018-01-7	Mancozeb
Maneb	88	0642	12427-38-2	Maneb
MCPA	82	0017	94-74-6 3653-48-3 20405-19-0 2039-46-5 19480-43-4 1713-12-8 1713-11-7 26544-20-7 2698-40-0	2-Methyl-4-chlorophenoxyacetic acid Sodium 2-methyl-4-chlorophenoxyacetate Diethanolamine 2-methyl-4-chlorophenoxyacetate Dimethylamine 2-methyl-4-chlorophenoxyacetate 2-Butoxyethyl 2-methyl-4-chlorophenoxyacetate Butyl 2-methyl-4-chlorophenoxyacetate Isobutyl 2-methyl-4-chlorophenoxyacetate Isooctyl 2-methyl-4-chlorophenoxyacetate Isopropyl 2-methyl-4-chlorophenoxyacetate
MCPP	88	0377	7085-19-0 1929-86-8 32351-70-5 1432-14-0 28437-03-2	Mecoprop Potassium 2-(2-methyl-4-chlorophenoxy)propionate Dimethylamine 2-(2-methyl-4-chlorophenoxy)propionate Diethanolamine 2-(2-methyl-4-chlorophenoxy)propionate Isooctyl 2-(2-methyl-4-chlorophenoxy)propionate
Metalaxyl	88	0081	57837-19-1	Metalaxyl
Metaldehyde	88	0576	108-62-3	Metaldehyde
Methamidophos	82	0043	10265-92-6	Methamidophos
Methidathion	88	0034	950-37-8	Methidathion
Methiocarb	87	0577	2032-65-7	Methiocarb
Methomyl	88	0028	16752-77-5	Methomyl

TABLE 8.1   List A under Phase 1 (*Continued*)

Registration standard	Date	Case	CAS No.	Chemical/Common name
Methoprene	82	0030	40596-69-8	Methoprene
Methoxychlor	88	0249	72-43-5	Methoxychlor
Methyl Bromide	86	0335	74-83-9	Methyl bromide
Methyl-Parathion	86	0153	298-00-0	Methyl parathion
Metiram	88	0644	9006-42-2	Metiram
Metolachlor	87	0001	51218-45-2	Metolachlor
Metribuzin	85	0181	21087-64-9	Metribuzin
Mevinphos	88	0250	7786-34-7	Mevinphos
Monocrotophos	85	0154	6923-22-4	Monocrotophos
Monuron	83	0045	150-68-5	Monuron
Monuron TCA	83	0045	140-41-0	Monuron trichloroacetate
Nabam	87	0641	142-59-6	Nabam
Naled	83	0092	300-76-5	Naled
Naptalam	85	0183	132-66-1 132-67-2	N-1-Naphthylphthalamic acid Sodium N-1-naphthylphthalamate
Napthalene	81	0022	91-20-3	Naphthalene
Napthaleneacetic Acid	81	0379	86-87-3 86-86-2 15165-79-4 25545-89-5 61-31-4 2122-70-5	1-Naphthaleneacetic acid 1-Naphthaleneacetamide Potassium 1-naphthaleneacetate Ammonium 1-naphthaleneacetate Sodium 1-naphthaleneacetate Ethyl 1-naphthaleneacetate
Nitrapyrin	85	0213	1929-82-4	Nitrapyrin
Norflurazon	84	0229	27314-13-2	Norflurazon
OBPA	81	0044	58-36-6	10,10'-Oxybisphenoxarsine
Oryzalin	87	0186	19044-88-3	Oryzalin
Oxamyl	87	0253	23135-22-0	Oxamyl
Oxydemeton-Methyl	87	0258	301-12-2	Oxydemeton-methyl
Oxytetracycline	88	0655	79-57-2 2058-46-0 7179-50-2	Oxytetracycline Oxytetracycline hydrochloride Calcium oxytetracycline
Paraquat Dichloride	87	0262	1910-42-5 2074-50-2	Paraquat dichloride Paraquat bis(methylsulfate)
PCNB	87	0128	82-68-8	Quintozene
Pendimethalin	85	0187	40487-42-1	Pendimethalin
Perfluidone	85	0195	37924-13-3	Perfluidone
Phenmedipham	87	0277	13684-63-4	Phenmedipham
Phorate	88	0103	298-02-2	Phorate
Phosalone	87	0027	2310-17-0	Phosalone
Phosmet	86	0242	732-11-6	Phosmet
Phosphamidon	87	0157	13171-21-6 297-99-4 23783-98-4	Phosphamidon (E)-Phosphamidon (Z)-Phosphamidon

TABLE 8.1 List A under Phase 1 *(Continued)*

Registration standard	Date	Case	CAS No.	Chemical/Common name
Picloram	88	0096	1918-02-1	Picloram
			6753-47-5	Triisopropanolamine picloram
			26952-20-5	Isooctyl picloram
			2545-60-0	Potassium picloram
			35832-11-2	Triethylamine picloram
				Isopropanolamine picloram
Potassium Bromide	84	0342	7758-02-3	Potassium bromide
Potassium permanganate	85	0220	7722-64-7	Potassium permanganate
Prometryn	87	0467	7287-19-6	Prometryn
Pronamide	86	0082	23950-58-5	Propyzamide
Propachlor	85	0177	1918-16-7	Propachlor
Propanil	87	0226	709-98-8	Propanil
Propargite	86	0243	2312-35-8	Propargite
Propazine	88	0230	139-40-2	Propazine
Propham	87	0283	122-42-9	Propham
Resmethrin	88	0421	10453-86-8	Resmethrin
			28434-01-7	Bioresmethrin
Rotenone	88	0255	83-79-4	Rotenone
				Derris resins other than rotenone
				Cube Resins other than rotenone
Simazine	84	0070	122-34-9	Simazine
Na & Ca Hypochlorite	86	0029	7778-54-3	Calcium hypochlorite
			7681-52-9	Sodium hypochlorite
Sodium Omadine	85	0209	15922-78-8	Omadine sodium
Streptomycin	88	0169	57-92-1	Streptomycin
			3810-74-0	Streptomycin sulfate
Sulfotepp	88	0338	3689-24-5	Sulfotepp
Sulfur	82	0031	7704-34-9	Sulfur
Sulfuryl Fluoride	85	0176	2699-79-8	Sulfuryl fluoride
Sulprofos	81	0076	35400-343-2	Sulprofos
Sumithrin	87	0426	26002-80-2	Phenothrin
Tebuthiuron	87	0054	34014-18-1	Tebuthiuron
Telone	86	0328	542-75-6	1,3-Dichloropropene
Temephos	81	0006	3383-96-8	Temephos
Terbacil	82	0039	5902-51-2	Terbacil
Terbufos	88	0109	13071-79-9	Terbufos
Terbutryn	86	0085	886-50-0	Terbutryn
Terrazole	80	0009	2593-15-9	Etridiazole
Tetrachlorvinphos	88	0321	22248-79-9	Tetrachlorvinphos
Thiophanate-Ethyl	85	0378	23564-06-9	Thiophanate
Thiram	84	0122	137-26-8	Thiram
TPTH	84	0280	76-87-9	Triphenyltin hydroxide

TABLE 8.1   List A under Phase 1 (*Continued*)

Registration standard	Date	Case	CAS No.	Chemical/Common name
Trichlorfon	84	0104	52-68-6	Trichlorfon
Trifluralin	87	0179	1582-09-8	Trifluralin
Trimethacarb	85	0112	2686-99-9	3,4,5-Trimethylphenyl methylcarbamate
			2655-15-4	2,3,5-Trimethylphenyl methylcarbamate
Vendex	87	0245	13356-08-6	Fenbutatin-oxide
Warfarin & its Na salt	81	0011	81-81-2	Warfarin
			129-06-6	Sodium warfarin
Zinc Phosphide	82	0026	1314-84-7	Zinc phosphide
2,4-D	88	0073	94-75-7	2,4-Dichlorophenoxyacetic acid
			3766-27-6	Lithium 2,4-dichlorophenoxyacetate
			2702-72-9	Sodium 2,4-dichlorophenoxyacetate
			2307-55-3	Ammonium 2,4-dichlorophenoxyacetate
				Alkanol* amine 2,4-dichlorophenoxyacetate *(salts of the ethanol and isopropanol series)
			2212-54-6	Alkyl* amine 2,4-dichlorophenoxyacetate *(100% Cl2)
			28685-18-9	Alkyl* amine 2,4-dichlorophenoxyacetate *(100% Cl4)
				Alkyl* amine 2,4-dichlorophenoxyacetate *(as in fatty acids of tall oil)
			5742-19-8	Diethanolamine 2,4-dichlorophenoxyacetate
			20940-37-8	Diethylamine 2,4-dichlorophenoxyacetate
			2008-39-1	Dimethylamine 2,4-dichlorophenoxyacetate
			53535-36-7	N,N-Dimethyloleylamine 2,4-dichlorophenoxyacetate
			3599-58-4	Ethanolamine 2,4-dichlorophenoxyacetate
			37102-63-9	Heptylamine 2,4-dichlorophenoxyacetate
			6365-72-6	Isopropanolamine 2,4-dichlorophenoxyacetate
			5742-17-6	Isopropylamine 2,4-dichlorophenoxyacetate
			6365-73-7	Morpholine 2,4-dichlorophenoxyacetate
			2212-59-1	N-Oleyl-1,3-propylenediamine 2,4-dichlorophenoxyacetate
			2212-53-5	Octylamine 2,4-dichlorophenoxyacetate
			2569-01-9	Triethanolamine 2,4-dichlorophenoxyacetate
			2646-78-8	Triethylamine 2,4-dichlorophenoxyacetate
			32341-80-3	Triisopropanolamine 2,4-dichlorophenoxyacetate
			55256-32-1	N,N-Dimethyl oleyl-linoleyl amine 2,4-dichlorophenoxyacetate
			1928-57-0	Butoxyethoxypropyl 2,4-dichlorophenoxyacetate
			1929-73-3	Butoxyethyl 2,4-dichlorophenoxyacetate
			1928-45-6	Butoxypropyl 2,4-dichlorophenoxyacetate
			94-80-4	Butyl 2,4-dichlorophenoxyacetate
			1713-15-1	Isobutyl 2,4-dichlorophenoxyacetate
			1928-43-4	Isooctyl(2-ethylhexyl) 2,4-dichlorophenoxyacetate
			53404-37-8	Isooctyl(2-ethyl-4-methylpentyl) 2,4-dichlorophenoxyacetate
			1917-97-1	Isooctyl(2-octyl) 2,4-dichlorophenoxyacetate
			94-11-1	Isopropyl 2,4-dichlorophenoxyacetate
			1320-18-9	Propylene glycol butyl ether 2,4-dichlorophenoxyacetate
2,4-DB	88	0196	94-82-6	4-(2,4-Dichlorophenoxy)butyric acid
			10433-59-7	Sodium 4-(2,4-dichlorophenoxy)butyrate

TABLE 8.1   List A under Phase 1 (*Continued*)

Registration standard	Date	Case	CAS No.	Chemical/Common name
			2758-42-1	Dimethylamine 4-(2,4-dichlorophenoxy)butyrate
			32357-46-3	Butoxyethanol 4-(2,4-dichlorophenoxy)butyrate
			6753-24-8	Butyl 4-(2,4-dichlorophenoxy)butyrate
			1320-15-6	Isooctyl 4-(2,4-dichlorophenoxy)butyrate
2,4-DP	88	0294	120-36-5	2-(2,4-Dichlorophenoxy)propionic acid
			53404-32-3	Dimethylamine 2-(2,4-dichlorophenoxy)propionate
			53404-31-2	Butoxyethyl 2-(2,4-dichlorophenoxy)propionate
			28631-35-8	Isooctyl 2-(2,4-dichlorophenoxy)propionate
6-12	81	0654	94-96-2	2-Ethyl-1,3-hexanediol

SOURCE: 54 FR (*Federal Register*) 7740, February 22, 1989.

correct the rejected application or request a waiver of permit conditions within 30 days of receipt of such notification.

The EPA administrator may issue an experimental-use permit only if it is determined that such a permit is needed to gather information that will be required for the formal registration of a pesticide. The use of a pesticide under an experimental-use permit, however, is subject to the terms and conditions of such a permit. Moreover, the use of the pesticide under this permit must be carried out only under the supervision of the EPA administrator.

The EPA administrator is also allowed to authorize any state to issue an experimental-use permit for a pesticide provided it is issued under appropriate terms and conditions. The EPA administrator may also issue an experimental-use permit to any public or private agricultural research agency or educational institution for a period not exceeding 1 year (or any other prescribed period).

8.8   Trade Secrets and Confidentiality

Section 10 of FIFRA includes a provision to safeguard against the public disclosure of trade secrets. The applicant desiring confidentiality of submitted data, in whole or in part, is required to

- Mark clearly those portions of the data that are considered trade secrets or commercial or financial information.

- Submit such marked data or material separately from the rest of the application.

The EPA administrator will not make public any information that may contain or relate to trade secrets or commercial or financial information obtained from an applicant as long as such confidentiality is properly requested.

The EPA administrator is generally bound by the applicant's own determination of confidentiality. If any release of information relating to the pesticide products is judged necessary by the EPA administrator to carry out the statutory obligations, a notice by EPA will be issued to the applicant. Such a notice to the applicant will include information relative to the public release of supporting data. This is required in order to allow the applicant to seek a declaratory judgment in the event of a dispute over such disclosure.

It has been EPA's position that any restricted use and release of supporting data and information should be limited only to certain matters, such as manufacturing, processing, and product formulations. EPA takes the position that information relating to product efficiency, hazard, and safety is fully open to public disclosure. The court decisions since the amendment of section 10 of FIFRA in 1978 have yet to reflect on this position unilaterally assumed by EPA.

8.9 Reregistration Fees

The 1988 FIFRA Amendments established specific reregistration and maintenance fees. These fees are separate from the original application fee. However, the 1990 FIFRA Amendments provided a reduction or waiver of the fee for minor agricultural use.

The reregistration fees are required to be paid in two stages. The first payment is during phase 2 of the reregistration process whereas the second payment is due during phase 3. According to the statute, the fees will fall in the range of $50,000 to $100,000 per active ingredient present in a pesticide for a major food or feed use. If the active ingredient is included in lists B, C, or D, the final fee will rise to $150,000.

In addition, the statute authorizes an annual maintenance fee of $425 per product for most pesticides during the period March 1, 1989, through 1997. This will apply to those registrants with 50 or fewer products along with a maximum ceiling of $20,000 for the maintenance fee per year. For registrants with more than 50 but not exceeding 200 pesticide products, the maintenance fee will be $100 per product, starting with the 51st product. For registrants with more than 200 products, there will be no maintenance fee for products starting with the 201st product.

EPA is authorized to adjust the maintenance fee in order to generate $14 million each year. The 1988 FIFRA Amendments, however, prohibit EPA from imposing any other type of user fee for pesticide registration until 1997.

8.10 Indemnification

Section 15 of FIFRA contains both general and specific indemnification provisions for registrants, end users, dealers, and distributors in commerce.

In a bold departure from the prior amendments to FIFRA, the 1988 FIFRA Amendments have eliminated payment to the manufacturers in the form of indemnity unless it is specifically appropriated by Congress. If so appropriated, the EPA administrator will make an indemnity payment to a registrant after notifying the registrant that its registration of a pesticide is being suspended in order to prevent an imminent hazard.

There are three general requirements for indemnities:

- The registrant is to be notified by EPA that the registration of a pesticide has been suspended to prevent an imminent hazard.

- The registration of the pesticide is to be canceled due to a final EPA determination that the use of that pesticide will create an imminent hazard.

- Any person who owned any quantity of such pesticide immediately before the notice to the registrant suffered losses by reason of such suspension or cancellation of the registration.

If the above criteria are satisfied, the EPA administrator will make an indemnity payment to the person who suffered such losses. However, the payment will not be made if the EPA administrator finds that such person had knowledge that the pesticide was not approved for registration by EPA and continued to produce such pesticide without giving a timely notice of such facts to EPA.

Under FIFRA, there are specific indemnification provisions for end users, dealers, and distributors. An end user will qualify for indemnification if

- The registrant is notified that EPA intends to suspend the registration of a pesticide or that an emergency order for suspension of the registration has been issued.

- The registration in question is suspended under emergency order and is canceled or its classification or label is changed upon appropriate public hearings and scientific review or following a voluntary cancellation and publication of a notice to that effect.

- The end user who owned any quantity of the pesticide in question immediately before the notice to the registrant solely for the purpose of using it as an end user and not for distributing, selling, or further processing for distribution or sale and suffered a loss due to the suspension or cancellation of the pesticide.

For wholesalers, dealers, and commercial distributors holding any quantity of suspended or canceled pesticides, reimbursements from the seller are limited to the costs of acquiring the pesticides from the seller, unless the seller, at the time of distribution or sale, provided in writing that the pesticide was not subject to reimbursement.

The costs of transportation are not subject to such reimbursement. However, a registrant who distributes or sells a pesticide directly to a person who is not an end user and who cannot use or resell any quantity of the pesticide, because of the suspension or cancellation of the pesticide, is required to reimburse that person for the cost of first acquiring that pesticide.

Thus, an indemnity payment to any person will generally be determined on the basis of cost of the pesticide. But such payment will exclude the cost of transportation to the person owning such pesticide immediately before the issuance of a suspension or cancellation notice to the registrant. Under no circumstances can such payment to any person exceed the fair market value of the pesticide immediately before the issuance of the EPA notice for suspension or cancellation.

Under this federal indemnification provision, it is the end user of a suspended or canceled pesticide who is eligible for reimbursement from the Justice Department's Judgment Fund, appropriated under section 1304 of Title 31 of the United States Code. As a result, such payments do not require specific congressional appropriation, but only a certification from the U.S. Attorney General. All other persons will have to seek reimbursements from the registrant, wholesaler, dealer or distributor in commerce. There is no statutory bar for reimbursement under an arm's-length contractual provision entered between a manufacturer or registrant and the wholesaler, dealer, or distributor.

8.11   Imports and Exports

Section 17 of FIFRA deals with pesticides and devices intended for import and export. In this context, the term *device* means any instrument or contrivance which is intended for trapping, destroying, repelling, or mitigating any pest or certain forms of plant or animal life, excluding human beings and microorganisms that are found in living humans or animals. For export purposes, all pesticides, devices, and active ingredients used in producing pesticides may be prepared or packed according to the specifications or directions of the foreign buyers as long as they conform to the statutory requirements for label and labeling and are not misbranded. In the event of an unregistered pesticide, the foreign purchaser is required to sign a statement acknowledging that the purchase is being made with the knowledge that such pesticide is not registered for use and thus cannot be sold in the United States. This must be done prior to the export of the pesticide. Additionally, a copy of the statement must be forwarded to an appropriate government official of the importing country.

A pesticide will be considered misbranded if

- Its labeling bears any statement, design, or graphic representation that is false or misleading.

- It is contained in package or other container or wrapping which does not conform to the standards established by EPA.

- It is an imitation of or is offered for sale under the name of another pesticide.

- Its label does not bear the registration number assigned to the establishment or facility where it was produced or manufactured.

- Any word, statement, or other information required by the statute or the regulations promulgated thereunder to appear on the label or labeling is not prominently or conspicuously displayed.

- The accompanying label does not contain directions for use which are necessary for effecting the intended use of the product.

- The label lacks a warning or cautionary statement which is necessary to protect public health and the environment.

- A pesticide is not properly registered under the statutory requirements, and its label does not prominently or conspicuously display the following: "Not Registered for Use in the United States of America."

Whenever any registration of a pesticide is suspended or canceled, the EPA administrator will transmit such notification to the governments of the importing countries and to appropriate international agencies through the Department of State.

For the purposes of importation of pesticides and devices from foreign countries, the Secretary of the Treasury is responsible for notifying EPA about the arrival of such imported products. Upon request, samples of such pesticides or devices will be delivered to the EPA administrator with appropriate notice to the consignee or owner of the imported items. This offers the consignee or owner an opportunity to appear before the EPA administrator and exercise applicable rights to introduce testimony pertaining to the imported products. An imported item may be refused admission to the United States if an examination of a representative sample indicates that it is adulterated or is misbranded or otherwise violates the statutory provisions. It may also be refused admission if it is found to be injurious to public health or to the environment. In such circumstances, the Secretary of the Treasury may refuse delivery of the imported product to the consignee and may also destroy the imported item.

The consignee cannot export such item within 90 days from the date of notice of admission refusal. The Secretary of the Treasury, however, may deliver the imported product to the consignee pending examination and decision in the matter or upon an execution of bond for the full invoice amount of the product together with any duty levied on it. All charges for storage, cartage, and labor on pesticides or devices which are refused admission or delivery to the consignee or owner will have to be paid by the consignee or owner. Any default in such payment will constitute a lien against any future importation by such consignee or owner.

8.12 Transportation, Storage, Disposal, and Recall

Section 19 of FIFRA contains the statutory provisions for transportation, storage, recall, and disposal that may occur as a result of a suspension or cancellation order. The EPA administrator is empowered to require a registrant or an applicant for registration of a pesticide to submit or cite data or relevant information regarding methods for safe storage and disposal of excess quantities of a pesticide. This requirement may apply for registration as well as reregistration. In addition, the EPA administrator may impose several requirements for the labeling of a pesticide. These include

- Requirements and procedures for the transportation, storage, and disposal of the pesticide

- Requirements and procedures for the transportation, storage, and disposal of the pesticide container

- Requirements and procedures for the transportation, storage, and disposal of the rinsate containing the pesticide

- Requirements and procedures for the transportation, storage, and disposal of any material used to contain or collect excess or spilled quantities of the pesticide

In addition to being suspended and canceled, a pesticide may also be recalled in order to protect public health or the environment. Such a recall may be either voluntary or mandatory.

In a voluntary recall, the EPA administrator will request the registrant to submit within 60 days a plan for voluntary recall of the pesticide. When such plan is approved, an order will be issued by EPA to the registrant to conduct its recall as planned. If the submitted plan by the registrant is determined inadequate to protect public health or the environment, the EPA administrator may hold an informal hearing.

In a mandatory recall, instead of requesting the submission of a plan by the registrant, the EPA administrator issues a regulation whereby a plan for the recall of the pesticide is prescribed. Such a regulation, if issued, may apply to any person who is or was a registrant, distributor, or seller of the pesticide or any successor-in-interest to such a person.

In essence, FIFRA incorporates a system of shared responsibility for storage, transportation, disposal, and recall of pesticides whose registrations have been suspended or canceled. EPA has been given broad authority to regulate pesticides that must be disposed of appropriately to protect public health and the environment.

8.13 Inspection and Enforcement

Section 9 of FIFRA authorizes representatives of EPA and representatives of authorized states to enter facilities or establishments at reasonable times for the purpose of inspection and obtaining samples. Such an entry is allowed to

- Any establishment or location where pesticides or devices are held for distribution or sale

- Any place where any suspended or canceled pesticide is being held for the purpose of determining its compliance with section 19 of FIFRA, which deals with the storage, disposal, transportation, and recall of pesticides

The government inspectors, upon entry, may obtain samples of the pesticides or devices which are packaged, labeled, and released for shipment including samples of the containers or labeling. However, prior to an entry for the purpose of such inspection, an authorized agency representative must present appropriate credentials and a written statement setting forth the reasons for the inspection to the owner, operator, or agent in charge of the establishment or other place where pesticides or devices are held for distribution or sale. Such a statement is required to provide an indication if a violation of the law is suspected. If no violation is suspected, alternate and sufficient reasons for the inspection must be provided.

Additionally, authorized agency representatives may obtain and execute warrants when there is reason to believe that statutory provisions of FIFRA have been violated. These warrants may authorize entry, inspection, and copying of records along with the seizure of any pesticide or device which is found in violation of the law.

Section 13 of FIFRA empowers the EPA administrator to issue an order prohibiting sale or use in conjunction with removal and seizure of a pesticide or device that is found to be in violation of any of the statutory provisions. The EPA administrator may issue a written or printed "stop sale, use, or removal" order to any person who owns, controls, or has custody of a pesticide or device which has been, or is intended to be, distributed or sold in violation of the law. Once such an order has been received, no person can sell, use, or remove the pesticide or device except in accordance with the provisions of the order. Any such prohibited pesticide that remains unsold or in original unbroken packages may be subject to seizure and disposal if a court so directs.

8.14 Penalties and Sanctions

Section 14 of FIFRA sets forth the provisions for both civil and criminal penalties. In general, any registrant, commercial applicator, wholesaler, dealer, or other distributor who violates any provision of the statute may be fined a civil penalty of not more than $5000 for each offense. For a private applicator, this civil penalty may not exceed $1000 for each offense. No civil penalty can be imposed, however, unless the person charged has been given proper notice and an opportunity for a hearing in the county, parish, or city where such person resides.

A registrant, applicant for a registration, or producer of a pesticide who knowingly violates any provision of FIFRA may also be subject to a criminal sanction. This can include a fine of not more than $50,000 or imprisonment for up to 1 year or both. For a commercial applicator of a restricted-use pesticide who knowingly violates any statutory provision, the criminal penalty can include a fine of not more than $25,000 or imprisonment of up to 1 year or both. For a private applicator charged with such a violation, criminal penalties may include a fine of up to $1000 or imprisonment of up to 30 days or both.

Section 26 of FIFRA authorizes primary enforcement responsibility for pesticide use violations to the states, provided such states

- Have adopted adequate pesticide use laws and regulations
- Have adopted and are implementing adequate procedures for the enforcement of such state laws and regulations
- Will keep appropriate records and submit such reports showing compliance to the EPA administrator

The caveat here is that such authorized states must have state pesticide laws no less stringent than what the statutory provisions of FIFRA call for.

8.15   Pesticide Residues in Food

One of the most thorny issues surrounding pesticides is their safe level in food and feed products. EPA has been examining ways in which a pesticide or pesticide product may be assessed for health risks. The debate is still continuing as to how much risk is acceptable for pesticides that may remain in food as residues. While no regulatory framework has been developed, it is important to remember, as indicated earlier, that FIFRA provisions reflect risk-benefit balancing. As a result, EPA will have to consider both risks and benefits associated with the continued use of a pesticide, including the likelihood that residues of such a pesticide may enter the human food chain or animal feedlot through crop harvesting and food processing.

The Federal Food, Drug, and Cosmetic Act (FFDCA) is designed to ensure food safety by prohibiting the sale of "adulterated food" that contains an unsafe "food additive." Note that while formulating the public health policy in 1958, Congress adopted Representative Delaney's amendment which prohibited any food additive that had been shown to induce cancer in humans or in laboratory animals. The Delaney Clause is included in section 409 of FIFRA which provides the limiting conditions of a substance that may be used as a food additive. The same yardstick may be used by EPA for regulating pesticides. It has been EPA's position that the Delaney clause precludes the setting of any pesticide tolerance in processed food if it is determined to be cancer-causing. This EPA position is based on the Delaney clause's establishing a "no-risk standard" for carcinogenic food additives. In this context, the Food and Drug Administration has the responsibility for monitoring the levels of pesticides in foods and food products. FDA is also the enforcer of food safety under the Federal Food, Drug, and Cosmetic Act. Section 348(c)(3)(A) of FFDCA deals with the issue of "safe use" of food additives which can also apply to the safe levels of pesticides in foods. Thus, under the current regulatory framework, pesticide residues in foods are regulated by both EPA and FDA. The Court of Appeals for the Ninth Circuit upheld the Delaney clause in *Les v. Environmental Protection Agency* by ruling that only Congress has the power to change the current ban against the use of pesticides that cause cancer. The court held that this is true even if the cancer risk is *de minimis*.

EPA's criteria for determination of pesticidal activity are set forth in 40 CFR section 153.125. In addition, EPA has developed a list of substances that have been determined to be pesticidally inert (i.e., do not contain any active ingredient). Table 8-2 is a reproduction of this EPA list.

8.16   Reporting and Record Keeping

Section 8 of FIFRA requires the EPA administrator to prescribe regulations requiring maintenance of records by producers, registrants, and applicants for

TABLE 8.2 Pesticidally Inert Substances

Substance	Uses	Substance	Uses
Acetone	Solvent.	Petroleum distillate, oils, hydrocarbons, also paraffinic hydrocarbons, aliphatic hydrocarbons, paraffinic oil.	Lubricant, solvent.
Alkyl* amino betaine (*46 percent C_{12}, 24 percent C_{14}, 10 percent C_{16}, 6 percent C_{10}, 7 percent C_8, 5 percent C_{18}).	Corrosion inhibitor, surfactant.	Polyoxyethylene sorbitol, mixed ethyl ester of.	Emulsifier.
Alkyl monoethanolamide	Emulsifier.	Polyvinylpyrrolidone	Emulsifier.
Aluminum chloride	Detergent.	Potassium bisulfate	Builder.
Aluminum hyroxybenzenesulfate sulfonate.	Emulsifier.	Potassium carbonate	Detergent.
Aluminum powder	Filler.	Potassium dodecylbenzenesulfonate	Anionic detergent.
Aluminum carbonate	Detergent.	Potassium laurate	Emulsifier.
Ammonium citrate	Sequestrant.	Potassium myristate	Emulsifier.
Ammonium lauryl sulfonate	Emulsifier.	Potassium N-(s-(nitroethyl)benzyl) ethylenediamine.	Emulsifier.
Ammonium oleate	Detergent, emulsifier.	Potassium phosphate, tribasic	Sequestrant.
		Potassium ricinoleate	Emulsifier.
Ammonium oxalate	Detergent.	Potassium toluene sulfonate	Detergent.
Amyl acetate	Diluent.	Potassium xylene sulfonate	Detergent.
Borax	Detergent.	Propanol (propyl alcohol)	Solvent, except in tinctures or where sole or major ingredient.
Butyl alcohol, tertiary	Solvent, odorant.		
Carbon	Carrier, absorbent.		
Castor oil	Emulsifier.		
Citric acid	Sequestrant.	Soap	Detergent.
Diethanolamine dodecylbenzene sulfonate.	Detergent.	Sodium acetate	Buffer.
		Sodium alkyl (100 percent C_9): benzene sulfonate	Detergent.
Sodium oleate	Emulsifier.		
Dimethyl phthalate	Perfume.	Sodium bicarbonate	Detergent.
Disodium monoethanolamine phosphate	Emulsifier.	Sodium carbonate	Detergent.
Dodecyl benzene sulfonic acid	Detergent.	Sodium chloride	Builder.
Essential oils	Perfume.	Sodium decylbenzene sulfonate	Detergent.
Ethanol (ethyl alcohol)	Solvent, except in tinctures or where sole or major ingredient.	Sodium diacetate	Sequestrant.
		Sodium dihydroxyethylglycine	Chelate, buffer.
		Sodium diisopropylnaphthalene sulfonate.	Detergent.
Ethanolamine	Emulsifier.	Sodium di(monoethanolamine) phosphate.	Emulsifier.
Ethanolamine dodecylbenzene sulfonate.	Detergent.		
Ethoxylated lanolin	Ointment base.	Sodium dodecylbenzene sulfonate (may be active as a sanitizer in dishwashing formulations).	Detergent.
Ethylenediamine	Emulsifier.		
Ethylenediaminetetraacetic acid (including all salts and derivatives).	Sequestrant.	Sodium dodecyl diphenyl oxide sulfonate.	Perfume.
Fumaric acid	Sequestrant.	Sodium glycolate	Sequestrant.
Gluconic acid	Buffer.	Sodium laurate	Detergent.
Isooctyl phenoxy polyethoxy ethanol	Surfactant.	Sodium N-lauroylsarcosinate	Detergent.
Isopropanol (isopropyl alcohol)	Solvent, except in tinctures, or where sole or major ingredient.	Sodium lauryl sulfate	Detergent.
		Sodium metasilicate	Detergent.
		Sodium N-methyl-N oleyltaurate	Emulsifier.
		Sodium mono and dimethyl naphthalene sulfonate.	Detergent.
Isopropyl myristate	Solvent.		
Juniper tar	Odorant.	Sodium oleate	Emulsifier.
Lauryl alcohol	Detergent, odorant.	Sodium phosphate	Emulsifier, buffer.
		Sodium salt of turkey red oil	Emulsifier.
Lauryl methacrylate	Emulsifier.	Sodium sesquicarbonate	Detergent.
Limonene	Odorant, perfume.	Sodium silicate	Detergent.
Magnesium chloride	Builder.	Sodium sulfate	Detergent.
Magnesium lauryl sulfate	Detergent.	Sodium sulfonated oleic acid	Emulsifier.
Magnesium silicate	Odor absorbent.	Sodium thiosulfate	Builder.
Menthol	Perfume.	Sodium toluene sulfonate	Detergent.
Methanol (methyl alcohol)	Solvent, except in tinctures, or where sole or major ingredient.	Sodium tripolyphosphate	Sequestrant.
		Sodium xylene sulfonate	Detergent.
		Tetrapotassium pyrophosphate	Sequestrant.
		Tetrasodium pyrophosphate	Sequestrant.
Methyl ethyl ketone	Solvent.	Toluene sulfonic acid	Emulsifier.
Methyl salicylate	Perfume, odorant.	1,1,1-Trichloroethane	Diluent.
Mineral oil, mineral seal oil, or white mineral oil.	Lubricant.	Triethanolamine	Emulsifier.
		Triethanolamine dodecylbenzene sulfonate.	Detergent.
Monoethanolamides of the fatty acids of coconut oil.	Emulsifier.		
		Triethanolamine laurate	Emulsifier.
Monosodium phosphate	Emulsifier, buffer.	Triethanolamine lauryl sulfate	Emulsifier.
Morpholine	Corrosion inhibitor.	Triisopropanolamine	Emulsifier.
Nonylphenoxypolyethoxyethanol	Surfactant.	Triisopropylamine	Emulsifier.
Octylphenol	Nonionic surfactant.		
		Trisodium phosphate	Detergent.
Oil of citronella	Perfume, odorant.	Turkey red oil	Emulsifier
		Undecylenic acid	Perfume.
Oil of eucalyptus	Perfume.	Xylene	Solvent.
Oil of lemongrass	Perfume.	Zirconium oxide	Dye.
Oleic acid	Solvent.		

registration regarding their operations as well as production of pesticides and devices. In this context, the statute specifically exempts the maintenance of records for financial data, sales data other than shipment data, pricing data, personnel data, and research data. The statute provides that all retailers, sellers, or distributors of a pesticide must notify EPA and other appropriate state and local authorities about any possession of a pesticide whose registration has been suspended or canceled. See App. 8A.

In addition, FIFRA includes several record-keeping requirements. In the past, these were required only of pesticide producers, but at present, under the statutory requirements, all registrants and applicants for registration must keep appropriate production and business records of the pesticides and devices they produce. Such records are to be made available to EPA upon request. Failure to submit required information pursuant to 40 CFR §§153.66, 153.69 and 153.78 is an actionable violation of section 6(a)(2) of FIFRA. For such a violation, EPA may recommend or seek to impose a civil and/or criminal penalty.

8.17   Applicators for Restricted-Use Pesticides

EPA may classify or not classify a pesticide product as set forth in 40 CFR part 152, subpart I. A pesticide may be classified for restricted use or for general use. A restricted-use pesticide may only be used by a certified applicator, or under the direct supervision of a certified applicator. Furthermore, EPA may, by regulation, prescribe restrictions with respect to the pesticide product's composition, labeling, packaging, use, or its distribution and sale, or to the status or qualifications of its user.

Section 11 of FIFRA describes the certification procedure for the use of restricted-use pesticides by applicators. The term *certified applicator* means any individual who is certified pursuant to the reregistration provisions of the statute set forth in section 4 of FIFRA. A certified applicator is authorized to use and supervise the use of any pesticide which is classified by EPA for restricted use.

This certification may be issued by a state whose plan for applicator certification has been approved by EPA. If the EPA administrator rejects a certification plan submitted by a state, EPA will provide such a state due notice and an opportunity to be heard before rejecting such a plan. Any plan for certification of the applicators of pesticides in a state that is without an approved plan must conform to the requirements set forth in the statute for state certification. Such certification plan is required to conform with the state plan that has been approved by EPA for certification purposes.

Appendix 8B reproduces EPA's list of pesticides earmarked for restricted use and set forth in 40 CFR §152.175. These listed pesticide products are classified for limited use by or under the direct supervision of a certified applicator.

8.18   Relationship with Other Federal Statutes

FIFRA is by no means the only federal statute to regulate pesticides. Besides FIFRA, pesticides may be regulated under several other federal statutes, which include but are not limited to

- Section 112 of the Clean Air Act Amendments of 1990 which deals with air toxics

- Sections 301 and 307 of the Federal Water Pollution Control Act of 1972, as amended, which impose permit requirements and application of certain treatment standards for wastewaters prior to their discharge

- Section 3004 of the Solid Waste Disposal Act, as amended by the Resource Conservation and Recovery Act of 1976 for activities involving pesticide storage and treatment as well as handling and disposal of pesticide wastes including pesticide containers

- Section 342 of the Federal Food, Drug, and Cosmetic Act of 1958 which set the tolerance limits for pesticide residues in food

- Section 6 of the Occupational Safety and Health Act of 1970 which set forth the hazard standards for the protection of agricultural workers from exposure to pesticides

- Section 1262 of the Poison Prevention Packaging Control Act of 1970, which authorizes the Consumer Product Safety Commission to declare a substance to be hazardous for bearing or containing a pesticide in order to protect consumers, including children.

Besides EPA, there are several other federal agencies which regulate or otherwise influence applicable federal regulations to control the production, sale, or use of pesticides in the United States, such as the Food and Drug Administration, the Department of Agriculture, the Occupational Safety and Health Administration, the Consumer Product Safety Commission, and the U.S. Fish and Wildlife Service. Nonetheless, the statute gives EPA the major regulatory responsibility over the pesticides.

8.19 Preemption of Local Legislation by FIFRA

Section 24(a) of FIFRA provides that a state may regulate the "sale or use of any federally registered pesticide or device" in the state. However, the statute permits this only if and to the extent that such state regulation does not permit any sale or use that is prohibited by FIFRA.

This statutory construction has given rise to legal debates whether FIFRA preempts local legislation of pesticides. The question whether FIFRA preempts local legislation has never been fully answered by the U.S. Congress. Addtionally, there has been certain shifting in EPA's position on FIFRA preemption. EPA currently allows certain flexibility to the state agencies which delegate their authority to the local government units. Moreover, in *Wisconsin Public Intervenor v. Mortier,* the Supreme Court held that FIFRA does not bar or exclude local governments from passing controls regarding the use of pesticides.

Apart from the above, cities, towns, counties, and other political subdivisions, by taking advantage of the home rule, have passed ordinances which impose local controls on pesticides. However, the propriety of such local legislation is being argued by legal scholars. Several recent court decisions have added to this controversy, including the Supreme Court's decision in *Papas v. Upjohn Co.* The *Papas* Court remanded a federal appeals court decision and vacated the court's earlier decision which held that FIFRA impliedly preempts tort claims based on inadequate labeling. The Supreme Court asked the appeals court to review the case in light of *Cipollone v. Liggett Group Inc.* In *Cipollone,* the Supreme Court held that congressionally mandated warnings on cigarette packages do not insulate tobacco manufacturers from state tort actions.

This has created additional confusion since section 24(b) of FIFRA indicates that a state may not impose or continue in effect any requirements for labeling or packaging which is different from or in addition to what is required under FIFRA.

Therefore, any final opinion on FIFRA's preemption must wait until Congress clarifies the current statutory provisions in sections 24(a) and 24(b) of FIFRA through appropriate amendments, or the Supreme Court renders a direct ruling on the subject of federal preemption of state and local legislation by FIFRA.

Recommended Reading

1. Janice L. Greene, *Regulating Pesticides: FIFRA Amendments of 1988,* Special Report, Bureau of National Affairs, Washington, 1989.

2. "Monitoring and Enforcement of Food Safety—An Overview of Past Studies," study by the staff of the U.S. GAO, GAO/RCED-83-153, Sept. 9, 1983.

Important Cases

1. *Papas v. Upjohn Co.,* No. 90-1837 (S. Ct. June 29, 1992).

2. *Cipollone v. Liggett Group Inc.,* No. 90-1038 (S. Ct. June 24, 1992).

3. *Wisconsin Public Intervenor v. Mortier,* No. 89-1905 (S. Ct. June 21, 1991).

4. *United States v. Dotterweich,* 320 U.S. 277, 64 S. Ct. 134, 88 L. Ed. 48 (1943).

5. *Mobay Chemical Corp. v. Costle,* 447 F. Supp. 811 (W.D. Mo. 1978), *appeal dismissed,* 439 U.S. 320, *reh denied,* 440 U.S. 940 (1979).

6. *Les v. Environmental Protection Agency,* No. 97-70234 (9th Cir. July 8, 1992).

7. *Chevron Chemical Co. v. Costle,* 443 F. Supp. 1024 (N.D. Cal. 1978).

8. *EDF v. Ruckelshaus,* 465 F. 2d 538 (D.C. Cir. 1972).

9. *EDF v. Ruckelshaus,* 439 F. 2d 584 (D.C. Cir. 1971).

10. *Riden v. ICI Americas, Inc.,* No. 89-0903-CV-W-1 (W.D. Mo. May 14, 1991).

11. *Dow Chemical Co. v. Blum,* 469 F. Supp. 892 (E.D. Mich. 1979).

APPENDIX 8A   Recommended Standard Format for the Notification of Possession of Canceled or Suspended Pesticide under FIFRA Section 6(g)

Company Name _____

Address _____

Contact Person _____

Contact Person Phone Number (___) ___ – ___

Signature of Responsible Company Official, Date _____

Pesticide Active Ingredient Subject To FIFRA §6(g) reporting: _____

Relationship(s) of the Company to the Pesticide: _____

(i.e., registrant, producer, exporter, distributor, retailer, commercial applicator, applicant for registration, applicant or holder of an experimental use permit.)

Location 1 of Pesticide

Address _____

EPA Reg. # _____
___ Units of ___ lbs/gal containers
___ Units of ___ lbs/gal containers
___ Units of ___ (___) containers

EPA Reg. # _____
___ Units of ___ lbs/gal containers
___ Units of ___ lbs/gal containers
___ Units of ___ (___) containers

EPA Reg. # _____
___ Units of ___ lbs/gal containers
___ Units of ___ lbs/gal containers
___ Units of ___ (___) containers

Total Quantity of Location 1:
_____ lbs
_____ gal
_____ (___)

Location 2 of Pesticide

Address _____

EPA Reg. # _____
___ Units of ___ lbs/gal containers
___ Units of ___ lbs/gal containers
___ Units of ___ (___) containers

EPA Reg. # _____
___ Units of ___ lbs/gal containers
___ Units of ___ lbs/gal containers
___ Units of ___ (___) containers

EPA Reg. # _____
___ Units of ___ lbs/gal containers
___ Units of ___ lbs/gal containers
___ Units of ___ (___) containers

Total Quantity of Location 2:
_____ lbs
_____ gal
_____ (___)

Location of 3 Pesticide

Address _____

EPA Reg. # _____
___ Units of ___ lbs/gal containers
___ Units of ___ lbs/gal containers
___ Units of ___ (___) containers

EPA Reg. # _____
___ Units of ___ lbs/gal containers
___ Units of ___ lbs/gal containers
___ Units of ___ (___) containers

EPA Reg. # _____
___ Units of ___ lbs/gal containers
___ Units of ___ lbs/gal containers
___ Units of ___ (___) containers

Total Quantity of Location 3:
_____ lbs
_____ gal
_____ (___)

SOURCE: "Proposed Rules," *Federal Register*, Vol. 56, No. 60, March 28, 1991.

APPENDIX 8B List of Restricted Use Pesticides

Active ingredient	Formulation	Use pattern	Classification[1]	Criteria influencing restriction
Acrolein	As sole active ingredient. No mixtures registered.	All uses	Restricted	Inhalation hazard to humans. Residue effects on avian species and aquatic organisms.
Acrylonitrile	In combination with carbon tetrachloride. No registrations as the sole active ingredient.	do	do	Other hazards—accident history of both acrylonitrile and carbon tetrachloride products.
Aldicarb	As sole active ingredient	Ornamental uses (indoor and outdoor).	do	Other hazards—accident history.
Allyl alcohol	No mixtures registered	Agricultural crop uses	Under further evaluation	
	All formulations	All uses	Restricted	Acute dermal toxicity.
Aluminum phosphide	As sole active ingredient. No mixtures registered.	do	do	Inhalation hazard to humans.
Azinphos methyl	All liquids with a concentration greater than 13.5 pct.	do	do	Do.
	All other formulations	do	Under futher evaluation	
Calcium cyanide	As sole active ingredient. No mixture registered.	do	Restricted	Do.
Carbofuran	All concentrate suspensions and wettable powders 40% and greater.	do	do	Acute inhalation toxicity.
	All granular formulations	Rice	Under evaluation	
	All granular and fertilizer formulations	All uses except rice	do	
Chlorfenvinphos	All concentrate solutions or emulsifiable concentrates 21% and greater.	All uses (domestic and non-domestic)	Restricted	Acute dermal toxicity.
Chloropicrin	All formulations greater than 2%	All uses	do	Acute inhalation toxicity.
	All formulations	Rodent control	do	Hazard to non-target organisms.
	All formulations 2% and less	Outdoor uses (other than rodent control).	Unclassified	
Clonitralid	All wettable powders 70% and greater	All uses	Restricted	Acute inhalation toxicity.
	All granulars and wettable powders	Molluscide uses.	Restricted	Effects on aquatic organisms.
	Pressurized sprays 0.55% and less.	Hospital antiseptics.	Unclassified	Acute dermal toxicity.
Cycloheximide	All formulations greater than 4%	All uses	Restricted	
	All formulations 0.027% to 4%	do	Under evaluation	
	All formulations 0.027% and less	Domestic uses	Unclassified	
Demeton	1 pct fertilizer formulation, 1.985 pct granular formulation.	All uses, including domestic uses.	Restricted	Domestic uses: Acute oral toxicity. Acute dermal toxicity. Nondomestic outdoor uses. Residue effects on avian and mammalian species.
Dicrotophos	All granular formulations, emulsifiable concentrates and concentrated solutions.	All uses	do	Acute dermal toxicity. Residue effects on mammalian and avian species.
	All liquid formulations 8% and greater	All uses	Restricted	Acute dermal toxicity; residue effects on avian species (except for tree injections).
Dioxathion	All concentrate solutions or emulsifiable concentrates[2] greater than 30%.			Acute dermal toxicity.
	Concentrate solutions or emulsifiable concentrates[2] 30% and less and wettable powders 25% and less.	Livestock and agricultural uses (non-domestic uses only).	Unclassified	
	All solutions[2] 3% and greater	Domestic uses	Restricted	Do.
	2.5% solution[2] with toxaphene and malathion.	All uses	Under evaluation	

Chemical	Formulation	Use	Classification	Reason
Disulfoton	All emulsifiable concentrates 65% and greater, all emulsifiable concentrates and concentrate solutions 21% and greater with fensulfothion 43% and greater, all emulsifiable concentrates 32% and greater in combination with 32% fensulfothion and greater.	...do	Restricted	Do.
	Non-aqueous solution 95% and greater	do	Restricted	Acute inhalation toxicity.
Endrin	Granular formulations 10% and greater	Commercial seed treatment.	Restricted	Acute dermal toxicity.
	All emulsions, dusts wettable powders, pastes, and granular formulations 2 pct and above.	Indoor uses (greenhouse)	do	Acute inhalation toxicity.
		All uses	Restricted	Acute dermal toxicity. Hazard to nontarget organisms.
EPN	All concentrations less than 2 pct.	do	do	Hazard to nontarget organisms.
	All liquid and dry formulations greater than 4%.	do	do	Acute dermal toxicity; acute inhalation toxicity; residue effects on avian species.
Ethoprop	Emulsifiable concentrates 40% and greater	Aquatic uses	do	Effects on aquatic organisms.
	All granular and fertilizer formulations	Aquatic uses	do	Acute dermal toxicity.
Ethyl parathion	All granular and dust formulations greater than 2 pct, fertilizer formulations, wettable powders, emulsifiable concentrates, concentrated suspensions, concentrated solutions.	All uses	Under evaluation	
		do	Restricted	Inhalation hazard to humans. Acute dermal toxicity. Residue effects on mammalian, aquatic, avian species.
	Smoke fumigants.			
	Dust and granular formulations 2 pct and below.	do	do	Inhalation hazard to humans. Other hazards—accident history.
Fenamiphos	Emulsifiable concentrates 35% and greater	do	do	Acute dermal toxicity.
Fensulfothion	Concentrate solutions 63% and greater, all emulsifiable concentrates and concentrate solutions 43% and greater with disulfoton 21% and greater, all emulsifiable concentrates 32% and greater in combination with disulfoton 32% and greater.	do	Restricted	Do. Acute inhalation toxicity.
Fluoroacetamide/1081	Granular formulations 10% and greater	Indoor uses (greenhouse)	do	Do.
	As sole active ingredient in baits. No mixtures registered.	All uses	Restricted	Acute oral toxicity.
Fonofos	Emulsifiable concentrates 44% and greater	do	do	Acute dermal toxicity.
	Emulsifiable concentrates 12.6% and less with pebulate 50.3% and less.	Tobacco	Unclassified	
Hydrocyanic acid	As sole active ingredient. No mixtures registered.	do	Restricted	Inhalation hazard to humans.
Methamidophos	Liquid formulations 40% and greater.	do	Restricted	Acute dermal toxicity; residue effects on avian species.
Methidathion	Dust formulations 2.5% and greater	All uses except nursery stock, safflower and sunflower.	do	Residue effects on avian species.
	All formulations		do	Do.
	All formulations	Nursery stock, safflower and sunflower.	Unclassified	

Active ingredient	Formulation	Use pattern	Classification [1]	Criteria influencing restriction
Methomyl	As sole active ingredient in 1 pct to 2.5 baits (except 1 pct fly bait).	Nondomestic outdoors-agricultural crops, ornamental and turf. All other registered uses.	Restricted	Residue effects on mammalian species.
	All concentrated solution formulations.	do	do	Other hazards-accident history.
	90 pct wettable powder formulations (not in water soluble bags).	do	do	Do.
	90 pct wettable powder formulation in water soluble bags.	do	Unclassified	
	All granular formulations.	do	do	
	25 pct wettable powder formulations.	do	do	
	In 1:24 pct to 2.5 pct dusts as sole active ingredient and in mixtures with fungicides and chlorinated hydrocarbon, inorganic phosphate and biological insecticides.	do	do	
Methyl bromide	All formulations in containers greater than 1.5 lb.	All uses	Restricted	Do.
	Containers with not more than 1.5 lb of methyl bromide with 0.25 pct to 2.0 pct chloropicrin as an indicator.	Single applications (nondomestic use) for soil treatment in closed systems.	Unclassified	
	Container with not more than 1.5 lb having no indicator.	All uses	Restricted	Do.
Methyl parathion	All dust and granular formulations less than 5 pct.	do	do	Other hazards-accident history. All foliar applications restricted based on residue effects on mammalian and avian species.
	Microencapsulated	do	do	Residue effects on avian species. Hazard to bees.
	All dust and granular formulations 5 pct and greater and all wettable powders and liquids.	do	do	Acute dermal toxicity. Residue effects on mammalian and avian species.
Mevinphos	All emulsifiable concentrates and liquid concentrates.	do	do	Do.
	Psycodid filter fly liquid formulations	do	do	Acute dermal toxicity.
	2 pct dusts	do	do	Residue effects on mammalian and avian species.
Monocrotophos	Liquid formulations 19% and greater.	do	do	Residue effects on avian species. Residue effects on mammalian species.
	Liquid formulations 55% and greater.	do	do	Acute dermal toxicity. Residue effects on avian species. Residue effects on mammalian species.
Nicotine (alkaloid)	Liquid and dry formulations 14% and above.	Indoor (greenhouse).	do	Acute inhalation toxicity.
	All formulations.	Applications to cranberries.	do	Effects on aquatic organisms.
	Liquid and dry formulations 1.5% and less.	All uses (domestic and nondomestic).	Unclassified	
Paraquat (dichloride) and paraquat bis(methyl sulfate).	All formulations and concentrations except those listed below.	All uses	Restricted	Other hazards. Use and accident history, human toxicological data.

Common name	Formulations	Uses	Classification	Criteria influencing restriction
	Pressurized spray formulations containing 0.44 pct Paraquat bis(methyl sulfate) and 15 pct petroleum distillates as active ingredients.	Spot weed and grass control.	do	Acute dermal toxicity.
	Liquid fertilizers containing concentrations of 0.025 pct paraquat dichloride and 0.03 percent atrazine; 0.03 pct paraquat dichloride and 0.37 pct atrazine, 0.04 pct paraquat dichloride and 0.49 pct atrazine.	All uses	Unclassified	Residue effects on avian species (applies to foliar applications only). Residue effects on mammalian species (applies to foliar application only).
Phorate	Liquid formulations 65% and greater.	do	Restricted.	Effects on aquatic organisms.
Phosacetim	All granular formulations	Rice.	do	Hazard to non-target species.
	Baits 0.1% and greater.	All uses	Restricted.	Residue effects on mammalian species. Residue effects on avian species.
Phosphamidon	Liquid formulations 75% and greater.	do	do	Acute dermal toxicity.
	Dust formulations 1.5% and greater.	do	do	Residue effects on mammalian species. Residue effects on avian species. Do.
Picloram	All formulations and concentrations except tordon 101 R.	do	do	Residue effects on mammalian species.
	Tordon 101 R forestry herbicide containing 5.4 pct picloram and 20.9 pct 2,4-D.	Control of unwanted trees by cut surface treatment.	Unclassified	Hazard to nontarget organisms (specifically nontarget plants both crop and noncrop).
Sodium cyanide [3]	All capsules and ball formulations.	All uses	Restricted.	Inhalation hazard to humans.
Sodium fluoroacetate	All solutions and dry baits	do	do	Acute oral toxicity. Hazard to nontarget organisms. Use and accident history.
Strychnine	All dry baits, pellets and powder formulations greater than 0.5 pct.	do	do	Acute oral toxicity. Hazard to nontarget avian species. Use and accident history.
	All dry baits, pellets and powder formulations.	All uses calling for burrow builders.	do	Hazard to nontarget organisms.
	All dry baits, pellets and powder formulations 0.5 pct and below.	All uses except subsoil.	do	Do.
	do	All subsoil uses.	Unclassified	
Sulfotepp	Sprays and smoke generators	All uses	Restricted.	Inhalation hazard to humans.
Tepp	Emulsifiable concentrate formulations	do	do	Inhalation hazard to humans. Dermal hazard to humans. Residue effects on mammalian and avian species.

331

Active ingredient	Formulation	Use pattern	Classification [1]	Criteria influencing restriction
Zinc Phosphide	All formulations 2% and less	All domestic uses and non-domestic uses in and around buildings.	Unclassified	
	All dry formulations 60% and greater.	All uses	Restricted	Acute inhalation toxicity.
	All bait formulations	Non-domestic outdoor uses (other than around buildings).	...do	Hazard to non-target organisms.
	All dry formulations 10% and greater	Domestic uses	...do	Acute oral toxicity.

[1] "Under evaluation" means no classification decision has been made and the use/formulation in question is still under active review within EPA.
[2] Percentages given are the total of dioxathion plus related compounds.
[3] (NOTE.—M–44 sodium cyanide capsules may only be used by certified applicators who have also taken the required additional training.)

Toxic Substances under TSCA

9.1 Introduction

There has been a steady proliferation of the manufacture, processing, and use of wide-ranging chemicals over the last several decades, particularly since World War II. While these chemicals generally play an important role in our lives and add significantly to the overall national economy and social benefits, they also pose health and environmental risks in varying degrees.

In 1971, the President's Council on Environmental Quality (CEQ) recommended that the administrator of the Environmental Protection Agency be empowered "to restrict the use or distribution of any substance which [the administrator] finds hazardous to human health or to the environment." After several attempts by Congress, the Toxic Substances Control Act (TSCA) of 1976 was enacted. By passing TSCA, Congress set in motion a statutory framework for evaluating the effects of all new chemical substances on human health, living organisms including microbes, and the environment.

TSCA is designed to improve our understanding of chemicals and chemical substances and to provide appropriate controls and safeguards to ensure public and environmental safety. Pursuant to TSCA, EPA is authorized to obtain from the industry appropriate data on products as well as their use, health effects, and other related health and safety matters that may concern chemicals, chemical substances, and chemical mixtures. Furthermore, EPA is authorized to require appropriate testing of these chemicals in order to gather necessary information. Under this statutory scheme, if warranted, EPA may regulate the manufacturing, processing, distribution in commerce (including import), use, and disposal of chemicals, new or old, and chemical mixtures. TSCA actually authorizes EPA with regulatory control over the manufacture of chemicals that may or may not turn out to be toxic and well before their entry into the environment. This is a significant departure from EPA's statutory authority under earlier legislation, e.g., the Federal Insecticide, Fungicide, and Rodenticide Act (FIFRA) which regulates pesticides inherently designed to be toxic.

EPA's authority under TSCA is extended over the toxic substances as provided in section 112 of the Clean Air Act (CAA), section 307 of the Federal Water Pollution Control Act (FWPCA), and section 6 of the Occupational Safety and Health (OSH) Act. TSCA arms EPA with significant new authority not included in other federal statutes to anticipate and address chemical risks and to regulate their manufacturing, commercial distribution, and ultimate use by and for the general public. As a result, EPA may undertake a testing program under TSCA to evaluate the characteristics of a chemical, such as chronic or acute toxicity, or to establish its health and environmental effects, including carcinogenic, mutagenic, teratogenic, behavioral, and synergistic effects. Pursuant to TSCA, an eight-member interagency committee of government experts on chemical substances advise the EPA administrator concerning chemicals that should be tested. In addition, manufacturers of new chemicals are required to give a premanufacturing notice to EPA prior to the manufacture of such chemicals.

The word *toxic* in the statute does not bear any specific reference to toxins or poisons or to any specific class of chemicals or chemical substances. Instead, the statute evaluates those situations which may be brought under its jurisdiction. In short, the broad objective of TSCA is to provide the necessary control before a chemical is allowed to be mass produced or becomes accessible to the general public or enters the environment. Simplistically, TSCA has a two-pronged approach. First, TSCA enables EPA to acquire information sufficient to identify and evaluate potential hazards from chemical substances. Second, TSCA allows EPA to regulate and, if necessary, restrict the production, use, distribution in commerce, and disposal of a chemical or chemical substance.

TSCA's authority as granted to EPA for the purpose of regulating disposal of toxic substances is sufficiently broad. As a result, EPA can regulate the disposal of a toxic substance into all the media of the environment, including underground waters, if such disposal presents an unreasonable risk of injury to health or to the environment. In short, the EPA administrator is authorized under TSCA to promulgate and enforce rules that are necessary to preserve public health and the environment from the dangers presented by toxic chemicals and chemical substances.

Under TSCA, EPA has been given broad discretionary power to take any regulatory measures deemed necessary to restrict chemicals suspected of causing any harm to humans and the environment. To date, only a few chemical substances, e.g., polychlorinated biphenyls (PCBs), asbestos, dioxins, and chlorofluorocarbons (CFCs), have been the prime focus of TSCA because of their known adverse health and environmental effects as well as public outcry about their production, use, and disposal. There have already been several recent attempts by Congress to amend TSCA and to bring a horde of other chemical substances within TSCA's jurisdiction. Senator Harry Reid (of Nevada) has announced that he intends to amend TSCA to place the burden of proving whether chemicals are safe on the industry. Hence, TSCA must be treated as not only a revolutionary statute, but also an evolutionary statute.

9.2 Definitions

Section 3 of TSCA provides statutory definitions of various important terms used in the textual language.

The term *chemical substance* means any organic or inorganic substance of a particular molecular identity. It may include (1) any combination of such substances occurring in whole or in part naturally or due to a chemical reaction and (2) any element or uncombined radical. It specifically excludes mixtures, pesticides, tobacco or tobacco products, radioactive materials, foods, food additives, drugs, cosmetics, and medical devices.

The term *new chemical substance* means any chemical substance which is not included in the chemical substance list compiled and published by EPA.

The term *manufacture* means to import into the customs territory of the United States, produce, or manufacture a chemical substance.

The term *mixture* means any combination of two or more chemical substances if the combination does not occur in nature and is not, in whole or in part, the result of a chemical reaction.

The term *process* means the preparation of a chemical substance or mixture, after its manufacture, for distribution in commerce (1) in the same form or physical state as it was received or in a different form or physical state from the one in which it was received, or (2) as part of an article containing the chemical substance or mixture.

The term *distribution in commerce* of a chemical substance means (1) the sale of the chemical substance, mixture, or article in commerce; (2) the introduction or delivery for introduction into commerce of the substance, mixture, or article; or (3) the holding of the substance, mixture, or article after its introduction into commerce.

Note that TSCA does not include a statutory definition of a *toxic substance*. The reason is simple. By passing TSCA, Congress authorized a reasonable and prudent regulatory framework to control the manufacture, processing, and distribution of various chemical substances and mixtures on the basis of environmental, economic, and social impact. Hence, the congressional intent behind the enactment of TSCA is not a one-sided consideration, but it srikes a balance among several competing interests.

9.3 Manufacturing and Processing Notification

One of the major provisions of TSCA is the premanufacture notification (PMN). Section 5(a) of TSCA requires manufacturers of a new chemical substance to give EPA at least 90 days notice prior to the manufacture of that chemical. With few exceptions, all new chemicals are subject to a PMN review prior to their manufacture. Any chemical which is not listed on an inventory of existing chemicals published by EPA will be considered new for the purpose of PMN. In addition, EPA may designate a particular use of an existing chemical under its significant new use rules (SNURs) based on the anticipated extent and type of exposure to human beings or to the environment by that existing chemical. A manufacturer or processor of an existing chemical for such significant new use must also submit a notice to EPA at least 90 days prior to the manufacture of the chemical for that use.

In both of the above situations, the EPA administrator may extend the 90-day premanufacture review period for an additional 90-day period for good cause. However, the reasons for a longer review period may be challenged in a court of law. Also, section 5(a) of TSCA requires the EPA administrator to consider certain relevant factors, including the following, before determining any significant new use of an existing chemical:

- The projected volume of manufacturing and processing of a chemical substance
- The extent to which a use changes the type or form of exposure of human beings or the environment to a chemical substance
- The extent to which a use increases the magnitude and duration of exposure of human beings or the environment to a chemical substance
- The reasonably anticipated manner and methods of manufacturing, processing, distribution in commerce, and disposal of a chemical substance

Thus, a significant new use may be found based on both the extent to which an existing chemical is manufactured or used and how such a chemical is commercially distributed or disposed of.

Notices in connection with the PMN should include the name of the chemical, its chemical identity along with the molecular structure, proposed or anticipated categories of use; an estimate of the quantity to be manufactured; the by-products that may result from the manufacture, processing, and disposal of the chemical; any other pertinent and known test data on health and environmental effects that the chemical, or any article containing such chemical, may have. Section 5(d) of TSCA requires that these notices be made available, subject to the provisions of section 14 of TSCA dealing with the disclosure of data, for examination by interested persons.

Within 5 days of receipt of such a notice, the EPA administrator is required to publish a notice in the Federal Register which

- Identifies the chemical substance for which a notice or data have been received
- Lists the uses or intended uses of such substance
- Describes the nature of the tests performed on such substance and the test data developed pursuant to section 4 of TSCA

In addition, at the beginning of each month, the EPA administrator is required to publish a list in the Federal Register of

- Each chemical substance for which a premanufacture notice has been received and for which the prescribed notification period has not expired

- Each chemical substance for which the prescribed notification period has expired since the last publication of such list in the Federal Register

For those chemicals on the priority list requiring special testing, the EPA administrator is required to publish in the Federal Register the reasons for not taking any action relative to production and use prior to the expiration of the notification period.

The EPA administrator may also issue a proposed order, to take effect after the expiration of the notification period, to prohibit or limit the manufacture, processing, distribution in commerce, use, or disposal of a chemical substance. The proposed order may also include measures to prohibit or limit any combination of such activities of a chemical substance.

This order applies to the case of a chemical substance for which there is insufficient information or lack of information to permit a reasoned evaluation of the health and environmental effects including risk of injury to health or the environment. If a manufacturer or processor of a chemical substance is subject to such a restrictive proposed order pending development of necessary information, such manufacturer or processor may file objections against the issuance of such order. Such a filing must specify the objections and state the grounds for such objections within 30 days of receipt of the notice.

If the EPA administrator issues a proposed order prohibiting or limiting the manufacture, processing, distribution in commerce, use, or disposal of a chemical substance, but such order does not take effect because of the timely objections filed, then the EPA administrator may not apply for an injunction from the U.S. district court where the manufacturer or processor is found, resides, or transacts business, in order to prohibit or limit such activities as they relate to that chemical substance. However, if the EPA administrator finds that there is a reasonable basis to conclude that these activities of a chemical substance present an unreasonable risk of injury to public health or the environment, then the EPA administrator may take an action including court injunction prohibiting or limiting such activities to the extent necessary to protect public health and the environment against such risks.

Exemptions are granted from the PMN requirements for certain chemicals. These exemptions include and apply to

- The categories of chemical substances listed on the inventory of existing chemicals
- Chemicals produced in small quantities solely for experimental purposes or for research and development (R&D)
- Chemicals used for test marketing
- Chemicals determined by the EPA administrator as not posing or presenting an unreasonable risk

In addition, a manufacturer or processor of a chemical substance may petition the EPA administrator for exemption from the PMN requirements when the chemical's existence is only for a short time and there is no likelihood of human or environmental exposure.

Note that for each PMN submitted to EPA by a manufacturer or processor, the application must include all test data about the chemical that are available to, or within the control of, the manufacturer or processor of that chemical.

9.4   Testing of Chemicals

As indicated earlier, the EPA administrator may require manufacturers or processors of potentially harmful chemicals to conduct certain tests to evaluate whether the intended use will result in an unreasonable risk of injury to health or damage to the environment. Section 4(a) of TSCA provides the testing requirements that the EPA administrator can impose. The costs of testing the chemical, however, are borne by the manufacturers or processors of that chemical.

Under TSCA, testing requirements are to be promulgated by EPA regulations, but only after an opportunity is provided for comments and a public hearing. The manufacturer or processor of a proposed new chemical may, however, petition the EPA administrator to develop testing standards for the chemical before any tests are run.

9.5   Hazardous Chemical Substances and Mixtures

Section 6(a) of TSCA authorizes the EPA administrator to prohibit or limit the manufacturing, processing, distribution in commerce, use, or disposal of a chemical substance or mixture where such activities or any combination of them presents or will present an unreasonable risk of injury to health or to the environment. In such a case, the EPA administrator may apply, by rule, one or more of the following requirements for adequate protection against such risk:

- A requirement that either (1) prohibits the manufacturing, processing, or distribution in commerce of such substance or mixture or (2) limits the amount of such substance or mixture which may be manufactured, processed, or distributed in commerce

- A requirement that (1) prohibits the manufacturing, processing, or distribution in commerce of such substance or mixture for (*a*) a particular use or (*b*) a particular use in a concentration exceeding the level specified by the EPA administrator, or (2) limits the amount of such substance or mixture which may be manufactured, processed, or distributed in commerce for (*a*) a particular use or (*b*) a particular use in a concentration exceeding the level specified by the EPA administrator

- A requirement that establishes that such substance or mixture or any article containing such substance or mixture be marked with or accompanied by clear and adequate warnings and instructions with respect to its use, distribution in commerce, or disposal or a combination of any of these activities

- A requirement that obligates the manufacturers and processors of such substance or mixture to make and retain records of the processes used and to monitor or conduct tests which are reasonable and necessary to ensure compliance with any rule under section 6 of TSCA
- A requirement that prohibits or otherwise regulates the manner or method of commercial use of such substance or mixture
- A requirement that prohibits or otherwise regulates the manner or method of disposal of such substance or mixture, or of any article containing such substance or mixture, by its manufacturer or processor or by any other person who uses or disposes of such substance or mixture or article for commercial purposes
- A requirement that directs manufacturers or processors of such substance or mixture to (1) give notice of any unreasonable risk of injury to distributors in commerce of such substance or mixture and to other persons in possession of such substance or mixture, (2) give public notice of such risk of injury, and (3) replace or repurchase such substance or mixture based on election by the person to which the requirement is directed

Pursuant to section 6(a) of TSCA, the EPA administrator is required to apply one or more of these requirements for appropriate protection against an unreasonable risk of injury to health or to the environment, using the least burdensome requirements. Any rule by the EPA administrator that limits but does not ban a chemical becomes effective when initially proposed in the Federal Register. This applies only if the EPA administrator determines that the chemical is likely to present an unreasonable risk of serious or widespread injury to health or to the environment during the rule making procedures. However, when a rule prohibits or bans outright the manufacturing or processing of a chemical, the EPA administrator is required to obtain a court order or injunction before the rule becomes effective.

If a chemical contains a hazardous contaminant due to the manufacturing processes involved, the EPA administrator may order the manufacturer to change the processes to avoid such contamination. But when a chemical contains a contaminant that poses an unreasonable risk of injury to health or the environment, the manufacturer or processor may be required to give public notice and to repurchase or recall such chemical.

The court may also be asked by the EPA administrator to allow and approve any necessary action against those chemicals that present an imminent and unreasonable risk of serious or widespread injury to health or to the environment.

We note here that section 6 of TSCA authorizes the EPA administrator to undertake and impose one or more specific requirements for regulating hazardous chemical substances and mixtures in order to protect against an unreasonable risk of injury to health or to the environment. It specifically includes polychlorinated biphenyls as hazardous chemical substances. However, this does not imply that PCBs are the only hazardous chemical substances or mixtures that may be regulated under TSCA.

9.6   Record-Keeping and Reporting Requirements

Section 8(a) of TSCA authorizes the EPA administrator to promulgate rules requiring manufacturers and processors or persons proposing to manufacture or process a chemical substance to report to the EPA administrator and maintain all appropriate records. Such a report should include certain specific information:

- The common or trade name, the chemical identity, and the molecular structure of each chemical substance or mixture
- The categories or proposed categories of use of each substance or mixture
- The total amount of each substance and mixture manufactured or processed, reasonable estimates of the total amount to be manufactured or processed for each of its categories of use, and reasonable estimates of the amount to be manufactured or processed for each of its future or proposed categories of use
- A description of the by-products resulting from the manufacture, processing, use, or disposal of each substance or mixture
- All existing data concerning the health and environmental effects of such substance or mixture
- The number of individuals exposed, reasonable estimates of the number who may be exposed to such substance or mixture in their places of employment, and the duration of such exposure
- The manner or method of disposal of each substance or mixture in the initial report and any change in such manner or method of disposal in the subsequent reports

Section 8(a) of TSCA provides exemption to small manufacturers or processors of a chemical substance or mixture from all or part of these reporting requirements. For example, manufacturers or processors of mixtures of small amounts of R&D chemicals are required to report and maintain records only if the EPA administrator finds them necessary for enforcement purposes. In the absence of a determination that reporting is necessary due to an unreasonable risk to health or to the environment, small manufacturers are exempt from reporting requirements except for those chemicals which are subject to proposed or final testing requirements or other regulatory obligations.

The EPA administrator, in turn, is required to publish a list of all existing commercial chemicals. This list is to contain all chemicals manufactured or processed for commercial purposes in the United States or imported into the United States.

EPA issued a final inventory update, set forth in the Federal Register of June 12, 1986, requiring manufacturers and importers of certain chemical substances listed in the TSCA Chemical Substances Inventory to report certain information about those chemicals. Subsequently, EPA proposed a comprehensive assessment information rule for the purpose of gathering information for risk assessments and strategy developments for 47 chemical substances.

EPA has formalized reporting forms for the purpose of submitting information as set forth in 40 CFR §710.39. Persons submitting information to update inventory data base under 40 CFR part 710, subpart B must use original copies of Form U. EPA form 7710-35, reproduced in App. 9A, is to be used by a manufacturer or importer of a chemical to provide preliminary assessment information to EPA. This form must be submitted for each manufacturing or importing facility. EPA form 7710-25, reproduced in App. 9B, must be used for premanufacture notice for new chemical substances.

Section 8(e) of TSCA includes the provision for "substantial-risk" notification. Pursuant to this, any person who manufactures, processes, or distributes in commerce a chemical substance or mixture and who obtains data that reasonably conclude that such substance or mixture presents a substantial risk of injury to health or the environment must immediately inform the EPA administrator about it. This is not a requirement when that person has actual knowledge that the EPA administrator has already been adequately informed.

On February 1, 1991, EPA promulgated a notice announcing the opportunity to register for EPA's TSCA section 8(e) compliance audit program (CAP) for "substantial-risk information" reportable to EPA. Pursuant to section 8(e) of TSCA, any information that reasonably supports the conclusion that a chemical poses a substantial risk to human health or the environment must be reported within 15 days from the date such information is obtained. CAP allowed a one-time voluntary compliance for the reporting requirement of TSCA. The registration period for CAP was closed on May 2, 1991.

In deciding whether new information reasonably supports a conclusion of substantial risk, two factors must be considered: the seriousness of the adverse effect and the fact or probability of its occurrence. However, these two criteria should be weighed differently depending upon the seriousness of the effect and the extent of the exposure. Any decision-making process for determining section 8(e) reportability for substantial risk of injury to health or the environment under TSCA should primarily focus on whether new toxicologic or exposure data offer "reasonable support for a conclusion of substantial risk." Hence, a decision to report under section 8(e) of TSCA should not focus on a definite conclusion of risk and, therefore, does not have to involve exhaustive health or environmental assessments or an evaluation of socioeconomic benefits that may be derived from the use of a chemical substance.

EPA's CAP had been developed to encourage industry reporting by setting forth certain guidelines. These guidelines identify EPA's enforcement responses in advance and allow those in the industry to assess their liability prior to election to participate. Registration in the CAP was not mandatory, but purely voluntary. However, participation in the CAP gave the manufacturer, processor, or distributor of a chemical an early opportunity to learn about EPA's enforcement policy pertaining to the notification of substantial risk requirements. Participation in the voluntary CAP had a special incentive as well—it set a ceiling of $1 million for all stipulated civil penalties that a participating person (including corporations) can face.

9.7   Polychlorinated Biphenyls

As indicated earlier, polychlorinated biphenyls are specifically included in section 6(e) of TSCA because of their extremely toxic effects. Reportedly, trace doses of PCBs have been documented to cause birth defects, hair loss, and even death in test animals. Exposures to PCBs and their by-products, polychlorinated dioxins and dibenzofurans, have been linked to a skin reaction known as chloracne, bleeding and neurological disorders, liver damage, spontaneous abortions, deformed babies, cancer, and death. While drafting TSCA, Congress mandated that PCB manufacture be banned and all existing uses be phased out. Section 6(e) of TSCA precludes all manufacturing of PCBs after January 1, 1978, other than in a totally enclosed manner. TSCA also authorizes the EPA administrator to require all persons who manufacture, process, distribute in commerce, or use any PCB after January 1, 1978, to do so only in a totally enclosed manner except in specific situations.

EPA's PCB regulations actually took effect in 1979, and the agency permitted use of PCBs in a totally enclosed manner and in concentrations lower than 50 parts per million. However, in *Environmental Defense Fund v. EPA,* the court held that EPA had no scientific basis or justification for such continued use and placed EPA on a timetable to issue new rules involving a final phaseout. EPA's rules promulgated later permitted continued uses of nonfood PCB electric equipment until October 1, 1988.

Historically, PCBs have been widely used as dielectric fluids in transformers and capacitors. EPA's regulations promulgated under TSCA and as set forth in 40 CFR §761.3 classify PCBs in several distinct categories. Each has a key definition and specific regulatory requirements.

PCB categories and key definitions

1. Transformers
   a. PCB transformers that contain over 500 parts per million of PCB in the transformer fluid
   b. PCB-contaminated transformers (or electric equipment) that contain PCB between 50 and 500 parts per million
   c. Non-PCB transformers that contain less than 50 parts per million of PCB
2. Capacitors
   a. Small capacitors that contain less than 3 pounds of dielectric fluid
   b. Large capacitors that contain 3 pounds or more of dielectric fluid. They can be either low-voltage (i.e., less than 2000 volts) or high-voltage (i.e., 2000 volts or greater) capacitors.
3. PCB article—any manufactured article that contains PCBs and whose surface directly contacts with PCBs
4. PCB container—any container that contains PCBs or PCB articles and whose surface directly contacts with PCBs
5. PCB-contaminated electric equipment—any electric equipment that contains PCBs in concentration of 50 parts per million or greater but less than 500 parts per million.

6. PCB item—any PCB article, PCB container, or PCB equipment which contains any PCBs by design or by accident

EPA's regulations are based on the presumption that any oil-filled electric equipment is PCB-contaminated electric equipment unless the test results indicate otherwise. There are specific exceptions to this presumptive rule which include circuit breakers, reclosers, and cable.

Temporary storage

There are two types of on-site storage allowed by EPA's regulations promulgated under TSCA: storage for up to 30 days and storage not longer than 1 year.

Temporary storage at a facility for up to 30 days, starting from the date of removal from service, is permissible for

- All nonleaking PCB articles and electric equipment
- Leaking PCB articles and electric equipment only if they are placed in containers along with sufficient sorbent materials
- PCB containers containing nonliquid PCBs
- PCB containers containing liquid PCBs if (1) the PCB concentration is less than 50 parts per million and (2) a formal spill prevention control and countermeasure (SPCC) plan has been prepared for the facility

PCB articles and containers may be stored for a longer period, but not exceeding 1 year if certain requirements as set forth in 40 CFR §761.65 are met. All such PCB articles and containers must be stored and inspected per these regulatory requirements. Also storage areas must be appropriately marked.

No regulatory requirements are to be met for the storage of PCB-contaminated electric equipment if such equipment has been completely drained of all free-flowing dielectric fluid.

PCB disposal

As indicated earlier, TSCA does not ban the use of PCBs as long as they are used in a "totally enclosed manner." Statutorily, this means any manner which will ensure that any exposure of human beings or the environment to a PCB will be "insignificant." EPA regulations as set forth in 40 CFR §761.60 require that a liquid PCB containing over 500 parts per million be disposed of either by specially designed high-temperature incinerators or by other alternate EPA-approved destruction methods, which will reduce PCBs to harmless components. When the PCB level is in the range of 50 to 500 parts per million, it may also be disposed of in a high-efficiency boiler or in a chemical waste landfill provided the PCB liquid is not an ignitable waste with a flash point of 140°F or below.

Contaminated soil, rags, or other nonliquid materials containing PCBs at a concentration of 50 parts per million or higher may be disposed of either in an approved incinerator or in a chemical waste landfill.

PCB transformers must be disposed of either in an approved incinerator or in a chemical waste landfill. In the latter case, the transformer must first be drained and rinsed per EPA procedures set forth in 40 CFR §761.60(b)(1)(i)(B).

Large PCB capacitors, however, are required to be incinerated except when they contain less than 500 parts per million of PCBs. In the latter case, they may be disposed of in a chemical waste landfill.

There is no regulatory requirement for the disposal of PCB-contaminated electric equipment as long as all the free-flowing dielectric fluid has been drained out.

Record keeping and other requirements

The specific EPA regulations set forth for PCBs do not exempt them from the general regulatory requirements promulgated by EPA under TSCA. In short, PCBs are singly targeted for additional requirements beyond the general ones for toxic substances. These additional requirements are very specific for PCBs and must be adhered to.

One such requirement is record keeping as set forth in 40 CFR sections 761.30, 761.65, and 761.180. Pursuant to these, any facility using or storing at any time (1) at least 45 kilograms (99.4 pounds) of PCBs in PCB containers, or (2) one or more PCB transformers, or (3) fifty or more large PCB capacitors must prepare an annual report for the previous calendar year by July 1. Such a report must contain

- Dates of removal of PCBs from service, placing PCBs into storage, and placing PCBs into transport vehicle for disposal

- Location of initial storage and disposal facility along with the name of the owner or operator of the facility

- Total quantities of PCBs still remaining in service and dates and quantities of (1) PCBs in PCB containers, (2) number of PCB transformers and the weight of PCBs in them, and (3) number of large PCB capacitors

For disposal and storage facilities of PCBs, an annual report for the previous calendar year must also be prepared by July 1. Such a report must contain

- Dates of receipt of PCBs for storage and disposal along with the facility identification and the name of owner or operator of the facility sending PCBs for that purpose

- Dates of disposal or transfer to another PCB storage or disposal facility along with the identification of specific types of PCBs stored or disposed of

- Total weight of PCBs in PCB containers and PCB transformers that have been (1) received during the year, (2) transferred during the year, and (3) retained at the end of the year

- Total number of PCB articles or electric equipment received, transferred, or remaining at the facility

In addition to the preparation of these annual reports, EPA regulations require all records to be kept for agency inspection and necessary maintenance. Furthermore, all PCB transformer fire-related incidents must be reported to the National Response Center (NRC) immediately, and all practical and safe measures must be undertaken to prevent any potential release of PCBs into the water. Details of PCB spill reporting are given in Chap. 10.

EPA regulations also require registration of PCB transformers with the local fire department and the exterior labeling of the locations of PCB transformers with the mark ML.

On February 5, 1990, EPA revised its regulations to reflect a greater degree of control of PCBs. This revision incorporates a number of provisions reflective of EPA's regulations based on the Resource Conservation and Recovery Act (RCRA). For example, the new EPA regulations require a PCB waste generator to obtain a TSCA identification number, use proper manifests, and retain all necessary documents.

9.8 Asbestos

The Toxic Substance Control Act Amendments of 1986 added a new Title II to the original law. This is known as the Asbestos Hazard Emergency Response Act (AHERA) of 1986. AHERA requires EPA to promulgate regulations defining response actions for school buildings that contain friable asbestos. The EPA regulations promulgated under AHERA and set forth in 40 CFR part 763 provide requirements for conducting appropriate inspections for asbestos-containing materials by each local education agency. Additionally, EPA regulations include response actions, operations and maintenance, training, management plans, and record-keeping requirements.

In essence, these regulations are directed to address the issue of asbestos for what it is. Asbestos has so far demonstrated its carcinogenic potential only in the form of fiber. At the same time, asbestos has wide-ranging and desirable properties, such as resistance to heat and chemical attack, high tensile strength, and flexibility. While debates in scientific and legal circles continue about the use of asbestos, EPA regulations thus far have taken an approach of compromise based on known health effects and widespread industrial and commercial uses of asbestos.

Definitions

The term *asbestos* does not refer per se to any distinct chemical or chemical substance or mineral. Rather, this name is attributed to several fibrous variations of naturally occurring silicate minerals that have a fibrous crystalline structure and display very similar physical and chemical characteristics. Typically, six such minerals comprise this "asbestos" family: chrysotile (serpentine), amosite (cummingtonite-grunerite), crocidolite (riebeckite), tremolite, anthophyllite, and actinolite. In their asbestos form, these minerals can split into fine fibers.

Asbestos mixture refers to a mixture which contains bulk asbestos or another asbestos mixture as an integral and important component. But the term does not include mixtures which contain asbestos as a contaminant or an impurity.

For environmental purposes, asbestos has been specifically defined by the Occupational Safety and Health Administration as a class of individual crystallites or crystal fragments that display a length of greater than 5 microns with a length-to-diameter ratio of 3:1 or greater and a maximum diameter of less than 5 microns.

In essence, any material that contains asbestos and thus can be characterized as an asbestos-containing material (ACM) is subject to the regulations with certain specific exceptions. EPA regulations as set forth in 40 CFR §763.65 deal with who must report. As a result, all miners or primary processors of asbestos, or importers of bulk asbestos in 1981 (the last year of such activities without the violation of the law), must complete and submit EPA form 7710-36. Appendix 9C reproduces this form. EPA form 7710-37, reproduced in Appendix 9D, must be used to report secondary processing and importation of asbestos mixtures.

Asbestos-related disease and contamination

The ill effects of asbestos on human health have been recognized and documented for some time. Asbestos-induced diseases can be in the form of (1) asbestosis (a restrictive lung disease), (2) lung cancer, (3) pleural and peritoneal mesothelioma (tumor of cells in the linings of the lungs and the stomach), and (4) other types of cancer. It is generally agreed that the asbestos-related diseases are caused by the inhalation of asbestos fibers. Measuring of asbestos emissions or fibers released must, by necessity, be accurate.

Contamination of building interiors by asbestos fiber generally occurs due to fallout, contact or impact, and reentrainment. In the case of fallout, the asbestos fiber is released into a continuous low-level but long-term process. This may happen without physical disruption or crumbling of the ACMs or sprayed-on materials containing asbestos. This results mainly from a long-term deterioration of the binding components. When such deterioration takes place, a physical contact or impact may result in a release of asbestos fibers. Fibers are thus released into the environment, especially into the building's interior air. In addition, the structural surfaces may be reentrained through air disturbance within the building that can cause resuspension of the fibers. Physical indoor activities, such as sweeping, dusting, and vacuuming, can cause resuspension of asbestos fibers for years.

Removal or encapsulation of ACMs can also contribute to asbestos contamination of buildings. In addition, demolition of structures that contain ACMs can result in high levels of airborne asbestos fibers. The EPA regulations require proper handling of all ACMs prior to the demolition of any building.

In this context, note that AHERA does not require total removal of ACMs. An asbestos-related hazard may be remedied by adopting one of several prescribed methods, such as enclosure or encapsulation, besides removal. As a result, where asbestos is not crumbling or is not found in a friable form, the ACMs may be left in place, using appropriate enclosure or encapsulation.

Current EPA regulations ban almost all uses of sprayed-on or friable asbestos. Disposal of these forms of asbestos is also fully regulated.

EPA regulations set forth in 40 CFR §§763.80 through 763.117 require local education agencies to identify friable and nonfriable ACMs in various schools, inspect and reinspect, sample and analyze, warn, manage, notify, and maintain records for such ACMs.

Regulatory requirements and authorized agencies

AHERA requires promulgation of appropriate regulations for all asbestos-related response actions. Additionally, regulations must address appropriate training for personnel handling ACMs, proper operations and maintenance plans, and record keeping.

Besides EPA, several federal agencies have the authority to regulate asbestos:

1. Occupational Safety and Health Administration is responsible for setting limits for occupational exposure to asbestos.

2. Food and Drug Administration is responsible for preventing asbestos contamination in foods, drugs, and cosmetics.

3. Department of Transportation is responsible for the regulation of packaging and shipment of asbestos.

4. Consumer Product Safety Commission is responsible for regulating the use of asbestos in consumer products.

Note here that section 9 of TSCA authorizes EPA to ask another agency to undertake regulation of a chemical substance or mixture which (1) presents or will present an unreasonable risk and (2) may be prevented or sufficiently rendered harmless by another agency. TSCA also allows the EPA administrator to use other federal laws administered by EPA, such as the Federal Water Pollution Control Act or the Clean Air Act, to protect against unreasonable risks posed by toxic substances unless it is determined that in the public interest, such risks should be singularly addressed under TSCA.

9.9   Dibenzo-*para*-dioxins and Dibenzofurans

Dioxins became infamous in this country upon EPA's discovery of massive dioxin contamination in Times Beach, Missouri, in 1983, followed by a mass exodus of local residents. The dioxin at Times Beach was found to be a by-product of the manufacture of 2,4,5-trichlorophenol (2,4,5-TCP) which was produced by the hydrolysis of tetrachlorobenzene (TCB) at high temperature. The soil at Times Beach contained 2,3,7,8-tetrachlorodibenzo-*p*-dioxin (TCDD). In March 1984, EPA issued its final rule incorporating its control of TCDD under the authority granted by TSCA.

When exposed to heat, PCBs create polychlorinated dioxins and dibenzo-furans, two of the most toxic substances known. The term *dioxin,* however, connotes a group of toxic chemicals, generally described as dibenzo-*para*-dioxins and dibenzofurans. EPA regulations as set forth in 40 CFR part 766 deal with the requirements for these chemical substances under the authority granted by TSCA. Simplistically, these regulations impose testing requirements pertaining to section 4 of TSCA. These tests are required to ascertain whether certain specific chemical substances are contaminated with halogenated dibenzodioxins (HDDs) and halogenated dibenzofurans (HDFs). And 40 CFR section 766.35(a) requires manufacturers and, in certain situations, processors of chemical substances to submit letters of intent to test, along with protocols to be followed for analysis of chemical substances, for the presence of HDDs and HDFs. Any request for exclusion or waiver from such testing must include certain specific conditions as set forth in 40 CFR section 766.32.

Not all chemical substances, however, are subject to this testing requirement. Chemical substances identified under 40 CFR section 766.25 are subject to the testing and reporting rule. Table 9-1 is a reproduction of EPA-listed chemical substances that are subject to this testing for HDDs and HDFs.

At this time, EPA requires that only manufacturers and importers of such listed chemical substances perform this testing. However, processors may be asked to test if manufacturers and importers fail to do so. All appropriate test information must be submitted to EPA. For this purpose, EPA has developed a dioxin/furan reporting form (EPA form 7710-51) which is reproduced in App. 9E. Small manufacturers are exempt from the reporting process but not from the test.

Also, 40 CFR section 766.38 provides the reporting requirement for precursor chemical substances. A precursor chemical substance is produced under conditions that do not yield HDDs and HDFs, but it has a molecular structure that is conducive to HDD or HDF formation under favorable reaction conditions. Pursuant to EPA's requirement, a manufacturer or importer of a chemical product that produces a precursor chemical substance must report to EPA no later than September 29, 1987. Small manufacturers and those manufacturers and importers who produce the precursor chemical substances at or below 100 kilograms per year and only for R&D purposes are not subject to this reporting requirement. Those who have to report, must provide EPA with process and reaction condition data on part II of EPA form 7710-51. A separate such form must be submitted for each chemical product reported. Additionally, the precursor chemical substance used must be identified for the purpose of this reporting.

9.10 Exports and Imports

Section 12 of TSCA authorizes the EPA administrator to regulate a chemical substance, mixture, or article manufactured or processed for export if that chemical substance, mixture, or article presents an unreasonable risk of injury to health or the environment of the United States. In addition, the EPA

TABLE 9.1   Chemical Substances for Testing for HDDs/HDFs

(1) *Chemical substances known to be manufactured between January 1, 1984 and date of promulgation of 40 CFR Part 766*

CAS No.	Chemical name
79-94-7	Tetrabromobisphenol-A.
18-75-2	2,3,5,6-Tetrachloro-2,5-cyclohexadiene-1,4-dione.
118-79-6	2,4,6-Tribromophenol.
120-83-2	2,4-Dichlorophenol.
1163-19-5	Decabromodiphenyloxide.
4162-45-2	Tetrabromobisphenol-A-bisethoxylate.
21850-44-2	Tetrabromobisphenol-A-bis-2,3-dibromopropyl ether.
25327-89-3	Allyl ether of tetrabromobisphenol-A.
32534-81-9	Pentabromodiphenyloxide.
32536-52-0	Octabromodiphenyloxide.
37853-59-1	1,2-Bis(tribromophenoxy)-ethane.
55205-38-4	Tetrabromobisphenol-A-diacrylate.

(2) *Chemicals not known to be manufactured between January 1, 1984 and the date of promulgation of 40 CFR Part 766*

CAS No.	Chemical name
79-95-8	Tetrachlorobisphenol-A.
87-10-5	3,4',5-Tribromosalicylanilide.
87-65-0	2,6-Dichlorophenol.
95-77-2	3,4-Dichlorophenol.
95-95-4	2,4,5-Trichlorophenol.
99-28-5	2,6-Dibromo-4-nitrophenol.
120-36-5	2[2,4-(Dichlorophenoxy)]-propionic acid.
320-72-9	3,5-Dichlorosalicyclic acid.
488-47-1	Tetrabromocatechol.
576-24-9	2,3-Dichlorophenol.
583-78-8	2,5-Dichlorophenol.
608-71-9	Pentabromophenol.
615-58-7	2,4-Dibromophenol.
933-75-5	2,3,6-Trichlorophenol.
1940-42-7	4-Bromo-2,5-dichlorophenol.
2577-72-2	3,5-Dibromosalicylanilide.
3772-94-9	Pentachlorophenyl laurate.
37853-61-5	Bismethylether of tetrabromobisphenol-A.
	Alkylamine tetrachlorophenate.
	Tetrabromobisphenol-B.

administrator may require testing of any chemical earmarked for export if such testing is necessary to determine whether there is an unreasonable risk to health or the environment within the United States. In view of these, any person who exports or intends to export any chemical that is subject to a regulatory order or action, testing, manufacturing or processing notification must notify the EPA administrator. The EPA administrator, in turn, is responsible for notifying the governments of the importing countries about the availability of such data.

In the case of imports, no chemical substance, mixture, or article containing a chemical substance or ingredient is allowed into the customs territory of the

United States when such substance, mixture, or article fails to comply with the applicable regulatory requirements under TSCA or is in violation of any statutory provision of TSCA.

9.11   Enforcement and Penalties

Section 11 of TSCA authorizes the EPA administrator, and any duly designated representative of the EPA administrator, to inspect any establishment, facility, or other premises in which chemical substances or mixtures are manufactured, processed, stored, or held before or after their distribution in commerce. However, such inspection cannot include financial, sales, pricing, personnel, or confidential research data unless specifically stated in an inspection notice. The EPA administrator is empowered to subpoena witnesses or receive documents and other necessary information in carrying out the administrative duties.

Section 16 of TSCA provides specific penalties and seizures for violation of any provision of TSCA. Simply stated, a civil penalty may be in any amount but not exceeding $25,000 for each violation. Any repeat or continuation of a violation beyond a single day will constitute a separate violation.

Any person who knowingly or willfully violates any provision of TSCA may be subject to a criminal penalty in addition to, or in lieu of, any civil penalty. For such violation, upon conviction, one may be liable to a fine of not more than $25,000 for each day of violation or to imprisonment for not more than 1 year or to both.

9.12   Preemption and Effect of State Laws

In general, there is no federal preemption of state laws by TSCA. TSCA will not affect the authority of any state or local political subdivision to establish and enforce regulations concerning chemicals except in the following situations:

- EPA has issued a testing requirement for a chemical substance.
- EPA has restricted the manufacture or processing or otherwise regulated a chemical substance.

However, a state may issue requirements which are identical and mandated by other federal laws or may prohibit the use of any chemical. In addition, the EPA administrator may grant an exemption to a state under certain conditions from certain federal regulatory requirements.

9.13   EPA's Current Biotechnology Policy

Recent biotechnological and genetic engineering advances have allowed scientists to manipulate the genetic makeups of certain microorganisms. This has created the opportunity to redefine their behavioral pattern for apparent benefits to the human society. As a result, genetic engineering or biotechnological advances are geared to reap benefits in such areas as medicine, food, agriculture, national defense, and the like.

These so-called benefits are not one-sided or riskproof. They also have potentials for creating major environmental risks and health hazards.

In recognition of the potential for serious harm to human health and the environment, a comprehensive interagency plan has already been undertaken by the federal government to regulate the emerging biotechnology industry. It is currently EPA's intention to use its broad authority under TSCA to regulate various biotechnology products, including biologically engineered microbes and other living organisms.

In December 1984, the Reagan administration issued a guidance document entitled "Proposal for a Coordinated Framework for Regulation of Biotechnology." This provided a glimpse of how the federal government viewed the development of regulations for the emerging biotechnology under existing federal statutes. Subsequently, on June 26, 1986, a document entitled "Coordinated Framework for Regulation of Biotechnology" was issued as a result of a combined effort by EPA, the Food and Drug Administration (FDA), the U.S. Department of Agriculture (USDA), OSHA, and the Department of Health and Human Services (DHHS). This framework endorsed the view that these federal agencies can rely on existing federal statutes to regulate various products of biotechnology.

This government framework indicated that EPA can regulate various biotechnology products by virtue of the statutory authority granted to it under TSCA and FIFRA.

TSCA does not specifically mention biotechnology products or the scope of EPA's authority to regulate and control these products. However, TSCA grants EPA broad authority to gather information, assess risks, and otherwise regulate chemical substances and mixtures that may present an unreasonable risk of injury to health or to the environment. EPA's policy seeks to bring various biotechnology products under its statutory authority and thus close the gap created by the emergence of biotechnology and genetic engineering.

However, EPA's expansive interpretation of its perceived authority over biotechnology and genetic products under TSCA has raised legal questions and scientific debate. EPA's reliance on the plain meaning of TSCA's definition of chemical substances based on any DNA molecule, however created, is extremely broad and thus controversial. This has prompted a reexamination of the congressional intent for the enactment of TSCA by sifting through the legislative history. In the absence of an amendment to TSCA or enactment of another federal statute by Congress, EPA's current policy to regulate biotechnology products under the general authority granted by TSCA will remain a hotly debated issue. However, the most conservative approach on the part of manufacturers, processors, importers, and distributors in commerce of biotechnology products is to meet the EPA regulations promulgated under

TSCA and other federal statutes, including the National Environmental Policy Act (NEPA), the Federal Food, Drug, and Cosmetic Act (FFDCA), the Resource Conservation and Recovery Act, FIFRA, and OSH Act.

Recommended Reading

1. Symposium on Toxic Torts, Institute for International Research, Law Division, New York, November 8–9, 1990.

2. The Eighth Annual PCB Regulatory Briefing, Executive Enterprises, Inc., New York, May 25–26, 1988.

Important Cases

1. *Dayton Independent School District v. U.S. Mineral Products Co.,* 906 F. 2d 1059 (5th Cir. 1990).

2. *Rollins Environmental Services (FS), Inc. v. Parish of St. James,* 775 F. 2d 627 (5th Cir. 1985).

3. *Environmental Defense Fund v. EPA,* 636 F. 2d 1267 (D.C. Cir. 1980).

4. *Dow Chemical Co. v. EPA,* 605 F. 2d 673 (3d Cir. 1979).

APPENDIX 9A   Manufacturer's Preliminary Assessment Information Report

IMPORTANT: Before completing this form, please read the accompanying instructions carefully.   O.M.B. No. 2000-0420: Approval Expires 3/31/84

| **EPA** U.S. ENVIRONMENTAL PROTECTION AGENCY
401 M Street, S.W.
Washington, D.C. 20460
**MANUFACTURER'S REPORT
PRELIMINARY ASSESSMENT INFORMATION**
This information is required under the authority of Section 8(a), Toxic Substances Control Act, 15 U.S.C. 2607. | **When completed send this form to:**
**Document Control Officer
Office of Pesticides and
Toxic Substances
U.S.E.P.A.
P.O. Box 2080
Rockville, Md. 20852** | CONTROL NUMBER   ●

PERIOD COVERED
FROM: Mo. I Yr.   TO: Mo.   I Yr. |

Section I — CERTIFICATION

TECHNICAL CERTIFICATION STATEMENT

I hereby certify that, to the best of my knowledge and belief, all information entered on this form is complete and accurate. I agree to permit access to, and the copying of records by, a duly authorized representative of the EPA Administrator, in accordance with the Toxic Substances Control Act, to document any information reported here.

Signature | Date

Name and title — *Please print or type*

CONCERNING EPA DISCLOSURE OF INFORMATION

Any person who submits information to EPA under the Preliminary Assessment Information Rule (40 CFR 712) should be aware of EPA regulations (40 CFR Part 2) which govern disclosure of such information. Those regulations provide that such person may, if he or she desires, assert a confidentiality claim covering part or all of the information submitted. Information covered by such a claim will be publicly disclosed by EPA only to the extent, and by means of the procedures, set forth in 40 CFR Part 2. However, if no such claim accompanies the information when it is received, EPA may make that information public without notifying the submitter.

CONFIDENTIALITY STATEMENTS

Information disclosed to EPA on this form may be claimed confidential by marking the appropriate boxes below. The person signing the Confidentiality Certification Statement attests to the truth of the following four statements concerning all information that is claimed confidential. Note that chemical substance identity may not be claimed confidential for this rule.

1. My company has taken measures to protect the confidentiality of the information, and it intends to continue to take such measures.

2. The information is not, and has not been, reasonably obtainable without our consent by other persons (other than governmental bodies) by use of legitimate means (other than discovery based on a showing of special need in a judicial or quasi-judicial proceeding).

3. The information is not publicly available elsewhere.

4. Disclosure of the information would cause substantial harm to our competitive position.

CONFIDENTIALITY CERTIFICATION STATEMENT

I hereby certify that the Confidentiality Statements on this form are true as to that information below for which I have asserted a confidentiality claim.

Signature | Date

Name and title — *Please print or type*

Section II — CHEMICAL IDENTIFICATION

▶ **Part A**

CAS No.

Chemical name (first 15 characters)

▶ **Part B**

Category name (first 15 characters)

Inventory Form C number

Section III — RESPONDENT IDENTIFICATION

☐ *MARK THIS BOX TO CLAIM THIS SECTION CONFIDENTIAL*

▶ **Part A — Plant Site — Physical location**

Name

Number and street

City

County

State | ZIP code

Dun and Bradstreet number

▶ **Part B — Mailing Address of:**

☐ Corporate Headquarters   ☐ Plant Site

Name

Number and street

City

State | ZIP code

Dun and Bradstreet number (for corporate headquarters only)

▶ **Part C — Technical Contact**

Name and title

☐ At headquarters

Telephone *(Area code/number)*

☐ At plant site

▶ **Part D — Acknowledgement**

EPA will send acknowledgement to - *Name and title*

EPA Form 7710-35

Section IV — PRELIMINARY ASSESSMENT INFORMATION

NOTE   ·Mark the box to the left of any item below to claim the answer to the item as confidential. Report all quantities in kilograms (1 kilogram = 2.2 pounds). Enter N/A for any item that does not apply to you; do not leave any blanks.

▶ **Part A — Plant Site Activities** — *Information in part A must be your best estimate from readily obtainable data.*
For items 3b, 3c, and 3d, specify the accuracy of your answers.

☐ **1**.Total quantity imported	k g	☐ **2**.Quantity manufactured for sale or use k g
☐ **3 a**.Quantity lost during manufacture (3b + 3c + 3d must equal 3a)	k g	**3 c**. Quantity in wastes treated to destroy the chemical . k g ± %
3 b.Quantity lost to the environment k g ± %		**3 d**. Quantity in wastes not treated to destroy the chemical k g ± %

Activity (1)	Process category (2)	Quantity (kilograms) (3)	Total worker-hours (4)	Total workers (5)
☐ **4**.Manufacture of the chemical	**a.** Enclosed			
	b. Controlled release			
	c. Open			
☐ **5**.On-site use as reactant	**a.** Enclosed			
	b. Controlled release			
Total Quantity _____ k g	**c.** Open			
☐ **6**.On-site nonreactant use of the chemical substance	**a.** Enclosed			
	b. Controlled release			
Total Quantity _____ k g	**c.** Open			
☐ **7**.On-site preparation of products	**a.** Enclosed			
	b. Controlled release			
Total Quantity _____ k g	**c.** Open			

☐ **8**.MANUFACTURER'S PRODUCTS — Report the quantity of the chemical substance that you prepare for each of the following.

INDUSTRIAL PRODUCTS (domestic)	**a.** Chemical or mixture	k g	CONSUMER PRODUCTS (domestic)	**d.** Chemical or mixture	k g
	b. Article with some release	k g		**e.** Article with some release	k g
	c. Article with no release	k g		**f.** Article with no release	k g

g. Products for export →   k g

▶ **Part B — Chemical Substance Processing by Customers** — *Information in part B must be accurate to within ± 50%.*

☐ **9**.CUSTOMERS' USES AND PRODUCTS — Estimate the quantity of the chemical substance that your customers use or prepare for each of the following.

INDUSTRIAL PRODUCTS (domestic)	**a.** Chemical or mixture	k g	CONSUMER PRODUCTS (domestic)	**d.** Chemical or mixture	k g
	b. Article with some release	k g		**e.** Article with some release	k g
	c. Article with no release	k g		**f.** Article with no release	k g

g. Products for export →   k g

h. Quantity of chemical consumed as reactant →   k g

i. Unknown customer uses →   k g

☐ **10**.MARKET NAMES — If you report your customers' uses as unknown (9i above) for more than 20% of the total quantity of chemical substance that you manufacture and import (20% of items 1 and 2 above), list the market names under which you distribute the chemical. (If you need more space, attach an additional sheet.)

a.	**c.**
b.	**d.**

☐ **11**.CUSTOMERS' PROCESS CATEGORIES — Based on your knowledge of general industry practices, estimate the quantity of chemical substance that you sell to customers as the chemical and that your customers further process in each of the following categories.

a. Enclosed processes	k g	**c.** Open processes	k g
b. Controlled release processes	k g	**d.** Unknown	k g

EPA Form 7710-35

APPENDIX 9B   Premanufacture Notice for New Chemical Substances

O.M.B. No. 2070-0012: Approval Expires 3-3-86

	AGENCY USE ONLY
⬥**EPA** United States Environmental Protection Agency	Date of receipt

PREMANUFACTURE NOTICE
FOR NEW CHEMICAL SUBSTANCES

When completed send this form to:

DOCUMENT CONTROL OFFICER
OFFICE OF TOXIC SUBSTANCES, TS-793
U.S. E.P.A.
401 M STREET, SW
WASHINGTON, D.C. 20460

Enter the total number of pages in the Premanufacture Notice ⟶

Document control number	EPA case number

GENERAL INSTRUCTIONS

You must provide all information requested in this form to the extent that it is known to or reasonably ascertainable by you. Make reasonable estimates if you do not have actual data.

Before you complete this form, you should read the "Instructions Manual for Premanufacture Notification" (**Instructions Manual**).

Part I. GENERAL INFORMATION

You must provide the chemical identity of the new chemical substance, even if you claim the identity as confidential. You may authorize another person to submit the identity for you, but your submission will not be complete and review will not begin until EPA receives this information.

Part II. HUMAN EXPOSURE AND ENVIRONMENTAL RELEASE

You may need additional copies of part II, sections A and B if there are several manufacture, processing, or use operations that you will describe in the notice. You should reproduce these sections as needed.

Part III. LIST OF ATTACHMENTS

You should attach additional sheets if you do not have enough space on the form to answer a question fully. In part III, list these attachments, any test data or other data, and any optional information that you include in the notice.

OPTIONAL INFORMATION

You may include in the notice any information that you want EPA to consider in evaluating the new substance. The **Instructions Manual** identifies categories of optional information that you may want EPA to review.

CONFIDENTIALITY CLAIMS

You may claim any information in this notice as confidential. To assert a claim on the form, mark (X) the confidential box next to the information that you claim as confidential. To assert a claim in an attachment, circle or bracket the information you claim as confidential. If you claim information in the notice as confidential, you must provide a sanitized version of the notice, including attachments, to EPA with your submission. For additional instructions on claiming information as confidential, read the **Instructions Manual.**

Indicate below the categories of information you have claimed as confidential in the notice.

1. ☐ SUBMITTER IDENTITY
2. ☐ CHEMICAL IDENTITY
3. ☐ PRODUCTION VOLUME
4. ☐ USE INFORMATION
5. ☐ PROCESS INFORMATION
6. ☐ PORTIONS OF A MIXTURE
7. ☐ OTHER INFORMATION

TEST DATA AND OTHER DATA

You are required to submit all test data in your possession or control and to provide a description of all other data known to or reasonably ascertainable by you if these data are related to the health and environmental effects of the manufacture, processing, distribution in commerce, use, or disposal of the new chemical substance. Standard literature citations may be submitted for data in the open scientific literature. Complete test data, not summaries of data, must be submitted if they do not appear in the open literature. Following are examples of test data and other data. You should submit these data according to the requirements of §720.50 of the Premanufacture Notification Rule (40 CFR Part 720).

Test data

- **Environmental fate data**

  Spectra (UV, visible, and infrared)
  Density of liquids and solids
  Water solubility
  Melting point/melting range
  Boiling point/boiling range
  Vapor pressure
  Partition coefficient, n-octanol/water
  Biodegradation
  Hydrolysis (as a function of pH)
  Photochemical degradation
  Adsorption/desorption to soil types
  Dissociation constant
  Other physical/chemical properties

- **Health effects data**

  Mutagenicity
  Carcinogenicity
  Teratogenicity
  Acute toxicity
  Repeated dose toxicity
  Metabolism studies
  Sensitization
  Irritation

- **Environmental effects data**

  Microbial and algal toxicity
  Terrestrial vascular plant toxicity (e.g., seed germination studies, growth inhibition)
  Acute and chronic toxicity to animals (e.g., fish, birds, mammals, invertebrates)

Other data

- **Risk assessments**
- **Structure/activity relationships**
- Test data not in the possession or control of the submitter

EPA Form 7710-25 (4-26-83)

CERTIFICATION

I certify that to the best of my knowledge and belief:

1. The company named in part I, section A, subsection 1a of this notice form intends to manufacture or import for a commercial purpose, other than in small quantities solely for research and development, the substance identified in part I, section B.

2. All information provided in this notice is complete and truthful as of the date of submission.

3. I am submitting with this notice all test data in my possession or control and a description of all other data known to or reasonably ascertainable by me as required by §720.50 of the Premanufacture Notification Rule.

		Confidential
Signature of authorized official	Date	
Signature of agent — *(if applicable)*	Date	

Part I — GENERAL INFORMATION

Section A — SUBMITTER IDENTIFICATION

Confidential

Mark (X) the "Confidential" box next to any subsection you claim as confidential.

1a. Person submitting notice

Name of authorized official	Title
Company	
Mailing address (number and street)	
City, State, ZIP code	

b. Agent *(if applicable)*

Name of authorized official	Title
Company	
Mailing address (number and street)	
City, State, ZIP code	

c. If you are submitting this notice as part of a joint submission, mark (X) this box. ⟶ ☐

2. Technical contact

Name	Title
Company	
Mailing address (number and street)	
City, State, ZIP code	Telephone	Area code	Number

3. If you have had a prenotice communication (PC) concerning this notice and EPA assigned a PC Number to the notice, enter the number ⟶ | Mark (X) if none ⟶ ☐

4. If you have submitted a test-marketing exemption (TME) application for the chemical substance covered by this notice, enter the TME number assigned by EPA ⟶ | Mark (X) if none ⟶ ☐

5. If you have submitted a bona fide request for the chemical substance covered by this notice, enter the bona fide request number assigned by EPA ⟶ | Mark (X) if none ⟶ ☐

6. Type of Notice — *Mark (X)*   ₁ ☐ Manufacture   ₂ ☐ Import

Part I — GENERAL INFORMATION — Continued

Section B — CHEMICAL IDENTITY INFORMATION

Mark (X) the "Confidential" box next to any item you claim as confidential.

	Confi-dential
Complete either Item 1 or 2 as appropriate. Complete all other items. If another person will submit chemical identity information for you, mark (X) the box at the right. ⟶ ☐ Identify the name, company, and address of that person in a continuation sheet.	
1. Class 1 or 2 chemical substances (for definitions of class 1 and class 2 substances, see the **Instructions Manual**) **a.** Class of substance — *Mark (X)* 1 ☐ Class 1 2 ☐ Class 2	
b. Chemical name (preferably CAS or IUPAC nomenclature)	
c. Molecular formula and CAS Registry Number (if known)	
d. For a class 1 substance, provide a structural diagram. For a class 2 substance — (1) List the immediate precursor substances with their respective CAS Registry Numbers. (2) Describe the nature of the reaction or process. (3) Indicate the range of composition and the typical composition (where appropriate). (4) Provide a representative structural diagram (if possible).	

☐ Mark (X) this box if you attach a continuation sheet.

Part I — GENERAL INFORMATION — Continued

Section B — CHEMICAL IDENTITY INFORMATION — Continued

2. **Polymers** (For a definition of polymer, see the **Instructions Manual**.)

	Confidential

a. Indicate the **lowest** number-average molecular weight composition of the polymer you intend to manufacture. Indicate the **maximum** weight percent of low molecular weight species below 500 and below 1,000 absolute molecular weight of that composition. Describe the methods of measurement or the bases for your estimates.

☐ Mark (X) this box if you attach a continuation sheet.

b. You must make separate confidentiality claims for monomer or other reactant identity, composition information, and residual information. Mark (X) the "Confidential" box next to any item you claim as confidential.

(1) — Provide the chemical name and CAS Registry Number of each monomer or other reactant used in the manufacture of the polymer
(2) — Indicate the **typical** weight percent of each monomer or other reactant in the polymer.
(3) — Mark (X) the identity column if you want a monomer or other reactant used at two weight percent or less to be listed as part of the polymer description on the TSCA Chemical Substance Inventory.
(4) — Indicate the **maximum** weight percent of each monomer or other reactant that may be present as a residual in the polymer as manufactured for commercial purposes.

Monomer or other reactant and CAS Registry Number (1)	Confidential	Typical composition (2)	Identity Mark (X) (3)	Confidential	Maximum residual (4)	Confidential
		%			%	
		%			%	
		%			%	
		%			%	
		%			%	
		%			%	
		%			%	

☐ Mark (X) this box if you attach a continuation sheet.

c. Provide a representative structural diagram of the polymer, if possible.

☐ Mark (X) this box if you attach a continuation sheet.

Part I — GENERAL INFORMATION — Continued

▶ **Section B — CHEMICAL IDENTITY INFORMATION — Continued**

3. Impurities

(a) — Identify each impurity that may be reasonably anticipated to be present in the chemical substance as manufactured for commercial purposes. Provide the CAS Registry Number if available. If there are unidentified impurities, enter "unidentified."

(b) — Estimate the **maximum** weight percent of each impurity. If there are unidentified impurities, estimate their total weight percent.

Impurity and CAS Registry Number (a)	Maximum percent (b)	Confi-dential
	%	
	%	
	%	
	%	
	%	
	%	
	%	

☐ *Mark (X) this box if you attach a continuation sheet.*

4. Synonyms — Enter any synonyms for the new chemical substance identified in subsection 1 or 2.

	Confi-dential

☐ *Mark (X) this box if you attach a continuation sheet.*

5. Trade identification — List trade names for the new chemical substance identified in subsection 1 or 2.

☐ *Mark (X) this box if you attach a continuation sheet.*

6. Generic chemical name — If you claim chemical identity as confidential, enter the generic chemical name that you developed with EPA during prenotice communication. If you have not developed a generic name with EPA, provide a generic name that reveals the specific chemical identity of the new chemical substance to the maximum extent possible. Read the TSCA Chemical Substance Inventory, Initial Inventory, Volume I for guidance on developing generic names.

☐ *Mark (X) this box if you attach a continuation sheet.*

7. Byproducts — Describe any byproducts resulting from the manufacture, processing, use, or disposal of the new chemical substance at sites you control. Provide the CAS Registry Number if available.

Byproduct (1)	CAS Registry Number (2)	Confi-dential

☐ *Mark (X) this box if you attach a continuation sheet.*

Part I— GENERAL INFORMATION — Continued

▶ Section C — PRODUCTION, IMPORT, AND USE INFORMATION

Mark (X) the "Confidential" box next to any item you claim as confidential.

1. **Production volume** — Estimate the **maximum** production volume during the first 12 months of production. Also estimate the **maximum** production volume for any consecutive 12-month period during the first three years of production.	Confi-dential
Maximum first 12-month production (kg/yr) / Maximum 12-month production (kg/yr)	

Maximum first 12-month production (kg/yr)	Maximum 12-month production (kg/yr)

2. Use Information

You must make separate confidentiality claims for the description of the category of use, the percent of production volume devoted to each category, the formulation of the new substance, and other use information. Mark (X) the "Confidential" box next to any item you claim as confidential.

a. (1) — Describe each intended category of use of the new chemical substance by function and application.

(2) — Estimate the percent of total production for the first three years devoted to each category of use.

(3) — Estimate the percent of the new substance as formulated in mixtures, suspensions, emulsions, solutions, or gels as manufactured for commercial purposes at sites under your control associated with each category of use.

(4) — Mark (X) whether the use is site-limited, industrial, commercial, or consumer. Mark more than one column if appropriate.

Read the **Instructions Manual** for examples.

Category of use (1)	Confi-dential	Production (percent) (2)	Confi-dential	Formulation (percent) (3)	Confi-dential	Mark (X) appropriate column(s) (4)				Confi-dential
						Site-limited	Indus-trial	Com-mercial	Con-sumer	
		%		%						
		%		%						
		%		%						
		%		%						

☐ *Mark (X) this box if you attach a continuation sheet.*

b. Generic use description

If you claim any category of use description in subsection 2a as confidential, enter a generic description of that category. Read the **Instructions Manual** for examples of generic use descriptions.

_____

_____

_____

_____

☐ *Mark (X) this box if you attach a continuation sheet.*

3. Hazard Information — Include in the notice a copy or reasonable facsimile of any hazard warning statement, label, material safety data sheet, or other information which will be provided to any person regarding protective equipment or practices for the safe handling, transport, use, or disposal of the new chemical substance. List in part III any hazard information you include.

☐ *Mark (X) this box if you attach hazard information.*

Part II — HUMAN EXPOSURE AND ENVIRONMENTAL RELEASE

▶ ## Section A — INDUSTRIAL SITES CONTROLLED BY THE SUBMITTER

Complete section A for each type of manufacture, processing, or use operation involving the new chemical substance at industrial sites you control.

Mark (X) the "Confidential" box next to any item you claim as confidential.

1. Operation description	Confi-dential
a. Identity— Enter the identity of the site at which the operation will occur.	

Name	
Site address (number and street)	
City, County, State, ZIP code	

If the same operation will occur at more than one site, enter the number of sites. ⟶
Identify the additional sites on a continuation sheet.

☐ *Mark (X) this box if you attach a continuation sheet.*

b. Type —
Mark (X)

1 ☐ Manufacturing   2 ☐ Processing   3 ☐ Use

c. Amount and Duration — *Complete 1 or 2 as appropriate*

	Maximum kg/batch	Hours/batch	Batches/year
1. Batch			
	Maximum kg/day	Hours/day	Days/year
2. Continuous			

d. Process description
(1) Diagram the major unit operation steps and chemical conversions.
(2) Provide the identity, the approximate weight (by kg/day or kg/batch), and entry point of all feedstocks (including reactants, solvents, and catalysts).
(3) Identify by number the points of release to the environment of the new chemical substance.

☐ *Mark (X) this box if you attach a continuation sheet.*

FORM EPA-7710-25 (4-26-83)   Page 7

Part II — HUMAN EXPOSURE AND ENVIRONMENTAL RELEASE — Continued

▶ **Section A — INDUSTRIAL SITES CONTROLLED BY THE SUBMITTER — Continued**

2. Occupational Exposure

You must make separate confidentiality claims for the description of worker activity, physical form of the new chemical substance, number of workers exposed, and duration of activity. Mark (X) the "Confidential" box next to any item you claim as confidential.

(1) — Describe the activities in which workers may be exposed to the new chemical substance. Include activities in which workers wear protective equipment.
(2) — Indicate the physical form(s) of the new chemical substance at the time of exposure.
(3) — Estimate the maximum number of workers involved in each activity.
(4) and (5) — Estimate the maximum duration of the activity for any worker in hours per day and days per year.

Worker activity (1)	Confi-dential	Physical form(s) (2)	Confi-dential	Maximum number (3)	Confi-dential	Maximum duration		Confi-dential
						Hrs/day (4)	Days/yr (5)	

☐ Mark (X) this box if you attach a continuation sheet.

3. Environmental Release and Disposal

You must make separate confidentiality claims for the release number and the amount of the new chemical substance released and other release and disposal information. Mark (X) the "Confidential" box next to each item you claim as confidential.

(1) — Enter the number of each release point identified in the process description, part II, section A, subsection 1d(3).
(2) — Estimate the amount of the new chemical substance released directly to the environment or into control technology (in kg/day or kg/batch).
(3) — Identify the media (air, land, or water) to which the new substance will be released from that release point.
(4) — Describe control technology, if any, that will be used to limit the release of the new substance to the environment. For releases disposed of on land, characterize the disposal method.
(5) — Identify the destination(s) of releases to water.

Release Number (1)	Amount of new substance released (2)	Confi-dential	Media of release (3)	Control technology (4)	Confi-dential

(5) Mark (X) the destination(s) of releases to water.   1 ☐ POTW (publicly owned treatment works)   3 ☐ Other – *Specify*
2 ☐ Navigable waterway

☐ Mark (X) this box if you attach a continuation sheet.

Part II — HUMAN EXPOSURE AND ENVIRONMENTAL RELEASE — Continued

Section B — INDUSTRIAL SITES CONTROLLED BY OTHERS

Complete section B for each type of processing or use operation involving the new chemical substance at sites you do not control.

*To claim information in this section as confidential, circle or bracket the **specific** information that you claim as confidential.*

Operation description

Describe the typical processing or use operation. Identify the unit operation steps which may occur during the operation. Estimate the number of sites at which the operation is likely to occur. Identify situations in which worker exposure to and/or environmental release of the new chemical substance may occur. Estimate the percent of new chemical substance as formulated in products manufactured for commercial purposes in the operation or as used in the operation. Estimate the number of workers exposed and the duration of exposure. Identify controls which limit worker exposure and environmental release if typically used. Identify byproducts which may result from the operation.

☐ *Mark (X) this box if you attach a continuation sheet.*

Part III — LIST OF ATTACHMENTS

Attach continuation sheets for sections of the form and test data and other data (including physical/chemical properties and structure/activity information), and optional information after this page. Clearly identify the attachment and the section of the form to which it relates, if appropriate. Number consecutively the pages of the attachments. In the column below, enter the inclusive page numbers of each attachment.

Mark (X) the "Confidential" box next to any attachment **name** you claim as confidential. Read the **Instructions Manual** for guidance on how to claim any information in an attachment as confidential. You must include with the sanitized copy of the notice form a sanitized version of any attachment in which you claim information as confidential.

Attachment name	Attachment page number(s)	Confi-dential

☐ Mark (X) this box if you attach a continuation sheet.   Enter the attachment name and number.

APPENDIX 9C   Report of Commercial and Industrial Uses of Asbestos

O.M.B. No. 2000-0478; Approval Expires June 30, 1985

⊕**EPA** United States Environmental Protection Agency	**When completed send this form to:** **U.S. Environmental Protection Agency** **P.O. Box 2070** **Rockville, MD 20852**

REPORTING COMMERCIAL AND INDUSTRIAL USES OF ASBESTOS

Please read the accompanying instruction booklet before completing this form. Note that a separate form must be submitted for each plant site, except that total imports or exports of bulk asbestos may be reported in a consolidated corporate report. The instructions provide further directions.

When this form is complete, enclose it in the preaddressed envelope provided. Should you have any questions, please contact the Industry Assistance Office toll-free at (800) 424-9065 or in Washington, D.C. at (202) 554-1404.

All persons who are solely secondary processors, importers of asbestos mixtures, or importers of articles containing asbestos components will report in the first phase on EPA Form 7710-37, titled "Secondary Processing and Importation of Asbestos Mixtures." These persons will report in this form, EPA Form 7710-36, only upon receipt of notice to that effect from EPA (see 40 CFR 763, part D).

A(1). RESPONDENT IDENTIFICATION

1. Company name | Plant site name

Address (Number and street)

City | County | State | ZIP code

2. Principal technical contact person | Name | Job title | Office telephone (Area code, number, extension)

3. If applicable — Record mine identification number ⸺

4. If applicable — Record the name and address of the parent corporation which is responsible for the fiscal management of the reporting site.

Name of parent corporation

Address (Number and street)

City | County | State | ZIP code

5. Record the total number of years during which the manufacture and/or processing of asbestos has been conducted at the reporting site. ⸺ Number of years

A(2). RESPONDENT ACTIVITY

6. Identify your activity from the definitions in the instruction booklet and mark (X) the box next to the category below that best describes your activity(ies). Mark (X) next to all activities that you are reporting on this form. You must at least complete the portions of this report that are indicated next to the box you mark, and you must complete other sections if you conduct that activity. For example, if at the same plant site you process bulk asbestos to produce an asbestos mixture and you also process an asbestos mixture that is made at another location, then you as a primary processor must also complete part D as a secondary processor.

All persons who are miners, millers, primary processors, or importers of bulk asbestos must report their asbestos activities at all plant sites.

Mark (X) your activity(ies) that is reported on this form.

☐ Mine and mill — Complete parts A, B(1), G, H, I, J, K [B(3), if applicable]
☐ Primary processor — Complete parts A, C, G, H, I, J, K [B(3), if applicable]
☐ Importer of bulk asbestos — Complete parts A, B(2) [B(3), if applicable]
☐ Secondary processor — Complete parts A, D, G, H, I, J, K
☐ Importer of asbestos mixtures — Complete parts A and E
☐ Importer of article(s) containing asbestos component(s) — Complete parts A and F

A(3). CERTIFICATION FOR CLAIMS OF CONFIDENTIALITY

You may claim as confidential any information you submit on this form. The instructions provide specific instructions on how to assert claims of confidentiality. For information which you submit in attachments to the form, provide a copy which clearly indicates the information you wish to claim confidential.

Space is provided at the end of each line to claim information on that line confidential.

For any information on this form you claim as confidential, you must certify that the information is confidential. The signature below will attest to the truth and accuracy of the following statements. All four statements must be true about any information you claim confidential.

1. My company has taken measures to protect the confidentiality of the information, and it will continue to take these measures.

2. The information is not, and has not been, reasonably obtainable by other persons (other than governmental bodies) by using legitimate means (other than discovery based on a showing of special need and in a judicial or quasi-judicial proceeding) without my company's consent.

3. The information is not publicly available.

4. Disclosure of the information claimed as confidential would cause substantial harm to my company's competitive position.

7. Signature of authorized official | Date

B(1). PRODUCTION OF BULK ASBESTOS

QUANTITY OF BULK ASBESTOS MINED OR MILLED

Enter, in short tons, the amount of bulk asbestos you produced (mined or milled) for 1979 through 1981.

Year	Report in short tons						Confidential
	Chrysotile	Crocidolite	Amosite	Anthophyllite asbestos	Tremolite asbestos	Actinolite asbestos	Mark (X)
1979							
1980							
1981							

B(2). IMPORTATION OF BULK ASBESTOS

QUANTITY OF BULK ASBESTOS IMPORTED

Enter, in short tons, the amount of bulk asbestos you imported for 1979 through 1981.

Mark (X) one: ☐ This is a corporate consolidated report. ☐ This is a plant site report.

Year	Report in short tons						Confidential
	Chrysotile	Crocidolite	Amosite	Anthophyllite asbestos	Tremolite asbestos	Actinolite asbestos	Mark (X)
1979							
1980							
1981							

B(3). EXPORTATION OF BULK ASBESTOS

QUANTITY OF BULK ASBESTOS EXPORTED

Enter, in short tons, the amount of bulk asbestos you exported for 1979 through 1981.

Mark (X) one: ☐ This is a corporate consolidated report. ☐ This is a plant site report.

Year	Report in short tons						Confidential
	Chrysotile	Crocidolite	Amosite	Anthophyllite asbestos	Tremolite asbestos	Actinolite asbestos	Mark (X)
1979							
1980							
1981							

EPA Form 7710-36 (6-12-82)

Page 2

C. PRIMARY PROCESSOR PRODUCTION

Enter the following information for each type of end product that you shipped from your plant site. Classify your end products according to the list below.

End product shipped (Enter one end product per line) (1)		Type of asbestos fiber (Use one line per mineral type) (2)	Year	Quantity of asbestos consumed (Short tons) (3)	Product shipments				Confidential Mark (X)
Code number	Generic name				Total annual production (4)		Value shipped (Thousands of dollars) (5)		
					Quantity	Unit of measure	Domestic	Exports	
		☐ Chrysotile ☐ Anthophyllite	1979						
		☐ Crocidolite ☐ Tremolite	1980						
		☐ Amosite ☐ Actinolite	1981						
	☐ Continued	☐ Chrysotile ☐ Anthophyllite	1979						
		☐ Crocidolite ☐ Tremolite	1980						
		☐ Amosite ☐ Actinolite	1981						
	☐ Continued	☐ Chrysotile ☐ Anthophyllite	1979						
		☐ Crocidolite ☐ Tremolite	1980						
		☐ Amosite ☐ Actinolite	1981						

(6) Estimate the percentage of the total value shipped from the plant site reported here as primary processor production in 1981 ——————————— ____ %

Asbestos Mixture Product Subcategories

Papers, felts, or related products — Unit of measure
- 01. Commercial paper — Short tons
- 02. Rollboard — Short tons
- 03. Millboard — Short tons
- 04. Pipeline wrap — Short tons
- 05. Beater-add gasketing paper — Short tons
- 06. High-grade electrical paper — Short tons
- 07. Unsaturated roofing felt — Short tons
- 08. Saturated roofing felt — Short tons
- 09. Flooring felt — Short tons
- 10. Corrugated paper — Short tons
- 11. Specialty paper (Specify generic name) — Short tons

Floor coverings
- 12. Vinyl-asbestos floor tile — Square yards
- 13. Asbestos-felt-backed vinyl flooring — Square yards

Asbestos-cement products — Unit of measure
- 14. A/C pipe and fittings — Short tons
- 15. A/C sheet, flat — 100 sq. ft.
- 16. A/C sheet, corrugated — 100 sq. ft.
- 17. A/C shingle — Squares

Friction materials
- 18. Drum brake lining (light-medium vehicle) — Pieces
- 19. Disc brake pads (light-medium vehicle) — Pieces
- 20. Disc brake pads (heavy vehicle) — Pieces
- 21. Brake block (heavy equipment) — Pieces
- 22. Clutch facings (all) — Pieces
- 23. Automatic transmission friction components — Pieces
- 24. Friction materials (industrial and commercial) — Pieces

Textiles — Unit of measure
- 25. Cloth — Pounds
- 26. Thread, yarn, lap, roving, cord, rope, or wick — Pounds

Other products
- 27. Sheet gasketing (other than beater-add) — Sq. yards
- 28. Packing — Pounds
- 29. Paints and surface coatings — Gallons
- 30. Adhesives and sealants — Gallons
- 31. Asbestos-reinforced plastics — Pounds
- 32. Insulation materials not elsewhere classified (Specify generic name) — (Specify)
- 33. Mixed or repackaged asbestos fiber — Short tons
- 34. Other (Specify generic name) — (Specify)

D. SECONDARY PROCESSOR PRODUCTION

Enter the following information for each type of end product that you shipped from your plant site. Classify end products and asbestos mixtures consumed according to the lists on this form.

End product shipped (Enter one end product per line) (1)		Total annual production (2)			Asbestos mixture consumed and form (Enter one mixture per line) (3)		Function of asbestos in product (4)	Year	Annual consumption of asbestos mixtures (5)			Confidential Mark (X)
Code number	Generic name	Year	Quantity	Unit of measure	Code number	Generic name / Form			Quantity	Unit of measure	Delivered cost (Thousands of dollars)	
		1979						1979				
		1980						1980				
		1981						1981				
	☐ Continued	1979						1979				
		1980						1980				
		1981						1981				
	☐ Continued	1979						1979				
		1980						1980				
		1981						1981				

(6) Estimate the percentage of the total value shipped from the plant site reported here as secondary processor production in 1981 ——————————— ____ %

Typical terms for products made from asbestos mixtures

Automotive and friction products
- 101. Drum brake lining (light-medium vehicle)
- 102. Disc brake pads (light-medium vehicle)
- 103. Disc brake pads (heavy vehicle)
- 104. Brake block (heavy equipment)
- 105. Clutch facings (all)
- 106. Automatic transmission friction components
- 107. Friction materials (industrial and commercial)
- 108. Custom automotive body filler
- 109. Transmissions
- 110. Mufflers
- 111. Radiator top insulation
- 112. Radiator sealant
- 113. Other (Specify generic name)

Appliances
- 114. Appliance, industrial and consumer (Specify generic name)

Construction products
- 115. Boiler and furnace baffles
- 116. Decorated building panels
- 117. A/C sheet
- 118. Flexible air conductor
- 119. Hoods and vents
- 120. Portable construction building
- 121. Roofing, saturated
- 122. Roof shingles
- 123. Wallboard
- 124. Wall/roofing panels
- 125. Other (Specify generic name)

Clothing
- 126. Aprons
- 127. Boots
- 128. Gloves and mittens
- 129. Hats and helmets
- 130. Overgarters
- 131. Suits
- 132. Other (Specify generic name)

Floor coverings
- 133. Vinyl-asbestos floor tile
- 134. Asbestos-felt-backed sheet vinyl flooring

Electrical products and components
- 135. Cable insulation
- 136. Electronic motor components
- 137. Electrical resistance supports
- 138. Electrical switchboard
- 139. Electrical switch supports
- 140. Electrical wire insulation
- 141. Motor armature
- 142. Other (Specify generic name)

Fire and heat shielding equipment and components
- 143. Arc deflectors
- 144. Fire doors
- 145. Fireproof absorbent paper
- 146. Heat shields
- 147. Molten metal handling equipment
- 148. Oven and stove insulation
- 149. Pipe wrap
- 150. Stove lining, wood and coal
- 151. Stove pipe rings
- 152. Sleeves
- 153. Thermal insulation
- 154. Other (Specify generic name)

Gaskets
- 155. Sheet gasketing, rubber encapsulated beater addition
- 156. Sheet gasketing, rubber encapsulated compressed
- 157. Compressed sheet gasketing (other)
- 158. Metal reinforced gaskets
- 159. Automotive gaskets
- 160. Other (Specify generic name)

Marine equipment and supplies
- 161. Caulks, marine
- 162. Liners, pond or canal
- 163. Marine bulkheads
- 164. Other (Specify generic name)

Paints, coatings, sealants, and compounds
- 165. Asphaltic compounds
- 166. Automotive/truck body coatings
- 167. Buffing and polishing compounds
- 168. Caulking and patching compounds
- 169. Drilling fluid
- 170. Flashing compounds
- 171. Furnace cement
- 172. Glazing compounds
- 173. Plaster and stucco
- 174. Pump, valve, flange, and tank sealing components
- 175. Roof coatings
- 176. Textured paints
- 177. Tile cement
- 178. Other (Specify generic name)

Textile and felt products
- 179. Aluminized cloth
- 180. Rope or braiding
- 181. Yarn, lap, or roving
- 182. Wicks
- 183. Bags
- 184. Belting
- 185. Blankets
- 186. Carpet padding
- 187. Commercial/industrial dryer felts
- 188. Draperies
- 189. Drip cloths
- 190. Fire hoses

Textile and felt products — Con.
- 191. Ironing board pads and insulation
- 192. Mantles, lamp or catalytic heater
- 193. Packing and packing components
- 194. Piano and organ felts
- 195. Rugs
- 196. Tape
- 197. Theater curtains
- 198. Umbrellas
- 199. Other (Specify generic name)

Miscellaneous products
- 200. Aerial distress flares
- 201. Acoustical products
- 202. Ammunition wadding
- 203. Asbestos-reinforced plastic products
- 204. Ash trays
- 205. Baking sheets
- 206. Blackboards
- 207. Candlesticks
- 208. Chemical tanks and vessels
- 209. Filters
- 210. Grommets
- 211. Gun grips
- 212. Jewelry making equipment
- 213. Kilns
- 214. Lamp sockets
- 215. Light bulbs (all types)
- 216. Linings for vaults, safes, humidifiers, and filing cabinets
- 217. Phonograph records
- 218. Pottery clay
- 219. Welding rod coatings
- 220. Other (Specify generic name)

E. IMPORTATION OF ASBESTOS MIXTURES

Enter the following information for each type of asbestos mixture that you imported in 1981. Classify your imports according to the lists in section C or D.

Asbestos mixtures (Enter one mixture per line) (1)		Asbestos fiber content (if known) (2)			Year	Total annual imports (3)			Confidential Mark (X)
Code number	Generic name	Type of asbestos		Quantity (Per unit)		Quantity	Unit of measure	Value of imports (U.S. dollars)	
		☐ Chrysotile ☐ Anthophyllite ☐ Unknown			1979				
		☐ Crocidolite ☐ Tremolite			1980				
		☐ Amosite ☐ Actinolite		☐ Unknown	1981				
	☐ Continued	☐ Chrysotile ☐ Anthophyllite ☐ Unknown			1979				
		☐ Crocidolite ☐ Tremolite			1980				
		☐ Amosite ☐ Actinolite		☐ Unknown	1981				
	☐ Continued	☐ Chrysotile ☐ Anthophyllite ☐ Unknown			1979				
		☐ Crocidolite ☐ Tremolite			1980				
		☐ Amosite ☐ Actinolite		☐ Unknown	1981				

EPA Form 7710-36 (6-12-82)

Page 3

U.S. ENVIRONMENTAL PROTECTION AGENCY	COMPANY NAME AND ADDRESS (Same as reported in Item 1 on page 1)
REPORT OF COMMERCIAL AND INDUSTRIAL USES OF ASBESTOS Continued	

F. IMPORTATION OF ARTICLE(S) CONTAINING ASBESTOS COMPONENTS

Enter the following information for each article that you imported in 1981. Classify your imports according to the lists in section C or D.

Article name (Enter one article per line) (1)		Total annual imports (2)				Asbestos component(s) (Use one line per component) (3)	Confidential Mark (X)
Code number	Generic name	Year	Quantity	Unit of measure	Value (U.S. dollars)		
		1979					
		1980					
		1981					
		1979					
		1980					
		1981					
		1979					
		1980					
		1981					

G. EMPLOYEES

Record the number of employees at the reporting site in each of the categories below. Count each employee in one category only.

	Number of employees	Confidential Mark (X)
(1) TOTAL number of employees at plant site (Sum of 2, 3, 4, and 5)		
(2) Number of production employees		
(3) Number of employees in shipping, receiving, and moving		
(4) Number of maintenance employees		
(5) Number of other employees		

H. SUMMARY OF CURRENT WORKER EXPOSURES

This section requires you to report, by category of production workers, the arithmetic mean of the 8-hour time weighted average (TWA) and ceiling concentration exposure levels for all employees at the reporting site. Report for all employees for whom you have determined a TWA and according to each production line making the reported products. The instructions include a worksheet for making necessary calculations (see Appendix B).

End product (Enter name) (1)		Respondent activity (2)	Production work category (3)	Number of employees (4)	8 hr. time weighted average				Ceiling concentration level (Range) (9)	Confidential Mark (X)
Code number	Generic name				Arithmetic mean of detectable measurements (5)	Standard deviation (6)	Total number of measurements (filters) (7)	Number of non-detectable measurements (8)		
Not applicable		Miners and millers	Mine operations (includes all employees working in mine, including transporters)							
			Mill operations (includes all employees working in production areas of mill)							
		Primary processors	Fiber introduction operations (includes blending, mixing, bag opening, fluffing, willowing, etc.)							
		Primary processors and secondary processors	Wet mechanical operations (includes machining, sawing, drilling, cutting, grinding, pulverizing, weaving, etc.)							
			Dry mechanical operations (includes machining, sawing, drilling, cutting, grinding, pulverizing, weaving, etc.)							
			Other production employees							
		Primary processors	Fiber introduction operations (includes blending, mixing, bag opening, fluffing, willowing, etc.)							
		Primary processors and secondary processors	Wet mechanical operations (includes machining, sawing, drilling, cutting, grinding, pulverizing, weaving, etc.)							
			Dry mechanical operations (includes machining, sawing, drilling, cutting, grinding, pulverizing, weaving, etc.)							
			Other production employees							
		Primary processors	Fiber introduction operations (includes blending, mixing, bag opening, fluffing, willowing, etc.)							
		Primary processors and secondary processors	Wet mechanical operations (includes machining, sawing, drilling, cutting, grinding, pulverizing, weaving, etc.)							
			Dry mechanical operations (includes machining, sawing, drilling, cutting, grinding, pulverizing, weaving, etc.)							
			Other production employees							

(10) Sampling and analysis methods

☐ Mark this box if you use the NIOSH method to sample airborne concentrations of asbestos fibers in the workplace and analyze measurements (DHEW (NIOSH) Pub. No. 79-127).

If you utilize other sampling and analysis methods, enter the appropriate codes for those methods listed in the instruction booklet under part I.

(11) Enter the lowest detectable level of the measurements that your company expects to attain by the sampling and analysis methods. (Measurements below this level should be considered as "nondetectable." for the purposes of this rule.)

I. MEASURING ASBESTOS EMISSIONS OR FIBER RELEASE

NPDES Permit number _____	Date

☐ Yes, enclosed is a separate report for this section concerning measuring asbestos emissions.   ☐ No separate report

EPA Form 7710-36 (6-12-82)

Page 4

J. WASTE AND DISPOSAL

Enter the following information for each end product you report in section B(1), C, and D. If you cannot provide information by end product, provide information according to the form of waste (column 2).

End product (1)		Form of waste (sludge, slurry, broke or scrap, baghouse fines, or specify other) (2)	Total annual quantity of asbestos waste (Short tons) (3)	Average percent asbestos (4)	Disposal site (5)				Method of disposal (6)	Confidential Mark (x)
Code number	Generic name				Type of land disposal facility (See instructions) (a)	Location (b)	Ownership (c)	Permitted hazardous waste facility? (d)		
□ End product report □ Form of waste report				%	□ Surface impoundment □ Waste pile □ Land treatment □ Land fill □ Seepage facility □ Injection well	□ On-site □ Off-site	□ Company □ Private □ Municipal	□ Yes □ No		
□ End product report □ Form of waste report				%	□ Surface impoundment □ Waste pile □ Land treatment □ Land fill □ Seepage facility □ Injection well	□ On-site □ Off-site	□ Company □ Private □ Municipal	□ Yes □ No		
□ End product report □ Form of waste report				%	□ Surface impoundment □ Waste pile □ Land treatment □ Land fill □ Seepage facility □ Injection well	□ On-site □ Off-site	□ Company □ Private □ Municipal	□ Yes □ No		

(7) You may describe in a separate enclosure any steps you take to ensure that waste does not release airborne fibers.
□ Yes, enclosed is a separate description of waste disposal practices.   □ No separate report

K. POLLUTION CONTROL EQUIPMENT

On separate lines enter information about each piece of air pollution control equipment used to control asbestos emissions at your plant site.   □ Confidential

Item number (1)	Type of equipment (See code) (2)	Gas stream volume and temperature (3)	Equipment size (Baghouse and ESP only) (4)	Estimated collection efficiency (Percent) (5)	Normal operating schedule (Hrs./year) (6)	Quantity collected annually (Pounds) (7)	Source of emissions (8)	Stack or chimney? (9)	Special problems? (10)
		Volume ___ Temperature ___ □ N/A	Square feet ___	% □ Designed □ Actual	Hrs./yr. ___	Pounds ___ □ Designed □ Actual		□ Yes □ No	
		Volume ___ Temperature ___ □ N/A	Square feet ___	% □ Designed □ Actual	Hrs./yr. ___	Pounds ___ □ Designed □ Actual		□ Yes □ No	
		Volume ___ Temperature ___ □ N/A	Square feet ___	% □ Designed □ Actual	Hrs./yr. ___	Pounds ___ □ Designed □ Actual		□ Yes □ No	
		Volume ___ Temperature ___ □ N/A	Square feet ___	% □ Designed □ Actual	Hrs./yr. ___	Pounds ___ □ Designed □ Actual		□ Yes □ No	
		Volume ___ Temperature ___ □ N/A	Square feet ___	% □ Designed □ Actual	Hrs./yr. ___	Pounds ___ □ Designed □ Actual		□ Yes □ No	
		Volume ___ Temperature ___ □ N/A	Square feet ___	% □ Designed □ Actual	Hrs./yr. ___	Pounds ___ □ Designed □ Actual		□ Yes □ No	

(11) Estimate the percent of plant exhaust air that is treated by the equipment listed above————————————————————   %

Continuation of item answers

APPENDIX 9D   Report of Secondary Processing and Importation of Asbestos

---

⊕EPA

US ENVIRONMENTAL PROTECTION AGENCY
REPORTING SECONDARY PROCESSING AND IMPORTATION OF ASBESTOS MIXTURES

PART I – COMPANY INFORMATION

COMPANY NAME	FOR EPA USE ONLY	
ADDRESS (Street, City, State & ZIP Code)	TECHNICAL CONTACT	TELEPHONE NO.
	IMPORTER ☐ PRINCIPAL	☐ AGENT

PART II – SECONDARY PROCESSOR END PRODUCT(S)

From the list in Section I, enter the asbestos end product produced. Opposite each product, list the asbestos mixture that you process, and the quantity of each mixture that you consumed in 1981.

END PRODUCT(S)		ASBESTOS MIXTURE(S)		QUANTITY OF ASBESTOS MIXTURE CONSUMED	
CODE	GENERIC NAME	CODE	GENERIC NAME	QUANTITY	UNIT OF MEASURE

PART III – IMPORTERS OF ASBESTOS MIXTURE(S) OR ARTICLE(S) CONTAINING ASBESTOS COMPONENTS

List the asbestos mixture(s) or article(s) that you import and the quantity of each item that you imported in 1981. Opposite each item, enter a description of the asbestos component in the mixture or article.

ASBESTOS MIXTURE(S) OR ARTICLE(S)		QUANTITY OF ASBESTOS MIXTURE(S) OR ARTICLE(S) IMPORTED		DESCRIPTION OF ASBESTOS COMPONENT(S) IN ARTICLE
CODE	GENERIC NAME	QUANTITY	UNIT OF MEASURE	

CERTIFICATION FOR CLAIMS OF CONFIDENTIAL BUSINESS INFORMATION

An authorized company official may claim any information reported on this form as confidential business information. To do this, the confidential information must be clearly circled with a red marker. In addition, an authorized company official must sign below to certify the truth and accuracy of the following four statements, which apply to all information that is claimed.

1. My company has taken measures to protect the confidentiality of the information, and it will continue to take these measures.

2. The information is not, and has not been, reasonably obtainable by other persons (other than governmental bodies) by using legitimate means (other than discovery based on a showing of special need in a judicial or quasi-judicial proceeding) without my company's consent.

3. The information is not publicly available elsewhere.

4. Disclosure of the information claimed as confidential would cause substantial harm to my company's competitive position.

SIGNATURE OF AUTHORIZED OFFICIAL	DATE

EPA Form 7710-37 (10-81)

APPENDIX 9E   Dioxin/Furan Reporting Form

United States Environmental Protection Agency
Washington, DC 20460

♦EPA   Dioxins/Furans Report

Form Approved
OMB No. 2070–0017
Approval expires   8/89

When completed, send this form to	For Agency Use Only
Document Control Officer Office of Toxic Substances, TS-793 US Environmental Protection Agency 401 M Street, SW Washington, DC 20460	
	Document Control Number | Docket Number

Part I — General Information

Section A — Submitter Identification

Confidential

Mark (X) the "Confidential" box next to any subsection you claim as confidential.

1a. Person Submitting Notice	Name of authorized official	Title
	Company	
	Mailing address (number and street)	
	City, State, and ZIP Code	

Section B — Chemical Identity Information (Use a separate form for each chemical reported.)

Mark (X) the "Confidential" box next to any subsection you claim as confidential.

1. Chemical name and CAS Registry Number

Part II — Process and Release Information

Section A — Flow Diagram

Mark (X) the "Confidential" box next to any subsection you claim as confidential.

Complete this section for each unit process. Provide a general process block flow diagram that identifies major unit operations and treatment processes and indicate the types and points of release of byproducts and residuals. (See example I attached.)

(1) Include intermediates, coproducts and byproducts produced by the process.

(2) Proide a block for each major unit operation (e.g., reactor, washer, filtration, air emission control, aeration lagoon, etc.) in the production process and in the residuals management process.

(3) Identify process input such as raw materials, reagents, and solvents by chemical or common name and CAS number, and indicate the point of introduction with arrows.

(4) For each unit operation in which the temperature is not ambient, specify temperature or temperature range in each block of the flow diagram.

(5) Specify operating pressure or pressure range in each block of the flow diagram for each unit operation in which the pressure is not atmospheric.

(6) Identify the composition of the reaction vessel wherever one is used (e.g., stainless steel, glass-lined).

(7) Number all points in the flow diagram from which the chemical substance will be released into the environment. (See example I)

☐ Mark (x) this box if you attach a continuation sheet

EPA Form 7710-51 (9-86)

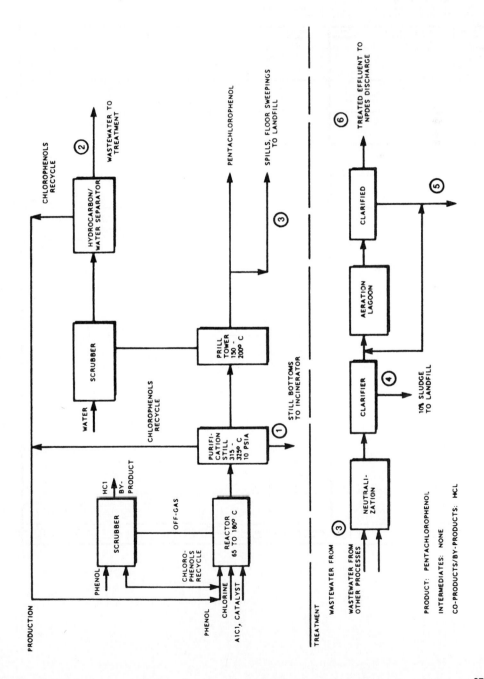

EXAMPLE I - PROCESS BLOCK FLOW DIAGRAM

PRODUCTION

PHENOL

CHLORINE
AlCl₃, CATALYST

PHENOL

CHLORO-
PHENOLS RECYCLE

SCRUBBER

OFF-GAS

HCl BY-PRODUCT

REACTOR
65 TO 180° C

CHLOROPHENOLS RECYCLE

WATER

SCRUBBER

CHLOROPHENOLS RECYCLE

HYDROCARBON/
WATER SEPARATOR

CHLOROPHENOLS RECYCLE

WASTEWATER TO TREATMENT ②

PURIFI-
CATION
STILL
315 -
325° C
10 PSIA

STILL BOTTOMS TO INCINERATOR ①

PRILL
TOWER
150 -
200° C

PENTACHLOROPHENOL

SPILLS, FLOOR SWEEPINGS
TO LANDFILL

③

TREATMENT

WASTEWATER FROM ③

WASTEWATER FROM
OTHER PROCESSES

NEUTRALI-
ZATION

CLARIFIER

④

10% SLUDGE
TO LANDFILL

AERATION
LAGOON

CLARIFIED

TREATED EFFLUENT TO
NPDES DISCHARGE ⑥

⑤

PRODUCT: PENTACHLOROPHENOL

INTERMEDIATES: NONE

CO-PRODUCTS/BY-PRODUCTS: HCL

371

Section B — Environmental Release and Disposal

You must make separate confidentiality claims for the release number and the amount of the substance released and other release and disposal information. Mark (x) the "Confidential" box next to each item you claim as confidential.

(1) — Enter the number of each release point identified in the process description, part II, Section A.
(2) — Estimate the amount of the chemical substance released directly to the environment or into control technology *(in kg/day or kg/batch)*.
(3) — Mark (x) this column if entries in columns (1) and/or (2) are confidential
(4) — Identify the media *(air, land, or water)* to which the substance will be released from the release point.
(5) — Describe control technology, if any, that will be used to limit the release of the substance to the environment. For releases disposed of on land, characterize the disposal method.
(6) — Mark (x) this column if entries in columns (4) and/or (5) are confidential
(7) — Identify the destination(s) of releases to water.

Release Number (1)	Amount of substance released (2)	Confi-dential (3)	Media of release (4)	Control technology (5)	Confi-dential (6)

(7) Mark (x) the destination(s) of releases to water   ☐ POTW *(publicly owned treatment works)*   ☐ Navigable waterway   ☐ Other *(specify)*   ☐ Mark (x) this box if you attach a continuation sheet

EPA Form 7710-51 (9-86)

Part III — Production, Import, and Use Information

Mark (x) the "Confidential" box next to any item you claim as confidential.

		Confi- dential
1.	**Production volume** — Report the production volume during the past 12 months of production. Also report the maximum production volume for any consecutive 12-month period during the past 3 years of production.	
	Past 12-month production *(kg/year)* Maximum 12-month production *(kg/year)*	

2. **Use Information** — You must make separate confidentiality claims for the description of the category of use, the percent of production volume devoted to each category, the formulation of the substance, and other use information. Mark (x) the "Confidential" box next to any item you claim as confidential.

(1) — Describe each category of use of the chemical substance by function and application.
(2) — Mark (x) this column if entry in column (1) is confidential.
(3) — Estimate the percent of total production for the past 3 years devoted to each category of use.
(4) — Mark (x) this column if entry in column (3) is confidential.
(5) — Estimate the percent of the substance as formulated in mixtures, suspensions, emulsions, solutions, or gels as manufactured for commercial purposes at sites under your control associated with each category of use.
(6) — Mark (x) this column if entry in column (5) is confidential.
(7) — Mark (x) whether the use is site-limited, industrial, commercial, or consumer. Mark more than one column if appropriate.
(8) — Mark (x) this column if entries in column (7) are confidential.

Read the Instructions Manual for examples.

Category of use (1)	Confi- dential (2)	Production (percent) (3)	Confi- dential (4)	Formulation (percent) (5)	Confi- dential (6)	Site- limited	Indus- trial	Com- mercial	Con- sumer	Confi- dential (8)

Columns (7): Mark (x) appropriate column(s)

☐ Mark (x) this box if you attach a continuation sheet.

3. **Hazard Information** — Include in the notice a copy or reasonable facsimile of any hazard warning statement, label, material safety data sheet, or other information which will be provided to any person regarding protective equipment or practices for the safe handling, transport, use, or disposal of the new chemical substance. List in part IV any hazard information you include.

☐ Mark (x) this box if you attach hazard information.

EPA Form 7710-51 (9-86)

4. Occupational Exposure — You must make separate confidentiality claims for the description of worker activity, physical form of the substances, number of workers exposed, and duration of activity. Mark (x) the "Confidential" box next to any item you claim as confidential.

(1) — Describe the activities in which workers may be exposed to the chemical substance. Include activities in which workers wear protective equipment

(2) — Mark (x) this column if entry in column (1) is confidential

(3) — Indicate the physical form(s) of the chemical substance at the time of exposure

(4) — Mark (x) this column if entry in column (3) is confidential

(5) — Estimate the maximum number of workers involved in each activity

(6) — Mark (x) this column if entry in column (5) is confidential

(7) and (8) — Estimate the maximum duration of the activity for any worker in hours per day and days per year

(9) — Mark (x) this column if entries in column (7) and/or (8) are confidential

Worker Activity (1)	Confidential (2)	Physical Forms (3)	Confidential (4)	Maximum number (5)	Confidential (6)	Maximum duration Hrs/day (7)	Maximum duration Days/yr (8)	Confidential (9)

☐ Mark (x) this box if you attach a continuation sheet

EPA Form 7710-51 (9-86)

Part IV — List of Attachments

Attach continuation sheets for sections of the form and optional information after this page. Clearly identify the attachment and the section of the form to which it relates, if appropriate. Number consecutively the pages of the attachments. In column (2) below, enter the inclusive page numbers of each attachment.

Mark (x) the "Confidential" box next to any attachment name you claim as confidential. Read the **Instructions Manual** for guidance on how to claim any information in an attachment as confidential.

Attachment name (1)	Attachment page numbers (2)	Confidential (3)

☐ Mark (x) this box if you attach a continuation sheet. Enter the attachment name and number.

Certification

I certify that to the best of my knowledge and belief:

1. The company named in part I, section A, subsection 1a of this form manufactures, imports, or processes, other than in small quantities for research purposes, the substance identified in part I, section B.
2. All information provided in this notice is complete and truthful as of the date of submission.

Signature of authorized official	Date	Confidential
Signature of agent *(if applicable)*	Date	Confidential

EPA Form 7710-51 (9-86)

Spill Reporting Requirements under Federal Statutes and Regulations

10.1 Introduction

Any release or spill, whether intentional or accidental, can trigger several statutory provisions and regulatory requirements, unless such a release qualifies as a "federally permitted" or, in the context of state laws, a "state-permitted" release. Spills can involve different substances and ingredients and are designated as such, e.g., oil spill, chemical spill, and hazardous or toxic substance spill.

Historically, however, it is the oil spill that received the congressional attention first because of the lasting impact created by several major oil spills in the rivers or oceans in the past. In recent years, congressional attention has been focused more on chemical spills involving various chemicals and chemical substances, particularly in the workplace. Most recently, with the enactment of the Comprehensive Environmental Response, Compensation, and Liability Act (CERCLA) and its progeny, the Superfund Amendments and Reauthorization Act (SARA), a great degree of attention has been paid to the releases or spills of hazardous substances. This reached a climax as the news of an unwanted and unplanned release of deadly methyl isocynate in Bhopal, India, broke out—a release that caused a disproportionate loss of human life, disfiguration, and other short- and long-term human miseries and ailments.

While all spills involving a hazardous substance or toxic chemical may cause extreme concern, in the practical sense they cannot be avoided altogether. This is due to the fact that manufacturing processes are not foolproof. Human error must therefore be taken into account in a contingency plan to deal with spills or releases. One can find some solace but not ambivalence in the philosophy which suggests that to err is human. Spills can occur for many reasons: negligence, accident, equipment failure, leaky container, poor maintenance, and intentional misconduct. In view of the possible devastating effect of a spill or release, particularly of certain extremely hazardous and toxic substances, several federal environmental laws impose strict reporting requirements. The intent here is not only to ensure a prompt cleanup, but also to determine what health or safety measures must be undertaken in order to minimize hazards to human health and the environment. In short, such reporting triggers measures that are part of government response and contingency plans.

Today, reporting requirements for spills or releases of oil, chemicals or chemical substances, and hazardous substances or materials are specifically covered in several federal environmental statutes and the regulations promulgated thereunder. In addition, similar requirements are included at the state and local levels.

Below is a list of the most important federal environmental statutes which deal with or address the issue of spills or releases:

- The Federal Water Pollution Control Act (FWPCA) and the Clean Water Act (CWA) along with its latest amendment, the Water Quality Act (WQA)
- The Comprehensive Environmental Response, Compensation, and Liability Act
- The Superfund Amendments and Reauthorization Act
- The Toxic Substances Control Act (TSCA)
- The Hazardous Materials Transportation Act (HMTA)
- The Resource Conservation and Recovery Act (RCRA)
- The Oil Pollution Act (OPA)

However, the general subject matter of these statutes does not necessarily include specific reporting requirements for that substance or material. For example, spill reporting quantities (RQs) for PCBs and other toxic pollutants are provided under CERCLA and not under TSCA which does not require EPA to specify RQs for toxic substances. Each of these federal statutes must be recognized and consulted separately. At the federal level, individual statutes often overlap. Moreover, they uniquely address and regulate certain specific types and aspects of spills and associated environmental concern.

The reporting obligation is also not the same in each of these statutory schemes. In most cases, the owner or operator of the facility from which a release or spill occurs is obligated to report. In certain cases, for example, pursuant to CWA, the reporting obligation shifts to the person in charge. In some other situations, the person actually causing the spill or the release is required to report.

Furthermore, many federal and state statutes have specific reporting requirements which must be met. Hence, while conformity with the federal reporting requirements may absolve a person from liabilities arising out of federal reporting obligations, that person may still be liable under the state laws and/or regulations promulgated thereunder for failure to report promptly and adequately at the state or local level.

Since many of the spills may eventually result in an explosion or fire, state and local laws and regulations may include specific spill reporting requirements. They are not preempted or automatically covered by the reporting requirements established by federal statutes and regulations. Hence, a prudent course of action should be based on complete spill reporting requirements in accordance with appropriate federal, state, and local laws and regulations.

10.2 Oil Spill

The Clean Water Act

Section 311(b)(3) of the Clean Water Act prohibits any unpermitted discharge of oil or hazardous substances into or upon the navigable waters of the United States, adjoining shorelines, or into or upon the waters of the contiguous zone. This statutory prohibition generally extends to the activities covered under the Outer Continental Shelf Lands Act of 1953 or the Deepwater Port Act of 1974. Additionally, such prohibited activities include those which may affect natural resources belonging to, appertaining to, or under the exclusive management authority of the United States. This prohibition, however, does not apply to certain permitted situations, e.g., discharge of oil which is permitted by the National Pollutant Discharge Elimination System (NPDES).

The oil spill prohibition is generally all-encompassing and highly restrictive since it is not linked to any specific quantity of oil. It has been interpreted by EPA that any discharge of oil, irrespective of amount, constitutes a statutory violation.

Section 311(a)(1) of FWPCA or CWA provides the following definition of oil:

"[O]il" means oil of any kind or in any form, including, but not limited to, petroleum, fuel oil, sludge, oil refuse, and oil mixed with wastes other than dredged spoil.

Section 311(a)(2) of CWA provides a very broad definition of discharge:

"[D]ischarge" includes, but is not limited to, any spilling, leaking, pumping, pouring, emitting, emptying or dumping.

However, this statutory definition of discharge excludes the following occurrences specifically:

- Discharges in compliance with a permit under section 402 of CWA
- Discharges resulting from one or more of permitted circumstances that are made part of a discharge permit issuance or modification under section 402 of CWA
- Discharges, continuous or anticipated intermittent, from a point source properly identified in a discharge permit or permit application under section 402 of CWA.

As emphasized earlier, under EPA's interpretation and current policy, any discharge of oil into the surface water or groundwater is a statutory violation unless such discharge is federally permitted. Hence, if there is a discharge of oil which, by all accounts, is minimal but not covered under a valid NPDES or an equivalent State Pollutant Discharge Elimination System (SPDES) permit, it will be a violation of the law.

EPA regulations as set forth in 40 CFR part 110 establish that any visible sheen of oil on the waters or any discoloration of the water surface is a presumptive or acceptable proof of discharge of oil. The courts have generally deferred to this interpretation by EPA and limited their decisions to the issues of liability for such discharge. A defense based on *de minimis* oil discharge or a discharge causing little or no harm has made little impact on the judiciary so far. The "sheen" test, according to these courts, creates a presumption of harm to the waters as a result of a spill or discharge. In a 1990 case entitled *Chevron, USA, Inc., v. Yost,* the Fifth Circuit Court of Appeals confirmed this trend by holding that EPA may prohibit *de minimis* discharges of oil or other hazardous substances in amounts that EPA considers harmful. Thus far, EPA has taken the position that those discharges of oil which cause a sheen may be "harmful to the public health or welfare of the United States." Nonetheless, this presumption by EPA may be rebuttable upon showing that no actual harm has been caused as a result of subsequent response actions.

With the passage of more recent federal environmental laws, particularly CERCLA, many of the original liability provisions of FWPCA have been supplemented and, to some extent, complemented. This has moved the focus from FWPCA to the statutory provisions of these later laws. However, note that certain requirements and liabilities based on FWPCA or its later amendments in the form of CWA and WQA have survived and are still applicable.

Section 311(b)(5) of FWPCA imposes a notification requirement on persons in charge of vessels and onshore and offshore facilities in the event of a spill of oil and hazardous substances (together, they are defined as *regulated substances*) into navigable waters, adjoining shoreline, or the contiguous zone. This notification must be made immediately after knowledge of a spill is gained. The notification or reporting requirement is triggered when the specified reportable quantities (RQs) are met or exceeded. EPA's Table 117.3 as set forth in 40 CFR section 117.3 provides the reportable quantities for hazardous substances under FWPCA or CWA, its later amendment. Table 1.1 reproduces EPA's Table 117.3. Note that the list of hazardous substances, the spills of which are to be reported under CWA, is not as comprehensive as that under CERCLA. But CERCLA does not impose any reporting requirement for an oil or petroleum spill. Additionally, CERCLA-based reporting involves releases of a hazardous substance into any media whereas CWA-based reporting is strictly for the release of oil or a hazardous substance into the navigable waters of the United States and adjoining shore lines. As indicated earlier, a discharge of any quantity of oil will trigger the notification requirement pursuant to the provisions of FWPCA or CWA.

For any federally unpermitted discharge of a regulated substance, CWA imposes a specific duty of notification. This reporting duty is imposed on any person in charge of a vessel or of an onshore or offshore facility with the knowledge of such discharge from the vessel or facility.

Hence, to meet the statutory obligation to report, both elements—responsibility and knowledge—must be present. If they are, the appropriate federal agency must immediately be notified. Currently, this notification must be made immediately to the National Response Center (NRC). If direct reporting to the NRC is not practicable, appropriate reports may be made to the U.S. Coast Guard or EPA.

Besides notification, the persons involved in the spill are required to undertake an appropriate cleanup following the spill or release. EPA may undertake this cleanup if the owner or operator fails to respond appropriately. In such a case, EPA can recover the total cleanup costs under the theory of strict liability. This theory of strict liability imposes liability without regard to fault. However, unlike absolute liability, limited defenses such as an act of God, act of war, and a third-party act or omission may be allowed as proper defense.

Furthermore, FWPCA includes a provision for citizen suits whereby any citizen may commence a civil suit for any violation after the violator has been given a 60-day notice. Interestingly, a citizen suit may also be brought against EPA for its failure to perform a nondiscretionary duty mandated under FWPCA in certain circumstances. Thus, the statutory provisions bind not only the violator but also EPA which can be duty-bound to enforce its regulations promulgated under the statute. In *Gwaltney of Smithfield, Ltd., v. Chesapeake Bay Foundation, Inc.,* the Supreme Court held that such citizen suits are permissible to remedy current, ongoing violations rather than past violations, unless a pattern or systematic practice of a series of violations can be proved.

The Oil Pollution Act

The OPA of 1990, which was signed into law on August 18, 1990, is perhaps the most comprehensive federal law dealing with various types of oil spills. OPA is basically a comprehensive oil pollution prevention, liability, and response statute that applies to all types of vessels and onshore and offshore facilities involved in the production, storage, transportation, and transfer of oil or petroleum. This law, to a large degree, is a congressional afterthought of the environmental pollution caused by the mammoth oil spill in Alaska by the Exxon ship *Valdez* in 1989. However, historically it is appropriate to note that Congress has been wary of oil pollution since the mid-1970s. In the early 1970s, several major oil spill incidents resulted from tanker accidents, break-ups, and leaks, on both U.S. and international waters. They caught the attention of the U.S. Congress which, until then, had paid very little attention to the environmental problem that could result from an oil spill on the river or ocean waters.

In essence, OPA has created a new liability and compensation scheme which goes beyond section 311 of FWPCA. FWPCA or its later versions, CWA or WQA, includes only a basic contingency plan for cleanup of oil spills. OPA has increased existing civil and criminal sanctions for offenses involving oil pollution to newer limits. These limits are set at extremely onerous levels not established under previous federal statutes. At the same time, OPA does not preempt the state laws pertaining to oil spills. Thus OPA leaves enough room for stiffer penalties or higher compensation schemes under applicable state laws.

Simplistically, OPA establishes strict liability for the owners and operators of vessels and onshore and offshore facilities. OPA defines *oil* very broadly to include oil of any kind and in any form. Thus, under OPA, the term *oil* includes but is not limited to petroleum, fuel oil, sludge, oil refuse, and oil mixed with other wastes except dredged spoil. It defines *recoverable damages* much more broadly than FWPCA or CWA. Under certain circumstances, the liability limits under OPA may be as high as ten times of that set forth in FWPCA or CWA. At the same time, it removes various liability limitation provisions which have thus far been available to the vessel owners on account of oil spills under the Limitation of Liability Act (LLA) of 1851, which prescribes the rule of limited liability. In addition, OPA does not preempt state laws which can set any limit for liability due to an oil spill. Moreover, the fact is that under OPA, there is a significantly deeper level of liability, the likes of which have not been seen earlier.

In this context, note that OPA also incorporates a funding mechanism, as CERCLA or Superfund does. It includes a $1 billion Oil Spill Liability Trust Fund (Trust Fund) to pay for various costs and damages, such as oil cleanup costs, operational and administrative expenses, and damages to natural resources.. The Trust Fund has been created by imposing a 5-cent-per-barrel tax on both foreign and domestic crude oil and specific petroleum products. Claims for recovery of removal costs for an incident made against this fund must be made within 6 years after the date of completion of all removal actions for that incident. Claims for recovery of damages, however, must be made within 3 years after the date when the injury and the discharge incident are reasonably discoverable.

In essence, OPA is created in the same mold as CERCLA. As a result, it incorporates many of the same concepts of CERCLA. OPA, however, is strictly applicable to those circumstances that involve an oil spill or pollution, whereas CERCLA applies to situations which specifically exclude oil or petroleum. Together, however, they address the statutory schemes for reporting of the releases or spills of regulated substances, which by definition include both petroleum and hazardous substances, except hazardous wastes as defined in RCRA.

Not surprisingly, OPA also defines *responsible parties* when a discharge or spill of oil is involved. These responsible parties are designated on the basis of the type of vessel or facility involved. Table 10.1 shows statutory designation of responsible parties.

Pursuant to OPA, each responsible party is liable for the removal costs and damages. The statute provides three specific defenses which are considered absolute: an act of God, an act of war, and an act or omission of a third party.

Other important provisions of OPA include certain minimum standards and operating requirements of oil or petroleum facilities and vessels:

- Double-hull requirements for vessels, newly constructed or upgraded to carry oil as cargo or cargo residue by the year 2015

TABLE 10.1   **Responsible Parties for Oil Spills**

Type of facility	Designated responsible parties
Vessels	Owner, operator, or demise charterer
Onshore facilities (excluding a pipeline)	Owner or operator
Offshore facilities (excluding a pipeline or facilities licensed as deepwater ports)	Lessee, permittee, or the holder of a right of use and easement
Deepwater ports	Licensee
Pipelines	Owner or operator

- Minimum standards and requirements for pressure-monitoring devices and warning devices against overfills

- Financial responsibility requirements for vessel and facility owners and operators to deal with a maximum but foreseeable oil spill

- Safety, licensing, and training requirements for operating personnel

- Notification requirements following an oil spill

Title IV, subtitle B of OPA imposes a statutory obligation on the owner or operator of a vessel or facility to respond immediately to a discharge or threat of a discharge of oil. The legislative intent behind OPA stressed that the responses to oil spills must be immediate and effective. As a result, there is a statutory requirement for prompt notification of an actual or potential oil spill. OPA significantly increases the penalties under CWA, as amended, for failure to notify appropriate federal agencies following a spill. On the basis of the statutory scheme in Title IV, subtitle C of OPA, any owner, operator, or person in charge of a vessel or facility who has the knowledge of an oil or hazardous substance discharge but fails to notify the appropriate federal agencies immediately may be liable for both civil and criminal penalties under section 311(b) of CWA. The civil penalty can include a fine of up to $25,000 for Class I and $125,000 for Class II penalties. In addition, one may be subject to imprisonment for up to 5 years pursuant to Title 18 of the United States Code. There are also steep penalties for a "knowing" discharge of oil or for a "knowing" endangerment caused by the discharge of oil. Criminal penalties for violations of OPA are governed by section 309(c) of CWA, as amended.

As discussed in Chap. 5, there is no uniform rule or interpretation as to what constitutes *knowing* for mental culpability (i.e., the scienter element) required in a routine criminal charge. Traditionally, the criminal codes of the states require an intent or mental culpability (i.e., *mens rea*) for a criminal offense that must be proved beyond a reasonable doubt. However, the federal environmental statutes do not contain specific language for the "knowing" element. As a result, the courts have been using their own interpretations of the statutory requirement, often relying on the legislative intent as a guide. The statute includes the term *unit of reportable quantity* which establishes what amount of oil, if spilled, will trigger the reporting obligation.

Section 4301 of OPA amends section 311(b)(5) of FWPCA for administrative penalties. It also includes an obligation on the part of the federal agency to immediately notify the appropriate state agency which may reasonably be expected to be affected by the discharge of oil or a hazardous substance.

As of now, EPA has yet to promulgate its regulations based on its statutory authority under OPA. As a result, a detailed regulatory scheme is unavailable. Meanwhile, since section 311 of the FWPCA or CWA has not been repealed by the enactment of OPA, EPA regulations promulgated under FWPCA or CWA must be complied with. This includes the notification requirements that may be triggered due to an actual oil spill or the threat of an oil spill. Interestingly, in *In re Exxon Valdez,* the federal district court held that oil spill is a maritime tort. Hence, as in the case of all tort claims, claims based on such oil spill must be decided on general principles of maritime law unless they are predicated on the theory of strict liability. In this context, maritime law does not allow recovery for economic losses in the absence of physical harm caused by the defendant's negligence.

10.3 Hazardous Substance Spill

CERCLA, as in the case of CWA, requires immediate notification if a CERCLA listed hazardous substance is released into any media in an amount equal to or greater than its RQ (see Table 7.2). In order to determine whether the RQ has been reached, one is required to consider any one 24-hour period. Moreover, if there are releases of one particular hazardous substance from several areas of one facility, all such releases must be added up for such determination. All reporting is to be made to the National Response Center (1-800-424-8802 or 1-202-426-2675) for any release of a hazardous substance that equals or exceeds its reportable quantity. Section 103 of CERCLA requires that any person in charge of a vessel or an onshore or offshore facility immediately notify the National Response Center of the release of a hazardous substance in an amount equal to or greater than its reportable quantity. This must be done as soon as that person has knowledge of such release. Hence, it is knowledge of a release which triggers the notification requirements; unknown releases do not function as a trigger. Also, a federally permitted release does not activate the reporting requirements irrespective of the quantity and types of hazardous substances released as long as the terms and conditions of the federal permit are not violated.

Section 102 of CERCLA requires EPA to establish the RQs for all hazardous substances. Details of EPA's regulations covering release reporting requirements are included in 40 CFR part 302. Each hazardous substance thus listed has a corresponding Chemical Abstracts Service (CAS) number as well as the RQ. If the hazardous substance happens to be a hazardous waste under RCRA, then the appropriate RCRA waste number is also provided in this list. The EPA-promulgated RQs for individual hazardous substances are given in Table 7.2.

As indicated earlier, unlike CWA where the spill is medium-specific in terms of surface waters, CERCLA spills include all the media—land, water (both groundwater and surface water), and air. As a result, the reporting duty pursuant to a CERCLA spill can be triggered if the spill or release is into groundwaters, surface waters, land surface or subsurface strata, and ambient air.

Not all RQs of hazardous substances are the same—they vary depending on their degree of hazardousness. Generally, EPA has designated five different quantities as the reportable quantity: 1 pound, 10 pounds, 100 pounds, 1000 pounds, and 5000 pounds; the first two are more common.

CERCLA also imposes both civil and criminal penalties for failure to comply with the reporting requirements for releases within the prescribed time. Even for those listed hazardous substances for which EPA has yet to establish specific RQs, reporting must be made using the statutorily established reportable quantity of 1 pound unless and until EPA's designation suggests otherwise. However, if a release is continuous and stable in quantity and rate, then it need not be reported every hour or every day. Instead, a one-time-only notice will suffice in those circumstances as long as the release of the hazardous substance involved is unchanged. Unfortunately, under the current regulatory framework of EPA, there is no legal definition of the term *continuous release*.

In addition to the release reporting obligation, section 103(c) of CERCLA requires notification to EPA of the existence of certain facilities from which there may be a known or suspected release:

> Within one hundred and eighty days after December 11, 1980, any person who owns or operates or who at the time of disposal owned or operated, or who accepted hazardous substances for transport and selected, a facility at which hazardous substances [as defined in section 101(14)(C) of CERCLA] are or have been stored, treated, or disposed of shall, unless such facility has a permit issued under, or has been accorded interim status under, subtitle C of the Solid Waste Disposal Act [RCRA], notify the Administrator of the Environmental Protection Agency of the existence of such facility, specifying the amount and type of any hazardous substance to be found there, and any known, suspected, or likely releases of such substances from such facility...

This imposes a tremendous burden on both present and past owners and operators of facilities. Any knowing failure to comply with this provision may incur both civil and criminal penalties.

Besides CERCLA, section 302 of the Emergency Planning and Community Right-to-Know Act (EPCRA) which makes up Title III of SARA has a specific reporting requirement. Section 304 of EPCRA requires owners or operators of certain facilities to report releases of extremely hazardous substances (EHSs) to state and local authorities under the emergency notification scheme. This notification is due immediately after the release of an EHS, in an amount that equals or exceeds its RQ, to the community emergency coordinator for the local emergency planning committee for any area likely to be affected by such release. This notification must also be made to the state emergency planning commission of the state likely to be affected by such release. Details of EPCRA notification are provided in Chap. 7.

In the context of RQ, note that section 102(a) of CERCLA authorizes EPA to adjust statutory RQs by appropriate regulations. EPA's current methodology for adjusting RQs begins with an evaluation of intrinsic physical, chemical, and toxicological properties, commonly known as *primary criteria,* of each hazardous substance. The primary criteria that are examined by EPA include

- Aquatic toxicity
- Mammalian toxicity (oval, dermal, and inhalation)
- Ignitability, reactivity, and chronic toxicity
- Potential carcinogenicity

The data for each hazardous substance are evaluated using various primary criteria. Therefore, each hazardous substance could receive several tentative RQ values based on its particular intrinsic properties. Under EPA's current scheme, the lowest of such tentative RQs ultimately becomes the primary-criteria RQ for that substance.

It is apparent, therefore, that RQs are not fixed or absolute. They may be revised based on the availability of new information pertaining to the potential harm to public health or welfare or the environment due to the release of hazardous substances. Typically, when a hazardous substance released into the environment degrades relatively rapidly to a less hazardous form by the natural degradative processes of biodegradation, hydrolysis, and photolysis (BHP), its initial RQ level in the scale of 1, 10, 100, 1000, or 5000 pounds is generally raised one level. Conversely, if a hazardous substance degrades to a more hazardous product following its release, the RQ of the original substance may be lowered by one or more levels, reflecting the RQ of a more hazardous substance that is formed after the release.

10.4 PCB and Other Chemical Spills

The statutory reporting scheme in section 8 of the Toxic Substances Control Act is not very explicit. However, EPA has concluded that section 8(e) of TSCA requires an immediate reporting of a spill involving a chemical substance if it presents a substantial risk of injury to health or the environment (i.e., a serious threat) unless such spill is reported under any other spill reporting requirements.

Section 8(e) of TSCA actually delineates the broad reporting requirements due to the manufacture, processing, or distribution in commerce of a toxic chemical or mixture. Pursuant to this provision, a manufacturer, distributor, or processor of a chemical is required to report to EPA whenever he or she or it becomes aware that the released chemical may pose a substantial risk of injury to public health or the environment.

EPA discussed its policy of TSCA reporting requirements in a 1978 report entitled "Statement of Interpretation and Enforcement Policy." Pursuant to this policy, EPA finds that TSCA provides a broader application of spill reporting requirements than other applicable federal statutes including CERCLA. Thus, EPA has decided that under TSCA, the spill reporting is not limited to specific chemicals but can be triggered whenever the release of any chemical poses a substantial risk to human health or the environment. Furthermore, EPA takes the position that under TSCA, there is no specific reportable quantity or threshold quantity which must be equaled or exceeded for imposing a reporting requirement, but that such reporting should be based on the sufficiency of a significant threat in any particular release of a chemical or chemical substance. Hence, following the reporting requirements under CWA or CERCLA is not enough. One must also determine whether the release poses a "substantial risk" to trigger reporting under TSCA.

So far, PCBs as a group have constituted the most important and talked about chemical substance under TSCA. Pursuant to TSCA, EPA has provided specific regulations to deal with PCBs.

Besides reporting requirements for the release or spill involving PCBs, EPA has set forth its PCB cleanup policy in connection with such spill or release in 40 CFR sections 761.120 to 761.135. This cleanup applies to PCB spills involving PCBs at concentrations of 50 parts per million or greater. Any PCB spill into the water, however, is considered on a site-by-site basis.

In addition, pursuant to 40 CFR part 761, subpart G, the spill cleanup is required for those spills which occur after May 4, 1987. For earlier PCB spills which have already been discovered, no additional cleanup is required if the spill has already been cleaned up in accordance with the requirements imposed by appropriate EPA regional offices. In essence, all earlier PCB spills are subject to EPA's discretion with respect to their cleanup or decontamination, as the case may be.

Cleanup requirements also differ based on whether more than or less than 1 pound of PCBs by weight has been spilled. Under current PCB cleanup policy of EPA, cleanup of high- and low-concentration spills involving 1 pound or more of PCBs by weight or 270 gallons or more of untested mineral oil will be considered complete if all the immediate requirements, cleanup standards, and sampling and record-keeping requirements as established by EPA in 40 CFR §761.125(c) are met.

Under current EPA policy, notifications of certain PCB spills are to be made by the responsible parties to the appropriate regional office of EPA. These PCB spills include

- Spills involving more than 10 pounds of PCBs (no PCB spill is to be reported to EPA if the spill involves 10 pounds or less of PCBs and if appropriate cleanup is undertaken)
- Spills that result in direct contamination of drinking water supplies, surface waters, or sewers
- Spills that directly contaminate grazing grass or vegetable gardens

As indicated earlier, in addition to the notification requirement, EPA regulations promulgated under TSCA include a cleanup requirement. These PCB response requirements can best be summarized in the following categories:

1. Spills greater than 1 pound and 500 parts per million or more in concentration of PCBs:

- Response requirements are immediate and must be undertaken within 24 hours (or 48 hours for PCB transformers) of knowledge of the spill.
- Notification must be made to EPA and NRC
- Posting of warning signs, area cordoning, record keeping, and maintenance of appropriate documents pertaining to visible contamination are required.
- The cleanup must include removal of all visible traces of fluid PCBs from the hard surfaces.
- Cleanup requirements differ depending on spill location, i.e., for cleanup of grounds at substations, 25 to 50 parts per million; for cleanup of solid surfaces at substations, 100 μg per 100 square centimeters; for nonrestricted access areas involving soil, 10 or 25 parts per million; and for solid surfaces in nonrestricted access areas, 10 or 100 μg per 100 square centimeters.

2. Spills of less than 1 pound and less than 500 parts per million in concentration of PCBs:
   - Double wash and/or rinse of the hard surfaces contaminated with PCBs
   - Cleanup of all indoor residential surfaces, other than vaults, to 10 micrograms per 100 square centimeters (by standard commercial wipe tests)
   - Cleanup of the visibly contaminated soil plus 1 lateral foot within the spill area by excavation and removal of excavated soil for appropriate disposal
   - Completion of the cleanup within 48 hours of the spill and certification that the cleanup was in compliance with applicable regulations

There are appropriate cleanup standards that must be conformed with. These cleanup standards are based on whether the spills involve restricted-access areas, unrestricted-access areas or outdoor electrical substations. The level of cleanup required in these situations reflects the anticipated level of risk to public health or to the environment.

Under section 8(e) of TSCA, the duty to notify for chemical spills presenting a substantial risk of injury to health or to the environment, including those of PCBs, does not automatically bind the individual employees of the company involved in manufacturing, processing, or distribution of the chemical in commerce. The employees may meet their regulatory duty of reporting if they notify the designated company official of the spill unless the company's established procedures include an affirmative duty to notify EPA. The designated or responsible company officials, however, have a personal responsibility to notify EPA as well as to ensure that all appropriate reports are provided to EPA. The "responsible party" who must notify the appropriate EPA regional office of a PCB spill and obtain guidance for appropriate cleanup measures under 40 CFR §761.125(a)(1) is either the owner of the PCB equipment, facility, or other source of PCBs or a designated agent of the owner.

EPA's PCB regulations, set forth in 40 CFR §761.120(e), do not affect spill reporting requirements or cleanup standards that are imposed under other federal statutes, including but not limited to CWA, RCRA, CERCLA, and

SARA. Under EPA's current policy, if these standards become competing and more than one requirement applies, the stricter standard must be met.

10.5 Spills of Miscellaneous Materials

Generally, discharge or a spill of any harmful chemical or chemical substance, besides PCBs, requires appropriate notification unless such discharge or spill or release is federally and/or state permitted.

At present, PCBs along with other toxic pollutants are listed under section 307(a) of the Clean Water Act. This list, as set forth in 40 CFR section 129.4, includes aldrin/dieldrin, DDT, DDD and DDE, endrin, toxaphene, benzidine, and PCBs. For reporting purposes, however, the toxic pollutants are considered as hazardous substances and must thus satisfy the spill reporting requirements based on CERCLA.

Unlike CWA and CERCLA, including its progeny SARA, RCRA does not include a specific provision for spill reporting. However, EPA has promulgated certain regulations based on its authority under sections 3002, 3003, and 3004 of RCRA. These EPA regulations, set forth in 40 CFR parts 260 through 268, contain specific provisions for the reporting of spills or discharges of hazardous wastes as defined under RCRA.

These RCRA-based spill reporting requirements differ from those based on CWA and CERCLA in several respects:

- RCRA-based regulatory provisions apply only to hazardous wastes.

- The reporting duty is not regulated or triggered by any specific reportable or threshold quantity of hazardous waste; instead it depends on a determination of whether the spill or discharge poses a threat to human health or the environment that warrants reporting.

For all spills or discharges of hazardous waste that are not transportation-related, the notification or report should be made to appropriate federal, state, and local agencies. Spills during transportation, however, should be reported to NRC. In this context, note that there are specific Department of Transportation (DOT) regulations set forth in 49 CFR parts 171 to 177 which were promulgated under the Hazardous Materials Transportation Act (HMTA). These regulations specifically address spills involving hazardous materials during transportation. All discharges or spills of hazardous waste that are transportation-related, irrespective of quantities, must be reported to NRC.

Other parties connected to hazardous wastes have different reporting obligations under the present regulatory framework. For example, generators of wastes and persons involved in treatment, storage, or disposal (TSD) of hazardous wastes must report those releases, fires, and explosions that threaten public health or the environment. The owners or operators of surface impoundment, landfills, and land treatment facilities must report if they find that hazardous wastes or hazardous constituents have entered the groundwater.

Section 104 of HMTA authorizes DOT to promulgate regulations dealing with hazardous materials. In this context *hazardous material* means a substance or material in a quantity or form which may pose an unreasonable risk to public health and safety or to property when transported in commerce. By this definition, hazardous materials include not only the listed hazardous substances but also other substances which pose an unreasonable risk to public health, safety, and property. These hazardous materials may include, but are not limited to, radioactive materials, explosives, etiologic agents, flammable and combustible liquids or solids, oxidizing or corrosive materials, compressed gases, and poisons. The lists of these hazardous materials and hazardous substances along with their RQs are set forth in 49 CFR §172.101. The specific threshold quantities are based on standards that must be used to determine if a spill poses a sufficient or significant threat to warrant any reporting to the appropriate agency. DOT's hazardous material table is indeed very comprehensive since it provides the hazard class, identification number, labeling, and packaging requirements for the listed hazardous materials. Reporting is required if any hazardous material is released and it poses a substantial threat to human health or the environment. However, under HMTA, this reporting is required only if the release occurs during transportation.

As indicated earlier, if there is a release of any such hazardous substance that equals or exceeds its RQ, it must be reported to NRC immediately. Two 24-hour telephone numbers are available for NRC: 800-424-8802 and 202-267-2675. If a hazardous waste or other hazardous material is discharged or spilled during transportation, and if it is determined by a state or local government that it must be immediately removed to prevent harmful consequences, then an authorized official of such government may allow its removal without a manifest.

Any release or leak from an underground storage tank (UST) holding a regulated substance (i.e., petroleum or a hazardous substance, but not a hazardous waste) is also covered under specific reporting requirements. However, such notification must be made to the regional office of EPA following an actual release or a suspected spill. Details of the regulatory requirements for USTs are discussed in Chap. 5.

10.6 Spill Reporting Forms

In almost all situations described above, an oral report in the form of a telephone call to NRC or to other designated agency locations should be made first. Under specific statutory and/or regulatory requirements, the caller is required to provide certain minimum information, e.g., time and date of spill, spill location, source and type of spill, and quantity spilled. In many instances, such oral reports must be substantiated by formal written reports.

Specific forms should be used for the purpose of reporting spill incidents to the appropriate federal agencies. In addition, individual states may require appropriate forms be used to report a particular type of spill or release.

On June 26, 1991, EPA promulgated its final rule on toxic chemical release reporting. This rule replaces EPA form R (EPA form 9350-1) and related instructions to report a toxic chemical release or spill. While the "range reporting" for notification of toxic chemical releases of less than 1000 pounds per year under EPCRA (or Title III of SARA) is retained as an option, the new EPA rule requires checking off of one of three revised ranges: 1 to 10, 11 to 499, and 500 to 999 pounds. Any facility that is confident of zero emissions or release because of no possibility of chemical releases to any one of the environmental media may enter "NA" in the release column.

Finally, it should be emphasized that there are specific exclusions to spill reporting requirements. These exclusions are not across-the-board type of exclusions, but apply strictly to the reporting requirements that relate to a specific statute and the regulations promulgated thereunder.

Figures 10.1, 10.2, and 10.3, developed by EPA, provide important and useful reporting information.

Recommended Reading

1. *Oil Spill Law Information Service,* Thompson Publishing Corp., Washington, April 1991.

2. Susan M. Cooke, *The Law of Hazardous Waste: Management, Cleanup, Liability, and Litigation,* vol. 2, Matthew Bender, New York, 1991.

3. *Spill Reporting Procedures Guide,* The Bureau of National Affairs, Washington, 1989.

4. "Asbestos-Containing Materials in School Buildings: A Guidance Document," EPA no. C00090, OPTS Docket 61004, 1979.

Important Cases

1. *Gwaltney of Smithfield, Ltd., v. Chesapeake Bay Foundation, Inc.,* 484 U.S. 49, 108 S. Ct. 376, 98 L. Ed. 306 (1987).

2. *Chevron, USA, Inc., v. Yost,* No. 89-3845 (5th Cir. Dec. 12, 1990).

3. *In re Exxon Valdez,* Civ. No. A89-095 (D.C. Alaska Feb. 9, 1991).

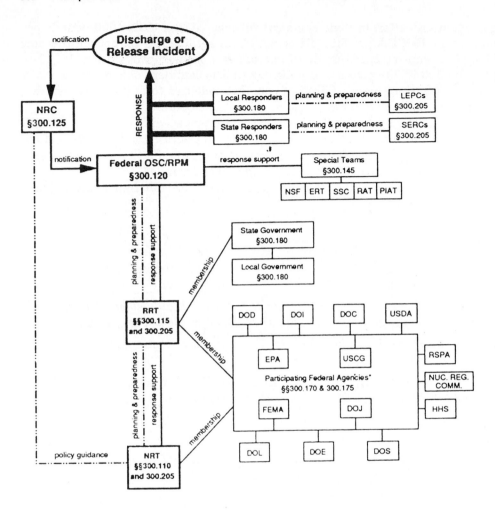

The same federal agencies participate on both the National Response Team (NRT) and the Regional Response Team (RRT). Federal agencies on the RRT are represented by regional staff. Abbreviations used in this figure are explained in §300.4.

Figure 10.1 National response system concepts.

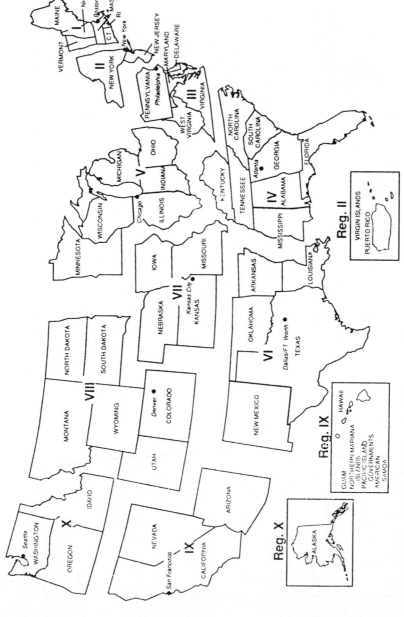

Figure 10.2 Standard regional boundaries for ten regions.

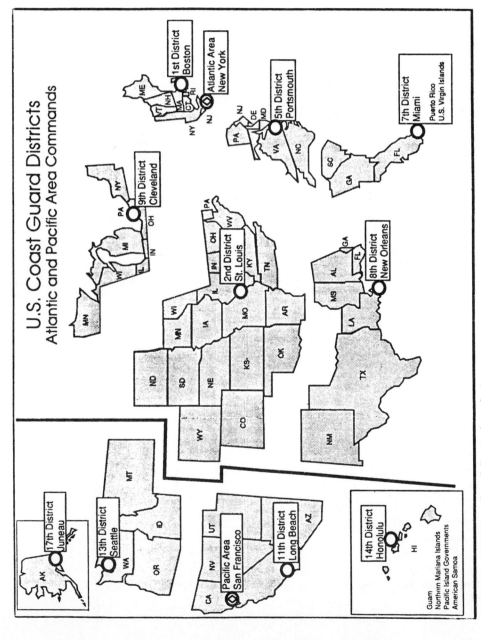

Figure 10.3 U.S. Coast Guard district boundaries.

Radioactive Waste— A Continuing Dilemma

11.1 Introduction

The 1940s ushered in the dawn of the atomic era in the United States. The discovery of nuclear fission paved the way for abundant generation of electricity from nuclear fuel. It also introduced medical tools and techniques, vast ranges of nuclear weapons, and nuclear submarines and underwater vessels. Such awesome display of nuclear fission and the wide range of use of various radioactive materials and isotopes have produced many millions of tons of nuclear or radioactive wastes which contain a vast array of radioactive elements. The half-lives of these radioactive elements may range from microseconds to millions of years.

Depending on the presence of such radioactive elements and their half-lives, the wastes may be termed low-level or high-level radioactive wastes. The proliferation of both low- and high-level radioactive wastes has caused a nightmare in terms of their safe management and control, particularly with respect to their proper disposal. In this context, note that while there are definite health and safety hazards associated with radioactive wastes, they are not necessarily more harmful or less harmful than other toxic or hazardous substances. Nevertheless, it is generally recognized that radioactive wastes must be properly handled and disposed of in order to lessen health hazards and promote public safety.

Unfortunately, proper disposal of both low- and high-level radioactive wastes is complex and expensive. Land burial of such waste is feasible only where certain geological and physical requirements can be met. Ocean burial also imposes certain restrictions including geopolitical considerations. The feasibility of incineration or other forms of burning or destruction of radioactive waste also has major technical limitations.

Since our first step into the atomic age, Congress has woefully failed to grasp the problems that could emerge from entering into the so-called atomic era. As a result, the regulatory agencies responsible for the management of radioactive wastes have been sidetracked by vague statutory framework and political considerations. This already difficult situation has been exacerbated by the public attitude highlighted by the "not in my backyard" syndrome. This has created a dilemma as well as a problem of gigantic proportions.

The Atomic Energy Act of 1954 directed the Atomic Energy Commission (AEC) to provide for the development and promotion of peaceful uses of nuclear energy, including the use of nuclear power for the generation of electricity. Very little was said in this federal statute about waste handling and disposal resulting from the use of atomic energy. During the 1970s and 1980s, several important federal statutes dealing with radioactive wastes were enacted. But the regulations promulgated thereunder have not been very effective owing to the lack of a coherent national policy with regard to the disposal of radioactive wastes.

Nonetheless, to understand the dilemma and challenges that confront proper management of radioactive wastes, it is important to look into the important provisions of several energy and environmental statutes that focus on radioactive wastes and their management. These are discussed later in this chapter.

11.2   Radioactive Wastes and Radioactivity

The hazards associated with radioactive wastes stem from their basic characteristics since radiation or radioactivity cannot be detected by sensory perceptions. Additionally, the effects of radiation are cumulative and may not have immediate evidentiary signs.

Radioactivity can be found in certain natural products, for example, uranium and thorium ores. Radioactivity can also result from several human activities such as mining, milling, processing, and use of radioactive ores and isotopes.

On a global basis, the International Atomic Energy Agency (IAEA), an intergovernment organization founded under the auspices of the United Nations, recognizes three types of nuclear or radioactive wastes: high-, intermediate-, and low-level radioactive wastes. The classification is based on several factors, such as temperature and source of the waste and number and half-lives of the isotopes or radioactive elements present in the waste. Generally, high-level radioactive wastes are generated during reprocessing of spent nuclear fuel and use of fuel rods. The intermediate-level radioactive wastes consist of nuclear reactor byproducts and associated materials, such as equipment and tools that become radioactive. The low-level radioactive wastes are typically generated from medical, commercial, and nonmilitary uses of radioactive isotopes and refuse matter that are contaminated by radioactive materials. In the United States, however, radioactive wastes are generally classified into two broad categories: low-level and high-level wastes. This classification reflects the criteria used by IAEA.

The general classifications of radioactive wastes do not necessarily coincide with all statutory and regulatory definitions. As a result, in the context of a particular statute or regulation, only the definitions provided in such statute or regulation should be taken into consideration. Traditionally, in the United States, two basic factors have been used to define radioactive waste in the absence of a formal definition. First, it must be produced, fabricated, manipulated, or regulated by people. Second, under certain situations, human activity must enhance the background or natural radiation. Interestingly, Congress has not attempted to regulate uranium and thorium mining under the Atomic Energy Act of 1954. These mining operations include naturally occurring radioactive materials.

Definition

In the absence of an established, comprehensive, and commonly understood definition of radioactive waste, it is important to have a basic understanding of radioactivity, since it is the center of all debates and public concerns. In the simplest sense, radioactivity occurs when the nucleus of an atom "decays" because of the release of alpha particles (i.e., helium nuclei) consisting of two protons and two neutrons, beta particles (i.e., electrons), or gamma rays (similar to x-rays). These three kinds of nuclear radiation may also come from radioactive materials found in nature.

All the isotopes (i.e., atoms of the same element that differ in mass or atomic weight) of certain naturally occurring elements, e.g., thorium, uranium, radium, and radon, are radioactive. But only a few isotopes of certain elements (e.g., carbon) are radioactive. In the latter case, only the nonstable isotopes give rise to radioactivity whereas stable isotopes remain nonradioactive.

High doses of radiation cause serious health effects which can be categorized as somatic and genetic effects. Somatic effects of radiation could increase the chance of cancer and shorten the lifespan of the person exposed to it. Genetic effects of radiation, however, may be passed on to the offspring of the exposed person by mutations of genes. It has been estimated that a person in the United States receives an average exposure of 180 millirems of radiation per year.

11.3   Federal Energy Statutes

Congressional attempts to control and manage the use of atomic energy in this country started with the enactment of the original Atomic Energy Act of 1946. Unfortunately, as indicated earlier, even the federal energy laws of the 1970s and the 1980s, while being true congressional attempts to address the issue of atomic energy, did not discuss at length the disposal practices of radioactive wastes.

In this context, note that the disposal of high-level radioactive wastes is the responsibility of the federal government. This is not the case with low-level radioactive wastes which require little or no shielding and no cooling. Most low-level radioactive wastes from nuclear power plants are typically solidified, put into drums, and placed at the bottom of trenches, about 20 feet deep, at a licensed disposal site. The collection, transportation, and disposal of these low-level radioactive wastes are regulated and monitored by the Nuclear Regulatory Commission and the Department of Transportation (DOT). Under the Low-Level Radioactive Waste Policy Act of 1980, discussed later in this chapter, each state is responsible for the disposal of its own low-level radioactive waste. The Environmental Protection Agency is responsible for providing radiation control guidelines to protect public health and the environment, but it does not have any direct control over the handling and ultimate disposal of either high- or low-level radioactive wastes unless such authority is specifically vested by a federal statute.

To get a true sense of the statutory developments for the control and management of radioactive wastes, however, one must start with the Atomic Energy Act of 1954.

Atomic Energy Act of 1954

The Atomic Energy Act (AEA) of 1954 amended the Atomic Energy Act of 1946 almost in its entirety. Since it is drastically different from the original statute, it is generally regarded as the baseline of U.S. atomic energy legislation. However, AEA hardly provides any specific textual language which can be used to determine how radioactive waste has to be handled and/or disposed of.

In reality, AEA and its later amendments, while dealing generally with the use, development, and control of nuclear energy, do not provide any specific authority to the regulatory agencies with regard to the control, handling, and disposal of radioactive waste. There are, however, specific provisions in §2074 and §2094 of this statute with respect to foreign distribution of special nuclear materials and source material and the role of the regulatory agency in such distributions.

During the period between 1947 and 1974, the Atomic Energy Commission had virtually sole responsibility over the U.S. nuclear energy activities. When it was dissolved by Congress in 1974, its activities were split between two new federal agencies, the Nuclear Regulatory Commission (NRC) and the Energy Research and Development Administration (ERDA). NRC was entrusted with nuclear power regulation while ERDA's role was restricted to research, promotion, and military use of nuclear energy. Later, ERDA merged with the Department of Energy (DOE) when the latter was created in 1977.

Currently, DOE is the administrative agency responsible for locating and developing high-level nuclear waste repositories while NRC continues with its original role of regulating nuclear power. This has created a vacuum. The congressional solution to the problem of radioactive waste has been essentially relegated to a search for political accommodation rather than setting specific guidelines to DOE for addressing various radioactive waste disposal problems. This problem is further compounded by a basic assumption made by NRC in connection with the civilian nuclear power reactor licensing proceedings. NRC has assumed that disposal facilities for spent nuclear fuel and high-level radioactive wastes will be automatically available when such a need arises. The determination of need has never been comprehensively approached. As a result, NRC's duties fell far short of developing a comprehensive solution to the burgeoning problem of radioactive waste disposal.

Past regulation promulgated under AEA is limited to only three kinds of nuclear materials and two kinds of nuclear facilities. The three kinds of regulated nuclear materials are source material, special nuclear material, and byproduct material. *Source material* is defined as (1) uranium, thorium, or any other material which is so determined by AEC (now by NRC) pursuant to the provision of 42 U.S.C. (i.e., United States Code) §2091 or (2) ores containing one or more of the foregoing materials in certain specific concentrations. The term *special nuclear material* means (1) plutonium, uranium (enriched isotope 233 or isotope 235), and any other material as determined by NRC, but does not include source material, or (2) any material artificially enriched by any such material, except source material. The term *byproduct material* means any radioactive material, other than special nuclear material, yielded in or made radioactive by exposure to radiation through the process of production or utilization of special nuclear material.

Although these definitions are arguably broad, AEC and NRC have taken the position that their regulatory jurisdiction starts only after uranium and thorium have been removed from their existing places in nature.

Uranium Mill Tailings Radiation Control Act

The Uranium Mill Tailings Radiation Control Act (UMTRCA) of 1978 was more of an afterthought by Congress to address public concern about the hazards of the uranium mill tailings wastes. Through its adoption of UMTRCA, Congress required EPA, in consultation with NRC, to provide a report to Congress which "identifies the location and potential health, safety, and environmental hazards of uranium mine wastes together with recommendations, if any, for a program to eliminate these hazards."

Note in this context that UMTRCA has two major titles. Title I establishes a program for remedial action by DOE at some twenty specifically identified inactive uranium mill tailings sites. Title I requires that the stabilization of these inactive sites conform to the prevailing EPA standards. Title II defines uranium and thorium mill tailings and their residuals as *byproduct material,* similar to the original definition provided in AEA. Pursuant to AEA, a byproduct material is a result of nuclear fission, such as cesium or strontium. In addition, Title II includes specific requirements relating solely to uranium and thorium mill tailings.

In essence, UMTRCA deals with the siting problems associated with uranium mill tailings rather than the entire gamut of radioactive wastes. However, one of the stated purposes of this federal statute is "to conduct, assist and foster development, use and control" of atomic energy so as to make the maximum contribution to the general welfare of the public. Hence, any pertinent provision of UMTCRA can be considered in regulatory decision making which deals with radioactive waste.

Low-Level Radioactive Waste Policy Act

The Low-Level Radioactive Waste Policy Act (LLRWPA), passed in 1980, specifies that each state is responsible for disposing of the low-level radioactive waste (LLRW) generated within its border. The sole exceptions are wastes generated as a result of defense activities or federal research and development activities.

Low-level radioactive waste can be in any one of the three physical forms (solids, liquids, and gases). Typical low-level radioactive wastes comprise protective clothing, filters and resin, solidified liquids, animal carcasses, contaminated soil, discarded equipment and tools, and scintillation wastes. LLRWPA basically encourages a regional concept of LLRW disposal. It urges the states to enter into compacts with each other to establish regional disposal sites.

Typically, LLRW is a material contaminated with radioactivity with a relatively short half-life, ranging from a fraction of an hour to several years. Unfortunately, however, the level of radioactivity in LLRW varies significantly, in spite of its name. In fact, the radioactivity level in certain LLRW is fairly high. This raises serious questions about the arbitrary categorization of radioactive wastes by Congress. Thus, any regulatory promulgation dealing with radioactive wastes under the statute is extremely difficult.

Historically, since 1962, federal policy has dictated that commercially generated LLRW should be disposed of in commercial, privately operated disposal sites. Prior to 1962, AEC accepted commercially generated LLRW at various AEC disposal sites. In this regard, LLRW should be distinguished from other forms of radioactive waste, such as spent reactor fuel or the high-level radioactive waste (HLRW) that results from reprocessing spent fuel, tailings from uranium mines and mills, and transuranic wastes that result from nuclear weapons fabrication and spent-fuel reprocessing. None of these wastes were permitted to be disposed of at the AEC sites.

This important distinction, however, makes little difference visually since LLRW cannot always be distinguished from HLRW or, for that matter, from other hazardous wastes. In addition, the issue of "mixed waste" (waste that is both radioactive and chemically hazardous) was not addressed at all in the original LLRWPA or its 1985 amendments, the Low-Level Radioactive Waste Policy Amendments Act (LLRWPAA) of 1985. Thus, the precise allocation of regulatory responsibility between NRC and EPA for any commercially mixed waste has yet to be fully resolved.

The enactment of LLRWPA by Congress was actually precipitated by the action of three states in 1979 when they made it clear that they would not bear the national burden for LLRW disposal forever. However, by passing LLRWPA in 1980, Congress essentially put the burden for establishing a regional LLRW disposal system back on the state governments and washed its hands of the problem altogether.

This statute, by embracing a simplistic concept, has failed to seek a uniform, nationwide federal solution to a national problem of gigantic proportions. By pushing the responsibility to the state level, Congress has bypassed the complexity of the issue and virtually abandoned its own responsibility. Unfortunately, to date there is not a single new site for LLRW disposal.

During later years, Congress has attempted to specify the details of implementing LLRWPA in other ways. In December 1985, Congress amended LLRWPA to allow the states without disposal facilities 7 more years beyond LLRWPA's original deadline of 1986 to develop them.

So far, only eight regional compacts for low-level radioactive waste management have been negotiated by the states and ratified by Congress through the Omnibus Low-Level Radioactive Waste Interstate Compact Consent Act of 1986. These eight regional compacts are

- The northwest interstate compact (Alaska, Hawaii, Idaho, Montana, Oregon, Utah, Washington, and Wyoming)
- The Rocky Mountain compact (Arizona, Colorado, Nevada, New Mexico, Utah, and Wyoming)
- The southeast interstate compact (Alabama, Florida, Georgia, Mississippi, North Carolina, South Carolina, Tennessee, and Virginia)
- The midwest interstate compact (Iowa, Indiana, Michigan, Minnesota, Missouri, Ohio, and Wisconsin)
- The central midwest interstate compact (Illinois and Kentucky)
- The central interstate compact (Arkansas, Iowa, Kansas, Louisiana, Minnesota, Missouri, Nebraska, North Dakota, and Oklahoma)
- The northeast interstate compact (Connecticut, New Jersey, Delaware, and Maryland)
- The southwestern compact (Arizona, California, and any eligible states)

Unfortunately, many of the compacts with more than two states have yet to choose a single site. To further complicate matters, a number of states are trying to go alone or in pairs, rather than risk a collaboration on a regional scale and thus alienate the citizens of the respective states. This has stymied the general premise of LLRWPA—that LLRW disposal is best handled on a regional basis. The pointed departure by some states which embraced the "single-state option" not only is a blow to the original congressional plan, but also highlights congressional failure to provide specific guidelines and to comprehend the long-term effect of not selecting new LLRW disposal sites at a federal or national level. As a result, the long-heralded 1986 deadline for approval of regional compacts by Congress has been a total bust.

Nuclear Waste Policy Act

The Nuclear Waste Policy Act (NWPA) of 1982 was passed by the 97th Congress to address the growing problem of nuclear waste. In an attempt to solve the inadequacy of the government and the industry to deal with accumulated radioactive wastes, NWPA established a comprehensive plan for collection and transportation as well as selection, construction, and operation of disposal sites (i.e., repositories) for high-level nuclear or radioactive waste. The high-level radioactive wastes, as discussed earlier, generally include used fuel or materials left following reprocessing of the spent (i.e., used) fuel.

A funding program is at the heart of NWPA. In essence, NWPA provides that the repository selection and construction activities are to be funded from the Nuclear Waste Fund which is to be derived from a levy on nuclear waste generators and owners. It includes several provisions which expressly authorize state grants for specific activities at different stages of the disposal process. For example, the grants may be available after a site has been selected, or when the President of the United States has chosen a candidate site, or when NRC has authorized construction, or when a state and DOE have entered a cooperative agreement. Unfortunately, NWPA does not include any specific provision that authorizes any state funding for preliminary site characterization studies.

In *Nevada ex rel. Loux v. Herrington,* the U.S. Court of Appeals for the Ninth Circuit held that NWPA is intended for distributing disposal costs among the generators and owners of radioactive waste, for promoting public confidence, and for protecting the public and the environment. As a result, the *Herrington* Court ruled that the costs of radioactive waste disposal "should be the responsibility of the generators and owners of such waste." The court also ruled that state and public participation in the planning of waste sites (i.e., repositories) must be an essential element in order to "promote public confidence" in the safe disposal of radioactive waste.

DOE is the administrative agency required under NWPA to locate and develop the nation's first high-level radioactive waste repository. Presently, nuclear weapons production waste is stored at military facilities, such as the Hanford Reservation in Washington. The spent fuel rods are stored in water-cooled pools or cooling ponds on-site at various nuclear power plants scattered throughout the nation.

This temporary storage of high-level radioactive wastes has been found to be both ineffective and dangerous. The waste may migrate off-site, resulting in serious environmental contamination, or it may fall into the hands of terrorist groups and thus be used for illegal and subversive activities. It has been estimated that by the year 2000, the United States will have about 87,000 tons of high-level radioactive wastes for disposal.

Unfortunately, only two waste repositories have been planned to date, and none has been constructed yet. Part of the problem lies with congressional failures to provide specific and adequate funding. In addition, in an attempt to orchestrate comprehensive waste disposal requirements, Congress made NWPA very long, complex, and confusing. Nevertheless, NWPA is truly the first legislative attempt to confront a complicated and politically unpopular national dilemma of how to design and implement permanent disposal sites for thousands of tons of high-level radioactive wastes, many of which may have a half-life exceeding several hundred thousand years.

11.4   Federal Environmental Statutes

An overview of important federal environmental statutes was presented in Chap. 1. It is important to recognize that the environmental statutes may also include use, generation, storage, transportation, treatment, and disposal of radioactive wastes, particularly those which are low-level radioactive wastes.

The following discussion of federal environmental statutes is limited to those matters related to radioactive wastes.

National Environmental Policy Act

The National Environmental Policy Act (NEPA), enacted in 1969, is a short, general statute. But it has a definite effect on the federal decision-making process. As indicated in Chap. 1, compliance with the provisions of NEPA is important to private interests if they involve federal permit decisions or federal construction grants and assistance. NEPA is also an important tool for private citizens, particularly when they have an environmental concern. Generally speaking, NEPA gives everyone a statutory right to force a review of federal decisions regardless of—and especially—when the federal agency involved does not have distinct environmental responsibilities. Thus, NEPA stands out as the first congressional attempt to establish a national policy to protect the quality and condition of the environment.

In essence, NEPA established a national policy that requires all federal agencies, including military agencies, to give full consideration to environmental effects while being involved in other plans and programs. To ensure that agencies implement this policy, NEPA directs that all agencies prepare and disclose an Environmental Impact Statement (EIS).

Unfortunately, NEPA's disclosure mandate conflicts with certain explicit needs in the nuclear programs, e.g., the military's need to prevent disclosure of military activities that involve national security. The Freedom of Information Act (FOIA), which is incorporated in NEPA by reference, can protect military secrets by exempting properly classified national security matters from disclosure. As a result, the courts have struggled to resolve the conflict between the environmental goals of NEPA and the need to protect national security by a veil of secrecy.

Title I of NEPA imposes specific obligations on all federal agencies. These agencies are required to consider the environmental impacts of their proposed actions and to account for them in their decision-making process. They are also required to make the EISs, prepared by the regulatory agencies, available to the President, the Council on Environmental Quality (CEQ), and the public. The EISs are, however, obligatory in connection with all major federal actions "significantly affecting the environment."

Title II of NEPA created the CEQ whose role is to provide general advice and assistance to the President of the United States in the preparation of an annual Environmental Quality Report (EQR), to research conditions and trends in the quality of the environment, and to evaluate the various programs and activities of federal agencies for consistency with NEPA policy.

In *Baltimore Gas & Electric Co. v. Natural Resources Defense Council, Inc.,* the Supreme Court upheld the validity of the NRC's table S-3 rule which in effect directed the nuclear licensing board to assign a zero value to the environmental impacts of the long-term storage of high-level nuclear waste. Besides deferring to expert agencies' decisions, the Court vigorously supported NEPA's mandatory requirement for federal agencies to "consider and disclose" the "significant environmental impacts" of any major federal action. The majority opinion of the Court held that NRC had fully considered the uncertainties concerning the long-term storage of nuclear wastes in the salt-bedded repositories and reasonably concluded that the uncertainties were insufficient to affect the outcome of any individual licensing proceedings.

In spite of this expressed deference by the Supreme Court to expert regulatory agencies in *Baltimore Gas,* it is perhaps wrong to suggest that the federal courts, particularly in the field of nuclear power regulation, will not question a federal agency decision. The facts in *Baltimore Gas* were unique because they involved highly complex technical considerations surrounding NRC's zero-release assumption. It is generally expected that while reviewing decisions by federal agencies in matters of great public concern, such as radioactive waste disposal, the courts will be engaged in a broad, formalistic review to ensure that the agencies are employing rational decision making and are adequately considering and disclosing the environmental impacts of such actions. Also while the courts are generally reluctant to disapprove of such agency actions on substantive grounds, in all probability they will enforce the procedural provisions of NEPA when required.

Resource Conservation and Recovery Act

Since its enactment in 1976, the Resource Conservation and Recovery Act has been amended several times, and each amendment has placed additional burdens on manufacturing and other industrial operations to promote safer solid and hazardous waste management programs. Regardless of the later amendments, subtitle C, which deals generally with hazardous waste, remains the centerpiece of RCRA. Besides the regulatory requirements for hazardous waste management, it spells out the mandatory obligations of generators and transporters of hazardous waste as well as of owners and operators of hazardous waste treatment, storage, and disposal facilities.

Before any action is planned under RCRA, it is essential to understand what constitutes a solid waste and a hazardous waste. These details of RCRA are given in Chap. 5. Note that the statutory definition of solid waste specifically excludes source, special nuclear, or byproduct material as defined by the Atomic Energy Act of 1954. By statutory construction of RCRA, if a material is not solid waste, it cannot be a hazardous waste. However, other types of radioactive wastes—particularly low-level radioactive wastes that are generated in health clinics, hospitals, R&D operations, and nuclear power plants which do not qualify as source, special nuclear, or byproduct material under AEA—are not statutorily exempted from the broad provisions of RCRA.

EPA regulations define a solid waste to be a hazardous waste if it exhibits any one of the four characteristics: ignitability, corrosivity, reactivity, and toxicity. Besides these four characteristics of hazardous waste, EPA has listed three types of hazardous wastes: hazardous wastes from nonspecific sources, hazardous wastes from specific sources, and discarded commercial chemical products and all off-specification species, containers, and spill residues.

In general, RCRA provisions apply to radioactive waste, particularly when they trigger one or more of the four specific characteristics. Although RCRA specifically excludes source, special nuclear, and byproduct material, it applies to the cleanup of other radioactive waste and commercial "mixed waste" containing radioactive elements. Section 3004(j) of RCRA prohibits storage of any hazardous waste at facilities generating mixed radioactive and hazardous wastes, unless such storage is solely for accumulation of certain quantities of hazardous waste necessary to facilitate proper recovery, treatment, or disposal. Section 3004(u) of RCRA requires compliance schedules for corrective actions for the cleanup of released hazardous waste or constituents (including mixed waste containing radioactive waste) from the TSD facilities as a condition for granting of a permit to operate such facility.

Comprehensive Environmental Response, Compensation, and Liability Act

CERCLA, popularly known as Superfund, was passed by Congress and signed into law on December 11, 1980. While RCRA basically deals with the management of hazardous wastes that are generated, treated, stored, or disposed of, CERCLA provides a response to the environmental releases of various hazardous substances. This also includes the release of various pollutants or contaminants into the air, water, or land. A CERCLA response (or liability) is triggered by the actual release or the threat of a release of "hazardous substance" or "pollutant or contaminant" into the environment. Details of CERCLA are provided in Chap. 6.

A release of any radioactive material that is considered to be hazardous in one medium is considered to be hazardous in all other media. CERCLA is not medium-specific. Consequently, certain air toxics, such as radon and several radioactive gases, which are hazardous under the Clean Air Act (CAA), are also hazardous substances under CERCLA. As a result, wherever any hazardous substance is released into any medium, e.g., soil, water, or air, this activates CERCLA.

Section 101 of CERCLA defines *release* to include any leaking, spilling, emitting, etc., whether intentional or accidental. There are two exceptions to this definition where CERCLA does not apply. The first exception includes releases of source, special nuclear, and byproduct material resulting from a "nuclear incident" that is subject to financial protection requirements established by NRC under section 170 of AEA. The second exception includes releases from uranium mill tailings sites which are being cleaned up by DOE under Title I of UMTRCA. These releases are exempt from CERCLA for the purpose of section 104 or any other appropriate response action.

CERCLA also contains limited exceptions which exempt certain releases of radioactive materials from its reporting requirements. The "federally permitted releases" (i.e., permitted under other federal laws) are exempted from the reporting requirements. Such exempted releases also include releases of source, special nuclear, or byproduct material.

Since the exemptions of CERCLA are based on legal rather than technical considerations, it is possible that two adjoining waste sites containing identical radioactive waste could be treated entirely differently under CERCLA—one site may be exempted from while the other may be included in CERCLA's coverage.

As stated earlier, releases of radioactive materials from nuclear incidents at private commercial nuclear plant sites which are subject to financial protection requirements under section 170 of AEA are exempt from CERCLA. However, releases of identical radioactive materials from nuclear incidents at DOE facilities that are covered by an indemnification provision authorized under the same section of AEA are not exempt from CERCLA. Therefore, it is also possible that radioactive waste sites containing identical radioactive wastes may be cleaned up under different standards depending upon whether CERCLA applies.

Finally, with respect to the releases of radioactive materials that are not exempted from CERCLA, EPA has the authority under CERCLA and Executive Order 12316 to require owners, operators, generators, and transporters of hazardous substances to clean up releases of radioactive materials from private and federal sites. All such parties are jointly and severally liable for the entire cleanup. In addition, EPA can issue administrative orders against such responsible parties and levy fines and penalties for noncompliance. Alternatively, EPA itself can expend monies from the Superfund for the cleanup of private sites and then reimburse the Superfund with monies obtained through civil lawsuits brought against the potentially responsible parties (PRPs). However, pursuant to §107(j) of CERCLA, any recovery of EPA's response costs from a federally permitted release is to be granted under other existing laws rather than CERCLA.

Superfund Amendments and Reauthorization Act

The Superfund Amendments and Reauthorization Act (SARA), enacted on October 17, 1986, is a progeny of CERCLA. It addresses environmental releases of hazardous substances from closed or abandoned hazardous waste disposal sites. The most revolutionary part of SARA is the Emergency Planning and Community Right-to-Know Act (EPCRA) covered under Title III. Details of EPCRA are given in Chap. 7.

The general provisions of SARA reinforce and/or broaden the basic regulatory program dealing with the releases of hazardous substances under CERCLA. Section 313 of SARA requires owners and operators of certain facilities that manufacture, process, or otherwise use one of the 329 listed chemicals and chemical categories to report all the environmental releases of these chemicals annually. This makes information about total annual releases of chemicals from industrial facilities available to the public.

Section 206 of SARA created a new §310 of CERCLA in order to deal with citizen suits. Section 310 of CERCLA allows citizen suits with respect to the cleanup of private and federal sites contaminated with radioactive substances. Also citizen groups, for the first time, can bring suits against EPA and other federal agencies for failure to comply with their nondiscretionary duties or for violation of standards, regulations, conditions, requirements, or orders issued pursuant to CERCLA. Such actions will, however, be excluded in connection with the cleanup of naturally occurring radon- and radium-contaminated sites.

SARA also contains new provisions with respect to indoor and naturally occurring radon. Title IV of SARA established the Radon Gas and Indoor Gas Quality Research Act of 1986. Pursuant to this, EPA is required to establish a program to obtain information on indoor air quality; to coordinate federal, state, and private research and development; and to assess appropriate federal government actions to mitigate the risks associated with indoor air quality.

In addition to Title IV, section 118 of SARA contains a number of provisions affecting naturally occurring radioactive materials and radium-contaminated soils. For example, pursuant to §118(b), no later than 90 days after enactment of the statute, EPA is required to provide a grant of $7.5 million from the Superfund to the state of New Jersey for the transportation of some 14,000 containers of radon-contaminated soil from residential areas in New Jersey since it is the subject of remedial action. Under CERCLA, these containers are to be transported to and temporarily stored at any site in New Jersey designated by the governor of the state.

Other environmental laws

The above-mentioned federal environmental statutes are by no means the only ones to reckon with. There are other federal environmental laws that may influence decisions on how to deal with radioactive wastes. In addition, there is a considerable degree of overlap and cross-reference among various environmental laws and, to some extent, with energy statutes, that authorize or require the cleanup of radioactive materials and waste. Besides certain specific provisions under several federal statutes, such as the Clean Air Act, the Clean Water Air, the Safe Drinking Water Act, and various other environmental and energy laws described above, radioactive waste management may be subject to certain regulatory requirements of the Department of Transportation and the Occupational Safety and Health Administration (OSHA). These regulations are authorized under the Hazardous Materials Transportation Act (HMTA) and the Occupational Safety and Health (OSH) Act.

In addition, various state cleanup laws directly or indirectly focus on the cleanup of radioactive wastes and radioactively mixed commercial or industrial wastes. Hence, any decision regarding radioactive waste management must take into account all applicable laws, federal and state, that are in any way connected with the issue of radioactive waste.

11.5   Federal Preemption and State Power

Pursuant to the explicit authority granted to Congress by the Constitution, a state law can be preempted under certain circumstances.

In *Pacific Gas and Electric Co. v. State Energy Resources & Development Commission,* the Supreme Court indicated the circumstances under which a state law may be preempted. First, a federal statute may expressly supersede the state law. Second, a federal statute may occupy a field where a state law is clearly inadequate and/or inappropriate, e.g., where a uniform national law is required and it is so pervasive as to "leave no room for the states." Third, a federal statute can preempt a state law where there is a direct conflict and the state law clearly defeats the congressional purpose.

In this context, the Supremacy Clause in the Constitution governs competing state and federal claims. According to the Court's holding in *Maryland v. Louisiana,* the Commerce Clause in the Constitution limits state laws, particularly when the state laws excessively burden interstate commerce. However, the state laws which deal with traditional police power of the state cannot be preempted.

In the field of radioactive waste, both federal laws and state laws may apply. While this sharing of power has been recognized by the courts in general, it is not clear how the Supremacy Clause and the Commerce Clause can restrict the state's legal authority to regulate the storage and/or transportation of radioactive waste. The joint regulatory power of federal and state agencies pursuant to expressed provisions of certain federal energy statutes creates considerable conflicts between the federal and state governments as each attempts to address and resolve the safety, health, and economic problems that so pervasively affect radioactive waste management. The environmental, health, and safety concerns, together with the emotional fears and the high stakes for all the parties involved, make the issue of radioactive waste virtually unmanageable. The present statutory schemes have created extraordinary tension and friction between the federal and state governments. While states seek to expand the scope of their police power by regulating the problems connected with the disposal of radioactive waste, the federal government seeks to limit the states' role, citing the need for a national uniform rule, thus seeking federal preemption.

The possible application of federal preemption in radioactive waste management has raised a number of major legal issues. The constitutional design of the Supremacy Clause to put a check on state power was eloquently stated by Chief Justice Marshall in *McCulloch v. Maryland.* The *McCulloch* Court held that the means of giving efficiency to the sovereign authorities vested by the people in the national government are those adapted to that end, fitted to promote, and having a natural relation and purpose with the objects of that government. According to this Court, the Constitution and the laws which are enacted in pursuance of it are the supreme laws of the land. The *McCulloch* Court held that there can be no doubt that the laws must be supreme, or otherwise they will be nothing.

Beginning with the landmark case *Gibbons v. Ogden,* the U.S. Supreme Court has demonstrated strong support for any national plan designed by the framers of the Constitution. The Supreme Court thus far has held and unhesitatingly applied the Supremacy Clause to preempt state action which obstructs or conflicts with a valid exercise of federal power. Unfortunately, even today, there is no definitive formula which the U.S. courts can uniformly apply to determine whether federal law preempts state action. The Supreme Court thus far has used a subjective analysis guided by (1) balancing of federalism and (2) whether the purpose and scope of the congressional act precludes state action. Such a case-by-case analysis by the Supreme Court has given no real clue as to how the regulatory activities of the states may be preempted or restricted, particularly where there is a strong public support for state involvement in safety and environmental matters typically associated with radioactive waste. This has provided virtually no basis for the states' role in radioactive waste management. As a result, the states and the public are in a quandary with respect to their legal scope of involvement in radioactive waste management.

In *Pacific Gas and Electric Co. v. State Energy Resources and Development Commission,* the Supreme Court applied the preemption doctrine to resolve the issue of nuclear waste disposal. The Court held that states may prohibit the construction of new nuclear plants without preempting federal authority until the federal government develops and approves a means to dispose of radioactive waste. This narrow construction by the Court has fueled more legal confusion and added to the ongoing tension between the federal government and the states; and it has put the radioactive waste management along with the future of nuclear energy industry in a tailspin. This underlying conflict is amplified because the Court stated *in dictum* that it would preempt any state prohibition on construction of nuclear energy plants "grounded in safety concerns" alone because the federal government has "occupied the entire field of nuclear safety." However, Justice Blackmun's concurring opinion leaves open the possibility that states may increase the breadth of their authority on safety as well as on economic grounds, prior to the construction and operation of a nuclear power plant. This has added further complexity to an issue already shrouded by a mystic veil.

The economic approach is a functional balance between the benefits of local flexibility and those of federal uniformity. However, it may not be the only area for state action. According to the legal scholars, the *Pacific Gas* decision sends a tacit message to the states that they may continue to address the growing safety, environmental, and economic problems associated with nuclear energy and radioactive waste disposal.

To the supporters of states' role in radioactive waste management, a 1992 Supreme Court decision brings a much needed boost. In *New York v. United States,* the Supreme Court held that the provisions of LLRWPA requiring a state to take title to the radioactive wastes generated within its borders is inconsistent with the allocation of power between the federal and state governments based on the Tenth Amendment. The Court also held that the statute's incentives for states to develop disposal sites or join regional compacts are consistent with the Constitution's allocation of power. However, the title-taking provision is not consistent because it coerces the states into carrying out the federal policies. The Court ruled that the choice given to the states between two unconstitutional alternatives—accept the ownership of waste or regulate it—is outside Congress's enumerated powers.

To the consternation of the nuclear energy industry as well as the federal and state governments, Congress has yet to clarify to what extent it is willing to recognize state authority over nuclear power and radioactive waste disposal. The failure by the federal government to establish a national policy to ensure a proper and safe method for the disposal of radioactive waste has often been criticized as a "nonpolicy of indecision and delay." Even an attempt by Congress to legislate a national solution to the problem of radioactive waste disposal has so far been pared down to a compromise. This ultimately resulted in the passing of LLRWPA without any strong guidance for the regulatory framework. Thus, as discussed earlier, LLRWPA has provided something for everyone, but only in the area of low-level radioactive waste. LLRWPA offered a functional division of state and federal responsibilities with regard to low-level radioactive waste only. In doing so, it surrendered the apparent authority over LLRW to the states. Even so, LLRWPA failed to explicitly suggest whether or when such state action may be preempted.

In *Northern States Power Company v. Minnesota,* the court held that the stringent state standards with regard to emissions from a nuclear power plant are preempted by the Atomic Energy Act and federal regulations. This is a leading case in the field of nuclear power plant construction and operation. Unfortunately, there is no other clue in the entire field of radioactive waste management to the legal extent of the state's role or its permissible power and authority.

11.6 Judicial Review and Remedy

Pursuant to the federal Administrative Procedure Act (APA), an act or an omission to act by a federal agency is subject to judicial review. In short, this opens up the possibility of challenging a federal rule or regulation or a decision by a federal agency and makes it subject to judicial review. The purpose of judicial review is to formally seek legal remedy available under specific circumstances.

As discussed earlier, in *Baltimore Gas & Electric Co. v. Natural Resources Defense Council, Inc.,* the Supreme Court upheld the validity of the NRC's table S-3 rule. The *Baltimore Gas* decision is significant not only for its holding but also for the unanimous opinion of the Court, cautioning lower federal courts against becoming overly involved in the technical determination of expert agencies.

The courts' deference to the administrative agencies, particularly when intricate technical and/or scientific issues are involved, is neither new nor uncommon. But the overly broad and powerful deference language in *Baltimore*

Gas has provided a sustaining legacy beyond a mere guideline for judicial review. It has, for all practical purposes, created a heightened notion of deference by the lower courts while reviewing agencies' decisions on broad nuclear matters, including radioactive waste. Practically this has made the power of the federal agencies lopsided and created a virtual status quo without the benefit of a thorough, independent, and neutral judicial review.

It remains to be seen what the long-term impact of *Baltimore Gas* will be in a situation involving judicial review of agency decision making. The later court decisions to date reveal that *Baltimore Gas* has generally caused the federal courts to move toward a reduced level of scrutiny of agency action, particularly in areas of scientific uncertainty.

The impact of *Baltimore Gas* has been highly pronounced in cases involving judicial review of agency proceedings pertaining to the nuclear power industry. These cases demonstrate the far-reaching effect of a Supreme Court warning. The current trend continues to illustrate the courts' general unwillingness to exercise a significant role in those operations that involve the nuclear power industry.

In *Three Mile Island Alert, Inc. v. NRC,* the Court of Appeals for the Third Circuit held that the NRC's decision to authorize certain restraints of the undamaged TM-1 reactor without conducting public hearings on certain safety issues was neither arbitrary nor capricious. In reviewing the NRC's restraint order, the court used an extremely narrow and deferential scope of review of NRC's determinations, particularly those the court found were within the expertise of NRC. Similarly, in *San Luis Obispo Matters for Peace v. NRC,* the D.C. Circuit Court held that the risk of a core meltdown and an earthquake occurring simultaneously at Diablo Canyon Nuclear Power Plant need not be included in an EIS because the probability of such an event is too remote and speculative.

In this context, note that in *Baltimore Gas* and other cases that followed it, the deferential treatment of a federal agency decision has been strictly limited to the field of nuclear power regulation. In essence, the narrow scope of judicial review seen so far has been pronounced only in matters involving nuclear power plant construction and operation.

While radioactive waste management is an area that affects the operation and management of nuclear power plants, it is certainly not limited to nuclear power plants. Hence, it is speculative to suggest that the trend started by *Baltimore Gas* will continue into all aspects of radioactive waste management.

In fact, in a few instances, the federal courts, while citing *Baltimore Gas,* have enjoined agency actions or asked the agency for further consideration of various environmental impacts. In *International Union, United Auto Workers v. Donovan,* the D.C. District Court asked OSHA to reconsider its denial of plaintiff's petition for the issuance of an emergency temporary standard in order to regulate exposure to formaldehyde in the workplace.

Thus, although no expansive judicial review appears on the horizon, some limited judicial review cannot be ruled out in the case of a federal agency action involving handling and disposal of radioactive waste.

11.7   Continuing Dilemma

Historically, nuclear energy has been a creation of the federal government and Congress. But the effects of nuclear power production, particularly the generation of vast quantities and types of radioactive waste, reach deeply into a territory traditionally regulated by the states. These traditional areas, may include, e.g., public health and safety, environmental protection, land use, power plant siting, and public utility planning.

As discussed earlier, the unique and often conflicting interests and roles of the federal government and of the states have created major and deep-seated tensions between them. The extraordinary hazards to the environment, public safety, and health arising from the ever-increasing production, transportation, use, storage, and disposal of radioactive waste and materials have elevated the conflict between the federal government and state interests to such levels of intensity as have not been experienced in any other area.

The ongoing debate of federal supremacy versus legitimate state interests is not limited to regulations involving radioactive waste. It often involves intricate and complex constitutional challenges based upon federal preemptive power. These conflicts are also being played out in the courts to test such primary issues as the scope and extent of congressional intent to preempt any conflicting state regulations. Recent court decisions and the very nature of radioactive waste have put the limelight on another new controversy—toxic tort and resulting liability. The controversy surrounding the generation, use, storage, and disposal of various types of radioactive waste and materials raises the question of the appropriateness of current jurisdiction of two federal agencies, NRC and EPA, along with various state health and environmental agencies which have received certain delegated authority from EPA pursuant to explicit provisions in some of the environmental statutes.

The continuing state participation in various administrative proceedings and judicial review has given rise to another new concern to the uncertainty over radioactive waste disposal problems—the propriety of state laws which impose a moratorium on the construction of nuclear plants. A case in point here is California's nuclear laws which were found unconstitutional in *Pacific Legal Foundation v. State Resources Conservation and Development Commission*. In this case, the Court of Appeals for the Ninth Circuit recognized the uncertainty regarding permanent disposal of radioactive materials and found that California could reasonably prove that nuclear power sources are uneconomical. Thus, the court skirted the preemption issue by finding a narrow area for state action.

In *Pacific Gas and Electric Co. v. State Energy Resources and Development Commission,* the Supreme Court dealt with the issue of whether a state statute dealing with radioactive waste disposal was preempted by federal law. The Court agreed with the view expressed by the Court of Appeals for the Ninth Circuit and accepted California's economic purpose as the rationale for enacting section 25524.2 of its Warren-Alquist Energy Act. The Court concluded that the state action in this situation falls outside the boundaries of federal preemption of nuclear safety regulation. Thus, the Court has opened the door to state legislative attempts to regulate certain aspects of the nuclear power industry.

The *Pacific Gas* decision has also raised a new question about punitive damages which are traditionally a part of state tort law. The previously unchallenged authority of the states over tort remedies was brought up more forcefully in *Silkwood v. Kerr-McGee Corp.* Despite the Supreme Court's careful distinction of safety and economics in *Pacific Gas,* the *Silkwood* Court held that the Atomic Energy Act did not preempt state laws from imposing punitive damages for injuries stemming from noncompliance with NRC's safety regulations. The Court upheld the state-defined punitive damages as though they formed the regulatory background for NRC's safety standards.

Thus, the combined effect of the *Pacific Gas* and *Silkwood* decisions has been a transformation of nuclear safety regulation from an exclusive federal domain to a joint enterprise among NRC, a federal regulatory agency, and a score of state and local regulatory agencies. This is a clear shift from the earlier preemption doctrine espoused by the Supreme Court. Moreover, recent court decisions indicate that preemption may not result until the state law *will necessarily* conflict—not *may possibly* conflict—with federal laws. This apparent reconciliation of dual statutory schemes by the Supreme Court may allow a state law to survive preemption when it has a different purpose or objective even though both federal and state laws regulate the same general area. This implies that the state laws, to survive, need not dwell solely on economic rationale, but may utilize other persuasive arguments to bolster their validity over the threat of preemption.

This judicial narrowing of the traditional application of preemptive power is seen as an open invitation to the floodgate of tort cases arising from various aspects of radioactive wastes, including but not limited to the question of radiation. As a result, the biggest emerging issue with radioactive waste management is the application of tort principles to find appropriate remedies. This has added a new dimension because of the potential personal injury cases that may stem from seemingly endless and continuously growing radioactive wastes.

The complexity of tort cases grounded on radiation or radioactivity can be found in current controversies and scientific debates concerning the safety issues of various nuclear programs and the growing suspicion of the public. Thus, a tremendous legal battle is brewing over how to establish a legal duty, to determine fault through the burden of proof, and to resolve causation before even reaching the question of assessing the damage and the appropriate remedy. Added to this enormous complexity are the problems associated with the dose estimation, frequency and relevancy, causation, and shifting of the burden of proof.

The ground is therefore ripe not just for traditional tort litigations stemming from personal injury cases, but also for making room for eventual toxic tort litigations arising out of radiation caused by generation, storage, treatment, handling, and disposal of radioactive wastes.

11.8 Emerging Issues

The issues involving radioactive waste management and disposal are complex and they arise from constitutional challenges to current dilemmas. This is due to the lack of a comprehensive, uniform, safe, and economic radioactive waste disposal program at the federal level. These challenges and dilemmas assume a herculean dimension if public suspicion and growing restlessness over the failures of Congress to resolve the crisis are taken into consideration. Typically, spent fuel rods from commercial power reactors are stored in cooling pools of water specially earmarked for that purpose in nuclear power plants. When they are sufficiently cooled off, the rods may be transferred either for reprocessing or for dry storage in steel or concrete vaults. Such storage is strictly an interim solution, and plans must be made for permanent disposal of radioactive wastes, particularly high-level radioactive wastes with half-lives of millions of years. Although scientists and engineers generally agree that the high-level radioactive wastes can be safely deposited within certain geological formations containing volcanic ash, salt, or granite which are stable for long periods, the public opposition to such disposal has become a major roadblock. This has resulted in a virtual standoff among the members of Congress who find it politically unsavory to act contrary to the wishes of the voting public.

Furthermore, reprocessing of spent fuel from reactors creates high-level radioactive waste, particularly during the initial stages of separation of uranium and plutonium from the spent fuel. So far, industries have proposed two solutions to handle and dispose of such wastes: vitrification and synroc (i.e., synthetic rock). In vitrification, melting powdered glass is mixed with dried and concentrated radioactive wastes. In synroc, the radioactive waste is "sealed" with a synthetic rock which is made from three titanate minerals and a metal alloy.

Challenging the need of a permanent repository, some have argued that these disposal processes need further documentation to substantiate their claims and that there is no pressing need for a deep repository. Instead, the radioactive wastes can be safely stored on the surface with available technical means. Also, the low-level radioactive wastes, which are hundredfold greater in volume than the high-level radioactive waste, can be effectively reduced by compaction, shredding, and incineration.

However, there is still scientific debate over how best to handle the so-called intermediate- or low-level radioactive wastes with high concentrations of radioactive isotopes. These wastes, unlike typical low-level radioactive wastes, require shielding, but they contain radioactive isotopes which have considerably lower half-lives than those present in high-level radioactive waste.

Another problem facing scientists is the mixed wastes which contain both radioactive wastes and nonradioactive hazardous wastes. While public opposition to scientific or technical solutions to low- and high-level radioactive waste disposal continues, the technical experts are now wondering how best to handle the mixed wastes.

Meanwhile, NRC has announced in its 1990 policy that nuclear power plants, hospitals, and other regenerators of radioactive wastes will be allowed to dispose of certain types of radioactive waste as ordinary trash. As a result, some of these wastes may be disposed of in landfills, sewers, and incinerators. In the past, NRC has granted disposal exemptions for some radioactive wastes on a case-by-case basis. Using this new "coherent framework," NRC claims that a classification of "below regulatory concern" for certain wastes (such as workers' clothing, animal carcasses from research laboratories, and packaging from reactor replacement parts) may allow disposal of certain radioactive wastes as ordinary trash or solid waste. The current requirement allows disposal of such wastes only in controlled, licensed waste repositories. By allowing this form of disposal in uncontrolled facilities, NRC claims that the much-needed money can be saved to focus on materials that "pose much more significant risks to the public."

According to NRC's current policy, only waste generated by commercial and research facilities may be deregulated provided such waste does not expose any person to a radiation dose exceeding 1 millirem per year, except for a "limited number" of people who may be exposed up to a dose of 10 millirem per year. In NRC's evaluation, a 1-millirem dose corresponds to one "annual risk of death from a radiation-induced cancer" out of about 5 to 10 million exposure cases. According to NRC, an average person in the United States receives a radiation dose of 360 millirems per year from natural causes, e.g., radon gas, medical treatment, and related procedures.

However, anticipating the resulting controversy of this NRC action, five states have already moved to bar its implementation within their borders. This has brought to the forefront the emerging issue of the urgent need for federal and state cooperation in solving the ever-increasing problem of radioactive waste disposal.

Another emerging issue is the Secretary of Energy's determination that NWPA does not require consultation with the states for recommending to Congress about the monitored retrievable storage (MRS) sites as an alternative to the long-term storage of high-level radioactive or nuclear waste. This is in direct opposition to permanent repositories which are authorized under NWPA. NRC has echoed DOE's determination by indicating that its adoption of any particular procedure to determine whether to concur with DOE's siting guidelines for MRS falls outside the rule-making requirements of APA and NWPA. While NRC's role is being upgraded by Congress from consultation to concurrence in the DOE's guidelines in order to create a special role for it, debates are still going on as to whether public rule-making procedures by NRC can be totally avoided in formalizing its procedures for determination of DOE's siting guidelines.

Meanwhile, proactive states, such as California, are going to the voters to pass sweeping toxic waste control initiatives which include radioactive wastes. Some of these states have been seriously considering public support for prohibiting the operation of nuclear power plants, thus eliminating environmental risks from such operations. This move is gaining momentum, particularly in those situations where the proponents also talk about economic benefits from such a ban. Under the dual argument of economic benefits and lessening of environmental risks, there may be enormous power shift to the states based on the 1983 Supreme Court decision which dealt specifically with the 1976 California law banning future nuclear power plant construction until permanent waste disposal sites are ready.

In light of the decision by the Supreme Court upgrading states' role in nuclear power plant construction and the concurrence of NRC in DOE's siting guidelines without following public rule-making procedures, a protracted battle between federal agencies and the states is almost certain. What is not certain, however, is how Congress will act to contain this highly emotional and controversial debate on state and federal roles in solving perhaps the most difficult waste disposal problem that the entire nation faces. Thus, one can only conclude that the dilemma involving radioactive waste may continue until and unless Congress launches a bold new program which defines specific roles of the federal and state governments in dealing with radioactive waste management.

Recommended Reading

1. EPA, Report to the Congress, "Potential Health and Environmental Hazards of Uranium Mine Wastes," June 22, 1983.

2. Nuclear Regulatory Commission, "NRC Task Force Report," NUREG-0212, 1977.

Important Cases

1. *New York v. United States,* No. 91-543 (S. Ct. June 19, 1992).

2. *Silkwood v. Kerr-McGee Corp.,* 104 S. Ct. 615 (1984).

3. *Baltimore Gas & Electric Co. v. Natural Resources Defense Council, Inc.,* 462 U.S. 87 (1983).

4. *Pacific Gas and Electric Co. v. State Energy Resources and Development Commission,* 461 U.S. 190 (1983), 103 S. Ct. 1713 (1983).

5. *Weinberger v. Catholic Action of Hawaii/Peace Education Project,* 454 U.S. 139 (1981).

6. *Maryland v. Louisiana,* 451 U.S. 725 (1981).

7. *Goldstein v. California,* 412 U.S. 546 (1973).

8. *Northern States Power Company v. Minnesota,* 447 F. 2d 1143 (8th Cir. 1971), *aff'd. mem,* 405 U.S. 1035 (1972).

9. *Perez v. Campbell,* 402 U.S. 637 (1971).

10. *Gibbons v. Ogden,* 22 U.S. 1 (1824).

11. *McCulloch v. Maryland,* 17 U.S. 316 (1819).

12. *Three Mile Island Alert, Inc. v. NRC,* 771 F. 2d 720 (3d Cir. 1985) *cert. denied,* _____ U.S. _____ , 106 S. Ct. 1460, *ruling denied,* 106 S. Ct. 2909 (1986).

13. *Pacific Legal Foundation v. State Resources Conservation and Development Commission,* 659 F. 2d 9903 (9th Cir. 1987).

14. *Nevada ex rel. Loux v. Herrington,* 777 F. 2d 529 (9th Cir. 1985).

15. *San Luis Obispo Matters for Peace v. NRC,* 751 F. 2d 1287 (D.C. Cir. 1984).

16. *International Union, United Auto Workers v. Donovan,* 570 F. Supp. 747 (D.D.C. 1984).

Remediation Technology: Selection from Several Alternatives*

12.1 Introduction

The raw materials which become a regulated waste when spilled, mishandled, or disposed of are ubiquitous in the industry. Historically, many of these substances have even been ingredients in common household cleaning products. Also, many substances have been used and disposed of for years prior to their assessment as potentially detrimental to human health or the environment. All these factors contribute to the high frequency of contamination when properties are transferred, facilities are decommissioned, and construction projects are initiated.

Contamination is most frequently found in the subsurface of various waste disposal sites. Typically, in the soils and groundwater beneath such waste disposal sites, contamination is not immediately visible and is also relatively stable. As a result, a subsurface contamination problem may persist undetected for decades after substances or materials were spilled. This chapter focuses on the technical approaches available for addressing myriad subsurface contamination problems.

Many unsuspecting property owners suddenly and often find themselves facing the complex and costly problem of subsurface contamination discovered as a result of common business transaction activities. This has prompted an increasingly frequent need for some form of contamination assessment. The most common source of contamination is a leak from a petroleum underground storage tank (UST) located at the site. Contamination also commonly results from historic handling and disposal practices associated with hazardous and toxic materials, such as degreasing solvents, metal-plating bath solutions, PCB-containing oils, cleaning fluids, and off-specification chemical products.

---

*This chapter was written by Glenn V. Batchelder, Vice President of Engineering, Groundwater Technology Incorporated, Norwood, Massachusetts.

12.2   Contaminant Migration

A typical scenario resulting from a gasoline leak is shown in Fig. 12.1. Upon release into the subsurface, liquid contaminants generally first encounter unsaturated soils, defined as soils located above the water table which are not saturated with water. The liquid contaminant flows downward though the soil pores along the path of least resistance. As the slug of contamination flows through the subsurface, it leaves behind a film of contamination that coats the soil particles. This film on the soil particles is referred to as *adsorbed contamination*. If a sufficiently large quantity of contamination is released, the free-flowing liquid contamination will eventually reach the water table as long as no impermeable barrier exists between such contamination and the groundwater.

Once the contamination reaches the groundwater, any water-soluble material will begin to dissolve in the groundwater; this phenomenon is referred to as *dissolved contamination*. Many contaminants such as gasoline, dry cleaning fluids, and degreasing solvents are only slightly soluble in water. As a result, initially the bulk of the material will remain in a separate liquid phase. As additional contamination reaches the water table, this separate-phase contamination spreads out on the surface of the groundwater.

In the case of materials which are less dense than water, such as most diesel oils and gasoline, the contamination spreads, floating on the water surface as shown in Fig. 12.1. With materials which are denser than water such as halogenated solvents, the contamination spreads on the surface and also sinks into the groundwater in separate-phase globules. These denser-than-water compounds often create a more complex migration problem such as the one illustrated in Fig. 12.2.

The fourth and final form in which contamination may exist in the subsurface is the vapor phase. If the contaminant is volatile, the adsorbed contamination will evaporate into the soil pores, resulting in a vapor phase stored in the subsurface matrix.

12.3   Remedial Assesssment

Any remediation strategy must address the potential existence, impact, and abatement of these four phases of contamination: adsorbed, dissolved, separate liquid, and vapor. An assessment is generally undertaken to determine the existence and extent of contamination prior to the beginning of a remedial action. An assessment commonly involves reviewing records on the past use of the property, analyzing soil samples from soil borings, analyzing water samples from monitoring wells, analyzing soil gas from vapor extraction points, and determining groundwater flow direction from monitoring well gauging. During the assessment, data are also collected to determine the impact of the contamination on potential receptors such as drinking water supplies. Based on the results of the assessment, the most cost-effective and expedient remediation strategy is developed to address the dual objective of minimization of risk and regulatory compliance.

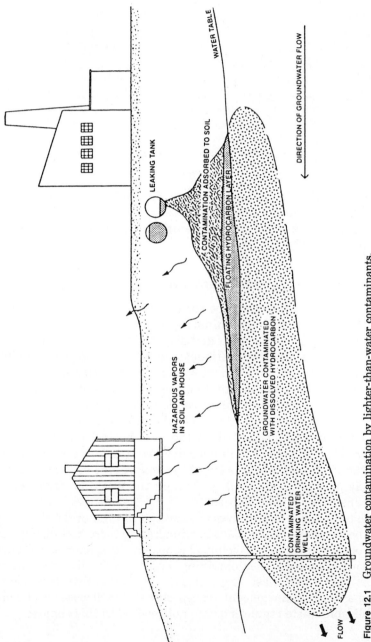

Figure 12.1 Groundwater contamination by lighter-than-water contaminants.

WATER TABLE

LEAKING TANK

CONTAMINATION ADSORBED TO SOIL

FLOATING HYDROCARBON LAYER

DIRECTION OF GROUNDWATER FLOW

GROUNDWATER CONTAMINATED
WITH DISSOLVED HYDROCARBON

HAZARDOUS VAPORS
IN SOIL AND HOUSE

CONTAMINATED
DRINKING WATER
WELL

FLOW

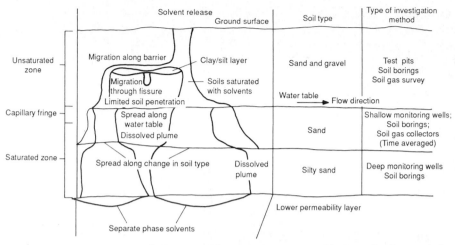

Figure 12.2 Migration and detection of heavier-than-water chlorinated solvents below ground.

While the contamination is assessed, it is advisable to conduct pilot studies to assess the applicability and potential cost of available technologies. Good pilot data are particularly useful in screening the cost-benefit ratios of different technical options objectively.

12.4   Comparison of Off-Site and On-Site Alternatives

One of the first issues which must be considered in developing a remediation plan is whether to treat the contamination on-site or to remove the contaminated material for off-site treatment and disposal. Many remediation plans include a combination of on-site treatment and off-site disposal. On-site treatment requires that either a treatment system be constructed at the site or a mobile unit be brought onto the site. In either case, this utilizes facility space for a finite time. In many cases, practical considerations make on-site treatment the only alternative, even given its inherent inconveniences.

Off-site treatment and disposal allow the contamination to be immediately removed from the site. One of the major disadvantages of off-site solutions is that the contamination or waste is out of the generator's control and therefore the waste may pose a future liability through the waste management activities of the transporter and/or disposer.

In this chapter, on-site alternatives are covered in greater detail than off-site alternatives because on-site solutions are inherently more technically complex. Off-site alternatives involve less technical considerations, such as selecting a reliable transporter and a disposer, obtaining a unit cost for disposal, and complying with the applicable waste regulations. The administrative logistics of prequalifying a waste management transporter and a disposer and complying with the applicable waste regulations are not covered in this chapter.

12.5   In Situ Remediation Technologies

On-site technological approaches can be divided into two categories: in-ground (in situ), and aboveground (ex situ). In situ technologies when compared to ex situ technologies have the advantage of minimizing the facility disturbances. Aboveground approaches provide more rapid and controlled remediation, but they in turn increase space requirements and material handling costs. The most commonly applied in situ technologies are groundwater extraction, soil vapor extraction, air sparging, and bioremediation. These in situ technologies are further discussed below.

Groundwater extraction

Groundwater extraction involves installing wells or trenches from which groundwater is pumped. When water is pumped from an extraction point such as a well or trench, the water level at that point is lowered, thereby creating a local depression in the water table, as shown in Fig. 12.3. This depression causes the water in the immediate vicinity to flow toward and ultimately into

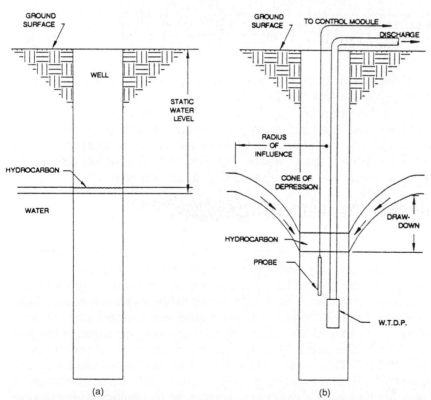

Figure 12.3   A hydrocarbon recovery well before (a) and after (b) installation of a water table depression pump.

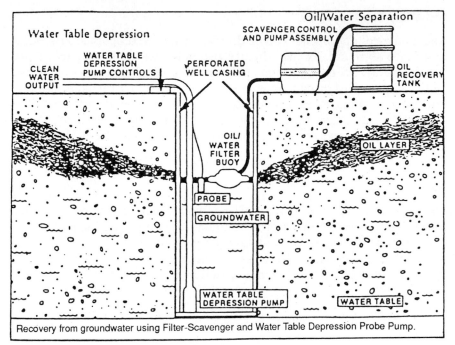

Figure 12.4 Typical two-pump recovery system.

the extraction point. The extraction points are strategically placed to capture dissolved contamination and prevent their further migration.

If separate-phase contamination which floats on water is present, the local depression of the water level will cause a contaminant to migrate toward the extraction point and accumulate. This allows the separate phase to be removed with a skimming system, as shown in Fig. 12.4. This technique is commonly applied when petroleum contamination is encountered.

Soil vapor extraction

In the soil vapor extraction method, extraction points are used to remove contamination from the subsurface. Soil vapor extraction is primarily used to address adsorbed volatile contamination, such as solvents and motor fuels. The extraction points, in the form of trenches and vertical perforated systems, are located in the unsaturated soils above the water table, as shown in Fig. 12.5. The points are generally placed in the zones of adsorbed contamination. A vacuum is applied to the extraction points, thereby drawing air through the soil pores into the extraction point. The mechanism of the removal process is illustrated in Fig. 12.6. This causes the adsorbed contamination to evaporate and subsequently be drawn into the extraction point. Soil vapor extraction is also commonly applied as an aboveground treatment technology where the extraction points are placed in piles of soil located on the ground.

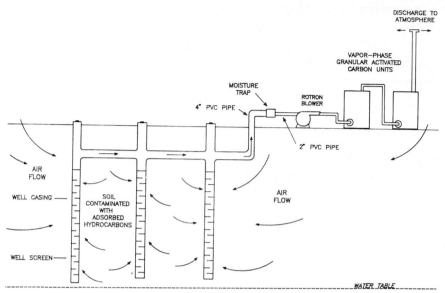

Figure 12.5 Soil ventilation system, vapor-phase carbon adsorption unit.

Air sparging

In air sparging, injection points are used instead of extraction points to remove subsurface volatile contamination. With air sparging, air is injected into the water table to strip dissolved contamination from the upper portion of the water table and the capillary fringe. Air sparging is one of the technologies illustrated in Fig. 12.7. The capillary fringe is the zone where the separate-phase contamination and the groundwater initially contact. This zone of the aquifer tends to be a continuing source of dissolved contamination and is difficult to access with soil vapor extraction and groundwater extraction. During the sparging process, the volatile contamination is transferred from the groundwater into the air. This air laden with contaminant vapors generated during the sparging process is normally drawn from the subsurface through a soil vapor extraction system.

Bioremediation

The final common in situ treatment technology is enhanced natural biodegradation of contaminants, generally referred to as *bioremediation*. This involves providing the indigenous microorganisms with oxygen and nutrients that they may require to efficiently degrade the contaminants. In the case of petroleum-related contaminants, the microorganisms degrade the contaminants into carbon dioxide and water, which are two environmentally benign end-products. The nutrients and oxygen are delivered to the zones of subsurface contamination through injection points such as infiltration galleries, as

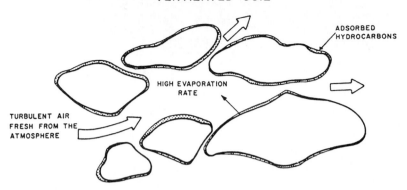

VENTILATED SOIL

ADSORBED
HYDROCARBONS

HIGH EVAPORATION
RATE

TURBULENT AIR
FRESH FROM THE
ATMOSPHERE

IN VENTILATED SOIL THE FRESH TURBULENT AIR MAXIMIZES THE EVAPORATION RATE.
THE SOIL VENT RAPIDLY REMOVES THE VOLATILIZED CONTAMINANT FROM THE SOIL.

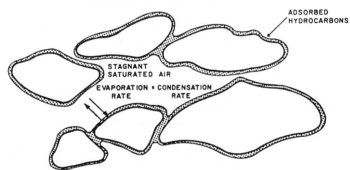

UNVENTILATED SOIL

ADSORBED
HYDROCARBONS

STAGNANT
SATURATED AIR

EVAPORATION = CONDENSATION
RATE   RATE

IN UNVENTILATED SOIL THE AIR TRAPPED IN THE CONTAMINATED SOIL BECOMES SAT-
URATED WITH THE CONTAMINANTS. THE ONLY MECHANISM FOR DECONTAMINATION IS A
SLOW DIFFUSION FROM PORE TO PORE.

Figure 12.6   Mechanisms for soil decontamination.

shown in Fig. 12.8. An infiltration gallery is a horizontally oriented distribution system where previously extracted groundwater is returned to the aquifer. The nutrients and oxygen are delivered to the subsurface through an infiltration gallery by metering them into the groundwater stream prior to its reinjection into the subsurface. The primary constraints on the applicability of bioremediation are that the contaminants must be biodegradable and accessible to a nutrient and oxygen delivery system.

The compounds which are most rapidly biodegraded are those which appear most familiar to the indigenous microorganisms such as petroleum products. Petroleum is created naturally and therefore is similar to the common carbon sources that the microorganisms consume for energy. This allows the micro-

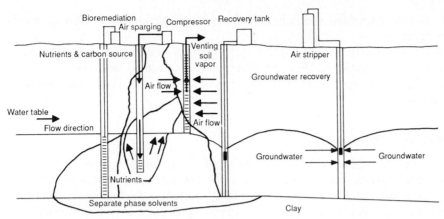

Figure 12.7 Remediation of heavier-than-water chlorinated solvents below ground.

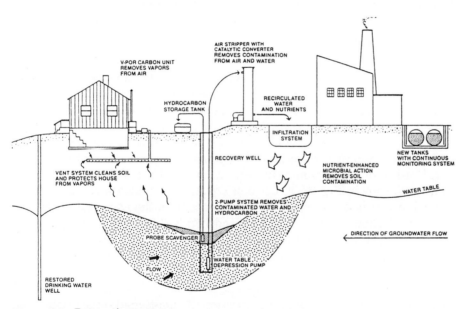

Figure 12.8 Restoration program.

organisms to easily and rapidly degrade almost all types of virgin or unadul-
terated petroleum contamination. Many of the other commonly encountered
contaminants, such as chlorinated solvents, are actually modified petroleum
compounds. Chlorinated solvents are those materials which are simply formed
by adding chlorine molecules to common types of petroleum molecules. As
more chlorine molecules are added to a particular form of petroleum, resulting
in a chlorinated petroleum compound, the compound becomes more difficult to

degrade. Other manufactured substances such as pesticides are also very difficult to degrade.

Another inhibitor to bioremediation is the presence of certain metals which poison the microorganisms, thereby slowing the biological activity.

Bioremediation is applied in situ and ex situ to both adsorbed and dissolved contamination.

12.6 Water Treatment Technologies

Groundwater extraction technology produces a wastewater stream as a byproduct. This wastewater normally has to be treated prior to discharge. The wastewater is typically discharged after treatment to a surface water directly or indirectly via a sanitary sewer or is discharged back into the subsurface. The permits required for these point source discharge options specify the degree of treatment required. The wastewater treatment often constitutes a major portion of the cost of operating a remediation system.

Water and wastewater treatment of organic contaminants

Water and wastewater treatment technologies can be divided into organic and inorganic methods. Organic contaminants include petroleum products, pesticides, polychlorinated biphenyls (PCBs), and solvents. Organic contaminants tend to have certain properties which make them treatable by air stripping, carbon adsorption, bioreaction, and gravimetric separation. Air stripping is one of the most common treatment technologies applied in which high-volatility, low-solubility contaminants are transferred out of the water into an airstream, as shown in Fig. 12.9. Carbon adsorption is also a common technology which involves transferring contaminants from the water onto the surface of carbon, as shown in Fig. 12.10. The carbon is then subsequently regenerated or disposed of. All these water treatment technologies are based on the principles of separation which exploit the relative insolubility of the contaminants.

When high-solubility contaminants such as methanol, acetone, or methyl tertbutylether (MTBE) are treated, the previously mentioned separation technologies are less effective. Destructive technologies such as biodegradation and ultraviolet (uv) ray or peroxide oxidation are more cost-effective on these high-solubility contaminants, as shown in Fig. 12.11. These destructive technologies reduce the contaminants to such environmentally compatible constituents as carbon dioxide, water, and chloride ions.

Water and wastewater treatment of inorganic contaminants

When inorganic contaminants such as heavy metals are addressed, an entirely different type of technology applies. The techniques used are generally

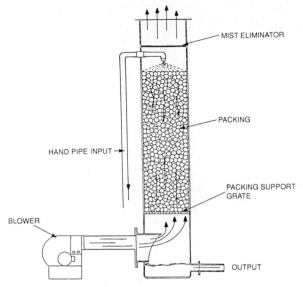

Figure 12.9   Air stripping tower.

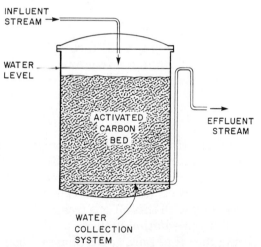

Figure 12.10   Liquid-phase carbon adsorption system.

based on separation technologies. Chemical precipitation is the most common approach to removing dissolved-metal contamination. This involves modifying the water chemistry to cause the dissolved metal to precipitate from solution, as shown in Fig. 12.12. The precipitated material is then collected as a sludge. Other common techniques are reverse osmosis and ion exchange. The waste stream generated from these technologies is a brine solution containing the contaminants in a concentrated form. This brine solution may be either disposed of as a waste or reduced to a lower-volume sludge through chemical

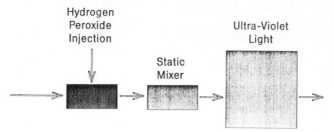

Figure 12.11 Chemical oxidation.

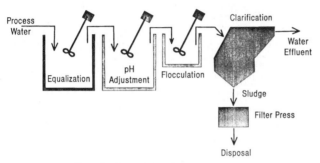

Figure 12.12 Chemical precipitation.

precipitation treatment or sludge conditioning. All three of the technologies described for treating heavy-metal contamination require substantial maintenance compared to the organic treatment technologies discussed above.

12.7 Off-Gas Treatment Technologies

Another common product of remediation system operation is an airstream contaminated with volatile organic materials, such as gasoline vapors or chlorinated hydrocarbons. These airstreams are the off-gas from air strippers or the soil gas removed from the subsurface during soil vapor extraction. The capital and operating costs associated with this off-gas treatment are often the most significant costs of an in situ remediation project.

The two approaches to treating the off-gas are vapor-phase carbon and thermal treatment. Vapor-phase carbon has the two distinct advantages of low energy costs and simplicity of operation. The major disadvantage of vapor-phase carbon is that a waste is generated which must be disposed of. This waste may be the extract from the regeneration process, or if no regeneration capabilities are utilized, the spent carbon itself must be disposed of. Further complicating the use of vapor-phase carbon is the recently imposed regulatory requirement which classifies many carbon regeneration systems as the RCRA treatment units, thus requiring RCRA part B permits.

The alternative to vapor-phase carbon is thermal treatment of the off-gas. Thermal treatment involves raising the temperature of the gas sufficiently to cause destruction of the contaminants. A modified version of thermal treatment is to pass the heated gas stream through a catalyst which allows the desired destruction to occur at a lower temperature. This reduces the energy costs for treating the gas. In either case, the advantage of thermal treatment is that the contaminants are destroyed, leaving behind no waste stream to be treated or disposed of. The only exception to this is halogenated hydrocarbons which produce small quantities of hydrochloric acid during combustion that must be scrubbed from the stream. The volume of waste material generated during this scrubbing activity requiring treatment or disposal is still substantially less than the by-products of carbon adsorption.

12.8 Comparison of In Situ and Ex Situ Techniques

All the technologies discussed to this point have been focused primarily on in situ treatment applications. When compared to ex situ techniques, in situ techniques have the advantages of low initial costs, treatment of inaccessible zones, minimization of surface disturbances, and relatively easy permittability. Generally in situ treatment requires longer to complete and has a greater uncertainty relative to performance than the ex situ technologies involving excavation of contaminated materials prior to treatment.

In evaluating ex situ treatment approaches, the following two major issues must be considered:

- Treatment technology. The objective of the treatment, the specific treatment approach, and where the treatment will be conducted must be evaluated.

- Final placement. Determination must be made of whether the excavated materials should be replaced on-site after treatment and, if the materials cannot be used for backfilling, of what available facilities may accept such material.

These two issues are inextricably intertwined in evaluating the cost-effectiveness and liability concerns associated with different approaches.

The technologies available for ex situ treatment are the same for both off-site and on-site treatment.

12.9 Ex Situ Treatment Technologies

The following are the most common ex situ treatment technologies:

- *Aboveground biotreatment.* This involves creating biotreatment soil piles which include oxygen and nutrient distribution systems to maximize the biodegradation activity. As with in situ bioreclamation, this technology is only applicable for treating those organic contaminants which are readily biodegradable. An example of this approach is shown in Fig. 12.13.

- *Aboveground venting or land farming.* This group of technologies is used to evaporate volatile organic compounds (VOCs) from the material. This tech-

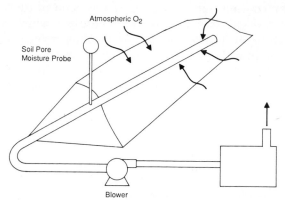

Figure 12.13 Aboveground biotreatment—windrow method.

nology is applied by exposing the soil to fresh air to maximize the rate of evaporation. Aboveground venting allows control of the contaminant emissions to the atmosphere whereas land farming is applied where air emissions are not a concern.

- *Low-temperature thermal desorption.* This technology involves heating the contaminated material to a temperature sufficient to rapidly volatilize the contaminants in a processing unit. The volatilized contaminants are then carried away from the material into an off-gas treatment system. This allows more rapid treatment or remediation of contaminated soil than soil venting does.

- *Incineration.* This technology is applied by heating the waste materials in the incinerator unit to temperatures sufficient to combust the organic contaminants and generally immobilize the inorganic contaminants. Incineration is relatively expensive compared to many of the contaminant-specific approaches. However, when wastes are treated with a wide range of contaminants, the cost-effectiveness of incineration improves. Also, under the federal land ban regulations, incineration is the prescribed technology for many wastes prior to land placement.

- *Solidification.* The intent of solidification technology is to chemically immobilize the contaminants by mixing the waste with a solidifying agent. This technology is particularly applicable to heavy-metal contaminants.

- *Soil washing.* This involves contacting soil with a washing fluid during which time the contaminant is transferred from the soil into the liquid. While this technology continues to be a promising concept, its applicability is still limited.

Other compound-specific technologies exist for contaminated soils, although most utilize some variation of the technologies described above. For evaluation of these technologies for on-site application, the special constraints must be considered, particularly the permitting prospects and the cost-effectiveness of constructing a temporary treatment facility. If the material is to be both treated and disposed of off-site, the prequalifications of the waste transporter and the disposer become the most critical portion of the technical evaluation process, with the technology selection issues becoming much less important.

In the overall evaluation of various technical options, an important consideration is the synergy between technologies, especially if in situ approaches are selected.

12.10 Strategy Development

The first step in developing an overall approach to addressing a contamination problem is to define the contaminant distribution, commonly known as a *mass balance*. By identifying the zones of the site and the medium in which the majority of the contaminant mass is contained, the remediation activities can be focused where they will have the greatest effect. In most cases, the majority of the contamination is retained in the area of the contaminant release. These areas, where the majority of the contamination is retained, commonly act as the continuing sources of groundwater contamination.

The mass of contaminant needs to be calculated for each phase: adsorbed phase, separate phase, dissolved phase, and vapor phase. Once the major mass sources of contamination have been identified, migration from those areas should be controlled and the sources abated. In determining the immediacy and degree of remediation efforts required from a liability control perspective, the risk of the contamination impacting any nearby receptors must be fully assessed.

The following are commonly impacted receptors:

- Dissolved phase and separate-phase contamination impacting on drinking water supply wells
- Dissolved phase and separate-phase contamination leaching into surface water
- Vapor phase and separate phase of volatile contamination causing vapors to accumulate in the basements of buildings
- Vapor phase and separate phase of volatile contamination migrating into the backfill of utility trenches

With the zones of highest contaminant mass identified and the risk to potential receptors assessed, a remediation strategy can be proposed which maximizes the impact of the contemplated remediation plans as a starting point in negotiating with the regulators. This can be extremely useful in arriving at an optimal corrective action plan and an agreed-upon closure standard.

Another critical consideration in selecting a remedial technology is determination of the accessibility of the contamination relative to the different technologies available. For example, volatile contamination located 40 feet below ground surface in a fine sand soil matrix would be most accessible to soil vapor extraction technology. Conversely, inorganic contamination located only 3 feet deep in the subsurface can be addressed most easily through excavation, barring any structures.

Recommended Reading

Bioremediation

1. Richard A. Brown and Richard Tribe, "Bioreclamation of Spills," *Hazardous Materials Management,* June 1990.

2. Scott B. Wilson, "Varieties of Bioremediation Technology," *Environmental Waste Management,* April 1990.

3. Paul M. Yaniga, "*In situ* Bioreclamation of Organic Waste," *Hazardous Waste Management,* April 1989.

4. "Microbe Muscle—Biotreatment Pioneers Are Breaking Down Political Technological Barriers to Hazwaste Remediation," *Environment Today,* January–February 1990.

5. "Bioreclamation Saves Groundwater Supply," *Public Works,* May 1988.

Chlorinated solvents

Richard A. Brown, Glenn V. Batchelder, and Richard W. Lewis, "Chlorinated Solvent Contamination: A Challenge to Find and Remove," *Chemical Processing,* January 1991.

General

1. Richard A. Brown and Kevin Sullivan, "Integrating Technologies Enhances Remediation," *Pollution Engineering,* May 1991.

2. Kenneth L. Manchen, "Rating Your Environmental Department," *Pollution Engineering,* April 1990.

Leaking underground petroleum storage tanks

1. Keith G. Angell, "After the Spill," *Environmental Compliance Sourcebook,* 1990.

2. "When Your Nightmare Comes True," *Journal of Petroleum Marketing,* June 1990.

Refineries

Richard A. Brown and Richard T. Cartwright, "Biotreat Sludges and Soils," *Hydrocarbon Processing,* October 1990.

Enforcement, Liability, and Compliance

13.1 Introduction

Environmental laws and regulations have been created to prevent and minimize what may be harmful to public health and the environment. The concern for public safety, health, and the environment has reached its zenith with the seemingly uncontrollable growth of toxic, radioactive, and otherwise hazardous chemicals, substances, and wastes which, taken together, may be grouped as hazardous materials. The legislative curb on industrial practices and programs has become significantly stringent since the 1970s when there was great public outcry against growing environmental pollution. Public criticism against lack of regulatory enforcement has remained highly vocal even in recent years for there is a continuous growth of hazardous materials.

So far, Congress has responded in kind to public cries for more environmental control by passing several major environmental statutes in recent years, e.g., the Toxic Substances and Control Act (TSCA); the Resource Conservation and Recovery Act; the Comprehensive Environmental Response, Compensation, and Liability Act; and the Superfund Amendments and Reauthorization Act. But Congress has not simply stopped by enacting these new environmental statutes. It has also added major and substantial amendments to some of the earlier environmental statutes, e.g., the Clean Air Act (CAA); the Clean Water Act (CWA); the Federal Insecticide, Fungicide, and Rodenticide Act (FIFRA); and the Safe Drinking Water Act (SDWA).

The most obvious result of these congressional actions is the tightening of enforcement of various statutory provisions. In turn, regulatory agencies have been given broad authority to take extreme measures against individuals and corporations for failure to comply. As a result, federal regulations promulgated under applicable federal statutes have incorporated stringent enforcement actions and liability provisions against the violators. In addition, in certain instances, the statutes themselves, including their amendments, have incorporated detailed enforcement measures for adoption by the regulatory agencies. This has left very little room for agency discretion and regulatory interpretations with significant latitude over such enforcement actions. As a result, the regulatory sanctions are not directed only toward the willful, grossly negligent, and intentional violators. In many situations, they even target those without any fault under the principle of strict liability.

The bottom line here is to restore public confidence. To achieve this difficult task, enforcement activities have begun to play an important role at the agency level. Gone are the days of passive cooperation, innocent bliss, and overt optimism. Instead, the agencies are doggedly pursuing their enforcement actions now. Not only has the number of enforcement staff been significantly increased, but also the agencies are seeking both civil penalties and criminal sanctions in a zealous fervor. The penalties are often being sought on the basis of per day or per violation. In addition, several agencies are overtly pursuing the "bad actor" policy against those violators who have had any record of past violations or are simply recalcitrants. The courts in turn are responding to strong prosecutorial advances by handing out stiff sentences, paying little or no attention to specific circumstances surrounding such violations.

Thus, environmental liability has taken on a completely new meaning. There are not just common law liabilities to worry about. Today, various statutory liabilities must be considered as well since the statutes themselves spell out the full range of civil and criminal penalties. And if this is not enough, most environmental statutes have either no statute of limitations or a very long one. This literally makes one liable for environmental violations for many years and, in certain cases, whole life. Worst of all, many of the newer environmental statutes and the regulations promulgated thereunder contain "strict liability" or "absolute liability" provisions, which offer little or no defense even if there is no fault on the part of the accused.

It appears that even the Securities and Exchange Commission (SEC) has joined in this foray. It is adopting an increasingly tougher position on public disclosure of environmental liabilities by corporations. If there is any "material adverse effect" due to environmental liabilities, a corporation is required to disclose such information in Form 10-K and in its annual report.

The ultimate message here is the need for strict compliance with the statutory provisions and applicable rules and regulations. Because of the expansive interpretation of statutory language by the courts in general and the concerted efforts by enforcement officials, both individuals and corporations will undoubtedly serve their interests best by avoiding potential liabilities through well-planned and well-heeded compliance programs of environmental laws and regulations.

13.2 Enforcement

Federal regulations promulgated to protect the public health and the environment are enforced under a number of federal statutes, many of which are characterized as environmental statutes. Although most of these federal environmental statutes go back only a few decades, the penalties on violators can be imposed based on standards set by regulations promulgated under any applicable federal or state law that is still a good law.

At the federal level, various civil penalties and criminal sanctions are typically imposed for environmental pollution of any of the three media (water, air, and land). Some of these penalties and sanctions may apply for pollution of a single medium and others for several media. For example, the Rivers and Harbors Act of 1899, popularly known as the "Refuse Act," as amended, and the Clean Water Act, as amended, contain statutory provisions for appropriate sanctions against water pollution, and the Clean Air Act, as amended, includes sanctions for violations of air quality standards as set forth in appropriate EPA regulations. However, several other environmental statutes such as RCRA, CERCLA, and SARA can impose penalties for a violation of the multimedia standards. In addition, there is a great deal of overlapping and cross-referencing of federal statutes. Hence, one may be subject to the provisions of several federal statutes and the regulations promulgated thereunder due to one single act or failure to act.

Nevertheless, not all federal statutes and the regulations promulgated thereunder provide the same sanctions. Depending on the statutory provisions and the regulations, both civil liabilities and criminal liabilities may vary depending on their roots and times of enactment. It is generally agreed that recent statutes or later amendments of older statutes are stricter in terms of both civil and criminal penalties. Also, as the federal regulatory agencies promulgate new standards in response to various statutory requirements, such penalties for violations will likely become stiffer.

Additionally, the regulatory agencies are gearing up for increased enforcement. The Department of Justice (DOJ) created its environmental crime unit only during the last decade. As a sign of what to expect, EPA has significantly expanded its enforcement staff. The Pollution Prosecution Act of 1990 requires the EPA administrator to increase the number of criminal investigators assigned to the Office of Criminal Investigations within EPA. It also requires the EPA administrator to provide fifty additional civil investigators by September 30, 1991, to develop and prosecute civil and administrative actions.

As stated earlier, federal environmental laws can be enforced civilly as well as criminally. Since their scope can vary from one to the other, it is extremely important to establish the statutory and regulatory sources of enforcement actions which may be imposed in several ways.

A brief review of the administrative actions and prosecutions that may be employed for known or suspected violations of federal environmental laws and regulations is provided below.

Inspections and search warrants

Agency inspectors carry out both routine and surprise inspections of facilities. Routine inspections are normally limited to information gathering pursuant to an agency approval or a permitting program. The surprise inspection, however, is generally undertaken if there is any suspicion or a tip is received about a possible violation of statutes or regulations that the agency is empowered to enforce.

These administrative inspections or searches are generally provided for in several environmental statutes. If not indicated, any inspection without a valid warrant may be denied. Furthermore, even if a warrantless inspection of a facility is authorized by an appropriate statute, its validity may be legally challenged on the Fourth Amendment's protection against unreasonable searches and seizures. This is important, particularly when a statutorily authorized inspection or search goes beyond the scope or limits of authority. Typically, inspection of financial records or trade secrets without an appropriate court order or warrant is outside the scope of agency inspections or searches. Unfortunately, most federal statutes do not provide appropriate standards that define the authority of the agencies in conducting a search or an inspection. This has resulted in several litigations, and the courts have carved out several exceptions to the warrant requirement. Notable among them is the "Biswell-Colonnade exception" which excuses the requirement of a warrant for an agency search. This is done by using the "open-fields" doctrine that removes the legitimacy of privacy of activities when they are taking place outdoors and not inside closed buildings or facilities. Nevertheless, any such warrantless search or inspection must be authorized by the statute itself. Otherwise, the agency inspection or search may be legally denied or may constitute a trespass. As a result, it is extremely important to have a good grasp of applicable statutory provisions before crying foul in a warrantless search or inspection.

If an agency inspection or search is conducted under a valid warrant, the warrant has to clearly provide the scope or limits of the search. If the warrant is not correctly executed, it can be legally challenged, particularly if the times and frequencies of inspections appear to be inappropriate.

Nonetheless, most of the recent environmental statutes authorize routine warrantless search and inspection by agency officials. Also, a statutory "right to entry" is not always limited to outdoor premises. The agency inspectors may legally enter various buildings and structures as well as any open area within the facility as long as the applicable statute grants such authority.

Administrative enforcement actions

There are several administrative enforcement actions which can be imposed by a regulatory agency as a result of an actual violation:

- Notices of violation
- Cease-and-desist orders
- Permit revocations
- Administrative consent orders for abatement and compliance

Procedurally and substantively, these methods of administrative enforcement are not the same in terms of correction and penalties. Some are less severe and can be settled easily by adopting minor changes and accepting agency consultations whereas others can be extremely severe as they may result in emergency orders including permit suspensions or revocations.

Civil penalties

One of the major deterrents against possible violation of applicable standards or regulations is the imposition of civil penalties. Traditionally, civil penalties were imposed by a court of competent jurisdiction only at the request of the agencies that were seeking compliance with regulatory requirements and discovered certain violations. Now there are statutes explicitly authorizing the agencies to enforce their regulatory programs developed under appropriate statutory provisions. Such enforcement can include an assessment of civil penalties through appropriate administrative processes.

Typically, civil penalties are imposed on the violators of applicable laws and regulations to make them comply. These civil penalties can, however, be either statutory or based on common laws. The common theme here is the existence of liability for environmental violations. Generally, the Department of Justice, the federal enforcement arm of the government, uses common laws whereas the federal environmental agencies, e.g., the Environmental Protection Agency, uses the statutory provisions to impose civil penalties on violators. It is important to note here that while seeking either a civil or a criminal penalty, EPA routinely refers such enforcement action to DOJ.

The civil penalties are not fixed in either situation. They can vary from minor slap-on-the-wrist type of fines to severe penalties. Also, fines for first-time offenders are considerably lower than those for repeat offenders of the same offense. The enforcement and regulatory agencies have been working together to develop and institute a bad actor policy. This not only will highlight the past history of environmental compliance or noncompliance records of a violator but also may subject such a violator to a harsher penalty than the nature of the violation deserves otherwise.

The statutes as well as the regulations promulgated thereunder typically provide a range of penalties. The statutory levels of civil penalties vary. They may be as high as $25,000 per day for the cited violation until its abatement.

However, such civil fines are not inclusive of other measures, such as damages and injunctory relief. Damages can be compensatory or punitive or both. Under certain statutes, the compensatory damages can be three times the harm or damage caused, although they are seldom imposed. Punitive damages are typically based on the nature of the violation and the severity of disregard and attitudes displayed by the violators for noncompliance. Also, punitive damages are awarded on a discretionary basis by a court.

Criminal sanctions

By and large, the common law remedies for environmental violations do not impose criminal sanctions. This is, however, not true for environmental statutes, particularly those of recent origin.

Almost all the federal environmental statutes since the 1970s, except the National Environmental Policy Act, have specific provisions for criminal sanctions. The underlying regulations, based on such authority, provide the full range of criminal sanctions that can be imposed on a violator.

The range of criminal sanctions, as in the case of civil penalties, varies from statute to statute. In most cases, the common thread of criminal sanctions is the traditional concept of *mens rea* (i.e., mental culpability), or intent of the wrongdoer. A few of these statutes, however, can impose criminal sanctions without taking into account the mental intent of the violator.

Generally, environmental statutes impose criminal sanctions for three types of actions: negligent actions, knowing (i.e., with knowledge) actions, and knowing endangerment actions. The severity of punishment follows in ascending order. The courts are not known to accept either ignorance of law or any overt or covert mistake in their understanding.

A violator's actual state of mind need not always be established by the enforcement agency in order to establish the "knowledge" element. Such knowledge may be presumed on the basis of several external factors, including the types of operation or the nature of the manufacturing facility and/or the objective standard of a "reasonable person." This is, however, an area of fervent litigation and controversy, and there is no simple rule of guidance.

Note that many federal environmental statutes do not preempt state laws. In fact, certain federal regulations contain specific provisions that allow their administration and enforcement by appropriate state agencies. These enforcement actions may be instituted following the state's Administrative Procedure Act, if there is one, or certain other applicable procedural requirements, such as notice and public hearing. For all such cases including federal enforcement actions, the rights accorded to all the defending parties must satisfy the due process of law guaranteed under the Fifth and the Fourteenth Amendments of the Constitution.

Apart from the above, certain statutes restrict or limit judicial review of the administrative actions. This is generally the case with the environmental statutes enacted since the 1970s. Nonetheless, the enforcement agency must prove that the defending party or parties have violated a specific provision of a statute or regulation which the agency is authorized to enforce.

13.3   Liability

Growing public awareness of environmental pollution is largely responsible for incorporation of broad environmental liabilities against individual and commercial or corporate polluters. In the past, polluters generally wished to stop and succeeded in stopping the federal government from bringing actions for environmental pollution by adroit use of constitutional doctrines and the framework of the Constitution itself, which has no explicit language for protection of the environment.

However, the present statutory scheme of Congress, largely influenced by public concern and criticism, has incorporated protection of the environment as one of its most persistent and consistent goals. The courts are also responding in kind by sifting through the legislative records for determining congressional intent in a given statute. It is no major surprise that in *United States v. Standard Oil Co.* the Supreme Court interpreted the provisions of the Rivers and Harbors Act of 1899, a so-called grandfather federal environmental statute, in an unprecedented and highly expansive way. The Court held that the statute forbids the discharge of refuse matter or industrial waste in the navigable waters of the nation. It should also be stressed that environmental liabilities are not restricted to toxic and hazardous substances or wastes, although at present they constitute most of the major environmental liabilities. Theoretically, any violation of a statute or regulations promulgated thereunder or that of federal or state common laws can make a violator liable.

This type of liability can take several forms, such as individual, corporate, and lender's liability. The nature or the type of liability depends on both the underlying action and the parties involved. It is appropriate to consider and scrutinize environmental liabilities in order to focus on the extent and gravity of potential liabilities.

Individual liability

No federal environmental statute imposes broader liability and sanctions on individuals and corporations than CERCLA. Section 107(a) of CERCLA lists four types of potentially responsible parties:

- Current owner or operator of a facility from which a release of a hazardous substance has actually occurred or is suspected

- Past owner or operator of a facility (including the person who owned, operated, or controlled the facility prior to its involuntary acquisition by a state or local government through bankruptcy, foreclosure, tax delinquency, abandonment, or voluntary acquisition by private parties) provided hazardous substances were disposed of at that facility during such ownership or operation

- Persons who generated or by contract, agreement, or other means arranged for disposal or treatment or arranged with a transporter for transport for disposal or treatment of a hazardous substance in a facility

- Persons who accepted any hazardous substances for transport to a disposal or treatment facility and also selected such facility for that purpose

The term *person* here can mean an individual, individuals, or a corporation, a partnership, and the like, as discussed in detail in Chap. 6. In addition, individuals may be liable for their own roles which may result in environmental damage and endangerment under the common laws and appropriate environmental statutes. Many of these environmental statutes include a double-barrel provision. Thus, individuals may also be liable in their functions as corporate officers or directors, corporate employees, and individual shareholders of a corporation if they actively participate in or are responsible for making decisions that directly affect or cause the release of hazardous substances or improper disposal of hazardous wastes. In many of these situations, individuals may be held directly liable, particularly under CERCLA. In such cases, the courts or even the plaintiffs may be reluctant to pierce the corporate veil.

Historically, individual liability for any wrongdoing is well established compared to corporate liability which is of recent statutory origin. Individuals remain liable under common laws for tortious conduct. Such tortious conduct can be found in environmental endangerment as well as damage to public health and safety. In addition, any individual violation of a specific provision of an appropriate federal environmental statute makes that individual liable for such conduct. For example, section 103 of CERCLA imposes a duty to notify. This duty rests on the person in charge of an offshore or onshore facility or a vessel with the knowledge of a release of a reportable quantity of a listed hazardous substance from such facility. Failure to report within the prescribed time can subject the individual in charge of the facility to a civil penalty of up to $10,000 or to a criminal sanction involving imprisonment for up to 1 year or to both. A similar civil liability—but no criminal sanction—can also be imposed on an individual for failure to meet the notification requirements under section 311 of the Federal Water Pollution Control Act (FWPCA) or its later amendments, the CWA and the WQA.

In the true sense, liability for environmental violations starts with individuals responsible for such violations, although such liability may not necessarily end there.

Corporate liability

The superimposition of corporate liability on the top of individual liability for environmental violations is more of an afterthought by legislators and regulators. This is reflective of the huge health, safety, and environmental damages and exorbitant costs of cleanup and remedial work. In short, its inclusion is logically based on the need to find "deep pockets" to pay for the resulting response costs.

The federal environmental statutes have specific provisions under which corporations may be held liable. In most cases, the parent corporation of a subsidiary may be found liable due to the activities undertaken by the subsidiary under certain circumstances. Thus, it is not uncommon for the courts to "pierce the corporate veil" to impose liabilities on the parent corporation. This may well result in the imposition of personal liabilities upon individual directors, officers, and shareholders.

The management and control of hazardous wastes and substances that are generated, transported, used, stored, or disposed of have become not just complex, but a lot more onerous to the corporations which must deal with them at various stages. Even reuse or recycling of hazardous wastes, which may appear to fit into the much touted congressional plan and regulatory objective of waste minimization, is inherently complex and not without its own pitfalls. The bottom line here is that any violation, no matter how unintended or accidental, can trigger some sort of liability. In most cases involving environmental liability, the statute of limitations holds little consolation for past violators.

As discussed in Chap. 6, under certain recent federal statutes, EPA can take legal action for recovery of costs within 6 years after the date of completion of the remedial action, regardless of when the actual disposal or release took place. If the cleanup is not considered complete, the statute of limitations has practically no fixed period to work as a stopper for an eventual remedial action or environmental liability unless one is seeking recovery of damages. In such a case, the claim must be presented within 3 years of the later of (1) the discovery of the loss due to the release of a hazardous substance or (2) the date of promulgation of the final regulations by EPA under section 301(c) of CERCLA for the assessment of damages to natural resources.

Corporate liability generally arises from various explicit and implicit duties and the subjective or objective roles of corporations. In addition, specific federal statutes have explicit or implicit provisions that give rise to corporate liability. Other individuals and/or corporate liabilities are discussed below.

Workplace liability. An employer has a general duty to provide a safe workplace for the employees along with suitable and safe equipment and other tools to work with. In addition, pursuant to common law principles, an employer is required to warn the employees about all defects and hazards that the employer knows of or should have known about while exercising reasonable care and diligence. If such defects or damages exist and the employees do not have any knowledge of them, the employer is also required to instruct the employees on how to avoid them safely.

The employer's duty to warn and to provide a safe workplace continues as long as the employer operates or maintains a workplace. Such duty is continual in nature and cannot be discharged by a one-time act or provision. Nonetheless, the employer is required to provide a reasonably safe—not foolproof or absolutely safe—workplace. The reasonableness standard of employer's degree of care actually depends on the type of workplace and the specific factors involved. Generally, the risk of injury to the employees is the yardstick by which the safety of a workplace is gauged. As a result, the type of warning and instructions to employees will vary, and the employer's liability may be lessened if such warnings and instructions are found to be effective and reasonable under the circumstances. This determination should be made by using an objective standard. According to one state court decision, an employer will generally not be liable if the employer "conducts its business in a manner conforming with the usage of others engaged in the same business under similar circumstances."

But another state court decision indicates that an employer's duty to warn is not pegged to the customs of one particular industry. The reason for this pointed departure by this court is that the industry practice is not always the measure of proper diligence or duty. Thus, in certain situations, higher standards than the customary practices within the industry may be imposed.

Hence, to avoid liability, an employer must exercise all reasonable care to provide a safe workplace. It is therefore appropriate to diligently seek out the hazards, both patent and latent, that exist in the workplace. Dependence on the industry custom for a determination of the duty to warn may not be adequate, particularly if the nature of the work is inherently dangerous and there is no industrywide custom to establish proper warnings.

At an extreme level, any work-related use, generation, transportation, storage, or disposal of hazardous and toxic substances and wastes, including by-products and off-specification species, may be construed as inherently dangerous and unsafe. It is therefore highly likely that an employer who is engaged in these operations will be bound to higher standards. In the absence of any specific statutory provision in the federal or state environmental laws or regulations, an employer is susceptible to liability for failure to meet the standards under the common law duty to warn. In such cases, the employer may not be shielded or its liability limited by the otherwise exclusive-remedy provision of the workers' compensation statutes for actions brought by employees because of workplace injuries. In general, an employee may resort to three principles of tort (i.e., intentional harm, dual capacity which goes beyond the employer-employee relationship, and third-party liability) to bypass such exclusive-remedy provisions.

Toxic tort liability.   The recent federal environmental statutes, such as RCRA, CERCLA, and SARA, do not address the issue of relief for personal injury or property damage stemming from exposure to hazardous and toxic substances or chemicals. As a result, plaintiffs alleging personal injuries or property damage resort to traditional tort laws based on judge-made laws founded in the principles of English common laws. The underlying principle of these common law torts, popularly known as toxic torts because of their implications of certain exposure to toxic substances, is that persons injured by the fault of others and not by fault of their own are compensable.

Simplistically, a tort is a civil wrong or conduct that causes personal injury or property damage to another. When certain conduct is unacceptable by society, society can condemn such conduct through the imposition of strict or even absolute liability under common law. What conduct is unacceptable is defined by the objective standard of a reasonable person.

This reasonableness standard is not, however, absolute or static. It changes with time and with the norms and expectations of the society. Belatedly, with the advent of strong public concern about hazardous and toxic substances and waste materials, the application of toxic tort is becoming more common. Under the principles of strict liability, corporations, like individuals, are becoming highly susceptible to various toxic tort liabilities due to personal injury or property damage that may be inflicted because of the way their businesses are conducted.

However, like all common-law–based injuries or damages, toxic tort allegations must be filed within a specific time from the exposure to the injurious agent, be it a hazardous or toxic substance or a waste material. Generally, if the statute of limitations runs out before the injury or damage is detected, the cause of action should not be toxic tort. However, this is indeed flexible since the statute of limitations may start to run from a different date than the date of actual exposure to an alleged harmful or hazardous substance, chemical, or waste that was not known or considered to be toxic at the time of its production, use, storage, transportation, or disposal.

Chapter 14 deals with the details of the toxic tort litigations and defenses.

Liability under RCRA. It has been pointed out in Chap. 5 that RCRA is a legislative outcome of the massive public concern about environmental hazards of solid and hazardous wastes. The general consensus prior to the enactment of RCRA was that management and control of hazardous waste or materials had become a serious problem and that legislation was required to ensure public health and safety. Through RCRA, Congress has set new standards and guidelines for industries on how to use, treat, store, or dispose of hazardous waste. RCRA also envisages the complete tracking of hazardous wastes from generation to disposal through a unique cradle-to-grave manifest program.

By enacting RCRA in 1976, for the first time Congress attempted to define solid and hazardous waste and thus to do away with the nagging debate over what is or is not a hazardous waste. However, RCRA does not put a stop to generation or use of hazardous wastes in various industrial processes. Instead, it puts the onus on everyone who has to deal with them and thus to follow certain requirements and set procedures. For example, RCRA regulations require permitting of treatment, storage, and disposal facilities which handle hazardous waste. It also sets forth numerous guidelines for state and federal inspections of TSD facilities for hazardous waste. Additionally, it sets the stage for various enforcement actions coupled with imposition of liabilities.

In an attempt to fine-tune the RCRA program, the original statute has been amended several times, the most recent and significant one being in 1984. The Hazardous and Solid Waste Amendments (HSWA) of 1984 resulted in more than seventy-five major changes and revisions in the original text. Almost all, however, are geared toward tightening of the compliance and enforcement programs. Furthermore, these amendments provide new definitions of solid waste along with an expanded listing of hazardous waste.

They also include additional enforcement provisions along with citizen suits and imminent-hazard actions which can impose restraints through injunctions. Generally, there has been no departure from the original purpose of RCRA, i.e., to reassure the public that the hazardous waste management practices are being conducted in a safe manner to protect public health and the environment. However, recent amendments have substantially increased the coverage and strict requirements of RCRA.

The basic tenet of RCRA philosophy still remains the acceptable norm of conduct on the basis of statutory and regulatory standards and guidelines encompassing generation, treatment, storage, transportation, and disposal of hazardous waste. Nevertheless, to ensure strict compliance, RCRA includes provisions for both civil and criminal penalties for the violators. Additionally, there are provisions for federal injunctive relief. For example, the EPA administrator may bring an action against a violator who is contributing to an "imminent and substantial endangerment" to health or the environment through activities involving handling, storage, treatment, transportation, or disposal of any solid or hazardous waste. In *United States v. Vertac Chemical Corp.,* the federal district court upheld EPA's interpretation of "imminent and substantial endangerment." EPA interpreted such a situation as one which poses a "risk of harm" or "potential harm" without the proof of actual harm. In addition, citizen suits may be brought against a violator of any standing administrative order or permit condition imposed under RCRA.

The enforcement program has also been broadened to allow individual states to oversee appropriate conduct by corporations and individuals involved. The majority of states to date have received RCRA's regulatory authority to enforce their own RCRA program.

As in the case of other recent federal environmental statutes, RCRA provides both civil and criminal penalties for

- Violation of the statute itself
- Violation of the EPA regulations promulgated under RCRA
- Violation of RCRA-based regulations as adopted by an individual state
- Violation of the RCRA permit requirements and the terms and conditions included there

Civil enforcement can be undertaken by an administrative consent order (ACO) or by bringing an action in a federal district court for injunctive relief. For criminal violations, however, the penalties may vary depending on the degree and nature of the violations. For example, fines of up to $50,000 per day and imprisonment for up to 5 years for knowing violations may be levied against first offenders, and these penalties are doubled for repeat offenders. A corporation found guilty under RCRA can be fined up to $1 million for creating a crime of "knowing endangerment."

Chapter 5 discusses what constitutes knowledge and knowing endangerment under RCRA. Surprisingly, in *U.S. v. Speach,* the Court of Appeals for the Ninth Circuit held that lack of requisite knowledge is a proper defense for transporting hazardous wastes to an unpermitted facility under RCRA.

Several courts, prior to the enactment of HSWA or the 1984 RCRA Amendments, held that the imminent-hazard provision of RCRA applied only to the current owners and operators of facilities and not to the off-site generators or past violators. The 1984 RCRA Amendments put a stop to such burgeoning decisions by specifically including past and present owners of facilities as well as transporters and generators of hazardous wastes under the general ambit of the imminent-hazard provision of RCRA.

Joint and several liability. Under the common law principles, there is no right of contribution among joint tort feasors. The majority of states have overcome this apparent gap of inequity by adopting a uniform contribution among joint tort-feasors act or similar state statute. Pursuant to such laws, if several persons are found jointly and severally liable and one ends up paying more than his or her proportional share in a divisible harm, that person may have a cause of action against other culpable parties for reimbursement of their respective proportional shares of the harm caused. Depending on the nature of the harm or damage inflicted, an apportionment of individual shares of harm is made.

This principle of contribution among joint tortfeasors has not gone unnoticed by federal courts. In this context, the federal district courts apply the laws of the state where such courts are physically located. As a result of the 1986 CERCLA Amendments by way of SARA, popularly known as the "Gore amendments," section 113(f) of CERCLA has incorporated this equitable apportionment. Pursuant to the Gore amendments, proportional contributions by joint tortfeasors are to be based on several factors. Notable among them are the

- Nature of the harm or fault
- Amount and nature (e.g., toxicity) of the hazardous substance released
- Migration potential of and risk of exposure to the released hazardous substance
- Degree and nature of responsibility by culpable individuals

In the context of traditional tort, joint and several liability is usually imposed

- When two or more persons act in concert
- Where defendants breach a common duty owed to a plaintiff
- Where there is a special relationship between the parties
- Where injuries to a plaintiff are indivisible

In the context of pollution-related violations, joint and several liability is generally imposed when the defendants are responsible for an indivisible harm to public health and safety or to the environment. This is indeed a question of fact and not of law. If it is found divisible, the court must apportion damages and will not impose joint and several liability on several defendants to tie up all such actors equally.

Some courts, however, have adopted a broader view of indivisibility in recent years. For example, in *Velsicol Chemical Corp. v. Rowe,* the court simply refused to consider the divisibility of the damages due to pollution since that would have imposed a tremendous burden of proof on the plaintiff to substantiate and separate the injuries caused by each defendant. Instead, the *Velsicol* Court, noting the modern trend, imposed joint and several liability on several defendants in absence of their ability to establish what portion of the total harm was caused by each. Several other courts have generally followed this modern common law rule of contribution by multiple tortfeasors which has been incorporated in Restatement (Second) of Torts, §875 (1977).

On the surface, CERCLA's liability standard merely states that owners and operators of facilities as well as transporters and generators of hazardous substances "shall be liable" for cleanup expenses, the cost of remedial action, and natural resource damage due to the release of any hazardous substance. A number of federal courts held that under §107 of CERCLA, liability may be joint and several if there are multiple contributors, the regulatory harm is indivisible, and the defendants cannot prove what portion of the total damage was caused by their individual contributions.

These courts generally suggest that CERCLA permits, but does not require, the imposition of joint and several liability on multiple defendants. If the imposition of joint and several liability appears harsh and unfair, these courts use their own discretions to determine and apportion damages. In *United States v. South Carolina Recycling and Disposal, Inc.,* the district court rejected the defendant's contention that apportionment should be based on volume since commingling of hazardous waste at the dump site made apportionment of individual contributions impossible. Instead, the court imposed a joint and several liability on each defendant. As of now, the courts generally have the ultimate say as to where and when joint and several liability should be imposed.

In *United States v. A & F Materials Co., Inc.,* the district court even rejected the test proffered by Restatement of Torts. Instead, it adopted an apportionment scheme of its own. While the court recognized its own authority to impose joint and several liability on the generator defendants, it emphasized that it was not necessarily obliged to do so. Instead, it favored an apportionment plan in those circumstances where generator defendants cannot prove what portion of the total damage was caused by their individual contributions of hazardous waste. The *A&F Materials* Court theorized its apportionment plan on the basis of the original CERCLA-related House bill which considered several criteria for apportioning damages involving multiple tortfeasors. These criteria include

- The ability of the parties to demonstrate that their contribution to a discharge, release, or disposal of a hazardous waste was distinguishable
- The amount of the hazardous waste involved
- The degree of toxicity of the hazardous waste
- The degree of involvement by the parties in the generation, transportation, storage, or disposal of the hazardous waste

On the other hand, in *United States v. Northeastern Pharmaceutical and Chemical Co., Inc.,* the court poignantly acknowledged the unclear legislative history on the issue of joint and several liability. However, by accepting the routine permissibility of joint and several liability, it held that since in the instant case only one generator, one transporter, and one landowner were involved and all were responsible for an indivisible harm, each defendant must be jointly and severally liable for the tort itself.

In *United States v. Alcan Aluminum Corp.,* the Court of Appeals for the Third Circuit held that a PRP is entitled an opportunity to prove that a harm is divisible and that there is a reasonable basis for apportionment of the cleanup costs. Note that the court's holding did not alter the proposition that if the harm is not divisible and therefore cannot be apportioned, joint liability will apply.

In summary, imposition of joint and several liability when two or more persons independently cause a distinct harm is a matter of discretion for most courts. Legal scholars tend to agree with this approach. Even EPA itself states that CERCLA may impose joint and several liability in most cases but not necessarily in all cases.

Strict liability. Within legal circles, it is generally accepted that the CERCLA liability standard is based on strict liability, although the statute does not contain such specific language for liability. This standard has been accepted by the majority of the courts throughout the country. These courts have generally relied on the legislative history of CERCLA as well as the standard of liability provided in §311 of the Federal Water Pollution Control Act and its later amendments in the Clean Water Act. Not surprisingly, EPA has steadfastly adhered to its policy, which is based on strict liability under CERCLA.

However, this does not imply that there can be no meritorious defenses against a charge of strict liability. In reality, there are limited defenses available in strict-liability cases since it is not an absolute liability. Even the undisputed victims of exposure to hazardous substances due to accidental or incidental releases may be denied recovery on the contention of such popular tort defenses as contributory negligence, assumption of risk, plaintiff's unusual sensitivity, defendant's public duty, intervening acts by third parties, and acts of war or God. Obviously, the strengths and weaknesses of these defenses may vary from jurisdiction to jurisdiction depending on appropriate state laws and regulations.

In a step-by-step approach, in determining whether strict liability may be imposed, the first step is to decide whether federal or state laws should apply. Arguably, state law should apply since the inactive hazardous waste disposal facilities are physically located within the confines of a state. Traditionally also, state law oversees property law. However, under CERCLA, federal courts have exclusive original jurisdiction for claims due to the release of a hazardous substance. In this context, even prior to the enactment of CERCLA, a number of courts interpreted section 311 of the Federal Water Pollution Control Act as a clear and convincing standard of strict liability. They used federal statutes for determining applicable standards.

It is presently widely accepted by both courts and legal scholars that CERCLA imposes strict liability. Unfortunately, as in other issues of CERCLA, the question remains less than fully resolved in the absence of a clear statutory provision or an appropriate decision by the Supreme Court. Notwithstanding the original strict-liability theory based on *Rylands v. Fletcher,* §§519 and 520 of Restatement of Torts (1938) and Restatement (Second) of Torts (1976) and the recently enacted federal statutes are probably the best available yardstick for the nation's courts to determine whether such a standard should be imposed in a particular situation. So far, most courts have resoundingly opted for strict liability when adjudicating liability under CERCLA.

Liability under CERCLA. The original CERCLA of 1980 provides a relatively short, general, and somewhat vague mandate for identification, investigation, cleanup, and remediation program as well as liabilities associated with inactive or abandoned hazardous waste disposal sites. This is perhaps due to the fact that CERCLA was a congressional attempt to address health and environmental issues that became nationwide concerns following the discovery of Love Canal in Niagara Falls, New York, and other significant and infamous abandoned or inactive hazardous waste disposal sites.

Nevertheless, CERCLA, with its later amendments taken as a whole, is considered a major federal environmental statute that imposes significant liability for discharge or release of hazardous substances into the environment. As stated earlier, section 107 of CERCLA imposes liability on persons, including corporations, in four distinct categories:

- Present owner or operator of a vessel or a facility
- Past owner or operator of a facility if disposal of any hazardous substances occurred during such ownership or operation
- Any generator who contractually or otherwise arranged for disposal or treatment or arranged with a transporter for transport for disposal or treatment of hazardous substances
- Any transporter who accepted to transport hazardous substances for disposal or treatment to a facility selected by that transporter

Private causes of action may be brought against these potentially responsible parties under joint and several liability as well as strict liability, as discussed earlier.

The vagueness and ambiguity in the original CERCLA of 1980, however, have given rise to serious debates and litigation. This is particularly intense with respect to issues dealing with the scope and standard of liability as well as the procedural requirements for government and private cost recovery actions.

In addition, a number of lower federal courts have tried to grapple with the issue of parent corporation liability under CERCLA. In these cases, the parent corporations have generally been held responsible for the actions of their subsidiaries. To do this, the courts have used the equitable powers to pierce the corporate veil. This is done to discourage parent corporations from setting up undercapitalized subsidiaries for the purpose of treatment, transportation, or disposal of hazardous waste. These decisions are generally based on findings of active participation or management of the activities of the subsidiaries by the parent corporation. However, in one federal case, *Joslyn Manufacturing Company v. T.L. James and Co., Inc.*, the Fifth Circuit Court of Appeals held that parent corporations are not liable for the actions of their subsidiaries. This court firmly rejected the argument based on management control by the parent corporation.

Generally, a corporation that has a controlling or management interest over the operations of another corporation or over its own subsidiary is liable under CERCLA. A dissolved corporation may also remain liable under CERCLA in certain circumstances, e.g., when it might not have distributed the proceeds of liquidation to the shareholders. The successor corporation that receives the shares of another corporation as a result of a merger may also be subject to CERCLA-based liability.

However, recent cases indicate that an acquiring corporation may not be liable for the activities of the acquired corporation under CERCLA if it acquires only the assets of the acquired corporation and not simply the shares. This exemption, however, will not apply if

- The acquiring corporation expressly or implicitly agrees to accept the liability of the acquired corporation.
- The asset purchase is a consolidation or merger between the two corporations.
- The acquiring corporation continues as the acquired corporation.
- The purchase-sale transaction of assets between the corporations is fraudulent.

While it is almost impossible to determine if and when the Supreme Court will decide on such hotly debated issues, it is prudent to structure a parent-subsidiary relationship that minimizes the potential for CERCLA liability.

CERCLA includes a broad definition of a *potential responsible party* (PRP) to determine who may be held responsible for the cleanup costs. The definition, discussed above, includes four categories of PRPs. The cleanup of an actual or threatened release of hazardous substances is the responsibility of these PRPs. If this is not accomplished, the federal government may take administrative action to remedy the release and then obtain reimbursement of all related costs from the appropriate responsible parties.

As indicated in Chap. 6, CERCLA deals mostly with the remedial actions rather than with the standards of conduct which are the vestiges of several other environmental statutes. Unfortunately, many key provisions of this statute are rather vague and somewhat ambiguous. Nonetheless, the remedial actions designated under CERCLA provide for both short-term response (i.e., mitigation of immediate sources of damage) and long-term remediation (i.e., permanent protection including confinement of releases of hazardous substances). As a result, remedial actions may include a host of activities ranging from providing alternative sources of drinking water to temporary relocation of the affected population. The National Contingency Plan (NCP) spells out the details of the cleanup actions that may be undertaken by either the government or private parties. EPA has developed its National Priorities

List (NPL) of inactive hazardous waste disposal sites based on the known and threatened releases of hazardous substances of such sites within the United States. Under the existing regulation, the federal government cannot undertake long-term remedial actions without meeting certain procedural requirements. Even if it does, it cannot recover the necessary costs of response from the responsible parties unless the site is listed on the NPL.

Besides this threat of cost recovery action, the federal government has potent enforcement tools under CERCLA. It can sue various PRPs under the provisions of sections 106 and 107 of CERCLA. Section 106(c) of CERCLA covers the imminent-hazard provision. Under CERCLA liability in section 107(a)(4), the responsible parties are liable for

- All costs of removal or remedial action incurred by a state or by the United States
- Any other necessary response costs incurred by any other person consistent with the NCP
- Damages for injury to or destruction of or loss of natural resources
- The costs of any health assessments or health effects study carried out by the Agency for Toxic Substances and Disease Registry (ATSDR) pursuant to §104(i) of CERCLA

Section 107(a) of CERCLA also authorizes the federal government to bring enforcement action against the PRPs to recover the response costs in cleaning up an inactive hazardous waste disposal site. According to the holding of the *Northeastern Pharmaceutical* Court, under the provisions of section 107 of CERCLA, the government is not required to prove that its response costs are consistent with the NCP. Rather, the burden of proof is on the defendant(s) to show inconsistencies of such response costs, as incurred, with the NCP. The same court, however, held that all nongovernment entities must prove that such reimbursement costs are necessary and consistent with the NCP under section 107(a) of CERCLA.

Note here that liability stemming from damage to natural resources is *strict* and does not require a showing of actual fault by the responsible party. Section 107(c) of CERCLA provides maximum limits of liabilities for a PRP. Section 107(b) of CERCLA specifies three statutory defenses where the damage can solely be attributed to

- An act of God
- An act of war
- An act or omission of a third party who is not an employee or agent of the defendant or whose act or omission is not a result of a contractual relationship with the defendant

These defenses are good only upon showing, by a preponderance of evidence, that the defendant(s) exercised due care and took appropriate precautions against foreseeable acts or omissions by any such third party as well as against the foreseeable consequences.

Liability under SARA. By enacting SARA in 1986, Congress did not significantly alter the basic liability provisions over cost recovery actions that are included in section 107(a) of CERCLA. However, a number of peripheral changes were made in CERCLA by way of SARA that can affect such cost recovery actions.

SARA added several new provisions to CERCLA. For example, two new important provisions deal with the substantive requirements for government cleanup work and the level of scrutiny to be used by the courts in evaluating such cleanup work. The basic benefit here is that all government cleanups must generally comply with the criteria provided in federal and state environmental statutes that are legally applicable or relevant and appropriate requirements (ARARs). In addition, the judicial review of a remedy undertaken by the federal government will be made on the administrative record of EPA unless such decisions are found to be arbitrary and capricious or otherwise not in accordance with the law.

Other important new provisions in CERCLA, due to the passing of SARA, include an equitable allocation of response costs among the defendants. In addition, they include the authority of the federal government to enter into settlement agreements with the PRPs. Other important provisions of SARA include the statute of limitations applicable to cost recovery actions, the nationwide service of process, and the government decision to sue PRPs before giving them a notice or an opportunity to undertake and perform necessary remedies. However, most importantly, SARA specifies how the government may seek contributions from the PRPs as an equitable remedy.

SARA has empowered EPA with broad administrative authority. It has reauthorized CERCLA for 5 years more and provided EPA with $8.5 billion to spend on various cleanup operations under the provisions of CERCLA during this 5-year period.

SARA also redefined the act of release of hazardous substances into the environment to include new costs that may be claimed against the PRPs. SARA has significantly increased the enforcement and liability provisions of CERCLA. A new section, section 109, established for the first time, stresses civil penalties for a number of violations, including violations of the terms-of-settlement agreement reached earlier. Another new section, section 117(e), authorizes public participation in remedial action plans and grants of up to $50,000 in technical assistance for the purpose of interpreting the nature of the hazard, remedial investigation/feasibility study (RI/FS), record of decision (ROD), remedial design, selection and construction of remedial action, operation and maintenance, or removal action to those who may be affected by a hazardous substance release. Additionally, the statute of limitations on filing personal injury lawsuits has been extended. This allows the tolling of the statute of limitations until the knowledge of the injury is gained.

Lender liability

Banks and other lending institutions may be liable as PRPs in government or private cost recovery actions due to the release or discharge of a hazardous substance into the environment. In two lower federal court decisions, the courts held that the lender may qualify as an *owner or operator* of an abandoned or inactive hazardous waste disposal facility. According to these courts, if a lender exercises significant or major control over the facility or completes the foreclosure of a mortgage lien on or security interest in the property contaminated with a hazardous substance, the lender will wear the shoes of the owner or operator. In a 1991 federal case *United States v. Fleet Factors Corp.,* the Supreme Court upheld an appeals court decision that a lender who is in a position to affect decisions about hazardous waste disposal made by a company receiving loaned money may be liable for that company's cleanup costs under CERCLA.

More and more, the courts are expansively redefining the term *owner or operator* in the context of PRPs under CERCLA. This puts a greater burden on the prospective lender to conduct a due-diligence environmental audit of the property to be purchased with the loaned money.

Strategies for possible negotiation and settlement of CERCLA liability are discussed in Chap. 15.

On April 29, 1992, EPA promulgated its final rule on lender's liability to define and interpret the provisions of sections 101(20) and 101(35) of CERCLA, as amended. Section 101(20)(A) of CERCLA exempts a person from CERCLA-based liability of an owner of a contaminated facility provided his or her or its indicia of ownership are held primarily to protect a security interest in that facility.

Unfortunately neither the statute nor the legislative history of CERCLA, as amended, provides any indication as to what can be the extent of "participation in management" involving a contaminated facility without violating this exemption. EPA's final rule attempts to define what is "security interest exemption." Moreover, it clarifies and specifies the range of activities that may be undertaken by a person for maintaining indicia of ownership primarily to protect a security interest.

Pursuant to EPA's rule, several activities of a lender (i.e., a holder in interest) are permissible which do not exceed the bounds of the security interest exemption under section 101(20)(A) of CERCLA. EPA's lender liability rule, for the first time, clarifies what activities a lender may undertake while preserving the "innocent landowner" defense by qualifying for the security interest exemption. It specifies the circumstances under which government or government-appointed lenders acquiring possession or control of contaminated facilities as conservators or receivers remain eligible as "involuntary owners" under section 101(35)(A)(ii) of CERCLA.

EPA's rule does not restrict the clause "primarily to protect a security interest" only to those situations where indicia of ownership are held primarily for investment purposes. The rule suggests that a holder may have other or secondary reasons for maintaining indicia of ownership. However, to avoid any possible fouling of the security interest exemption, the primary reason for holding indicia of ownership of a contaminated facility must be the protection of the holder's security interest in that facility.

The rule provides examples of several actions or activities that will constitute "participation in management" as well as those which will not be considered as participation in management. According to EPA, the following actions by a lender will not constitute participation in management:

- Actions at the inception of a loan or other transaction, for example, an environmental audit or inspection or cleanup of a facility or to comply or come into compliance with any applicable law or regulation

- Actions, such as loan policy and workout, that are consistent with holding indicia of ownership primarily to protect a security interest prior to foreclosure, including but not limited to, cleanup of a facility during the term, complying or coming into compliance with applicable federal, state, or local environmental laws, rules, and regulations during the term of the security interest, securing or exercising authority to monitor or inspect a facility in which indicia of ownership are maintained or taking other actions to adequately police the loan or security interest.

- Actions involving response actions under section 107(d)(1) of CERCLA or under the direction of an on-scene coordinator

The rule clearly allows policing and workout activities by a holder (viz. a private and government lending institution or entity, receiver, or conservator) without triggering any participation in management as long as such activities occur prior to foreclosure. A holder or a lender is not immune from CERLCA-based liability if he or she or it arranges for disposal or treatment of a hazardous substance or accepts for transportation and disposal of hazardous substances at a facility of his, her, or its selection following foreclosure. However, a lender may liquidate, maintain business activities, undertake any response action under section 107(d)(1) of CERCLA or under the direction of an on-scene coordinator and take measures to preserve, protect, or prepare a contaminated facility, the secured asset, for its eventual sale or disposition.

Liability in bankruptcy

The environmental laws which are designed to prevent and/or remedy an environmental contamination appear to run counter to bankruptcy laws that provide a debtor with a "fresh start" by an automatic stay of debts. Under the Bankruptcy Code, only an individual debtor may receive a discharge of environmental or other claims in bankruptcy. The Bankruptcy Code defines a claim as a right to payment or to an equitable remedy in the event of a breach of performance. Liabilities under CERCLA or other environmental laws constitute claims that may be present or contingent. However, a claim may be denied if the right to payment or to an equitable remedy is speculative or contrary to the provision of the Bankruptcy Code. Also, if allowed, an environmental claim is subject to additional considerations. Typically, such a claim is considered unsecured or contingent, thus making the claim somewhat tenuous.

Moreover, a state law that directly conflicts with the provisions of the Bankruptcy Code may be preempted if the enforcement of the state law is too onerous to interfere with the bankruptcy adjudication. In *Ohio v. Kovacs,* the Supreme Court held that the liabilities for response costs are automatically stayed or suspended as a result of bankruptcy. The Court, however, did not address the issue when a claim based on environmental obligation arises for bankruptcy purposes. The Court held that the cleanup order by the state of Ohio had actually been converted to an obligation to pay for the cleanup which is dischargeable in bankruptcy. In the 1991 case of *Jensen v. California Department of Health Services,* the Ninth Circuit Court of Appeals held that a claim based on environmental cleanup costs arises when the conduct of a debtor in bankruptcy creates the threat of environmental contamination and not upon the expenditure of the response costs. In *In re Chateaugay Corp.,* the Second Circuit Court of Appeals affirmed a lower court ruling which held that a debtor's reorganization effectively discharges both past and future contamination cleanup costs that arise from prebankruptcy activities.

However, the liability of the bankruptcy trustee is not extinguished if there is a release of hazardous substance from a facility that comes under the trustee's control due to the voluntary or involuntary entry of the debtor into bankruptcy. By accepting the existing assets and liabilities of the debtor, the bankruptcy trustee becomes responsible for necessary cleanup and remediation work. According to the court's decision in *Midlantic National Bank v. New Jersey Department of Environmental Protection,* a bankruptcy trustee may not abandon a contaminated property in contravention of state laws which are designed to protect public health and the environment.

In a 1992 case, *In re CMC Heartland Partners,* the Court of Appeals for the Seventh Circuit held that cleanup obligations of an owner of a contaminated facility under CERCLA run with the land. The court ruled that such obligations are not discharged in owner's bankruptcy. According to this court's holding, if and when an owner of a contaminated facility emerges after bankruptcy, the CERCLA-based liability also returns.

Note that the issue of CERCLA liability in a bankruptcy situation is still very tepid since the narrow holding of the *Kovacs* Court. The uncertainty caused by this holding and the judicial effort to balance environmental goals and bankruptcy resolutions will continue unless there is a Supreme Court decision or congressional act indicating when a claim arises in a bankruptcy.

13.4 Compliance

Simplistically, the best way to avoid environmental liabilities is to avoid all statutory and regulatory violations. This can be done by meeting all the statutory and regulatory requirements and complying with them at all times.

For a due-diligence review of compliance requirements, it is important that all aspects of environmental considerations including applicable federal, state, and local laws and the regulations promulgated thereunder be fully analyzed and reviewed. Typically, the following steps are included in a well-planned due-diligence review:

- Environmental audit of the facility
- Soil contamination investigation
- Contamination assessment including investigation of groundwater and surface water quality
- Risk assessment of operational activities
- Corrective action programs undertaken and successfully completed

As indicated earlier, environmental liabilities can be imposed on almost all living entities, be they an individual, a partnership, or a corporation, since the statutory definition of *person* literally includes every living individual and on-going enterprise. While there is no specific or single rule as to how best to avoid potential liabilities, there are certain minimal standards or acts which should be included in all well-defined and properly planned compliance programs.

Environmental audit

A proper and well-planned environmental audit is an important tool in establishing and maintaining continued compliance with environmental laws and regulations. Furthermore, it allows one to seek and identify whether there are any actual or potential violations of any aspect of the operation of a facility that are not in compliance with applicable environmental laws and regulations. Controversies exist about the soundness of undertaking an environmental audit since by doing this, one may risk possible criminal charges if certain violations are disclosed. EPA's formal policy and the Department of Justice's enforcement policy released on July 1, 1991, have not helped in minimizing this apprehension. Nevertheless, through such an environmental audit, steps can be taken to protect the facility's owner or operator as well as the facility's seller, lender, and buyer who may be subject to liability because of the sale or purchase of a contaminated facility. This is important even for a person who is seemingly innocent. For example, to qualify for an "innocent-landowner" defense, a person has to establish his or her or its exercise of due care in view of all relevant facts and circumstances. Pursuant to the statutory provisions, an innocent landowner may assert this defense if only the property is found to be contaminated prior to the purchase and the contamination is solely due to the acts of the seller with whom the purchaser has no contractual relationships for liability of cleanup.

Section 101(35)(B) of CERCLA provides a list of items that a prospective purchaser of land or property should establish to assert the third party or innocent-landowner defense. This list generally includes all appropriate inquiries, at the time of acquisition, into the previous ownership and uses of the property consistent with good commercial or customary practice. To determine CERCLA-based liability, the courts will consider

- Any specialized knowledge or experience on the part of the purchaser or defendant
- The relationship of the purchase price paid to the value of the property if it were uncontaminated
- Commonly known or reasonably ascertainable information about the property
- The obviousness of the presence or likely presence of contamination of the property
- The ability to detect such contamination by appropriate inspection

According to EPA's guidelines issued on June 6, 1989, as contained in "Guidance on Landowner Liability," there is no uniform standard of inquiry by prospective landowners. Based on EPA's hierarchy of standards, commercial transactions are subject to the highest level of prepurchase inquiries with respect to existing contamination of a property. The courts in general have not accepted the innocent-landowner defense if the prepurchase inquiries were inadequate, there were lapses of due care by the purchaser subsequent to the discovery of contamination, a contractual relationship existed between the seller and the purchaser, or the purchaser has certain degree of culpability for the discovered contamination. In *Sanford Street Local Development Corp. v. Textron Inc.,* the federal district court held that a significantly discounted price for a contaminated site can constitute an arrangement for disposal of hazardous substances.

As discussed earlier, liabilities can arise from a number of federal environmental statutes as well as applicable state and local laws. Each provides substantial penalties; some involve civil penalties, others involve criminal sanctions, and still others include both civil and criminal sanctions. Under the provision of joint and several liability included in most of the recent federal environmental laws, the federal government can receive the entire costs of cleanup and remediation from a single liable party. Hence, appropriate care must be exercised for all environmental considerations.

Violations can take the form of failure to purchase applicable and necessary environmental permits or to meet their specific terms and conditions or general violations of environmental statutes or regulations. Such violations can be construed under specific statutory provisions or under the common law if personal injury or property damage results.

Under the common law principles of negligence, trespass, private and public nuisance, and strict liability, compliance with regulations or permit conditions may not be an adequate defense if a personal injury or property damage is experienced by a third party.

For these reasons, a typical environmental audit may be carried out in several phases. A phase 1 audit includes a general physical inspection of the facility, interviews with management, employees, past owners, operators, lessees and tenants, and a historical review of past use of the property, permits issued, and documents contained in the agency's files. Phase 2 includes a thorough investigation of the facility including, but not limited to, collection of soil and groundwater and surface water samples, installation of monitoring wells, stack tests, and in-depth review of all past operations and practices. Phase 3 generally involves actual cleanup and remediation activities. Phase 3 is, by its very nature, generally undertaken only at the behest of a regulatory agency or court order.

While an environmental audit can be done in several formats, almost all such audits should include certain minimal elements. A typical checklist of such elements or considerations is shown in Table 13.1. For real estate transactions, an environmental audit checklist should be comprehensive enough to include all environmental considerations to qualify as a due-diligence review prior to such transactions.

As indicated earlier, there has been serious concern over the potential benefits and possible pitfalls of an environmental audit. Whether the findings of an internal audit have to be disclosed to regulatory agencies has been widely argued by legal scholars. Doubts have been cast since EPA's policy statement on environmental audits issued on July 9, 1986, does not contain any specific provision stating that the findings of an audit will not be used for the purposes of an enforcement action. However, on July 1, 1991, the Department of Justice issued a policy document encouraging self-policing and voluntary disclosure of violations by the regulated community. The benefit here is that of possible leniency of sanctions if voluntary, timely, and complete disclosures as well as appropriate efforts to remedy past noncompliance are made. However, if one discloses the findings of a voluntary audit that reveals noncompliance and one fails to adequately address and remedy such compliance, such a disclosure will result in a great potential for severe enforcement measures.

Miscellaneous compliance programs

Apart from an environmental audit, a good compliance program may include several other activities:

- Establishing compliance schedules
- Understanding and complying with permit conditions
- Keeping and maintaining accurate records
- Prompt reporting of any spill, discharge, or release, whether accidental or intentional
- Instituting a good housekeeping plan
- Using labels, placards, or other warning signs, if warranted
- Establishing reporting requirements and designating appropriate individuals to contact regulatory agencies and other required departments
- Correcting all violations quickly and effectively, including occasional and unexpected equipment or plant upsets

TABLE 13.1 Checklist for Environmental Audit or Assessment

Location of site _____

SIC code and nature of operation _____

Owner of site _____

Operator of site _____

Site address _____

Name, address, and phone number of inspector or reviewer _____

Date of site inspection or date of interview _____

	Yes	No	Additional comments/ remarks/location
General Observations, Natural Resources, and Physical Layout			
1. Staining or discoloration of any area	_____	_____	_____
2. Absence of vegetation or appearance of dead vegetation	_____	_____	_____
3. Hills, mounds, or depressions	_____	_____	_____
4. Old structures, foundations, buildings which look presently unused and somewhat dilapidated	_____	_____	_____
5. Signs of flowing or standing liquid	_____	_____	_____
6. Any discoloration or malodor from waters, soils, storage piles, etc.	_____	_____	_____
7. Transformers—PCB-labeled or no label at all	_____	_____	_____
8. Open or damaged drums, if any, and types	_____	_____	_____
9. Manholes, drainage ditches, culverts, catch basins, etc.	_____	_____	_____
10. Stockpiling of materials and types	_____	_____	_____
11. Unpaved storage area and parking lots	_____	_____	_____
12. Fuel storage tanks			
a. Aboveground	_____	_____	_____
b. Underground	_____	_____	_____
13. Waste treatment plants, lagoons, etc.	_____	_____	_____
14. Insulation materials suspected of asbestos	_____	_____	_____
15. Fill or vent pipes that indicate the existence of storage tanks underground	_____	_____	_____
16. Unusual or noxious odors (solvent, petroleum, other hydrocarbons)	_____	_____	_____
17. Lack of knowledge or record of previous operations on-site	_____	_____	_____
18. Other items or concerns of environmental contamination, past and present	_____	_____	_____
19. Regulatory compliance record of past and current owners and operators of the site	_____	_____	_____
20. Records of complaint by neighbors and other interested parties, including federal, state, and local authorities	_____	_____	_____
21. Safety and environmental training program for personnel	_____	_____	_____
22. Passive attitude of management toward environmental compliance	_____	_____	_____

TABLE 13.1 Checklist for Environmental Audit or Assessment (*Continued*)

	Yes	No	Additional comments/ remarks/location
Water Discharge			
1. Discharge of process water	_____	_____	_____
2. Discharge of cooling water—contact or noncontact	_____	_____	_____
3. Inoperative or poorly maintained sampling equipment	_____	_____	_____
4. Lack of flow measurement device for plant effluents	_____	_____	_____
5. Broken or nonexistent dikes, berms, or other secondary containment	_____	_____	_____
6. Storm drains in liquid bulk transfer area	_____	_____	_____
7. Improperly installed or total absence of oil-water separators	_____	_____	_____
8. Nonexistence of on-site laboratory or QA/QC facility	_____	_____	_____
9. Runoff from storage piles	_____	_____	_____
10. Improper or invalid NPDES permit and monitoring requirements	_____	_____	_____
11. Lack of or outdated SPCC plan	_____	_____	_____
12. Expiration of an NPDES or SPDES permit for discharge of wastewater	_____	_____	_____
13. Any violation of the terms and conditions of such permit	_____	_____	_____
14. Any sign of dredge or fill activity at site	_____	_____	_____
Air Emissions			
1. Nonregistered vents or exhaust stacks	_____	_____	_____
2. Lack of knowledge about emission points and releases from these points	_____	_____	_____
3. Lack of stack monitoring or compliance data	_____	_____	_____
4. Improper or broken bags or particulate collection devices	_____	_____	_____
5. Expired certificates or permits to operate	_____	_____	_____
6. Dusty indoor environment	_____	_____	_____
7. Lack of air pollution control equipment	_____	_____	_____
8. Noncompliance with terms and conditions of operating permit	_____	_____	_____
Solid or Hazardous Waste			
1. Poor or sloppy housekeeping	_____	_____	_____
2. Evidence of past spills	_____	_____	_____
3. Unlabeled or ruptured drums, containers, etc.	_____	_____	_____
4. Dated or missing MSDS	_____	_____	_____
5. Mixing of waste materials, including trash, without any viable separation process	_____	_____	_____
6. Lack of shipping records or transportation documents	_____	_____	_____
7. Incomplete analysis of waste generated, stored, or transported off-site for disposal	_____	_____	_____
8. Lack of knowledge of what constitutes hazardous and nonhazardous waste	_____	_____	_____
9. Leaky, dented, or corroded tanks or containers	_____	_____	_____
10. Types of chemicals used, stored, or disposed of	_____	_____	_____
11. An expired TSD facility or a solid waste management facility permit	_____	_____	_____
12. Any violation of such permit if it is current	_____	_____	_____

- Motivating employees to act in full compliance with the applicable laws and regulations
- Cooperating fully with government agencies or their contractors in all cleanup activities
- Committing top management to full environmental compliance

Limitation of real estate transactions liability

Exposure to liability arising from real estate transactions can be limited through careful allocation of liabilities in applicable contract agreements for purchase or sale of property. These considerations can involve

- Hold harmless, indemnification, or other warranty provisions in the purchase and sale or transfer agreements
- Setup of an appropriate escrow fund for cleanup and remediation requirements and performance bonds
- Complete disclosure of past violations and selling the property in an as-is condition
- Transfer of only the assets, but retaining or sharing liability and undertaking appropriate remedial actions
- Provision of an option to repurchase the property under certain specific conditions
- Obtaining adequate insurance coverage against environmental liability, particularly for sudden or accidental releases

Note here that not all these measures are foolproof and the parties may remain liable for some of or all the environmental shortcomings or mishaps. For example, using the as-is provision, the seller of a property may not get around the statutory liability that is imposed on the owner or operator of an inactive hazardous waste disposal facility. An owner or operator of a facility may be subject to an "establishment" or "transfer" or "disclosure" clause in certain jurisdictions unless past violations are fully disclosed and explicitly assumed by the new owner or operator. Nevertheless, by allocating specific liabilities, the seller or the buyer may be able to reduce potential liability significantly.

In some jurisdictions, private parties can contract away their liabilities. In such situations, the inclusion of the as-is clause in the purchase and sale agreement makes good sense for the seller of the property. The buyer, however, can bolster her or his innocent landowner defense by undertaking a comprehensive environmental assessment of the property prior to closing. Such an assessment may also allow the potential buyer to formulate a sound basis as to the extent of contamination and liability which may arise from such existing contamination, thus revealing what terms of the deal may make it appropriate or acceptable.

Recommended Reading

1. Susan M. Cooke, *The Law of Hazardous Waste: Management, Cleanup, Liability and Litigation,* vol. 3, Matthew Bender, New York, 1991.

2. Ridgway M. Hall and David R. Case, *All About Environmental Auditing,* 1st ed., Federal Publication, Inc., Washington, 1990.

3. Frank P. Grad, *Treatise on Environmental Law,* vol. 1A, Matthew Bender, New York, 1987.

4. C. Chadd and L. Bergeson, *Guide to Avoiding Liability for Waste Disposal,* 1986.

5. Joseph F. Dimento, *Environmental Law and American Business,* Plenum Press, New York, 1986.

Important Cases

1. *Midlantic National Bank v. New Jersey Department of Environmental Protection,* 106 S. Ct. 755 (1986).

2. *Ohio v. Kovacs,* 105 S. Ct. 705 (1985).

3. *United States v. Standard Oil Co.,* 384 U.S. 224 (1966).

4. *In re CMC Heartland Partners,* No. 91-3005 (7th Cir. July 2, 1992).

5. *U.S. v. Speach,* No. 90-50708 (9th Cir. June 29, 1992).

6. *United States v. Alcan Aluminum Corp.* No. 91-5481 (3d Cir. May 14, 1992).

7. *Jensen v. California Department of Health Services,* BAP No. EC 90-1655-AsMeo (9th Cir. Mar. 15, 1991).

8. *In re Chateaugay Corp.,* No. 90-5024 (2d Cir. Sept. 6, 1991).

9. *United States v. Fleet Factors Corp.,* 901 F. 2d 1550 (11th Cir. 1990), *cert. denied,* 58 U.S.L.W. 2713 (1991) ("Fleet Factors II") *aff'g.* 724 F. Supp. 955 (S.D. Ga. 1988) ("Fleet Factors I").

10. *Kayser-Roth Corp., Inc., v. United States,* 910 F. 2d 24 (1st Cir. 1990).

11. *Smith Land & Improvement Corp. v. The Celotex Corp.,* 851 F. 2d 86 (3d Cir. 1988), *cert. denied,* 57 U.S.L.W. 3471 (Jan. 17, 1989).

12. *United States v. Monsanto Co.,* 858 F. 2d 171 (4th Cir. 1988).

13. *Levin Metals Corp. v. Parr-Richmond Terminal Co.,* 799 F. 2d 1312 (9th Cir. 1986).

14. *United States v. Northeastern Pharmaceutical and Chemical Co., Inc.* (NEPACCO), 810 F. 2d 726 (8th Cir. 1986), *cert. denied,* ____ U.S.____, 108 S. Ct. 146 (1987).

15. *New York v. Shore Realty Corp.,* 759 F. 2d 1032 (2d Cir. 1985).

16. *Sanford Street Local Development Corp. v. Textron Inc.,* No. 1:90-CV-582 (W.D. Mich. Aug. 8, 1991).

17. *State of Idaho v. Banker Hill Co.*, 635 F. Supp. 665 (D. Idaho 1986).

18. *United States v. Maryland Bank & Trust Co.*, 632 F. Supp. 573 (D. Md. 1986).

19. *United States v. Mirabile*, 632 F. Supp. 573 (E.D. Pa. 1985).

20. *Aminoil, Inc., v. United States*, 599 F. Supp. 69 (C.D. Cal 1984).

21. *United States v. A&F Materials, Co., Inc.*, 578 F. Supp. 1249 (S.D. Ill. 1984).

22. *United States v. South Carolina Recycling and Disposal, Inc.*, 653 F. Supp. 984 (D.S.C. 1984).

23. *United States v. Vertac Chemical Corp.*, 489 F. Supp. 870 (E.D. Ark. 1980).

24. *Velsicol Chemical Corp. v. Rowe*, 543 S.W. 2d 337 (Tenn. 1976).

25. *Rylands v. Fletcher*, L.R.3H 330 (1868).

Toxic Torts—Litigation and Defense

14.1 Introduction

Public awareness of environmental concerns in the United States reached a frenzied level after the discovery of various hazardous substances and waste in Love Canal, Niagara Falls, New York, and dioxin in Times Beach, Missouri. This heightened level of concern has not ebbed as more stories are being revealed. New nuances and twists of past improper disposal practices involve toxic chemicals, hazardous substances, and hazardous wastes. This has given rise to a new concept of private action against the tortious conduct of the violators. This is indeed the genesis of toxic tort litigation in the United States.

The concept of tortious conduct, however, is an import from England where tort law was introduced through common or judge-made law. The premise of tort law is simple. If one suffers personal injury or property damage due to the fault of others, the injured person should be compensated by the perpetrators who caused the injury or damage.

Not every type of injury or damage can give rise to compensation. To be compensated, the wrongful or tortious conduct must be an act that is condemnable or otherwise unacceptable by the society. Typically, the act or conduct giving rise to the injury or damage is judged by the objective standard of reasonableness. Liability due to a tortious act or conduct will arise only when such an act or conduct is determined to be unreasonable under the prevailing societal standards. Thus, this standard may vary depending on the time and place. A tort may also arise when one fails to meet a "duty of care" established under the principles of societal duty.

This generalized concept of legal duty to avoid causing harm to others through an act of commission or omission, as the case may be, is separate and distinct from criminal law, where a criminal act is statutorily defined rather than established by the standards or norms eschewed by society. Thus, a cause of action, based on tort principles, commonly gives rise to only civil action against the culpable party or parties.

The concept of *toxic tort,* as the term implies, is a reflection of societal expectations. These expectations may involve the use, manufacture, storage, treatment, and disposal of toxic chemicals and chemical substances in this nation. In fact, such expectations throughout this nation imply increasing public concern over the steady proliferation of chemicals and chemical substances in every aspect of life.

As the term implies, toxic tort relates to a tortious conduct that is linked to toxic chemicals or substances. Toxic tort claims may emanate from varied causes and sources ranging from exposures to asbestos fibers and various chemical substances to toxic air pollutants, hazardous substances, and adulterated or misbranded drugs. The claims may involve a single plaintiff or multiple (sometimes hundreds) plaintiffs and a similar number of defendants. From a practical point of view, a cause of action for toxic tort is generally related to the release or discharge of toxic chemicals or substances, resulting in certain personal injury and/or property damage. It is therefore no surprise that toxic tort, like all other actions involving tortious conduct, is ingrained in such common law principles as negligence, trespass, breach of warranty, misrepresentation, strict liability, and the like. As a result, most of the toxic tort cases are very similar to the well-established product liability cases in terms of their makeup. Likewise, toxic tort claims may include compensatory damages for physical injury, property degradation, or serious emotional distress. They may also include punitive damages as a future deterrent.

Note that a number of federal environmental statutes, such as the Clean Water Act (CWA), the Clean Air Act (CAA), the Toxic Substances Control Act (TSCA), the Safe Drinking Water Act (SDWA), and the Occupational Safety and Health (OSH) Act have specific liability provisions which can be applied in a toxic tort case. In addition, a number of state laws have general as well as specific liability provisions for situations which can be labeled as toxic tort cases. Nevertheless, the typical springboard of liability in a toxic tort case is traceable to common law principles.

The introduction of toxic tort cases to the United States can be attributed to the early asbestos-related lawsuits. With the passage of time, toxic tort cases are becoming as important as such celebrated cases as agent orange, dioxin, and those cases based on release and eventual exposure to various dangerous chemicals and chemical substances.

Toxic torts have a rather short history in spite of their premise based on century-old legal principles. With growing public concerns and a greater public awareness of health and safety problems arising from the releases of toxic and dangerous chemicals and chemical substances into the environmental media (i.e., air, water, and land), it is now anybody's guess whether toxic torts will be a major aspect of environmental litigation. It is generally agreed that such litigation is now and will continue to be inherently complex because of the paucity of scientific and medical data regarding the safety and health hazards of various chemicals and chemical substances. In addition, the difficulty in establishing the causal relationship between the release or exposure of such chemicals and the health affliction or damage suffered by the individuals makes any quick settlement almost a dream.

14.2   Liabilities in Toxic Torts

As indicated earlier, all tort laws, including the very concept of toxic tort, are founded in common law principles of fault or civil wrong. Hence, any liability that stems from a toxic tort may be recognizable under certain statutory provisions dealing with fault or common law theories of tort, based upon some underlying principles of fault and resulting harm.

In this context, note that common law, unlike statutes, does not have any fixed or written text. Historically, common law has been developed from societal expectations and reasons based on notions of fair play, conscience, and justice. By nature, common law principles are flexible since they reflect the expectations, habits, mores, standards, principles, or policies of a given society. Therefore, any civil wrong that is not approved or accepted by a society may give rise to a tort liability in that society. Historically, common law liabilities may arise from a number of incidents of tortious conduct, e.g., trespass, nuisance, and negligence.

Theories of liability in toxic tort cases, however, may include not only federal and state statutes or the legal principles based on common law theories, but also the theory of strict liability. The latter applies when the society finds certain acts or omissions to act totally unacceptable irrespective of the existence or nonexistence of fault. In short, strict liability is an extreme type of liability where the society's sanctions are firm, rigid, and unequivocal. In this context, any lack of a firm basis for recovery under the theories of liability espoused by common law is not necessarily a death knell to claims based on toxic tort. Common law principles and the liabilities that arise as a result of their violations are generally flexible, and the courts may consider any new theory of liability based on acceptable legal principles.

These toxic tort liabilities under common law principles are discussed further below because of their importance in toxic tort cases. Various federal environmental statutes that may be successfully invoked in a toxic tort case were covered in earlier chapters.

Negligence

Under the original English law of torts, the theory of liability is based upon "actual intent" and "actual personal culpability." Through years of development of societal standards, along with moral overtones, the injured parties were allowed to bring a cause of action arising out of negligence by the culpable party. However, to recover, the injured party had to show that the act not only was negligent, but also was inflicted directly on the victim.

The legal acceptance of negligence as a formal cause of action, however, removed the additional requirement of direct injury. Under the modern theory of negligent tort, indirect injuries stemming from negligent acts are equally formidable.

Negligence, as the plain meaning of the word implies, can be associated with either the commission of an act or the omission of an act. The latter arises when there is a legal duty to act. Furthermore, to give rise to any liability, the negligent act or failure to act must also result in certain damages or consequences in the form of personal injury or property damage.

Under the current legal theory of negligence, several elements must be satisfied to constitute a cause of action or liability:

- A legal duty to use reasonable care that is in conformity with the standards of conduct by a reasonable person
- A failure to meet or conform to such standards, thereby resulting in a breach of legal duty
- A causal connection between the negligent conduct and the resulting injury which falls within the concept of "proximate cause"
- An injury or actual loss to another party, the so-called victim

This legal duty of care has a significant connotation since it also includes the concept of the duty to warn. Hence, a facility that manufactures, uses, treats, stores, or disposes of toxic chemicals or hazardous substances and wastes may be required not only to take all possible steps for their proper handling, but also to warn the public about their presence and associated dangers. This is important in understanding the legal duty of care that may require a proper notification or warning. In *Ashland Oil, Inc. v. Miller Oil Purchasing Co.,* the court found that the selling of hazardous waste oil as benign oil falls far short of what constitutes the standards of ordinary care.

Note that in those cases where it is difficult to establish the defendant's duty of care, a plaintiff generally resorts to claims of *negligence per se*, thus linking the cause of action to some form of negligence based on legal standards.

The broad concept of negligence and the liability derived from it are firmly embedded in the "reasonable" person or objective-standard rule. One can be found liable under this rule if the evidence establishes that there was lack of conformity to this rule at the time of injury or loss and that the proximate cause of such injury or loss was the negligent act. However, for any recovery, the plaintiff is also required to establish a resulting injury or damage as a result of the negligent act.

Section 291 of Restatement (Second) of Torts (1964) provides a legal understanding of an unreasonable conduct and a negligent act:

> Where an act is one which a reasonable man would recognize as involving a risk of harm to another, the risk is unreasonable and the act is negligent if the risk is of such magnitude as to outweigh what the law regards as the utility of the act or of the particular manner in which it is done.

Section 292 of Restatement (Second) of Torts (1964) lists various factors that should be considered in determining one's utility of conduct for the purpose of establishing negligence. These include

- The social value attached by the law to the interest that is to be advanced or protected by one's conduct

- The extent of the chance that such interest will be advanced or protected by the particular course of conduct

- The extent of the chance that such interest can be adequately advanced or protected by another and by a less dangerous course of conduct

Based on the above, there is a common law duty of care not only for the manufacturing and use of toxic chemicals and substances. This duty of care under the principles of common law is also applicable to their proper storage, treatment, transportation, and disposal. It is also not uncommon for the courts to find a breach of the duty of care, in connection with the federal or state environmental statutes and the regulations promulgated thereunder. In essence, if there is any negligent release or discharge of toxic chemicals and hazardous substances, a toxic tort based on the theory of negligence may be brought; and if there is any damage or injury, the courts may award appropriate recovery against the culpable parties. This can happen irrespective of the fact that such culpable parties are in full compliance with applicable state and federal environmental statutes and the regulations promulgated thereunder, including specific permit conditions. The irony of the flip side to this situation is that when one violates the statutory or regulatory provisions, such acts of noncompliance may be used as *prima facie* evidence of negligence and ensuing liability.

Trespass

Under common law principles, when one invades another's rights to land or chattel (i.e., a thing) pursuant to a negligent or an intentional act, trespass is committed and the other party may recover for the injury that results. In a case of trespass, physical invasion of the plaintiff's land or property, caused directly by the defendant or because of the defendant's involvement, is a necessary element.

From the viewpoint of releases of toxic chemicals and hazardous substances into the environment, almost all trespass actions will arise from trespass to land or trespass to person rather than to chattel. Also, the intention or motive of the wrongdoer is not an essential element to establish a cause of action. However, since this tort protects one's exclusive rights of possession of land, a physical invasion into such possessory right is a general requirement. Moreover, if a trespassory invasion is established, it is hardly a defense that such invasion is socially warranted or beneficial to the plaintiff. Unlike the cause of action for negligence, one is liable even if one trespasses in good faith and with reasonable care.

However, a trespass will not result when one enters someone else's land with permission, invitation, or a license to enter. Under certain circumstances, an entry will be considered privileged, e.g., the entry of police or fire marshal or any other law enforcement personnel. However, such a privileged entry is generally limited not only by purpose but also by time and space. Hence, any stay on somebody's land beyond what is necessary in light of factual circumstances may constitute a trespass, even for enforcement or regulatory personnel.

Typically, in environmental cases involving a release or discharge of toxic chemicals or substances, causes of action based on both trespass and negligence are pleaded in spite of the fact that they are legally distinguishable, even under common law principles. Since the elements involving these two different torts are often found to coexist in most of the environmental releases of toxic chemicals or hazardous substances, they are found to coexist in almost all pleadings. For plaintiffs, since the courts have gotten used to such forms of pleadings, it may be wise to bring a cause of action based on both negligence and trespass. For defendants, however, it is important to refute such pleadings by showing that all the elements necessary for each of these tort actions are not present, hence one or both causes of action should be ignored by the court.

Nuisance

Nuisance is perhaps the most common but least understood cause of action under the common law principles of tort. Part of the reason for this dubious distinction is that the word *nuisance* is shrouded in legal confusion. While it is customarily associated with a specific type of conduct, legally nuisance is associated with a kind of interest that is invaded, damaged, or harmed. In strict legal parlance, nuisance relates to a particular type of liability rather than to a particular tort.

Under common law principles, one may use one's property, real or personal, in any manner that one sees fit as long as such use of property is made in a reasonable manner. A nuisance is caused when one's use of one's property becomes unreasonable, improper, or unlawful, thus resulting in some injury, annoyance, or inconvenience to the lawful rights of another. To recover in an action of nuisance, however, one's injury, annoyance, or inconvenience has to be material (i.e., significant).

There are two types of nuisance: public nuisance and private nuisance. In a public nuisance, an individual has to suffer a special damage that is not experienced by others in the same community. Section 821B of the Restatement (Second) of Torts (1977) defines a public nuisance as an "unreasonable interference with a right common to the general public." Such public rights may include safety, health, and property. To sustain a public nuisance action based on environmental pollution, the plaintiff has to establish some special or peculiar injury that is not suffered by the public at large within the community.

Private nuisance, however, results in a tort action when there is an unreasonable and substantial interference with the use or enjoyment of the property interest of another. The underlying difference between public nuisance and private nuisance is that the former type relates to public rights whereas the latter deals with personal rights in connection with use or enjoyment of land. Unlike trespass, for a nuisance tort action by the owner or possessor of land, a physical entry onto that land by the culpable party is not required.

In this context, note that in the United States public nuisance is generally statutory. Tort actions involving public nuisance are specifically addressed in federal and state statutes. Generally, such statutory provisions are largely included in the form of specific criminal provisions.

The courts may impose temporary or permanent injunctions against nuisance on equitable grounds. Section 371 of Restatement (Second) of Torts (1964) provides the general rule of liability whereby a possessor of land is liable for "physical harm to others outside of the land" because of an activity that involves an "unreasonable risk of physical harm" to such individuals. Hence, a cause of action based on nuisance is generally most effective when the defendant can be shown as the owner or operator of a hazardous waste disposal facility.

The often quoted phrase *coming to a nuisance* deserves special understanding. This is actually a defense ploy to an action of nuisance. It suggests that the complainant moved into an area where the alleged nuisance that is being complained of preexisted and hence the complainant's move to that area was made with full knowledge of such nuisance. The underlying premise of this defense is that since the complainant knew of this nuisance prior to the move, the complainant already accepted it and accounted for its existence, say, by appropriate adjustments in rent, lease, or purchase price of the complainant's real property. This defense is not generally accepted by the majority of the courts which adhere to the rule that any continued nuisance against one's use and enjoyment of property is not to be tolerated. However, since the courts will "balance the equities" in a tort action of nuisance, installation of appropriate control equipment in mitigating a nuisance action can be a major factor in courts' decisions. In short, "reasonableness" is the central piece of controversy in such an action.

In toxic tort cases, private nuisance actions are as common as public nuisance actions. In many tort actions involving private nuisance actions arising due to the presence of toxic chemicals, hazardous substance, or waste, presently the courts are carefully weighing the usefulness and applicability of the requirement of a special damage or injury that is not suffered by others to sustain such actions. This special consideration is indeed a difficult and questionable matter because of the very long latency period or manifestation of a disease, say, cancer, due to an exposure to toxic chemicals, hazardous substances, or hazardous waste.

Strict liability

In an action involving strict liability, there is no requirement for an allegation by a plaintiff that the defendant's act was unreasonable. As the term implies, liability in such a case becomes strict or literally "certain." As a result, it has been characterized as *liability without fault,* or something akin to *absolute liability*. According to the well-known tort expert and legal scholar

Dean Prosser, to establish strict liability, there is no need to show the following two important elements of tort liability:

- An intent by the defendant to interfere with a legally protected interest of the plaintiff without any legal justification
- A breach of a duty by the defendant to exercise reasonable care

Section 519 of the Restatement (Second) of Torts (1976) provides that one carrying on an "abnormally dangerous activity" is liable to others for their personal injury or property damage resulting from such activity. This liability will attach irrespective of the fact that the defendant has exercised utmost care to prevent such harm. In general, the courts use the strict liability standards when ultrahazardous activities are involved. In *Ashland Oil, Inc., v. Miller Oil Purchasing Co.,* the court found the chemical waste disposal company strictly liable because it was engaged in an ultrahazardous or abnormally dangerous activity.

Section 520 of the Restatement (Second) of Torts (1976) states that the following factors should be considered in order to determine an abnormally dangerous activity:

- Existence of a high degree of risk of some harm to a person, land, or chattel belonging to others
- Likelihood that the resulting harm will be great
- Inability to eliminate the risk of harm by the exercise of reasonable care
- Extent to which the activity is not a matter of common use
- Inappropriateness of the activity to the place where it is carried on
- Extent to which the usefulness or value of the activity to the community is outweighed by its dangerous attributes

In the 1990 case *Indiana Harbor Belt Railroad Co. v. American Cyanamid Co.,* the Seventh Circuit Court of Appeals held that the shipment of hazardous chemicals by rail through an urban area is not an abnormally dangerous activity under Illinois law. As a result, the court found that the defendant, a manufacturer of acrylonitrile, was not strictly liable. However, the court reinstated the plaintiff's claim for negligence in this toxic tort case.

From the plaintiff's point of view, the case is always stronger if the facts can substantiate claims for other types of torts besides a claim for strict liability.

In toxic tort cases, the allegation is generally based on the theory that manufacturing, use, storage, treatment, transportation, or disposal of toxic chemicals, hazardous substances, or wastes, as the case may be, is an abnormally dangerous activity. Hence, as the theory goes, if any harm is caused to another by such an activity, the perpetrator must be strictly liable. However, whether such an activity is abnormally dangerous is a question of law, and it is solely for the court to decide. In this context, note that strict liability is a legal doctrine that has been embraced by the courts as a result of social engineering rather than statutory provisions.

Additionally, the courts are not bound to reach a decision on the basis of the factors listed by the Restatement and Restatement (Second) of Torts since the factors have no statutory basis. However, the factors have great influence on the courts' decisions. The courts, both state and federal, have generally used them in their decision-making process. Hence, the factors espoused by the Restatement of Torts, first or second version, should not be discounted lightly.

Alternate liability

The first celebrated case involving alternate liability theory is *Summers v. Tice,* where two hunters simultaneously fired their guns, but only one hit the plaintiff. There was no proof as to which of the two hit the plaintiff. The court found both defendants liable in the absence of absolute proof against either defendant.

The premise of alternate liability is the principle that wrongdoers should not go unpunished because the plaintiff cannot prove which one of several possible defendants actually caused the harm.

Section 433B of the Restatement (Second) of Torts (1964) provides the theory of alternate liability while espousing the burden of proof in a tortious conduct. In short, alternate liability may be found

- Where the combined tortious conduct of two or more defendants brings harm to a plaintiff and one or more of such defendants seeks to limit liability on the grounds of divisibility of harm that allows apportionment among such defendants

- Where the conduct of two or more defendants is tortious, the harm to the plaintiff is proved to have been caused by only one of such defendants, and there is uncertainty as to which defendant actually caused the harm

The courts have been very careful and selective in using this theory of tortious conduct. So far, this has been applied in a handful of cases involving identical product manufacturers where injury to the plaintiff could not be directly or indirectly linked to any one manufacturer.

Enterprise liability

The theory behind enterprise liability is of recent origin. It has been applied mostly in cases involving product liability, particularly for drugs, medical devices, and various consumer products. Its genesis can be traced to a case in California, *Syndel v. Abbott Laboratories.* In this product liability case, the suit was brought against eleven major drug manufacturers of diethylstilbestrol (DES). The plaintiff argued that she suffered because her mother took the drug DES while she was pregnant. But the plaintiff could not establish the actual manufacturer of the drug DES that her mother took during the pregnancy. It turned out that there were more than 200 DES manufacturers at that time. However, the plaintiff established that the eleven defendants in the case were responsible for manufacturing 90 percent or more of the DES that was available during the time of her mother's pregnancy.

The *Syndel* Court held that under the rule of causation, once the plaintiff meets the burden of proof showing that a substantial share of the market was held by certain defendants, such defendants in turn have to prove that they are not responsible for the plaintiff's injury. Thus the theory of market share or enterprise liability became a new concept in tortious conduct.

Through the years, the courts have recognized enterprise liability as a specific form of tort. During its development as a separate tort, the courts have laid down certain specific elements which had to be met to establish enterprise liability. At this time, the courts are generally prone to consider the following factors to determine enterprise liability:

- Injury or illness was due to a fungible product manufactured or otherwise made available by all the defendants cited.
- Proof as to the specific manufacturer of the product that actually caused the plaintiff's injury or illness is lacking.
- All the defendants joined in the action have produced and sold the product in question at the same time.
- Injury or illness complained of has a causal relationship to the product manufactured by the defendants.
- All the named defendants manufactured and sold the same product in a manner that is unreasonably dangerous.
- The defendants joined in the lawsuit actually represent a substantial share of the market for the fungible product that is the cause of the plaintiff's injury or illness.

Under the enterprise liability theory, each of the joined defendants is liable for the plaintiff's injury or damage to the extent of the particular defendant's market share of the product in question. Using this theory, one may recover if one has been exposed to a particular toxic or hazardous substance or waste by-product that is generated by several manufacturers. To recover, the plaintiff has to establish that there is a causal relationship between the toxic or hazardous substance and the injury complained of and that the defendants named in the lawsuit represent a substantial share of the market for that substance.

14.3   Scientific and Medical Evidence

As indicated earlier, to recover from a tortious conduct, mere allegation is not enough. One has to establish the existence of the required elements for sustaining the cause of action and the relief sought.

In cases involving toxic torts, the proof of causation is extremely complex and difficult in view of the long latency periods for the manifestation of injury, e.g., cancer. This is compounded by the fact that in certain situations the underlying statutes of limitations are relatively short.

Under the traditional tort law, the plaintiff has to prove by preponderance of evidence that there is a causal connection between the act or omission of the defendant and the plaintiff's injury. This causal connection includes both the *cause in fact* and the *proximate cause* of the injury that is being complained of. Both elements must be satisfied before the question of damage can be resolved. In this context, *cause in fact* means an act or an omission to act that is both a sufficient and a necessary condition for the injury or damage suffered by the plaintiff. Proximate cause, also known as the *legal cause,* a term generally used by the courts, is actually a device for limiting the scope of defendant's liability based on those consequences having some reasonable relationship to the risks created by the defendant.

Traditionally, the courts have relied on two tests—the "but-for" test and the "substantial factor" test—to determine whether there is a cause in fact. In the but-for test, the determination is whether the harm would not have occurred but for the act or conduct of the defendant. In the substantial-factor test, the inquiry is whether the defendant's act or conduct was a substantial factor in bringing about the harm suffered by the plaintiff.

The proof of causation under these two tests is not always simple or straightforward, particularly in toxic tort cases. While certain toxic chemicals or substances have "signature" injuries which closely relate the injury to an exposure to a particular chemical, most are not so easy to establish. Typical signature injuries are generally associated with exposures to asbestos and heavy metals. However, if the defendant has other health complications, e.g., chronic illness, even these signature injuries become masked or significantly more complex to identify and to establish. For cases involving exposure to toxic chemicals or hazardous substances, there are too many factors, internal and external, which can obscure any specific claim.

Once the cause in fact is established, the plaintiff has to establish the proximate or legal cause of the injury or damage. At this stage, the question centers on whether the defendant's act or conduct was a sufficiently significant cause for the plaintiff's injury or damage so as to find the defendant legally responsible for such harm.

In complex toxic tort cases, the courts are relying more and more on medical and scientific evidence for proof of causation. Rule 702 of the Federal Rules of Evidence allows courts to determine whether expert testimony bearing scientific, technical, or other specialized knowledge should be admitted. Since most of the jurors as well as the judges are not trained in complex scientific or medical theories and procedures, the courts have been increasingly relying on qualified experts to help them understand how a potential harm can be inflicted as a result of a particular exposure to a toxic chemical or hazardous substance.

Typically, such qualified experts rely on scientific methods and data, including bioassays, epidemiological studies, and the sophisticated risk assessment tools, to arrive at an expert opinion. As a result, these qualified experts must have very specialized training and qualifications. They can be toxicologists, epidemiologists, health physicists, or physicians who can relate the chemical, physical, and biological nature of a toxic chemical or hazardous substance to an alleged exposure and the resulting harmful condition. In complex toxic tort cases, the courts are increasingly allowing the jury to decide whether an expert testimony is credible enough to decide the causal connection.

In this context, note that pursuant to a mandatory requirement under CERCLA, a new federal agency, the Agency for Toxic Substances and Disease Registry (ATSDR), has been created by the federal government. ATSDR has been given certain specific responsibilities as set forth in section 104(i) of CERCLA:

- To establish and maintain, in cooperation with the states, a national registry of diseases or illnesses as well as persons exposed to toxic substances

- To establish and maintain an inventory of appropriate health or medical literature, research, and studies on health effects of toxic substances

- To maintain a complete listing of areas closed to the public or otherwise restricted in use because of contamination by toxic substances

- To provide medical care and to conduct tests for exposed individuals in public health emergencies

- To carry out periodic surveys and screening programs to determine the relationship between exposure to toxic substances and the diseases or illnesses of exposed persons

Under SARA, the role of ATSDR has been largely expanded to include new activities. As a result, the Environmental Protection Agency has been seeking the assistance of ATSDR to perform a comprehensive "health assessment" due to the existence of an inactive hazardous waste disposal site. This is being done prior to the inclusion of a site on the National Priorities List (NPL).

Some other important activities by ATSDR are becoming increasingly important to toxic tort litigants:

- Formulation and revision of a priority list of at least 100 hazardous substances most commonly found at the NPL sites that pose the most significant potential threat to human health due to their known or suspected toxicity

- Preparation of toxicological profiles of such listed substances, including analyses of current data which may link exposure to these substances to adverse health effects

- Provision of health consultants to federal, state, and local governments upon request

- Initiation of research to develop new information on health effects for hazardous substances if current data are insufficient

- Detection, assessment, and evaluation of the effects of hazardous substances on human health through basic research involving advanced techniques

As a result, toxic tort litigants are relying more and more on ATSDR reports for health risk assessment in order to establish a causal connection between certain illnesses and exposure to specific toxic chemicals and hazardous substances. However, in toxic tort cases, the courts generally let the jury decide whether to hold expert testimonies as credible. The courts may preclude offers of scientific proof on complex matters that are either too controversial within the established scientific and/or medical community or at the nascent stage of a new scientific theory. Nevertheless, toxic tort litigants are generally allowed to introduce expert testimony although it is not based on "particularistic evidence."

14.4 Defenses

Depending on the causes of action, several defenses are possible in toxic tort cases. Also, since the courts generally balance the equities in tort or civil cases, the courts carefully consider the relative interests of the parties who may suffer in absence of an equitable remedy.

Nevertheless, the same defenses do not apply to all toxic tort cases. For example, in the case of strict liability which makes one culpable even though there is no fault, the defense is extremely limited.

In cases involving negligence, the traditional defenses are assumption of risk and contributory negligence by the plaintiff, both of which focus on the act or conduct of the plaintiff for the loss or injury suffered. In the first situation, the legal doctrine is based on the principle that if the plaintiff knew of the danger and still assumed the risk, she or he should not be permitted to benefit or recover from the loss or harm experienced. In this case, the plaintiff's act is considered to be an intervening, superseding cause that makes the defendant's negligence no longer a proximate cause for the injury or damage. There is an exception to this rule only when there is a greater intervening cause, such as the saving of another human life.

The defense based on contributory negligence is based on the concept that one should not benefit from one's wrong and that for recovery, one must have "clean hands." However, this defense also has certain limitations, e.g., the "avoidable-consequence" rule or the "last-clear-chance" rule. Under the avoidable-consequence rule which pertains to the issue of damages arising out of a negligent act, one cannot recover for injuries or losses if one could have safely avoided them by acting as a reasonable person. The last-clear-chance rule comes into play when both the defendant and the plaintiff are guilty of creating a perilous condition but the plaintiff could not and the defendant could avoid the consequences by resorting to other available means. In this case, the negligent defendant is liable, unlike the situation involving an avoidable consequence.

For trespass actions, the most potent defense is based on a landowner's permission, invitation, or an appropriate license to enter the land. In a nuisance action, however, some limited defenses are available including the controversial claim of "coming to a nuisance" in certain jurisdictions. Also, as indicated earlier, good faith efforts by a defendant in averting an injury or damage are routinely taken into consideration by the courts.

As indicated earlier, in a strict liability case, there is no defense based on lack of fault. Also, strict liability for certain activities is imposed by the statutes themselves. The statutory provisions, however, generally do not abrogate or do away with the defense of contributory negligence unless the statutes have explicitly abolished such a defense.

In this context, note that certain statutes have adopted comparative-negligence doctrines that apportion the loss between the plaintiff and the defendant or the defendants, as the case may be. Such apportionment is based on their share of negligence that resulted in the final injury or damage.

As in all other types of litigation, the statutes of limitations are also an appropriate defense in the toxic tort cases if and when they apply. In most environmental cases involving causes of action based on toxic torts, the statute of limitations does not start to run until the plaintiff actually has or should have had knowledge of the injury or loss suffered.

For government agencies, there is an added defense based on the principle of sovereign immunity. Historically, all governments are immune from toxic tort liabilities. This is based on the common law principle that the king or the queen can do no wrong. However, through the years, Congress has waived certain immunity from liability against the tortious conduct carried out by the federal government as an employer. One such congressional action is the ultimate passage of the Federal Employers' Liability Act (FELA). Similarly, state legislators have significantly reduced the levels and types of immunity that can protect state governments from tort liability.

In short, the available defenses in toxic tort cases are quite limited. Hence, a prudent defensive strategy should include other options, e.g., pollution liability coverage through insurance, a fast settlement, and, in the case of a litigation, impressing upon the jury and the judge the unreasonableness and impossibility of performance to meet applicable environmental standards. The latter defense may also include other elements, e.g., third party liability, the need for allocation of liability, and the lack of public policy as a guide. These defenses, however, are of limited merit since most toxic torts have an overtone of strict liability. As a result, other options that can indemnify or protect one from the severity of toxic tort liability should be looked into very carefully.

14.5 Insurance Coverage

At present, there is no single or uniform protection against environmental liabilities including the occurrences or episodes that fall in the category of toxic torts. Historically, the insurance industry offered a standard *comprehensive general liability* (CGL) coverage for protection against bodily harm and property damage.

In recent years, the insurance companies have been disputing the availability of environmental liability coverage through insurance policies that provide CGL. Their arguments are basically twofold:

- Government actions for cleanup and remediation costs, e.g., under CERCLA, do not constitute claims for "damages."
- These claims are essentially for equitable relief which the CGL policy does not cover.

Most of the current insurance policies also include a general exclusion clause from environmental liabilities covered. Typically, the insurance coverage is triggered when certain events or occurrences happen during the policy period of the coverage. However, for claims involving environmental and toxic torts, the answer is not simple. The underlying reason is that there have been a significant number of toxic tort cases since the 1970s and CERCLA-related cases since the 1980s. With increasing public awareness and government crackdowns, insurance carriers are finding out that suddenly they have become the ultimate victim of growing environmental pollution.

This awareness actually started in the 1960s when the environmental statutes at the federal level started to get tough on environmental violators. Of course, this awareness or low-level concern of the insurance carriers turned into a nightmare for the insurance industry as a whole in the late 1970s and early 1980s when several major federal environmental statutes were passed with clear indication of cleanup and remediation liability for the violator. The insurance carriers reacted defensively to these developments of the liability provisions of environmental laws as their costs of indemnification started rapidly mounting. Getting caught after the passage of the environmental laws and regulations, the insurance industry basically reacted after the laws were enacted. As a result, the position of the insurance industry on liability coverage for environmental pollution shifted gradually in several distinct stages:

1. Up until the late 1960s, the insurance carriers covered all cases of bodily injury and property damage under the CGL coverage.

2. Starting in the late 1960s, the insurance carriers focused their attention on the depth and breadth of pollution exposure and incorporated a pollution exclusion clause in the CGL. This essentially excluded coverage for pollution from the CGL except in the case of a "sudden and accidental" discharge, dispersal, release, or escape of a pollutant.

3. During the early 1980s, the insurance industry adopted yet another new position by denying coverage for any environmental pollution under its CGL policy. Instead, certain insurance carriers started offering a distinct and separate pollution coverage at an extremely high premium.

However, the courts have not fully endorsed the limitations for pollution exclusion as are put into effect by the insurance industry. As a result, this has become one of the most hotly debated issues. Unfortunately, the courts have not provided much direction since they are almost evenly spread in their interpretation of the insurance coverage under the CGL policy. As a result, this has become a real quagmire for both insurers and insureds.

One common area of most insurance litigation involves the exception provision to the exclusion clause in standard CGL. This is a question of indemnity and a duty to defend the insured for an injury or property damage that is caused suddenly by an accident. Under this provision, any environmental liability that arises due to a continuous release and knowing or intentional discharge of toxic or hazardous substances is not covered under the CGL policy since the cause of the injury or damage is intentional and has been or should have been known to the property owner.

Upon consolidation of three lower court cases in Michigan, the Michigan Supreme Court ruled that the insurance companies do not have to pay for pollution due to the standard practices of the insured company or if such pollution extends over a period as short as two weeks. Although this is the decision of a state court, not a federal court, this decision may be particularly influential in the outcome of similar cases in the future because of its ruling by a state's highest court. Recent state court decisions in various jurisdictions suggest no uniform approach by the state courts to the extent or scope of coverage under CGL. A significant number of federal courts, while applying state laws, have also produced diametrically opposite decisions; some have upheld the temporal meaning of the "sudden and accidental" clause, while others have relied on the ordinary meaning of *damages*. This may make forum shopping an important consideration for a desirable outcome.

This question of insurance coverage may be an important consideration in toxic tort cases. Generally, most of the toxic torts involve a sudden and accidental release of toxic chemicals or hazardous substances. If such releases are continuous, it is highly likely that the enforcement officials will take appropriate statutory actions including possible shutdown of a facility.

Nevertheless, there is no guarantee that one may be protected by a liability coverage under the CGL policy in a toxic tort situation. The whole issue of environmental liability and the applicability of insurance coverage, particularly under the CGL coverage, is hotly debated all over the country.

Additional discussion of insurance policies and their coverage is included in Chap. 16.

Recommended Readings

1. Michael Dore, *Law of Toxic Torts,* vols. 1 and 2, Clark Boardman Company, New York, 1991.

2. "Toxic Torts, The 1990 Symposium," Institute for International Research, New York, Nov. 8–9, 1990.

3. W. Page Keeton, Dan B. Dobbs, Robert E. Keeton, and David G. Owen, *Prosser and Keeton on The Law of Torts*, 5th ed., West Publishing Co., St. Paul, Minnesota, 1984.

Important Cases

1. *Christopherson v. Allied-Signal Corp.,* 902 F. 2d 362 (5th Cir. 1990).

2. *Indiana Harbor Belt Railroad Co. v. American Cyanamid Co.,* No. 89-3703 (7th Cir. Oct. 18, 1990).

3. *Brock v. Merrell-Dow Pharmaceuticals Inc.,* 874 F. 2d 307 (5th Cir. 1989) *cert. denied,* 110 S. Ct. 1511 (1990).

4. *In re Agent Orange Product Liability Litigation,* 611 F. Supp. 1223 (S.D.N.Y. 1985), *aff'd.* 818 F. 2d 187 (2d Cir. 1987).

5. *Allen v. United States,* 588 F. Supp. 247 (1984).

6. *Ferebee v. Chevron Chemical Co.,* 736 F. 2d 1529 (D.C. Cir. 1984).

7. *Ashland Oil, Inc. v. Miller Oil Purchasing Co.,* 678 F. 2d 1293 (5th Cir. 1982).

8. *Syndel v. Abbott Laboratories,* 26 Cal. 3d 588, 607 P. 2d 924, *cert. denied,* 49 U.S. 912 (1980).

9. *Summers v. Tice,* 33 Cal. 2d 80, 199 P. 2d 1 (S. Ct. Ca. 1948).

10. *Rylands v. Fletcher,* L.R.3 H.L. 330 (1868).

Negotiation and Settlement Strategies

15.1 Introduction

The proliferation of new comprehensive federal statutes and the incorporation of new provisions in existing statutes by major amendments have made the federal environmental laws something to reckon with by all concerned. Moreover, the Freedom of Information Act (FOIA) allows not only lawyers but everyone including the journalists and the public at large to have access to the information collected and compiled by regulatory agencies on environmental matters. The only restriction to such a sunshine provision applies to the confidential matters that are statutorily protected. This protection is accorded due to the concern for confidentiality and the trade secrecy that may be involved. However, pursuant to 40 CFR part 2, subpart B, any business confidentiality claim is subject to a final confidentiality determination by EPA's legal office. EPA's scrutiny will be based on several substantive criteria, the burden of establishing them rests on the claimant.

This makes the industrial operations subject to almost constant scrutiny and open surveillance. In Chap. 13, the scope of environmental liability is discussed at length to show how the traditional concepts of fault, causation, and damage have been radically expanded in recent years. In addition, various corporations and business enterprises that typically own or operate industrial or manufacturing facilities have generally been saddled with the "deep-pocket" ailment based on the presumption of their ability to pay. Under the unwritten deep-pocket rule, both regulatory agencies and the public at large generally target those who reportedly have deeper pockets among various potentially responsible parties (PRPs). In short, the deep-pocket rule is routinely connected to liability, irrespective of actual or potential contribution to a fault since it offers financial promises or ability to pay. Under this concept, not all PRPs who are equally guilty may end up paying equally for their culpable actions.

Unfortunately, the environmental statutes or the regulations promulgated thereunder have not attempted to rectify this inequitable share of guilt or liability. In fact, the enforcement policies of the regulatory agencies are largely guided by a strong overhanded approach against owners or operators of facilities with deep pockets. While such actions make sensational news, the entire liability scheme suffers from a sense of injustice.

From the perspective of damage control and the need to undertake a prompt cleanup due to a significant or serious environmental contamination, time is of utmost importance. Proper protection of public health and the environment is indeed a socially imperative and desirable goal. The ultimate objective and the general design of all environmental laws, including common laws, are to prevent wrongs and to undertake appropriate remedial measures. In short, our present sense of environmental justice has a built-in concept of environmental control, cleanup, and remediation.

The new wave of federal environmental statutes, as discussed in Chap. 1, came as a result of public awareness and concern in the late 1960s and early 1970s. Since that time, it has been the general goal of these statutes to protect both public health and the environment by including provisions for tough sanctions. The regulations promulgated by the agencies under these statutory authorities have established not only the standards of conduct of owners and operators of facilities through various federal permitting schemes, but also an expanded scheme of liability stretching from routine civil and criminal penalties to the ultimate cleanup and remedial obligations. This has resulted in an unwritten merger of both statutory and regulatory schemes with the omnipotent deep-pocket rule.

The reality of this apparent tactic should not be overlooked by industrial or business communities. If there is a need for business strategy, appropriate environmental responses as well as cleanup requirements and their resulting costs should be the focus of corporate planning and concerns. For its ultimate survival and success, each business or corporate entity must have a well-planned strategy for environmental management and crisis control. This proactive or visionary approach requires both a broad corporate policy and a comprehensive strategic plan to avoid, mitigate, and respond to any conceivable environmental problem or shortcoming. Therefore, it is critical not to limit corporate planning to minimum steps for compliance with environmental statutes and regulations. Instead, an ongoing business entity may benefit from an incorporation of the processes that involve effective negotiation and settlement into the business plans. This is extremely important for nonroutine environmental damages that may result from previously unforeseen, unintended, and accidental mishaps.

15.2 Environmental Cleanup and Government Response

CERCLA authorizes cleanup and remediation actions by the government as well as private parties. The basic thrust under CERCLA is not only to allow the federal or state governments to undertake appropriate cleanups or remediation, but also to let the PRPs themselves undertake such actions following a release or threat of a release of any hazardous substance. This acts in conjunction with the strict or no-fault liability concept attributed to CERCLA. Under this concept, a third party may undertake and claim appropriate cleanup or remediation costs for environmental release of a hazardous substance as long as such claims do not preclude a claim for personal injury or property damage. Such claims may also be asserted against government funds, e.g., the Superfund, set up for this purpose.

Section 104 of CERCLA authorizes the federal government to take any appropriate action in response to actual release or a potential threat of a significant release into any one of the three environmental media (i.e., air, water, and land). Such release may be of

- Any hazardous substance as defined under CERCLA

- Any pollutant or contaminant that may present an imminent and substantial danger to either public health or the environment

CERCLA also specifies certain procedural requirements for both government and private cleanup and remedial action.

The Superfund Amendments and Reauthorization Act modified some of the government activities that may be undertaken pursuant to the response action authorized under section 104(a) of CERCLA. As a result, the federal government may act only after determining that necessary cleanup will not be undertaken by the owner or operator of the facility or by any of the PRPs.

Under the no-fault or strict liability concept of CERCLA, PRPs may be liable to the government and/or private parties for a number of environmental damages and associated cleanup costs. These costs may include

- The response costs incurred by the federal or state government

- The response costs incurred by private parties, excluding personal injury or property damage

- The costs resulting from damages to natural resources that are the property of federal or state governments

The authorized activities of the federal government under CERCLA include two major types of response actions: removal actions and remedial actions.

As discussed in detail in Chap. 6, the removal actions are generally temporary or short-term in nature. They include, but are not necessarily limited to, cleanup or removal of hazardous substances released onto land and water and monitoring of both the environmental contamination and the effectiveness of the cleanup. They may also prescribe restricted use of or access to the contaminated area.

Under a serious episode of environmental release of a hazardous substance, the removal action may include temporary evacuation of the people from affected areas. But removal actions are generally not continued beyond 1 year from the date of initial response to an actual or threatened release of a hazardous substance or a cost of $2 million. These limits, imposed by section

104(c)(1) of CERCLA, may be exceeded if continuance of removal actions is "appropriate and consistent" with the remedial actions to be undertaken later.

Remedial actions, however, are considered more permanent or long-term response actions. They can include

- Excavations or dredging of contaminated soil
- Pumping out of contaminated groundwater
- On-site treatment of contaminated soil and/or water
- Repair and/or removal of leaky underground and aboveground storage tanks and other containers
- Adoption of an appropriate spill prevention control and countermeasure (SPCC) plan to safeguard against future discharges

The federal government is authorized to undertake remedial actions by using remedial funds available under CERCLA's funding provisions. However, it must satisfy several procedural requirements including a remedial investigation and feasibility study (RI/FS) before such undertaking. Moreover, CERCLA authorizes the use of federal remedial funds only at those inactive hazardous waste disposal sites that have been listed on the National Priorities List. In addition, section 104(c)(3) of CERCLA, included by way of SARA, requires the federal government to enter into an agreement with the appropriate state government in order to incorporate both financial commitment and active participation of the state in the contemplated remedial action.

Any action undertaken by the federal government or that of a third party interested in recovering remedial costs from the PRPs must be consistent with the National Contingency Plan (NCP). As discussed in Chap. 6, the premise of the NCP is to ensure that the response or remedial options are fully evaluated under established procedures so that they become cost-effective.

SARA also focuses on the most controversial issue of remediation, i.e., how clean is clean. It is, therefore, appropriate to use a risk-based approach for developing cleanup criteria under CERCLA. The general aim of SARA is to undertake remedial actions based on appropriate treatment, including permanent destruction of the hazards associated with hazardous substances. Off-site transportation and disposal, and land disposal in particular, while not banned outright, are considerably restricted and least favored.

15.3 Cost Recovery Actions

Federal and state governments as well as private third parties can recover both removal and remedial costs that meet the NCP criteria. As discussed in Chap. 13, there is broad government discretion in deciding which PRPs should be brought to action. Under the principle of joint and several liability, cost recovery actions by the governments may be influenced by the deep-pocket rule rather than by the degree of culpability.

In general, there is a large group of expenditures or recoverable costs by the government under CERCLA by way of SARA. They may include

- Costs related to health assessment
- Costs of preparing nonbinding allocations of responsibilities (NBARs)
- Costs of removal actions
- Costs of indemnification of parties or contractors engaged in response actions
- Interests that may be accrued on recoverable costs
- Legal costs including court costs and attorneys' fees

Section 107 of CERCLA authorizes recovery of response costs incurred by private parties. They may include a wide range of response actions by private parties. However, to be recoverable, as indicated earlier, these response costs must be consistent with the NCP. The regulatory requirements for a private party response costs which are consistent with the NCP are set forth in 40 CFR part 300, subpart H. In this context, it should be emphasized that any private party cleanup must satisfy certain preconditions and qualifications in order for the cost to be recoverable. One important aspect in any private party undertaking is the requirement for providing an opportunity for public comments regarding the response action selected.

Chapter 16 deals with various elements and issues of cost recovery actions.

15.4 Provisions for Negotiations and Settlements

The original CERCLA did not expressly provide for joint and several liability. Nor did it contain any specific framework for contribution. The courts have generally interpreted that section 107 of CERCLA appears to establish strict liability for costs due to response actions and for damages to environmental resources. In *New York v. Shore Realty Corp.,* the court actually went further and held that a listing on the NPL is not a requirement for a liability action under section 107 of CERCLA.

The requirement of a site's listing on the NPL for CERCLA's section 107 action is obviously an area of intense debate since there is no statutorily established guiding rule. Even when the courts attempt to apply the rule of the Restatement (Second) of Torts by placing the burden of proof of apportionment on each PRP instead of a blanket adoption of joint and several liability using the legislative history of CERCLA as a guide, the courts find it extremely difficult to grant partial summary judgment.

Lack of a clear statutory language imposing strict liability and the difficulty of assessing and apportioning liability among the PRPs have also been major stumbling blocks. Perhaps this is why the actual apportionment of responsibilities for response and remedial costs has become a matter of negotiation and dispute resolution. However, these negotiations which often lead to final settlements are not limited to the PRPs only. EPA, being at the center of such disputes that demand environmental damage control due to the release of hazardous substances, holds the ultimate key to a final resolution of the dispute.

To be just and fair, the final resolution has to be acceptable to both the PRPs and various agencies of the federal government. Note that CERCLA does not preempt the states from imposing additional liabilities on the PRPs. However, section 114(b) of CERCLA bars multiple recoveries for claims or damages arising from CERCLA as well as other federal and state statutes. Hence, if a final settlement for response and remedial costs or claims for environmental damages can be properly and fully negotiated, further compensatory costs by the PRPs can be prevented. As a result, a lot may ride on negotiated settlements. Therefore, they require carefully planned strategies and evaluation of various options.

Government settlement

On July 8, 1991, EPA published a notice about its model CERCLA RD/RA consent interim decree (model consent decree) that may be used in remedial design/remedial action (RD/RA) settlements with PRPs. Additionally, for a proper negotiation by the federal government, EPA and DOJ jointly published a settlement policy guideline, "Hazardous Waste Enforcement Policy," on February 5, 1985. This is now largely codified in the statutory provisions of the 1986 CERCLA Amendments which were enacted in the form of SARA. These new settlement and negotiation provisions set forth in section 122 of CERCLA grant a great deal of discretionary authority to the federal government.

Unfortunately, EPA's model consent decree (MCD) does not offer generous incentives to a PRP who may be a prospective settlor. The MCD appears to shift the balance in favor of EPA with minimal rewards to the PRPs who may settle. The MCD expressly reserves the right for the government to force the settling PRP to perform additional response actions that may be necessary to "meet the performance standards or to carry out the remedy selected" in EPA's record of decision (ROD). Thus, EPA retains the discretion regarding additional response actions which remain undefined with respect to their eventual scope. Hence, a settling PRP may have no control over the selection, nature, and costs of response actions that may "do the job." Moreover, a "dispute resolution" has to be achieved through exchange of position papers between the settling PRP and EPA. Unfortunately, it is EPA that will make the final decision.

Some of the important mandatory and discretionary settlement provisions include the following:

- Under section 122 of CERCLA, the government has discretion over whether to enter into settlement agreement with any PRP as it has outlined in its guidance manual (however, it mandates the federal government to explain its settlement position in writing to the PRPs).

- The federal government is required to issue a notice of anticipated remedial action and solicitation of participation to the PRPs.

- The federal government can enter into settlement agreement for CERCLA's section 104 response actions with the PRPs by pursuing either administrative consent order or judicial decree.

- The government must use a moratorium for 90 days on CERCLA's section 104(b) actions involving remedial investigation/feasibility study (RI/FS) or other duties after providing notice and information in order to allow the PRPs make good-faith proposals for response actions.

- The government is required to observe a moratorium on response actions under section 104(a) of CERCLA or any enforcement action under section 106 of CERCLA for 120 days after providing notice and information with respect to such actions in order to enable proposals from the PRPs.

- The federal government is encouraged to negotiate *de minimis* settlements at the early stages of negotiations with the PRPs.

- All government agreements to undertake or finance CERCLA's section 106 remedial actions must be in the form of a consent decree.

- The federal government is authorized to accept mixed funding settlements.

- The federal government is expressly authorized to enter into covenants not to sue the settling *de minimis* PRPs for future liability.

- All cost recovery settlements that exceed $500,000 must be approved by the U.S. Attorney General.

- The federal government may provide preliminary nonbinding allocations of responsibilities (NBARs) to the PRPs, but such NBARs are neither judicially admissible nor reviewable.

The model consent decree provides boilerplate language to expedite the settlement process. However, it is still subject to negotiations by and between EPA and the PRPs. The model consent decree will generally be used for fashioning a quick settlement by EPA. EPA's regional offices will actually be providing the PRPs with a proposed consent decree based on this model as well as site-specific considerations. Moreover, EPA's regional offices may work with the Department of Justice to draft and negotiate settlement terms that go beyond the model consent decree.

A decision by the federal government to use or not use the procedures outlined in §122 of CERCLA for settlement purposes is not subject to judicial review. In this context, the states are expressly allowed under both CERCLA and SARA to play a significant role in any settlement process involving response actions. As a result, the states are fully authorized to actively participate in the selection of cleanup and remedial criteria and methods. Hence, these settlement provisions may also be followed by the state governments if and when they are involved.

EPA's settlement policy involving selection of a remedy or remedial action under the RI/FS process is based on applicable or relevant and appropriate requirements (ARARs). Any remedy to satisfy CERCLA requirements has to meet three specific standards:

- The remedy, in the form of a cleanup and to the maximum extent practicable, must include permanent solutions as well as alternative treatment methods or resource recovery technologies.

- The remedy must be cost-effective.
- The remedy must be consistent with the National Contingency Plan, including the requirements for public participation.

Pursuant to section 121(b) of CERCLA, remedial actions involving treatment which permanently and significantly reduces the volume, toxicity, or mobility of the hazardous substances, pollutants, and contaminants are preferred to remedial actions not involving such treatment. The remedial actions selected must provide a degree of cleanup which at a minimum ensures protection of human health and the environment.

As indicated earlier, EPA's current settlement policies are included in a memorandum entitled "Interim CERCLA Settlement Policy." This is strictly a policy guideline, and it does not have the legally binding effect of legislation. However, for all negotiations for possible settlement with the federal government, this is indeed an appropriate starting or reference document. In general, this guideline suggests several important points:

- EPA will seek to recover all costs or a substantial proportion of the costs that may be incurred for an appropriate remedy or cleanup of a contaminated site.
- EPA will consider providing some form of contribution protection in a settlement decree to protect the settling PRPs.
- EPA will seek to retain the right to reopen the consent decree.

EPA's guideline, however, does not define what is a substantial proportion of recovery. In an unwritten rule, EPA at one time insisted on not accepting any settlement offer that did not amount to a recovery of at least 80 percent of the total remedial costs. This was, however, not an official EPA policy. Since the late 1980s, EPA has issued several policy statements which attempt to obliterate the old, unwritten 80 percent recovery rule. At present, in any settlement that may be negotiated with EPA, several factors should be taken into consideration: the number of PRPs, amount and nature of contamination or wastes disposed of, types of remedial options, evidence received or accumulated, ability of the PRPs to pay, contribution of waste by each PRP, and possible outcome in the event of a litigation.

EPA's guideline for contribution protection and the scope of its release from further liability is, at best, noncommittal and deliberately vague. Generally, EPA will agree to a stipulation in the form of a consent decree. The judgment against a nonsettling party will be reduced if there is any settlement with other PRPs. This reduction in payment will, however, be only to the extent necessary to extinguish a settling party's liability to the nonsettling party. But, unlike a typical settlement of a private tort action, EPA may not provide a full and complete release of liability to the PRPs interested in settling with EPA. Instead, the scope of EPA's release in the consent decree will be guided by its degree of confidence in the remedy negotiated and finally accepted. Additionally, as indicated above, EPA can reopen a consent decree if

- EPA subsequently discovers new contamination previously undetected at a site for which a settlement has been negotiated and this newly discovered contamination poses an imminent and substantial endangerment to public health, welfare, or the environment.

- EPA subsequently obtains additional scientific data after the selected remedy is negotiated and undertaken and such data suggest that the site conditions pose an imminent and substantial endangerment to public health, welfare, or the environment.

EPA has, however, recognized that in any settlement with a private party that requires significant or substantial expenditure of funds by the private party in connection with the response action, the private party will require a general release or a covenant not to sue. It is not totally lost to EPA that no private party may voluntarily undertake to spend a large sum of money for a response action involving cleanup and/or remediation unless EPA offers to settle for good. Section 122 of CERCLA, added by way of SARA, established a new procedure which may promote and facilitate settlement negotiations between EPA and the prospective settling PRPs.

In this context, note that section 122(f)(1) of CERCLA includes a discretionary covenant not to sue a settling party concerning any liability to the government as a result of a release or threatened release of a hazardous substance. This may even be extended to future liability in the case of *de minimis* PRPs if the release or threatened release of a hazardous substance has been addressed by a remedial action, off-site or on-site. However, in such a case, four conditions must be met:

- The covenant not to sue is in the public interest.

- The covenant not to sue will expedite the response action consistent with the NCP.

- The PRP is in full compliance with the terms and conditions of a consent decree under section 106 of CERCLA for response to the actual or threatened release.

- The response action has been approved by the President of the United States.

Section 122(f) of CERCLA contains additional conditions for future liability. A settling party, unless she or he is involved in a *de minimis* settlement, can be subject to future liability arising from the release or threatened release of hazardous substances if such liability stems from conditions unknown at the time of a remedial action. As a result, a covenant not to sue will generally include this exception clause.

Historically, EPA's settlement policy underwent several changes because of the roadblock EPA faced while trying to reach settlements with the PRPs. Prior to 1985, EPA actually followed three distinct and separate approaches in negotiating settlements under CERCLA. In the first two years after the passage of CERCLA and until mid-1983, EPA followed a pro-industry stance and negotiated and executed several so-called sweetheart deals with the industry. During the period from mid-1983 to early 1985, EPA tried to change its image by aggressively pursuing a no-nonsense strategy and embarking on major litigation against the industry.

Neither of these strategies was successful, and EPA became the target of public criticism and ridicule. As a result, since early 1985 EPA introduced its third, middle-of-the-road strategy for settlement with PRPs. This strategy, unlike the previous ones, is based on pragmatic considerations and is portrayed as somewhat of an open policy. Pursuant to this strategic plan, EPA published its first interim CERCLA settlement policy in 1985, in order to offer the industry certain broad guidelines for private party settlements. The biggest contribution of this interim settlement policy has been a reduction of industry's uncertainty and anxiety over EPA's approaches toward settlement. Also, EPA, for the first time, openly embraced a settlement framework which incorporates a certain degree of government flexibility toward settlement with the PRPs.

With regard to the covenant not to sue a settling party, EPA also made specific strides. Here also EPA showed its flexibility by adopting a sliding scale for releases from liability.

Many of the policies formulated in EPA's settlement policy guidelines "Interim CERCLA Settlement Policy" have somehow found their way into certain statutory provisions of CERCLA. Unless there is a conflict with the specific statutory provisions, EPA is not barred from using this policy guideline, which sets its goal at receiving 100 percent of response costs. This includes the administrative costs. However, it should be stressed that this goal, set by EPA, is not necessarily the be-all and end-all of negotiations. Like all other negotiations between negotiating parties, there is a give-and-take aspect to EPA's current settlement policy.

As a guideline, the current EPA settlement policy document sets forth ten site-specific criteria for EPA's decision whether to settle for less than 100 percent of its costs:

- Amount of hazardous substance or waste contributed by each PRP
- Nature of the hazardous substance or waste contributed by each PRP
- Strength of evidence that traces the hazardous substance or waste at the site to the settling PRPs
- Ability of the settling PRPs to pay
- Risks of litigation if trial is sought
- Public interests in a settlement as opposed to litigation
- Economic considerations in obtaining a present sum certain
- Inequities and their aggravating factors
- Precedent value of a settlement if reached
- Future of the case after a settlement is reached

Considerations based on these factors are essentially flexible although EPA's position may be hardened if there is a threat of an imminent and substantial endangerment to public health or the environment due to the presence of a contaminated site. Also, EPA's published policy guideline attempts to unmask the mystery that surrounded the prior 80 percent recovery rule. According to this document, 80 percent recovery of the costs may not be acceptable as there are cases where 80 percent or more of cost recovery will not meet the substantial recovery test.

However, pursuant to another EPA guideline document, "Interim Guidelines for Preparing Non-Binding Allocations of Responsibility," issued on May 28, 1987, EPA may prepare an NBAR if such a request is made by a significant percentage of the PRPs with respect to a particular site. Such an NBAR will be prepared during the RI/FS process and will be based on the settlement criteria discussed above.

Section 104(e) and section 122(e)(3)(B) of CERCLA allow EPA to obtain information that is necessary, even by using its subpoena power, for preparing an NBAR. However, this authority is limited to the purpose of determining, selecting, or taking response actions. Under certain circumstances, where the costs of remediation are divisible, the NBAR may even allocate the costs among the PRPs.

Private settlement

CERCLA does not forbid a private settlement between a PRP and a private third party that took upon itself to clean up or remediate a contaminated site. This is, however, extremely rare since a typical cleanup cost runs into millions of dollars. Also, since CERCLA provides cost recovery by third parties against government funds (i.e., Superfund) set up for this purpose, this proposition has very little meaning in the real world.

15.5 Negotiating Strategy

A good negotiating strategy should be based on a full understanding of the position of the government and must carefully consider the factors which are statutorily inflexible as well as those which are historically flexible. A good starting point is to review EPA's policy guidelines for negotiating settlements with PRPs. It is important to consider whether a particular PRP can qualify under the *de minimis* settlement rule, for this offers considerable flexibility.

Nevertheless, a PRP in a given situation must also take certain stock of his or her own position. Several factors are worth considering in formulating a position for negotiation with EPA:

- The nature and quantity of hazardous substances or wastes for which the PRP is responsible

- The actual role of the PRP with regard to the contaminated site, since historically EPA has put the greatest burden on waste generators and least on waste transporters

- The efficacy, reliability, and permanence of cleanup goals
- The level of likely achievement of cleanup goals based on available remedial options
- The possibility of prompt implementation of a selected remedy
- The degree and types of closure and postclosure monitoring warranted
- The degree of innovation as opposed to existing proof with regard to the proposed technology for remedial action
- Need for additional regulatory approvals, permits, or licenses
- The degree of actual or potential hazards to public health, welfare, or the environment based on scientific and medical evidence and reliable health studies

Obviously, the stakes in negotiating a just and proper settlement with the government are high. The more information is available, the easier it is to decide how firm or flexible the strategic posture of a PRP should be in reaching a settlement with the government. It is also important to appreciate the flip side of the coin. CERCLA requires EPA to receive a "substantial offer" from the PRPs before EPA can forestall any of its contemplated actions. In addition, a PRP must take note of EPA's authorized power under CERCLA, as amended by SARA, to reject even a substantial offer. In essence, a PRP must weigh the merits of a defense against the PRP's actual and/or potential liability under a CERCLA section 106 order and the possibility of treble damages that may accrue before the PRP's defenses can be formally heard and eventually decided by a court. Any challenge to an EPA action based on a CERCLA section 106 order by labeling it "arbitrary, capricious, or otherwise not in accordance with law" must include strong meritorious defenses including sufficient cause by the PRP to refuse to comply with a CERCLA section 106 order.

The legislative history of CERCLA may provide useful information for a PRP's strategy in a given situation. The 1980 Senate debate over the meaning of the critical phrase *sufficient cause* provides an excellent account of how the legislators viewed a meritorious defense to an EPA order. A number of courts actually looked into Senator Stafford's remarks to find a judicial interpretation of what constitutes sufficient cause.

In addition, the courts have also looked into the legislative history surrounding SARA to find an objective standard of reasonableness to determine the propriety of an EPA unilateral order pursuant to section 106 of CERCLA. While a PRP must weigh available defenses to a section 106 order carefully, based on considerations of several factors discussed earlier, a careful search and reading of legislative histories surrounding both CERCLA and SARA may provide additional defenses.

15.6   *De Minimis* Settlement

In its interim guidance on settlements, EPA recognizes the necessity and appropriateness of a *de minimis* settlement. As the term implies, a *de minimis* settlement may be negotiated with those PRPs who contributed to or are otherwise found responsible for minimal amounts of hazardous substances at a contaminated site. A *de minimis* settlement is also based on the theory that the hazardous substances or wastes thus contributed do not result in a serious or significant threat to public health or the environment. This determination is based on not only their quantity, but also careful consideration of their nature. Section 122(g) of CERCLA provides that a *de minimis* settlement with a PRP is appropriate if

- The amount of the hazardous substances contributed by the PRP to the facility is minimal.
- The toxic or other hazardous effects of the substances contributed by the PRP to the facility are minimal.

Traditionally, EPA has defined the term *de minimis* on a site-by-site basis. This is also recognized in EPA's June 19, 1987, directive "Interim Guidance on Settlement with *de Minimis* Waste Contributors under Section 122(g) of SARA." The main objective here is to reach a quick *de minimis* settlement with the PRPs so that the government resources may be effectively preserved and directed toward those PRPs responsible for substantial and significant contamination of other sites.

This tacit recognition by EPA of the need to conserve and direct its limited resources against the major PRPs allows considerable leverage for PRPs in formulating their negotiating strategies. Also note that there is no specific definition of what constitutes *de minimis* contribution by a PRP. EPA regional offices use their own policies in such matters. Several EPA regions use a volume of 1 to 2 percent waste as a starting point for discussion with possible *de minimis* PRPs. Obviously, there is no hard and fast rule. Hence, other PRPs should not be dissuaded from trying to qualify as a *de minimis* PRP if their contributions can be shown to be minimal, insignificant, or otherwise insubstantial as compared to the total volume or the nature of the hazardous substances or wastes present at a contaminated site.

Historically, EPA has not entertained any *de minimis* settlement prior to its completion of the PRP search for a specific site and its completion of a RI/FS. In addition, all *de minimis* settlements are finalized only after a 30-day public comment period. Moreover, as indicated earlier, if the total response costs for a particular site exceed $500,000, the *de minimis* settlement has to be approved by the Department of Justice (DOJ).

The statutory language for *de minimis* settlement requires EPA to settle with *de minimis* PRPs as promptly as possible. They are to be encouraged whenever such settlements are practicable and in the public interest. However, the statute allows *de minimis* settlement when only a minor amount of response costs is involved and when the contribution by the PRP is minimal in comparison to other hazardous substances or wastes present at the site.

From a practical point of view, this is not a simple resolution. EPA does not and cannot always determine what is a "minor" amount of response costs or the extent of total volume, hazards, and toxicity of the hazardous substances and wastes deposited in a contaminated site at an early stage of its involvement. Hence, to take a safer position, EPA may seek substantial premiums before agreeing to an early bailout by the *de minimis* PRPs. EPA's *de minimis* settlement concept is founded on the theory that a PRP who wishes to enter an early settlement and obtain a complete release of liability should pay a premium above and beyond the PRP's proportional share of contribution based on the nature and amount of hazardous substances or wastes generated, transported, or disposed of. Note also that section 122(g)(2) of CERCLA authorizes but does not require EPA to provide the settling *de minimis* PRP a complete release from liability. Thus, unless a *de minimis* PRP is on top of things, the PRP may settle for terms unfavorable or worse—plainly unjust.

However, to settle or not to settle early is a Catch-22 decision. If a PRP waits to reach an early cash-out settlement, the PRP may continue to incur not only additional costs but also a cloud of uncertainty about the final outcome. For many corporations, this may be a public relations nightmare. On the other hand, any quick settlement by paying a premium price may irk the shareholders of the corporation. In essence, there is one message here. All PRPs, including *de minimis* PRPs, must weigh all options before deciding on a settlement.

15.7   Litigation versus Settlement

Settlement has been touted as a possible, and often better, alternative to protracted and expensive litigation. In many situations, a fair settlement is indeed a far better choice than litigation.

However, while CERCLA or SARA allows EPA to enter a covenant not to sue, it is not close-ended. As discussed earlier, section 122(f)(6) of CERCLA states that most of EPA's covenants not to sue must also include certain exceptions or provisions for reopening of a case for future liability against a settled PRP.

Such an exception or a reopener clause is generally provided for situations where a future liability "arises out of conditions which are unknown." This is true even when the President certifies that a remedial action has been completed and the covenant not to sue has been effective. Hence, a release that may be obtained through a settlement process pursuant to EPA's policy under CERCLA and SARA is far short of a general or complete release that one can expect in a typical tort settlement.

It is therefore a game of chess between the government and the PRPs to arrive at an appropriate scope of release as a result of a negotiated settlement. In spite of careful drafting, there may be enough openings which might subject PRPs who have settled to new EPA orders or threats of litigation. This is particularly important in light of the statutory provision in section 121 of CERCLA over the irksome and difficult question of how clean is clean or how to establish an appropriate cleanup standard. From a pragmatic point of view, the question has not been settled yet and may never be settled because of new ongoing scientific and medical discoveries relating to toxicity and dose-response relationship, health effects, and ultimate disease manifestation. Nevertheless, section 121 of CERCLA addresses the cleanup standards that must be met by appropriate remedial actions.

Hence, for PRPs in general, all options must be weighed carefully and without haste. One obvious option, in spite of its spiraling costs and uncertainties of outcome, is litigation. In many situations, a judicial resolution of a dispute involving site contamination may be more desirable in spite of the costs because of the opportunity of a day in court and the avoidance of zealous EPA negotiators who may not always be limited to the facts of a particular case.

A proper strategy under CERCLA, as stressed above, must include a variety of action plans as well as litigation. During the formulating stage, it is essential to decide on the pros and cons of all action plans considered as viable options.

Note that SARA has altered some of the past thinking about typical options under CERCLA. Notably, pursuant to SARA, the past rules of CERCLA litigation have undergone significant changes. Most of these changes are in the area of judicial review.

As a result of SARA, a new section 113(h) has been added to CERCLA. This precludes any judicial review of EPA's decisions for removal or remedial actions or administrative orders under section 104 of CERCLA except in certain situations which involve

- An action enforcing an administrative order or for an injunctive relief under section 106(a) of CERCLA or to recover a penalty for violation of such order

- A cost recovery action under section 107 of CERCLA

- An action under section 310 of CERCLA in the form of a citizen suit alleging that the removal or remedial action under section 104 or section 106 of CERCLA violates the statutory requirements

- An action under section 106 of CERCLA in which the United States has moved to compel a remedial action

- An action under new section 106(b)(2) of CERCLA for reimbursement of costs

SARA also modified the scope of judicial review. Pursuant to SARA, judicial review of remedy is limited to the administrative records available before EPA. No judicial review of the adequacy of an EPA response action taken or ordered is permissible beyond such administrative orders or records. In other situations, applicable principles and standards of judicial review will guide whether supplemental records or documents should be taken into account. In addition, SARA expressly adopted the "arbitrary and capricious" standard of review for EPA's response actions. As a result, the courts may uphold an EPA decision regarding the selection of a response action in the name of the President of the United States. This is, however, permissible only as long as an opposing party cannot establish, based on the administrative records, that the decision was arbitrary and capricious or otherwise not in accordance with the law.

Recommended Reading

1. "Personal Liability for Environmental Violations—Avoiding and Defending Civil Suits and Criminal Prosecutions," Litigation and Administrative Practice Services, No. 409, Practicing Law Institute, New York, 1991.
2. C. R. Schraff and R. E. Steinberg, *RCRA and Superfund,* vol. 1, Shepard's/McGraw-Hill Inc., Colorado Springs, 1990.

Important Cases

1. *Outboard Marine Corp. v. Thomas,* 773 F. 2d 883 (7th Cir. 1985), *vacated,* 107 S. Ct. 638 (1986).
2. *New York v. Shore Realty Corp.,* 759 F. 2d 1032 (2d Cir. 1985).
3. *NRDC v. Reilly,* C.A. No. 88-3199 (D.D.C. June 14, 1989).
4. *United States v. Chem-Dyne Corp.,* Consolidated Civil Action Nos. C-1-82-840 etc. (S.D. Ohio 1985).
5. *United States v. Metate Asbestos Corp.,* Civ. 83-309 GLO RMB (D. Ariz. 1985).

Response Costs, Claims, and Recovery

16.1 Introduction

In today's world, it is commonplace to learn about the discovery of an abandoned hazardous waste disposal site or the release of a hazardous substance that threatens public life and the environment. While the public may be getting used to the tantalizing details surrounding such discoveries, the resulting releases do not have anything routine in terms of their environmental impact. In fact, they trigger all applicable statutory and regulatory provisions, and each of such episodes may require certain response actions, including cleanup and remediation.

As more studies are conducted and revelations about the dangers of toxic chemicals and hazardous substances are made, the total cost of cleanup and remediation is skyrocketing. This has created a domino effect. Insurance companies not only are drastically modifying their traditional coverage provisions under the comprehensive general liability (CGL) or the umbrella liability (UL) policies, but also are setting a huge premium for special environmental coverage. In addition, the Environmental Protection Agency and various state and local regulatory agencies are constantly readdressing and redefining the often disputed issue of how clean is clean.

The total of all these is the ballooning of response costs whether undertaken by government units or by private parties, including the potentially responsible parties (PRPs). This is wreaking havoc among the PRPs who must somehow pay for all the appropriate response costs. For the PRPs, the arsenal of sanctions and the array of incentives in connection with site cleanup and remediation are extensive. Unfortunately, however, the financial means of many PRPs are often extremely limited.

By necessity, PRPs require a full evaluation of all available options and strategies. It is quite easy for an enforcement agency to literally throw the book at any PRP, invoking joint and several liability, irrespective of the PRP's degree of culpability. However, it is not so simple for a PRP to undertake an expensive response or settlement action.

Chapter 6 discusses the PRP liability provision of section 107 of the Comprehensive Environmental Response, Compensation, and Liability Act. Section 107 of CERCLA authorizes recovery of costs incurred in connection with response actions by both the U.S. government and private third parties. In fact, section 107 of CERCLA authorizes legal actions for the recovery of response costs from PRPs. In short, CERCLA has explicitly created a public (i.e., government) and a private right of action against PRPs, especially for cost recovery. Also, recovery of costs by private third parties in the way of claims against the government funds, such as the Superfund, is allowed when the PRPs cannot be identified or are bankrupt or otherwise insolvent.

Note that insolvency of a PRP cannot be invoked, unless the claim is discharged in bankruptcy, to relieve a culpable party from the response costs. The stigma or the possibility of a future enforcement action against an insolvent but not bankrupt PRP can last indefinitely unless barred by the statute of limitations.

In this context, it should be emphasized that the initial naming or labeling of a person as a PRP by a regulatory agency is seldom based on thorough research or exhaustive analysis of documents. The possibility of being included as a PRP is very high even through indirect transactions or mere references in documents. Also, the regulatory agencies have no specific statutory requirements or criteria to deal separately with *de minimis* PRPs as compared to major PRPs. So far, EPA has been flexible in its settlement with *de minimis* PRPs. However, this is not a hard-and-fast rule, and the regional offices of EPA have broad discretionary powers. As a result, EPA can also use very rigid standards for settlement discussions with *de minimis* PRPs.

While the above possibilities are very real, the PRPs are not without options that may lessen the ultimate burden under the federal environmental laws. It is, therefore, always smart to look for that silver lining in dark clouds to make the best of an otherwise difficult situation.

16.2 Response Costs

Chapter 6 discusses in detail various response actions pursuant to actual or threatened release of a hazardous substance under CERCLA. In addition, it may be necessary to undertake certain response actions when there is a spill or discharge of petroleum or oil under the Federal Water Pollution Control Act (FWPCA) of 1972 or its later amendment, the Clean Water Act (CWA), as amended. Furthermore, as discussed in Chap. 5, the Resource Conservation and Recovery Act (RCRA) calls for a corrective action program for dealing with released contaminants from a treatment, storage, or disposal (TSD) facility of hazardous wastes.

In plain dictionary English, any cost that is incurred in connection with a response is a response cost. However, there is a specific statutory and/or regulatory definition of what can be considered a response cost. In legal parlance, as has been stressed in earlier chapters, one must pay heed to the statutory or regulatory definitions, since they are terms of art and thus cannot be defined otherwise, regardless of the common sense meanings of those terms.

Definitions

Clean Water Act. Section 311(a)(8) of the CWA sets forth the following statutory definition of *removal:*

> "[R]emove" or "removal" refers to the removal of the oil or hazardous substances from the water and shorelines or the taking of such other actions as may be necessary to minimize or mitigate damage to the public health or welfare, including, but not limited to, fish, shellfish, wildlife, and public and private property, shorelines, and beaches.

The statute does not contain any definition of what constitutes a response action or response cost. However, section 311(c)(1) of CWA authorizes the President of the United States to act to remove or arrange for the removal of any oil or hazardous substance that is actually discharged or has the potential for discharge into or upon the navigable waters of the United States, including adjoining shorelines and waters of the contiguous zone, at any time. The only time this may not be carried out is when such removal can be done properly by the owner or operator of the vessel or the onshore or offshore facility from which such a discharge takes place.

Section 311(f)(4) of CWA states that the cost of removal of oil or a hazardous substance for which the owner or operator of a vessel or an onshore or offshore facility is liable includes any response cost or expense incurred by the federal or any state government. All costs incurred in restoring or replacing the natural resources damaged or destroyed due to a discharge of oil or a hazardous substance are appropriate response costs. Section 311(f) of CERCLA specifies that the owner or operator of a vessel or a facility from which oil or hazardous substance is discharged in violation of the no discharge provision of section 311(b)(1) of CWA is liable to the federal or state government for the actual costs of removal.

However, CWA puts a statutory ceiling to liability arising from such removal costs. Pursuant to section 311(f)(1) of CWA, this liability may not exceed $125 per gross ton of an inland oil barge or $125,000, whichever is greater. For any other vessel, the liability rises to $150 per gross ton of the vessel or $250,000 for a vessel carrying oil or hazardous substances as cargo, whichever is greater. In case of a willful negligence or willful misconduct, the owner or operator of the vessel will be liable to the federal government for the full amount of the removal cost. In such a case, the government will have to prove that the discharge that occurred was within the privity and knowledge of the owner.

For the owner or operator of an onshore or offshore facility, the liability for removal costs owed to the federal government under section 311(f)(2) of CWA may not exceed $50,000,000 unless the discharge is due to willful negligence or willful misconduct within the privity and knowledge of the owner. In the latter case, the liability may cover the full amount of such removal costs irrespective of their amount.

Section 311(f)(3) of CWA provides limited defense to the owner or operator of a vessel, or an onshore or offshore facility, against the liability that may arise from the costs of removal. These defenses include

- An act of God
- An act of war
- Negligence on the part of the U.S. Government
- An act or omission of a third party
- Any combination of the foregoing clauses

These defenses can be asserted regardless of whether an act or omission was negligent. While the statute does not specifically state whether the liability of the owner or operator of a vessel or an onshore or offshore facility due to the discharge of oil or a hazardous substance is strict, it has all the trappings of a strict liability because of these limited defenses. Also, the federal government may bring the liability action for removal costs against the owner or operator of a facility in any court of competent jurisdiction.

Section 311(o)(2) of CWA specifically allows any state or political subdivision to impose any of its requirements or liability for discharge of oil or a hazardous substance into any waters within such state. There is no federal preemption in such situations.

Comprehensive Environmental Response, Compensation, and Liability Act. Section 101(23) of CERCLA defines what is removal:

> "[R]emove" or "removal" means the cleanup or removal of released hazardous substances from the environment, such actions as may be necessary taken in the event of the threat of release of hazardous substances into the environment, such actions as may be necessary to monitor, assess, and evaluate the release or threat of release of hazardous substances, the disposal of removed material, or the taking of such other actions as may be necessary to prevent, minimize, or mitigate damage to the public health or welfare or to the environment, which may otherwise result from a release or threat of release. The term includes, in addition, without being limited to, security fencing or other measures to limit access, provision of alternative water supplies, temporary evacuation and housing of threatened individuals not otherwise provided for, action taken under section 104(b) of...[CERCLA], and any emergency assistance which may be provided under the Disaster Relief Act of 1974.

Note that the statute defines *respond* or *response* to mean remove, removal, remedy, and remedial action. Section 101(24) of CERCLA defines the last two terms as follows:

> "[R]emedy" or "remedial action" means those actions consistent with permanent remedy taken instead of or in addition to removal actions in the event of a re-

lease or threatened release of a hazardous substance into the environment, to prevent or minimize the release of hazardous substances so that they do not migrate to cause substantial danger to present or future public health or welfare or the environment. The term includes, but is not limited to, such actions at the location of the release as storage, confinement, perimeter protection using dikes, trenches, or ditches, clay cover, neutralization, cleanup of released hazardous substances or contaminated materials, recycling or reuse, diversion, destruction, segregation of reactive wastes, dredging or excavations, repair or replacement of leaking containers, collection of leachate and runoff, onsite treatment or incineration, provision of alternative water supplies, and any monitoring reasonably required to assure that such actions protect the public health and welfare and the environment. The term includes the costs of permanent relocation of residents and businesses and community facilities where the President [of the United States] determines that, alone or in combination with other measures, such relocation is more cost-effective than and environmentally preferable to the transportation, storage, treatment, destruction, or secure disposition offsite of hazardous substances, or may otherwise be necessary to protect the public health or welfare...

In short, *remedy* or *remedial action* implies a permanent action in contrast to *remove* or *removal* which is more of a temporary action. However, the terms are not mutually exclusive. A remedy or remedial action may follow a remove or removal action if the situation so warrants. Any of these four actions can constitute a necessary response action based on the circumstances surrounding an actual release or a potential release of a hazardous substance. Section 106 of CERCLA allows undertaking of an appropriate response or abatement action by the federal or a state government in the public interest.

Section 107 of CERCLA imposes liability on the owner or operator of a vessel or a facility for such response costs. Section 107(a)(4) of CERCLA, as stated in Chap. 13, imposes liability on four types of PRPs. The liability for these PRPs includes the "full and total costs of response and damages" if

- The release or threat of release of a hazardous substance was the result of willful misconduct or willful negligence within the privity or knowledge of such person.

- The primary cause of the release was a violation within the privity or knowledge of such person, of applicable safety, construction or operating standards or regulations.

- Such person fails or refuses to provide all reasonable cooperation and assistance requested by a responsible public official in connection with response activities under the National Contingency Plan.

Otherwise, the liability of the owner or operator of a vessel or facility from which a release or threat of a release occurs or that of any other PRP for the release of a hazardous substance is limited. Section 107(c)(1) of CERCLA provides the ceilings for such liability for specific situations involving release of a hazardous substance. Under section 107(c)(1) of

CERCLA, the liability of an owner or operator or other responsible person for each release of a hazardous substance from any vessel, other than an incineration vessel, carrying any hazardous substance as cargo or residue is $300 per gross ton or $5,000,000, whichever is greater. For any other vessel not carrying any hazardous substance as cargo or residue and which is not an incineration vessel, the liability is $300 per gross ton or $500,000, whichever is greater. For an incineration vessel or a facility, liability for such release is the total of all response costs plus $50,000,000 for any damages. In addition, if a person who is liable for a release or threat of release of a hazardous substance fails, without sufficient cause, to properly provide removal or remedial action upon a Presidential order, that person may be liable to the federal government for punitive damages in an amount at least equal to, but no more than three times, the amount of any costs incurred by the government by dipping into the Superfund.

As indicated in Chap. 13, the regulatory cap of a response action as set forth by EPA is $2,000,000 and a time limit of 12 months. These limits can be exceeded only when there is an emergency or when EPA determines that continued response action is both appropriate and consistent with the remedial action planned.

As indicated in Chap. 6, CERCLA also allows assertion of claims by private third parties for response costs against the Superfund set up by the federal government. However, there is a general controversy because of lack of uniform decisions by the nation's courts as to what response costs are recoverable by the government and by a private third party.

In this context, note that PRPs themselves may carry out a response action that is EPA-approved. But such action must be undertaken promptly and properly so as to avoid penalties and other sanctions. Under section 112 of CERCLA, any third party, other than a PRP, who voluntarily undertakes a response action in the form of removal or remedial action must seek reimbursement for the response costs from the owner, operator or guarantor of the vessel or facility from which a hazardous substance has been released. If the claim is not satisfied within 60 days of presentation, the claimant may present the claim to the Superfund established under CERCLA. For recovery, all CERCLA response costs must be consistent with the National Contingency Plan. Moreover, for recovery of response costs, claims must be asserted before the statutes of limitation run out (details are provided in Chap. 6).

16.3 Recovery of Response Costs

As indicated earlier, there is a general controversy as to whether the recovery of response costs, be it a removal or a remedial action, by a private party can be as comprehensive as that of a government recovery. Part of this controversy stems from the textual language of the applicable statutes.

There are several federal statutes, discussed below, which contain specific cost recovery provisions.

Clean Water Act

Section 311 of CWA addresses discharges of oil and hazardous substances, whether accidental or intentional. Besides a spill reporting requirement, the statute requires the person causing the discharge either to contain and clean up such spills or discharges or to pay for the cost of cleanup.

Section 311(f) of CWA, as discussed earlier, puts a cap on the liability of an owner or operator of a vessel or facility unless it is a result of willful negligence or willful misconduct. However, this statutory provision addresses the issue of liability only in connection with cost recovery by the government. The statute does not contain any specific language with respect to the scope or extent of cost recovery by third parties.

A number of lower courts granted a cost recovery action brought by private third parties involving a spill or release of oil or a hazardous substance. The general thrust of these decisions is that CWA prohibits any discharge of oil or hazardous substances. The statute does not specifically bar a cleanup or response action by a third party.

Comprehensive Environmental Response, Compensation, and Liability Act

Unlike CWA, CERCLA contains specific provisions for cost recovery actions by both government and private third parties. Section 107 of CERCLA establishes that the PRPs are liable for cleanup and/or remediation costs. However, it does not include any explicit language with respect to the standard of liability. Instead, CERCLA indicates that the standard of liability will be what can be imposed under the Federal Water Pollution Control Act (FWPCA) of 1972. Section 107(a) of CERCLA specifically provides that four types of PRPs will be liable for the response costs. Pursuant to the statutory provisions, the response costs due to an actual or threatened release of a hazardous substance can include

- All costs of removal or remedial action incurred by the federal or a state government which are not inconsistent with the NCP
- Any other necessary costs of response incurred by any other person consistent with the NCP
- Damages for injury to, destruction of, or loss of natural resources, including the reasonable costs of assessing such injury, destruction, or loss resulting from such a release
- The costs of any health assessment or health effects study carried out under section 104(i) of CERCLA

Thus, the statute explicitly states that any government, federal or state, can recover all response costs as long as they are consistent with the NCP. However, the statute does not clearly indicate whether a third party who undertakes a removal or remedial action on its own may also recover all response costs. Instead, the statute talks about "other necessary costs" that are consistent with the NCP for private third party cost recovery. Due to this seemingly different use of language in the statute, the courts are divided over the recoverability of attorneys' fees and particularly whether such fees can qualify under CERCLA's response costs.

In most of these cases, the courts have relied heavily on the legislative history and congressional deliberations behind the enactment of CERCLA. Some even looked for congressional intent behind the passing of FWPCA because of an explicit reference to FWPCA in CERCLA. Surprisingly, however, even such a similar approach or legislative research has led to dissimilar—and in a few situations, strikingly different—decisions.

The legislative history generally indicates that Congress intended the same standard of liability in all CERCLA-based actions. More precisely, the legislative history suggests a congressional intent to apply common law principles to issues not specifically included in the statutory provisions. However, this is camouflaged by vague statutory language and general congressional reticence.

Due to the explicit statutory language, the courts thus far have had no problem in allowing a cost recovery by the federal government for a wide range of removal and remedial activities. In *United States v. NEPACCO,* the court held that section 107 of CERCLA allows the government to recover response costs which may include

- Investigations, monitoring, and testing to identify the extent of the release or threatened release of hazardous substances
- Investigations, monitoring, and testing to identify the extent of danger to public health or welfare or the environment
- Planning and implementation of a response action
- Recovery of the costs associated with the above actions
- Costs of enforcement of the statutory provisions of CERCLA, including the costs incurred for the staff of EPA and the Department of Justice

Thus, a government recovery of response costs may include those costs directly and indirectly associated with the response action, e.g., legal costs and attorneys' fees. In the 1991 case, *United States v. Parsons,* the Court of Appeals for the Eleventh Circuit sent a chilling message when it held that the federal government may recover four times its response costs under section 107(c)(3) of CERCLA. To arrive at this ruling, the *Parsons* Court held that punitive damages are included in government cost recovery for cleanup or remedial actions. Moreover, according to this court, section 107(c)(3) of CERCLA permits government cost recovery of actual remedial costs plus treble damages when responsible parties fail to undertake, without "sufficient cause," an appropriate response action pursuant to an EPA administrative order. Interestingly, in another 1991 case, *Barmet Aluminum Corp. v. Reilly,* the Sixth Circuit Court of Appeals affirmed the lower court's decision that preenforcement challenges to EPA's response actions are barred by the limitation of judicial review provision of section 113(h) of CERCLA.

It, therefore, appears that the courts are willing to allow EPA a broad power or deference in connection with its response actions. Also, the courts are willing to approve recovery of all costs for such response action as long as the agency comports with the statutory provisions.

Furthermore, in the 1991 decision, *Independent Petrochemical Corp. v. Aetna Casualty & Surety Co.,* the Court of Appeals for the District of Columbia Circuit held, by applying Missouri law, that government-mandated cleanup costs are covered within the CGL policy. According to the *Aetna* Court, actions under the statutory provisions of CERCLA and RCRA are essentially claims for restitution. As a result, such claims were perceived by the court to be equitable in nature. Thus, the court ruled that environmental cleanup costs which have been pursuant to a government order constitute "damages" under the CGL.

However, the courts are reaching different decisions with regard to private third party recovery of response costs. They are generally in agreement that response costs associated with the first three of the above-listed five items are recoverable in a private party cost recovery action. However, there is no uniformity with regard to the recoverability of administrative, enforcement, or legal costs.

A few courts, in early private party cost recovery actions, allowed recovery for medical expenses, relocation costs, and property damages. However, in recent years, several courts denied private party recovery of costs pertaining to medical monitoring, loss of income, and damage to property. But in the 1991 case of *Key Tronic v. United States,* the federal district court allowed a broad-based recovery, including prejudgment interest, attorneys' fees, PRP search costs, and costs of negotiating a consent decree with EPA. However, this liberal construction of the cost recovery provisions based on the definition of "enforcement-activities-related" response cost to removal and remedial action under section 101(25) of CERCLA is not accepted by many other courts. For example, in *Wehner v. Syntex Corp.,* the court rejected the plaintiff's claim for even medical monitoring costs. Other courts have disagreed on the cost of alternate water supplies, permanent (but not temporary) relocation cost, and attorneys' fees.

Of all these, the issue of attorneys' fees has been most hotly debated by the parties. Not surprisingly, the court decisions are widely apart on this issue. In *Gopher Oil Co., Inc. v. Union Oil Co. California,* the Court of Appeals for the Eighth Circuit upheld the lower court's determination that in an "as is" sale fraudulently induced by a seller, the seller is liable for 100 percent of the cleanup costs of the contaminated facility sold. The court further held that reasonable attorney's fees, excluding work related to the fraud claim, are recoverable by a private party in a CERCLA-based cost recovery action. In *General Electric Co. v. Litton Industrial Automation Systems,* the Court of Appeals for the Eighth Circuit also held that attorneys' fees are recoverable in a CERCLA-based response action. However, in *Fallowfield Development Corp. et al. v. Strunk,* the federal district court held that attorneys' fees are not costs that are enforcement-activities-related and thus are not recoverable. This obvious disparity and divergence of opinions is best reflected in two well-known court decisions which are considered leading cases for diametrically opposed viewpoints. In *Pease & Curren Refining, Inc. v. Spectrolab, Inc.,* the court liberally construed the statutory provisions of cost recovery and held that attorneys' fees are recoverable by private parties. However, in *T & E Industries, Inc. v. Safety Light Corp.,* the federal district court followed a conservative approach by embracing the "American rule." Thus, the *Safety Light* court held that a private party may not recover attorneys' fees unless they are specifically included either in the statute or in the contract.

A few courts, however, treated all expenses, including attorneys' fees, in a CERCLA-based action as equitable claims. In *United States v. Hardage,* the federal district court, however, denied to accept broad equitable claims by private parties seeking cost recovery based on recoupment, restitution, and common fund or benefit theory. Interestingly, in *Idaho v. Hanna Mining Co.,* the Ninth Circuit Court of Appeals stated *in dictum* that CERCLA does not state whether attorneys' fees may be awarded for actions for natural resource damages.

Obviously, attorneys' fees and several other response costs including, but not limited to, medical monitoring costs, relocation expenses, and costs for alternate water supplies are not definite or specifically protected items under the statute for private party cost recovery. The ongoing legal debate will not be resolved unless Congress amends the current vague and somewhat ambiguous statutory language that contains an explicit provision only for government recovery, or until the U.S. Supreme Court finally decides what response costs qualify under the statute for private party recovery. Until one of these two things occurs, the parties will be susceptible to the vagaries of court decisions based on liberal or conservative interpretation of the current statutory language.

16.4   Cost Recovery Claims under CERCLA

For a recoverable response cost under CERCLA, the statute provides that certain specific elements must be satisfied to constitute a claim. These elements apply to both government and private third party cost recovery claims. They also apply to both types of CERCLA response actions—removal actions and remedial actions. While the statute does not specify what constitutes a mature or uncontroverted claim for recovery or even what limitations may apply, the statute is clear with respect to the elements of a cost recovery claim:

- There must be a release or a threatened release of a hazardous substance.
- Such release or threatened release must cause the incurrence of response costs.
- The costs incurred must be necessary costs of response.
- These response costs must be consistent with the NCP.

For recovery of remedial costs by a private party, the incurrence of such costs must be consistent with the NCP while satisfying the other three elements. For participation in response action and cost recovery under section

107(a) of CERCLA by others, 40 CFR section 300.700(c)(3) provides the following:

> For the purpose of cost recovery under section 107(a)(4)(B) of CERCLA: (i) A private party response action will be considered "consistent with the NCP" if the action, when evaluated as a whole, is in substantial compliance with the applicable requirements in paragraphs (c)(5) and (6) of this section, and results in a CERCLA quality cleanup; (ii) Any response action carried out in compliance with the terms of an order issued by EPA pursuant to section 106 of CERCLA, or a consent decree entered into pursuant to section 122 of CERCLA will be considered "consistent with the NCP."

This indicates that all response costs are NCP-consistent if the cleanup or remedial action meets the terms of a government order or consent order. In the absence of such government order or a consent decree, all response costs, to be recovered, must be consistent with the NCP. As discussed in Chaps. 6 and 13, this does not mean 100 percent compliance, but "substantial compliance" with the NCP procedures.

There are established NCP procedures as developed by EPA. The NCP procedures are not uniform for all response actions. The NCP procedures for removal actions are simpler than those for remedial actions. Details of federal regional plans and the scope of the NCP are set forth in 40 CFR part 300, subpart A. Note that EPA added a new section in 1990 to clarify and elaborate on the NCP procedures. Also, the NCP procedures are not limited to response actions in connection with releases or substantial threats of release of hazardous substances. The NCP also applies to discharges or substantial threats of discharge of oil to or upon the navigable waters of the United States, adjoining shorelines for the contiguous zone, and the high seas beyond the contiguous zone that may affect the natural resources of the United States. The operation responses required in different phases of oil removal are set forth in 40 CFR part 300, subpart D whereas 40 CFR part 300, subpart E provides regulatory methods and criteria for determining the extent of response for releases of hazardous substances.

Pursuant to section 107(a) of CERCLA, a CERCLA-based claim for cost recovery can be asserted by any person (including private parties and public or government agencies) who incurs costs in a response action. However, to constitute a cause of action based on a claim for cost recovery, all applicable elements of a cost recovery claim under CERCLA must be satisfied.

In addition, as with any other claim, a claimant must bring a cause of action before the statute of limitations runs out. The statutes of limitations that may apply in a CERCLA-based action are discussed in detail in Chap. 6. It is generally agreed that the statute of limitations does not start running until the response action, whatever it may be, is completed.

16.5 Insurance Coverage and Indemnification

Chapter 14 describes the recent turmoil over insurance coverage for pollution. The insurance industry has significantly amended its old coverage for pollution under the CGL policy three times since the mid-1960s. These amendments in insurance policy language first included "pollution exclusion" clauses; later, they required the adoption of special environmental liability coverage under a separate policy or a special rider, thus dropping pollution coverage from the CGL or umbrella liability coverage altogether.

Historically, many industrial operations, particularly those owned by corporations, generally relied on their CGL policies for a bailout in the event of claims brought against them in a typical cost recovery action pursuant to a bodily injury or property damage. However, in recent years, the insurance companies are denying their duty to defend and indemnify their insureds by quoting generally the pollution exclusion clause of the CGL. As a result, such drastic limitations and exclusions by the insurance companies, particularly the applicability of their pollution exclusion clauses, have given rise to much litigation.

Part of the problem stems from continued amendments or revisions of existing policy coverages and adoption of new policy coverage language under the CGL. As indicated in Chap. 14, until 1966, the insurance industry as a whole adopted a seemingly broad policy coverage under its CGL provision. In 1966, the CGL policy was amended to incorporate coverage based on an "occurrence" rather than on an "accident," for public relations reasons. Claims based on personal injury and property damage were thus covered under the CGL as long as they arose from an "occurrence." This basically changed the all-risk coverage in the CGL policies providing defense and indemnification for bodily injury and property damage from an accident to an occurrence category. It did not, however, change the coverage by limiting it to certain specified risks.

The *occurrence* under the CGL was defined as

> An accident, including continuous or repeated exposure to conditions, which results in bodily injury or property damage, during the policy period, and that was neither expected nor intended.

The term *occurrence* was thus focused on an accident but included continuous or repeated exposure to conditions which can result in bodily injury or property damage "during the policy period." However, in spite of this apparent broad coverage, it is particularly difficult to satisfy the provision in those cases that involve bodily injury and/or property damage diagnosed or revealed after many years because of the long latency or manifestation period. In many of these situations, the original insurance company does not provide CGL coverage for subsequent years when the manifestation or the lawsuit commences. Furthermore, there is no easy way to determine the exact time of occurrence of such an accident that is the cause of the bodily injury or property damage being complained of. However, the courts have generally held that coverage under the CGL may be denied only when the insured acted negligently or intentionally and as a result the bodily injury or property damage was expected or intended.

To overcome such unplanned and very expensive insurance coverage, most of the insurance carriers first added a pollution exclusion clause to the standard CGL policy in 1973. The pollution exclusion clause states that the coverage does not apply to cases of bodily injury or property damage that is due to the discharge or release of toxic chemicals, waste materials, or other irritants, pollutants, and contaminants into any environmental medium (land, air, and water). Most notably, however, this pollution exclusion clause itself included an exclusion or exception clause which exempted its application to cases where such discharge or release is "sudden and accidental." Thus, due to the inclusion of these double-negative clauses, the CGL policy coverage extended to only those pollution accidents which were both sudden and accidental.

This simplistic construction of policy coverage over pollution-related incidents has caused major controversies since the CGL policies do not define what is *sudden* or *accidental*. It has been left to the courts to define these terms and their applicability in a particular situation. Not surprisingly, the courts have interpreted them widely, often with diametrically opposite and conflicting interpretations.

In general, the decisions of the courts have differed widely based on whether they embraced or discarded a temporal meaning of the phrase *sudden and accidental*. Any attribution of a temporal meaning to this phrase tends to connote "immediate and unexpected," thus bringing into play the time-based application of the pollution exclusion clause under the CGL policy.

To quash the growing controversy and widely different decisions by the courts in various jurisdictions, the insurance industry attempted to change things in 1986. This time, the pollution exclusion clause in the CGL policy was drastically amended to include an absolute or "super pollution exclusion" clause. The ultimate goal of this draft was to deny any pollution coverage under the CGL policy. Instead, the insurance carriers offered a separate pollution coverage which allows only those losses based on claims made only during the period when the policy is in effect. Any claim brought after the policy period, irrespective of when the pollution was actually caused, is apparently excluded from this policy coverage. Thus, there may not be any lingering or indefinite claim based on an old or expired policy.

In reality, this has created more confusion and debate within the legal circles. The courts are reacting to this changing policy coverage by insurance carriers in different fashions. This has created a clear division or watershed between the so-called liberal courts and conservative courts. Moreover, the cases are still being litigated over the appropriateness of the sudden-and-accidental provision of the original pollution exclusion clause.

Some of the courts have even resorted to a reliance on equitable principles of law to determine the applicability of the pollution exclusion clause. This has resulted in markedly contrasting results. In *Maryland Casualty Company v. Armco, Inc.,* the Fourth Circuit Court of Appeals held that the response costs under CERCLA are equitable in nature. Based on this, the *Armco* Court held that CERCLA response costs are not covered within the policy coverage. However, a majority of courts in other jurisdictions have not followed similar analysis for policy coverage. Instead, they have looked outside the legal or technical meaning of *damages at law*. They have relied on the ordinary meaning of *damages at law* and held that all the costs that the insured has to pay because of the legal liability under CERCLA are response costs. The Second, Third, Ninth, and District of Columbia Circuit Courts have held that CERCLA response costs are "damages" and are therefore covered by the CGL policies. These courts, however, relied upon applicable state law to determine what constitutes a damage.

This interpretation is, however, being contested along with the broader issue of when CGL policy covers an insured for the response costs charged against such insured. Historically, insurance carriers have adopted the position that the term *damages* in the CGL policy does not include cleanup costs under CERCLA. Hence, they hold that they are obligated to pay only that money which constitutes a legal damage for the insured and not any other monetary liabilities, such as cleanup costs, that are grounded on equitable principles. This distinction between legal damage and equitable damage has also affected the decisions of the courts. In two well-publicized state decisions, one in Michigan and another in New Jersey, the state appellate courts expressed two exactly opposite opinions in similar factual situations.

Hence, this debate will not come to rest unless the Supreme Court finally decides on the current controversy over the extent of CGL policy coverage. An alternate solution is for Congress to amend the statutory language of CERCLA or other applicable federal law to make its intent explicit. Meanwhile, for claimants and opponents, it is an open game and both sides are better off by putting forward all their claims and oppositions based on principles of law as well as equity.

Note that EPA has the legal authority under CERCLA to seek cost recovery from an insurance carrier, as a PRP, for cleanup costs. Section 108(c)(2) of CERCLA provides for direct actions against "any guarantor providing evidence of financial liability" in connection with a facility from which a release or threatened release of a hazardous substance is expected. It is generally undisputed that insurance companies step into the shoes of a guarantor when they insure a person. This can become a potent weapon for EPA if it can convince the court that government-mandated cleanup costs are damages under the CGL policy.

EPA has already initiated several actions against the insurance carriers to recover various CERCLA response costs. In *Aetna Casualty and Surety Co., Inc. v. Pintlar Corp.,* the Court of Appeals for the Ninth Circuit held that under Idaho law, EPA's claim against a PRP triggers the insurer's duty to defend under the CGL policy. The court ruled that the plain meaning of the word *damages* in the CGL policy includes CERCLA response costs from the perspective of an ordinary person. In another 1991 case, the federal Court of Appeals in Washington held that the government may recover from the insurer for response costs owed by a company that went bankrupt. The court rejected the insurers' claim that cleanup costs are not covered under the CGL policies. In *Potomac Electric Power Co. v. California Union Insurance Co.,* the federal district court held that a company which sold the CGL policy is liable for the costs incurred in responding to the release of hazardous PCBs as a result of the dismantling of electric transformers by a third party even though the utility sold them as scrap metal. This court based its ruling that under the District of Columbia law, the term *damages* includes environmental cleanup costs.

These decisions certainly add more furor to the ongoing debate about the extent of coverage under CGL policies even with a specific pollution exclusion clause. Since insurance companies are regulated by state insurance commissions, state laws will largely dictate the outcome of disputes between the insureds and the insurers concerning the extent of coverage and the likelihood of indemnification under the CGL policies.

16.6 Claims under Common Law

Neither CWA nor CERCLA explicitly bars any claim for response costs based on common law principles. These common law principles may include trespass, negligence, private and public nuisance, ultrahazardous activity, and other tortious conduct.

These principles of common law tort are discussed in Chap. 14. It is important to remember that any claim for recovery can be initiated based on common law or federal or state statutes. They are not mutually exclusive. However, to be successful, in all such actions, the claimant must meet all the elements of a cause of action that supports the claim.

16.7 Other Statutory Claims

Thus far, the claims for cost recovery by both government and private third parties appear to be largely based on CERCLA. In this context, it is important to recognize that RCRA grants broad authority to the EPA administrator for waste management, resource recovery, and resource conservation. Chapter 5 discusses the subject of EPA's corrective action authority under the 1984 RCRA Amendments, also known as the Hazardous and Solid Waste Amendments (HSWA) of 1984.

This raises the possibility of cost recovery for corrective actions undertaken pursuant to a regulatory requirement or order. EPA has yet to promulgate its regulations detailing the scope of the corrective action program. So far, it has provided only an outline of the scope of its corrective action authority.

In this context, we emphasize that any cost for a RCRA-based corrective action due to the releases of contaminants from a treatment, storage, or disposal facility can be compared to the costs based on CERCLA response actions. In *Chemical Waste Management, Inc. v. Armstrong World Industries, Inc.,* the court went further and held that an owner or operator of a TSD facility for hazardous wastes can recover corrective action costs under section 107(a)(3) of CERCLA. The court ruled that "RCRA and its regulation do not preclude" recovery of CERCLA response costs by the owner or operator of a TSD facility which received RCRA interim status. This suggests that a recovery of costs for corrective action under RCRA may indeed be sought under the cost recovery provision of CERCLA. In another federal case, *Colorado v. Department of the Army,* the court held that the statutory schemes of RCRA and CERCLA are not mutually exclusive and that Congress intended CERCLA to operate "independently of and in addition to" RCRA.

The statutory definition of *hazardous substances* under CERCLA includes, among its listed contaminants or pollutants, all hazardous wastes as identified under or listed pursuant to section 3001 of RCRA.

It is somewhat premature to determine what will be the final regulatory requirements for cost recovery in a corrective action for the TSD facilities of hazardous waste. Legal attempts to recover costs incurred in a corrective action program are very real possibilities. When that happens, more than likely the courts will be at the center stage of controversies again for determination of insurance coverage by the CGL or other liability policy.

Recommended Reading

1. *Environmental and Toxic Tort Claims—Insurance Coverage in 1991 and Beyond,* Course Handbook Series no. 579, Practising Law Institute, New York, 1991.

2. Susan M. Cooke, *The Law of Hazardous Waste: Management, Cleanup, Liability, and Litigation,* vol. 2, Matthew Bender, New York, 1991.

3. "Toxic Law Reporter," The Bureau of National Affairs, Inc., Washington, June 12–Sept. 25, 1991.

4. Jerold Oshinsky, Roger Warin, and Catherine J. Serafin, *Environmental Insurance,* Federal Publications Inc., Washington, 1991.

Important Cases

1. *Independent Petrochemical Corp. v. Aetna Casualty and Surety Co.,* No. 89-5367 (D.C. Cir. Sept. 13, 1991).

2. *In re Chateaugay Corp.,* No. 90-5024 (2d Cir. Sept. 6, 1991).

3. *United States v. Parsons,* No. 90-8779 (11th Cir. July 22, 1991).

4. *Aetna Casualty and Surety Co., Inc. v. Pintlar Corp,* Nos. 89-35286 and 89-35287 (9th Cir. Nov. 7, 1991).

5. *Barmet Aluminum Corp. v. Reilly,* 927 F. 2d 289 (6th Cir. 1991).

6. *General Electric Co. v. Litton Business Systems, Inc.,* 715 F. Supp. 949 (W.D. Mo. 1989), *aff'd. sub nom, General Electric Co. v. Litton Industrial Automation Systems,* 920 F. 2d 1415 (8th Cir. 1990), *cert. denied,* No. 90-1221 (S. Ct. Mar. 25, 1991).

7. *Idaho v. Hanna Mining Co.,* 882 F. 2d 392 (9th Cir. 1989).

8. *Continental Insurance Cos. v. Northeastern Pharmaceutical and Chemical Co. (NEPACCO),* 842 F. 2d 977 (8th Cir. 1986), *cert. denied,* 109 S. Ct. 66 (1988).

9. *Maryland Casualty Company v. Armco, Inc.,* 822 F. 2d 1348 (4th Cir. 1987), *cert. denied,* 484 U.S. 1008 (1988).

10. *Ashland Oil, Inc. v. Miller Oil Purchasing Co.,* 678 F. 2d 1293 (5th Cir. 1982).

11. *Key Tronic v. United States,* No. CS-89-694-JLQ (E.D. Wash. Mar. 19, 1991).

12. *Gopher Oil Co. v. Union Oil Co. of California,* No. 91-1159 (8th Cir. Jan. 28, 1992).

13. *Potomac Electric Power Co. v. California Union Insurance Co.,* No. 88-2091 (D.D.C. Sept. 30, 1991).

14. *Pease & Curren Refining Inc. v. Spectrolab Inc.,* 744 F. Supp. 945 (C.D. Cal. 1990).

15. *Fallowfield Development Corp. et al. v. Strunk,* 1990 WL 52745 (E.D. Pa. Apr. 23, 1990).

16. *United States v. Hardage,* 750 F. Supp. 1444 (W.D. Okla. 1990).

17. *State of Colorado v. United States Department of the Army,* No. 86-C-2524 (D. Colo. Feb. 24, 1989).

18. *T & E Industries Inc. v. Safety Light Corp.,* 680 F. Supp. 696 (D.N.J. 1988).

19. *Wehner v. Syntex Corp.,* 681 F. Supp. 651 (N.D. Cal. 1987).

20. *Chemical Waste Management, Inc. v. Armstrong World Industries, Inc.,* 669 F. Supp. 1285 (E.D. Pa. 1987).

21. *United States v. NEPACCO,* 579 F. Supp. 823 (W.D. Mo. 1984).

Index

ABOUT THE AUTHOR

Somendu B. Majumdar is a senior attorney in the law firm of Devorsetz Stinziano Gilberti & Smith, P. C. in Syracuse, New York. Previously, he was VP-Technical Director of Aiken-Murray Corporation and the manager of environmental engineering and Director, International Marketing for the Airco Carbon Division of Airco, Inc. He has worked with several engineering/consulting firms including Metcalf & Eddy, Inc., Ebasco Services, Inc. and United Engineers & Constructors, Inc. in various technical and management positions. Dr. Majumdar received his B.S., M.S., and Ph.D. degrees in chemical and environmental engineering. He also holds MBA and JD degrees and attended the University of Maine, New York University, and Fordham University School of Law. He is a licensed professional engineer in several states and a member of the bar in New York, New Jersey, Connecticut, and Washington, DC.